HEAT EXCHANGERS

SELECTION, RATING, and THERMAL DESIGN

Sadık Kakaç
Hongtan Liu

Department of Mechanical Engineering
University of Miami
Coral Gables, Florida

CRC

CRC Press
Boca Raton Boston London New York Washington, D.C.

Acquiring Editor: Robert B. Stern
Project Editor: Debbie Didier
Cover design: Denise Craig
PrePress: Carlos Esser

Library of Congress Cataloging-in-Publication Data

Kakaç, S. (Sadık)
Heat exchangers : selection, rating, and thermal design / Sadık Kakaç, Hongtan Liu.
p. cm.
Includes bibliographical references and index.
ISBN 0-8493-1688-X
1. Heat exchangers. I. Liu, Hongtan. II. Title.
TJ263.K25 1998
621.402'5--dc21

97-24314
CIP

International Standard Book Number 0-8493-1688-X
Library of Congress Card Number 97-24314
Printed in the United States of America 1 2 3 4 5 6 7 8 9 0
Printed on acid-free paper

Preface

Heat exchangers are vital in power producing plants; process and chemical industries; heating, ventilating, air-conditioning, and refrigeration systems; and cooling of electronic systems. A large number of industries are engaged in designing various types of heat exchange equipment. Courses are offered at many colleges and universities on thermal design under various titles.

There is extensive literature on this subject; however, the information has been widely scattered. This is a systematic approach to be used as an up-to-date textbook based on scattered literature for senior undergraduate and first year graduate students in mechanical, nuclear, aerospace, and chemical engineering who have taken introductory courses in thermodynamics, heat transfer, and fluid mechanics. This systematic approach is also essential for the newcomers who are interested in industrial applications of thermodynamics, heat transfer, and fluid mechanics; and for designers and operators of heat exchange equipment. This book focuses on the selection, thermohydraulic design, design processes, and rating and operational problems of various types of heat exchangers.

The new criterion by the Accreditation Board on Engineering and Technology (ABET) requires engineering design across the curriculum. Therefore, one of the main objectives of this textbook is to introduce thermal design by describing various types of single-phase and two-phase flow heat exchangers, detailing their specific fields of application, selection, and thermohydraulic design and rating; and showing thermal design and rating processes with worked examples and end-of-chapter problems including student design projects.

Much of the text is devoted to double-pipe, shell-and-tube, compact, gasketed-plate heat exchanger types, condensers, and evaporators. Their design processes are described, and thermal–hydraulic design examples are presented. Some other types, mainly specialized ones, are briefly described without design examples. Nevertheless, thermal design factors and methods are common to all heat exchangers regardless their function.

The text begins with the classification of heat exchangers according to different criteria. Chapter 2 provides the basic design methods for sizing and rating of heat exchangers. Chapter 3 is a review of single-phase forced convection correlations in ducts. A large number of experimental and analytic correlations are available for heat transfer coefficient and flow friction factor for laminar and turbulent flow through ducts. Thus, it is often a difficult and confusing task for a student and even a designer to choose appropriate correlations. In this chapter, recommended correlations for the single-phase side of heat exchangers are given with worked examples. Chapter 4 discusses pressure drop and pumping power for heat exchangers and their piping circuit analysis.

One of the major unresolved problems in heat exchanger equipment is fouling; design of heat exchangers subject to fouling is presented in Chapter 5. The thermal design methods and processes for double-pipe, shell-and-tube, compact, and gasketed-plate heat exchangers are presented in Chapters 6, 8, 9, and 10 for single-phase duties, respectively. The important design correlations for the design of two-phase flow heat exchangers are given in Chapter 7. With this arrangement, students and newcomers in industry will achieve better understanding of thermal design and will be better prepared to understand the thermal design of condensers and evaporators that is introduced in Chapter 11. Appendices A and B provide thermophysical properties of various fluids, including the new refrigerants.

In every chapter, worked examples to illustrate the thermal design methods and procedures are given. Although the use of computer programs is essential for thermal design and rating of these exchangers, for students and newcomers manual thermal design analysis is essential in the initial learning period. Fundamental design knowledge is needed before one correctly uses computer design software and develops new reliable and sophisticated computer software for rating to obtain an optimum solution.

The end-of-chapter problems including student design projects are selected to enhance the design applications. A solution manual accompanies the text. Additional problems are added to the solution manual that maybe helpful to the instructors of the course.

Design of a heat exchange equipment requires explicit consideration of mechanical design, economics, optimization techniques, and environmental considerations. Information on these topics is available in various standard references and handbooks, and from manufacturers.

Several individuals have made valuable contributions to this book: E. M. Sparrow and A. Bejan reviewed the manuscript and provided helpful suggestions. We gratefully appreciate their support.

Sadik Kakaç has edited several books on the fundamentals and design of heat exchangers to which many leading scientists and experts made invaluable contributions, and is thankful to them. Both authors are greatly indebted, especially to the following individuals whose contributions to the field of heat exchangers made this book possible: Kenneth J. Bell, David Butterworth, John Collier, Paul J. Marto, Mike B. Pate, Ramesh K. Shah, and J. Taborek.

The authors express their sincere appreciation to their students, who contributed to the improvement of the manuscript by their critical questions. The authors wish to thank Claudia O. Gavrilescu and Liping Cao who helped during the classroom teaching and various stages of this project. Special thanks also go to T. Zhou for her assistance in the preparation of the manuscript.

Finally we wish to acknowledge the encouragement and support of our lovely wives who made many sacrifices during the writing of this text.

Sadık Kakaç
Hongtan Liu

The Authors

Sadık Kakaç is a Professor of Mechanical Engineering at University of Miami, Coral Gables, Florida. He received the Dipl. Ing. (1955) in Mechanical Engineering from Technical University of Istanbul; his S.M. (1959) in Mechanical Engineering and his S.M. (1960) in Nuclear Engineering, both from Massachusetts Institute of Technology. In 1965, he received his Ph.D. in the field of heat transfer from Victoria University of Manchester, U.K. He was elected as a member of the Turkish Scientific and Technical Research Council in 1972 and the general secretary of the Turkish Atomic Energy Commission in 1978, and represented Turkey in a number of scientific endeavors abroad. He is a recipient of the Alexander von Humboldt Research Award for Senior U.S. Scientists (1989), the Science Award of the Association of Turkish–American Scientists (1994), and the ASME Heat Transfer Memorial Award (1997). He is a fellow of the American Society of Mechanical Engineers (ASME). Formerly, Dr. Kakaç was a Professor of Mechanical Engineering at the Middle East Technical University in Ankara, Türkiye. Dr. Kakaç has focused his research efforts on steady-state and transient forced convection in single-phase flows and on two-phase flow instabilities in flow boiling. He is the author or coauthor of 5 heat transfer textbooks and more than 165 research papers; and edited 15 volumes in the field of thermal sciences and heat exchanger fundamentals and design, including *The Handbook of Single-Phase Convective Heat Transfer.*

Hongtan Liu is an Assistant Professor of Mechanical Engineering at University of Miami, Coral Gables, Florida. He received his B.S. degree (1982) in Mechanical Engineering from Shenyang Civil and Architectural Engineering Institute, and both his M.S. (1989) and Ph.D. (1993) degrees in Mechanical Engineering from University of Miami. He spent one year as a research fellow at the Institute of Thermodynamics A, Technical University of Munich in Germany. During the period of 1996–1997, Dr. Liu was a Senior Research Engineer at Energy Partners, Inc., where he was mainly responsible for the research in fluid flow, heat and mass transfer in fuel cells. As a graduate student, he received a Graduate Student Academic Excellence Award (1993) and was listed in Who's Who Among Students in American Colleges and Universities (1991–1992). As a faculty member, he received the Eliahu I. Jury Excellence in Research Award from the College of Engineering, University of Miami (1995). Dr. Liu has focused his research efforts on fuel cells, solar energy applications, boiling two-phase flow instabilities, and hydrogen energy systems. He is the author or co-author of more than 30 research papers.

Contents

Chapter 4 Heat Exchanger Pressure Drop and Pumping Power

Chapter 5 Fouling of Heat Exchangers

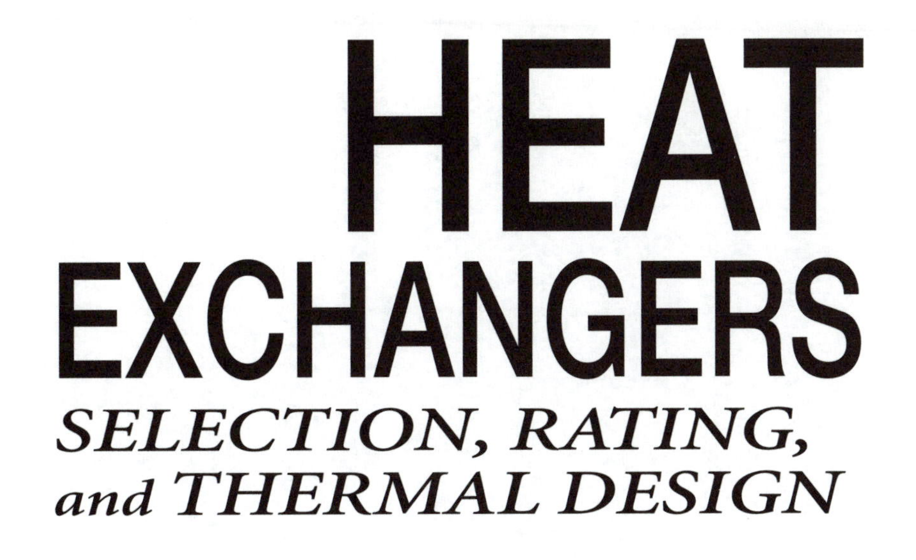
HEAT
EXCHANGERS
SELECTION, RATING,
and THERMAL DESIGN

1

Classification of Heat Exchangers

1.1 Introduction

Heat exchangers are devices that provide the flow of thermal energy between two or more fluids at different temperatures. Heat exchangers are used in a wide variety of applications. These include power production; process, chemical, and food industries; electronics; environmental engineering; waste heat recovery; manufacturing industry; and air-conditioning, refrigeration, and space applications. Heat exchangers may be classified according to the following main criteria:[1,2]

1. Recuperators and regenerators
2. Transfer processes: direct contact and indirect contact
3. Geometry of construction: tubes, plates, and extended surfaces
4. Heat transfer mechanisms: single phase and two phase
5. Flow arrangements: parallel, counter, and cross flows

The preceding five main criteria are illustrated in Figure 1.1.[1]

1.2 Recuperation and Regeneration

The conventional heat exchangers shown diagramatically in Figure 1.1a with heat transfer between two fluids is called a recuperator, because the hot stream A recovers (recuperates) some of the heat from stream B. The heat transfer is through a separating wall or through the interface between the streams as in the case of direct contact type of heat exchangers (Figure 1.1c). Some examples of the recuperative-type exchangers are shown in Figure 1.2.

In regenerators or in storage-type heat exchangers, the same flow passage (matrix) is alternately occupied by one of the two fluids. The hot fluid stores the thermal energy in the matrix; during the cold-fluid flow through the same passage later, energy stored will be extracted from the matrix. Therefore, thermal energy is not transferred through the wall as in a direct transfer type of heat exchanger. This cyclic principle is illustrated in Figure 1.1b. While the solid is in the cold stream A it loses heat; while it is in the hot stream B it gains heat (i.e., it is regenerated). Some examples of storage-type heat exchangers are rotary regenerator for preheating the air in a large coal-fired steam power plant, gas turbine rotary regenerator, fixed-matrix air preheaters for blast furnace stoves, steel furnaces, open-hearth steel melting furnaces, and glass furnaces.

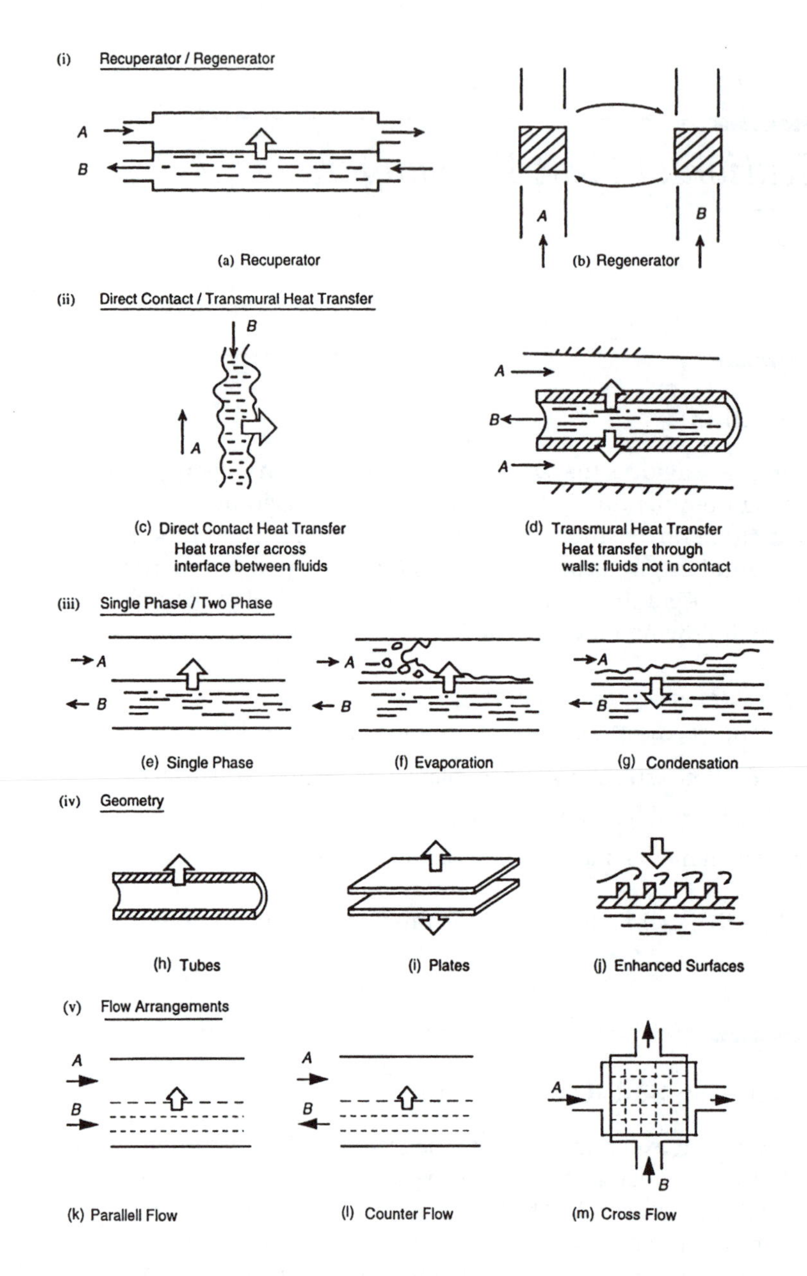

FIGURE 1.1
Criteria used in the classification of heat exchangers. (From Hewitt, G. F., Shires, G. L., and Bott, T. R. [1994] *Process Heat Transfer*, CRC Press, Boca Raton, FL.)

Regenerators can be classified as follows:

1. Rotary regenerator
2. Fixed-matrix regenerator

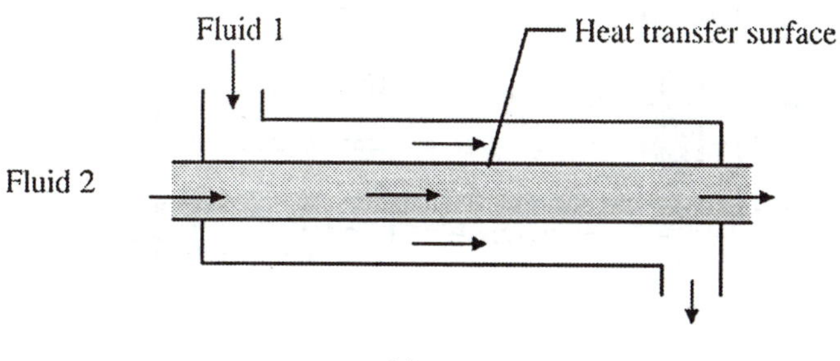

(a)

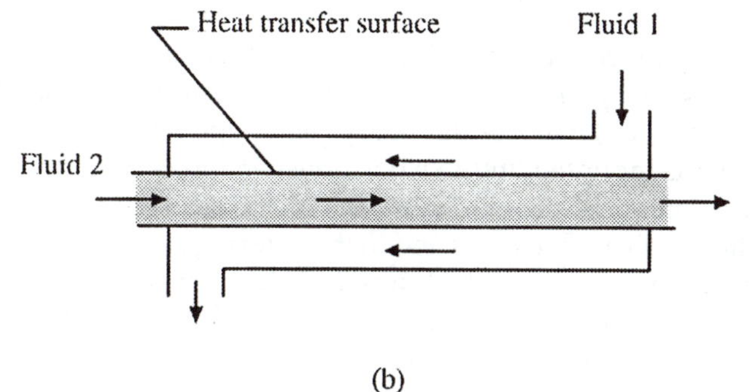

(b)

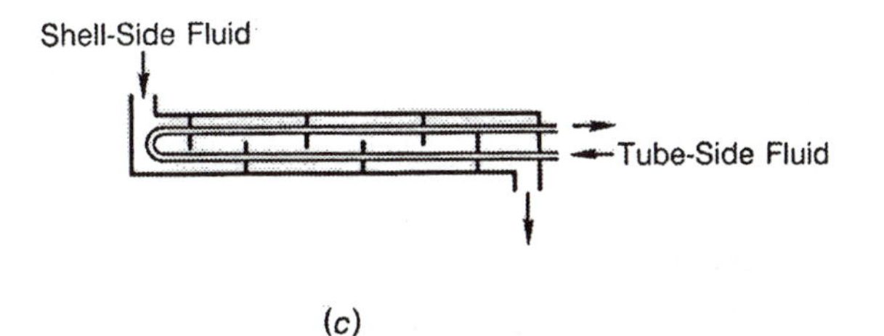

(c)

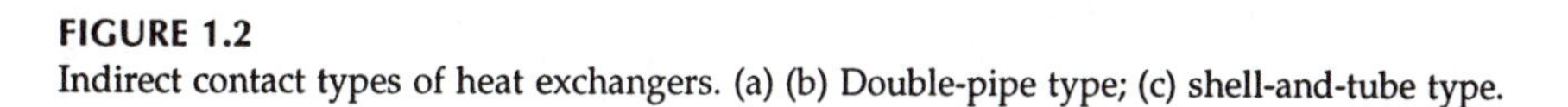

FIGURE 1.2
Indirect contact types of heat exchangers. (a) (b) Double-pipe type; (c) shell-and-tube type.

Rotary regenerators can be further subclassified as:

1. Disk type
2. Drum type

The disk-type and drum-type regenerators are shown schematically in Figure 1.3. In a disk-type regenerator, the heat transfer surface is in a disk form and fluids flow axially. In a drum type, the matrix is in a hollow drum form and fluids flow radially.

These regenerators are periodic flow heat exchangers. In rotary regenerators, the operation is continuous. To have this, the matrix moves periodically in and out of the fixed

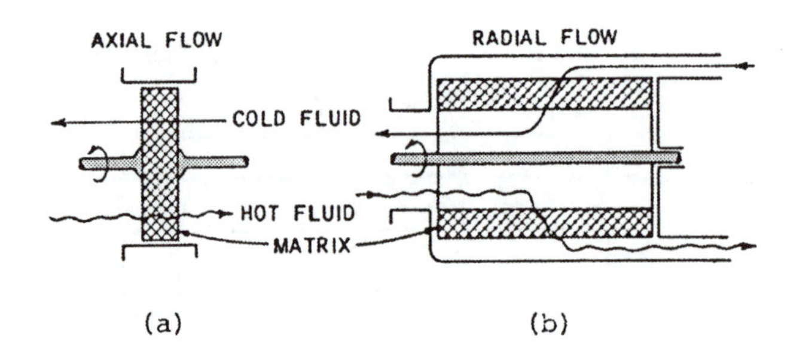

FIGURE 1.3
Rotary regenerators. (a) Disk type; (b) drum type. (From Shah, R. K. [1981] In *Heat Exchangers — Thermo–Hydraulic Fundamentals and Design*, Wiley, New York. With permission.)

stream of gases. A rotary regenerator for air heating is illustrated in Figure 1.4. There are two kinds of regenerative air preheaters used in convectional power plants:[3] the rotating-plate type (Figures 1.4 and 1.5) and the stationary-plate type (Figure 1.6). The rotor of the rotating-plate air heater is mounted within a box housing and is installed with the heating surface in the form of plates as shown in Figure 1.5. As the rotor rotates slowly, the heating surface is exposed alternately to flue gases and to the entering air. When the heating surface is placed in the flue gas stream, the heating surface is heated; and then when it is rotated by mechanical devices into the air stream, the stored heat is released to the air flow. Thus, the air stream is heated. In the stationary-plate air heater (Figure 1.6), the heating plates are stationary, while cold-air hoods — both top and bottom — are rotated across the heating plates; the heat transfer principles are the same as those of the rotating-plate regenerative air heater. In a fixed-matrix regenerator, the gas flows must be diverted to and from the fixed matrices. Regenerators are compact heat exchangers and they are designed for surface area density of up to approximately 6600 m^2/m^3.

1.3 Transfer Processes

According to transfer processes, heat exchangers are classified as direct contact type and indirect contact type (transmural heat transfer).

In *direct contact* type heat exchangers, heat is transferred between the cold and hot fluids through a direct contact between these fluids. There is no wall between hot and cold streams, and the heat transfer occurs through the interface between two streams as illustrated in Figure 1.1c. In direct contact-type heat exchangers the streams are two immiscible liquids, a gas–liquid pair, or a solid particle–fluid combination. Spray and tray condensers (Figure 1.7) and cooling towers are good examples of such heat exchangers.[4,5] Very often in such exchangers, heat and mass transfer occur simultaneously. In a cooling tower a spray of water falling from the top of the tower is directly contacted and cooled by a stream of air flowing upward (Figures 11.14 and 11.15).

In an *indirect contact* type heat exchanger, the heat energy is exchanged between hot and cold fluids through a heat transfer surface (i.e., a wall separating the fluids). The cold and hot fluids flow simultaneously while heat energy is transferred through a separating wall as illustrated in Figure 1.1d. The fluids are not mixed. Examples of this type of heat exchanger are shown in Figure 1.2.

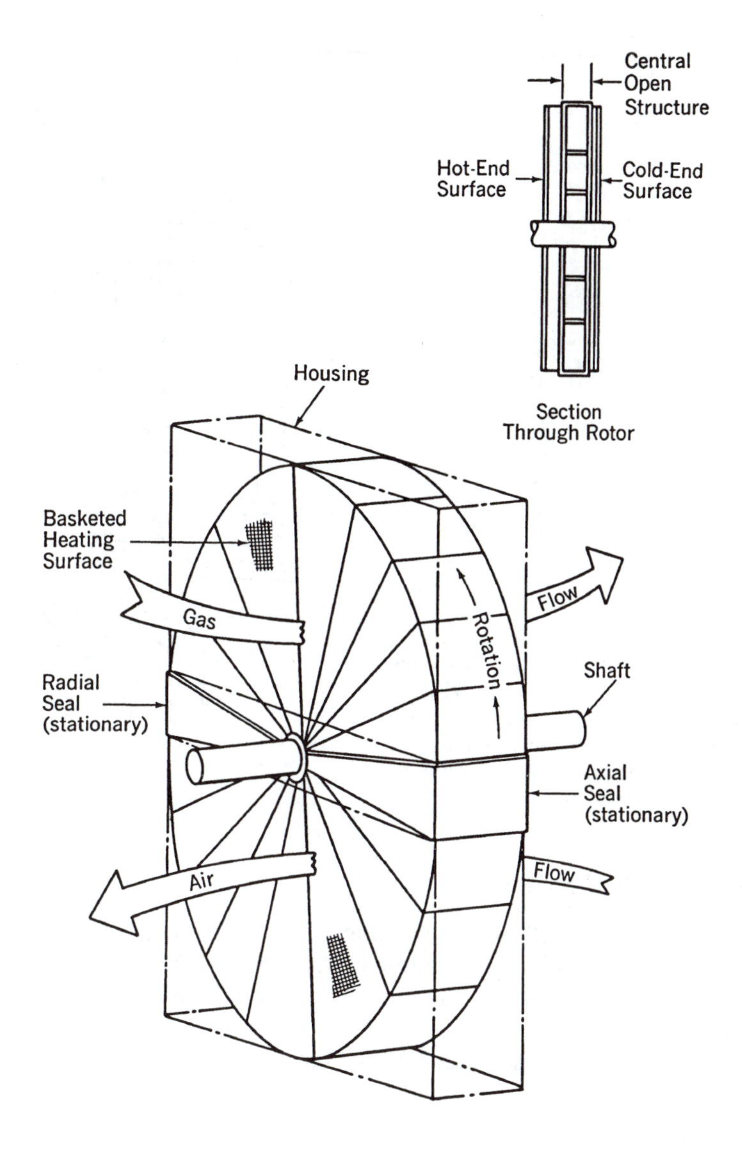

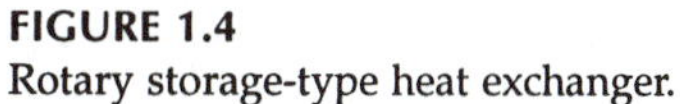
FIGURE 1.4
Rotary storage-type heat exchanger.

Indirect contact- and direct transfer-type heat exchangers are also called *recuperators*. Tubular (double-pipe, shell-and-tube), plate, and extended surface heat exchangers; cooling towers; and tray condensers are examples of recuperators.

1.4 Geometry of Construction

Direct transfer-type heat exchangers (transmural heat exchangers) are often described in terms of their construction features. The major construction types are tubular, plate, and extended surface heat exchangers.

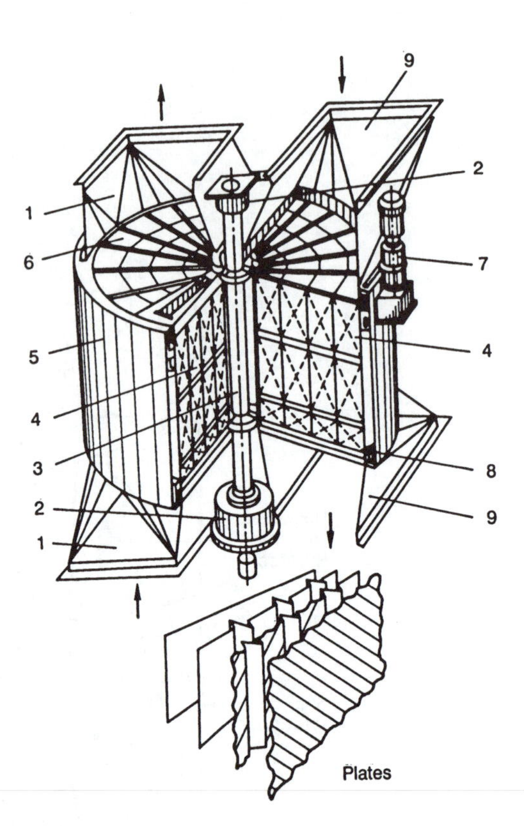

FIGURE 1.5
Rotating-plate regenerative air preheater in a large coal-fired steam power plant. (1) Air ducts; (2) bearings; (3) shaft; (4) plates; (5) outer case; (6) rotor; (7) motor; (8)sealings; (9) flue gas ducts. (From Lin, Z. H. [1991] In *Boilers, Evaporators and Condensers*, Wiley, New York. With permission.)

1.4.1 Tubular Heat Exchangers

These heat exchangers are built of circular tubes. One fluid flows inside the tubes and the other, on the outside of the tube. Tube diameter, number of tubes, tube length, pitch of the tubes, and tube arrangement can be changed. Therefore, there is a considerable flexibility in their design.

Tubular heat exchangers can be further classified as follows:

1. Double pipe
2. Shell and tube
3. Spiral tube type

(1) Double-Pipe Heat Exchangers

A typical double-pipe heat exchanger consists of one pipe placed concentrically inside another of larger diameter with appropriate fittings to direct the flow from one section to the next, as shown in Figures 1.2 and 1.8. Double-pipe heat exchangers can be arranged in various series and parallel arrangements to meet pressure drop and mean temperature difference requirements. The major use of double-pipe exchangers is for sensible heating or cooling of process fluids where small heat transfer areas (to 50 m^2) are required. This configuration is also suitable when one or both fluids are at high pressure. The major disadvantage is that these exchangers are bulky and expensive per unit transfer surface. Inner tubing may be single tube or multitubes (Figure 1.8). If the heat transfer coefficient is poor in the annulus, axially finned inner tube (or tubes) can be used. Double-pipe heat exchangers are built in modular concept (i.e., in the form of hairpins).

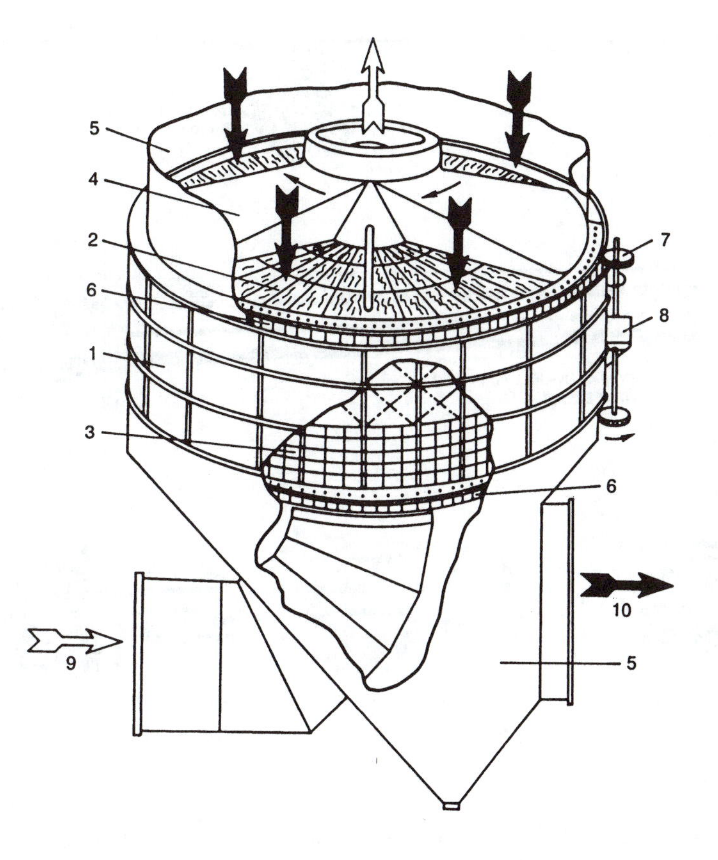

FIGURE 1.6
Stationary-plate regenerative air preheater. (1) Outer case; (2) plates; (3) plates in the lower temperature region; (4) rotating air ducts; (5) flue gas ducts; (6, 7) drive; (8) motor and drive-down devices; (9) air inlet; (10) gas exit. (From Lin, Z. H. [1991] In *Boilers, Evaporators and Condensers*, Wiley, New York. With permission.)

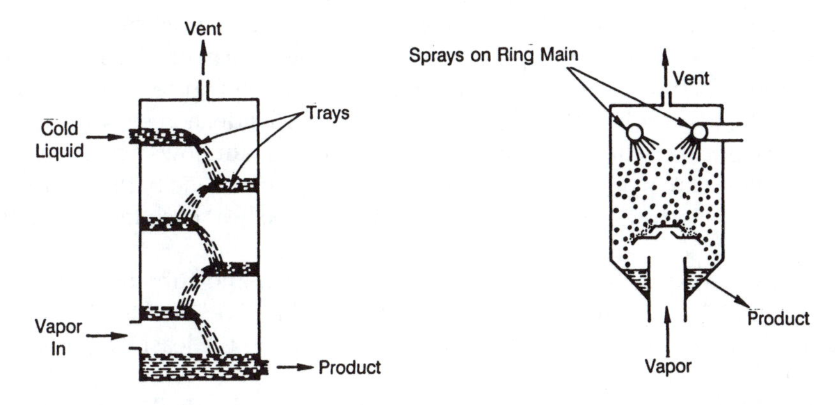

FIGURE 1.7
Direct contact type of heat exchangers. (a) Spray condenser; (b) tray condenser. (Adapted from Butterworth, D. [1988] In *Two-Phase Flow Heat Exchangers: Thermal–Hydraulic Fundamentals and Design*, Kluwer, Dordrecht, The Netherlands.)

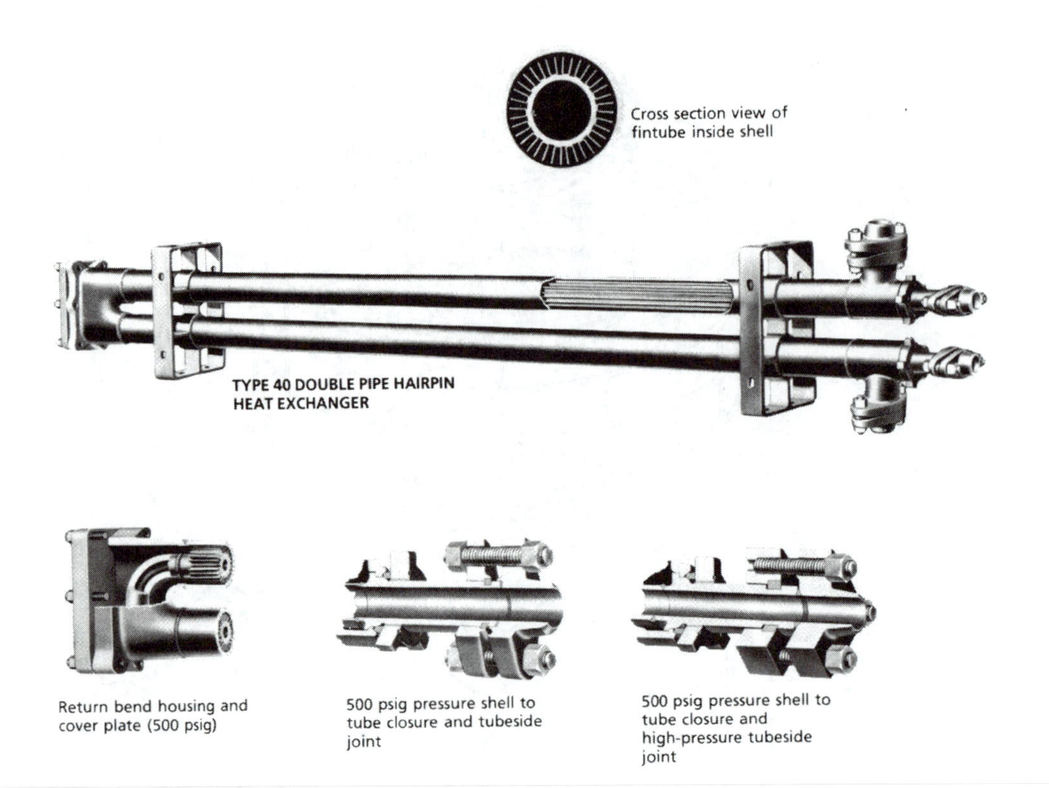

FIGURE 1.8
Double-pipe hairpin heat exchanger with cross-section view and return bend housing. (Courtesy of Brown Fintube.)

(2) Shell-and-Tube Heat Exchangers

Shell-and-tube heat exchangers are built of round tubes mounted in large cylindrical shells with the tube axis parallel to that of the shell. They are widely used as oil coolers, power condensers, preheaters in power plants, and steam generators in nuclear power plants; and in process and chemical industry applications. The simplest form of a horizontal shell-and-tube type of condenser with various components is shown in Figure 1.9. One fluid stream flows through the tubes while the other flows on the shell side, across or along the tubes. In a baffled shell-and-tube heat exchanger, the shell-side stream flows across between pairs of baffles, and then flows parallel to the tubes as it flows from one baffle compartment to the next. There are wide differences between shell-and-tube heat exchangers depending on the application.

The most representative tube bundle types used in shell-and-tube heat exchangers are shown in Figures 1.10, 1.11, and 1.12. The main design objectives here are to accommodate thermal expansion, to furnish ease of cleaning, or to provide the least expensive construction if other features are of no importance.[6]

In a shell-and-tube heat exchanger with fixed tube sheets, the shell is welded to the tube sheets and there is no access to the outside of the tube bundle for cleaning. This low-cost option has only limited thermal expansion, which can be somewhat increased by expansion bellows. Cleaning of the tube is easy (Figure 1.10).

The U-tube is the least expensive construction because only one tube sheet is needed. The tube side cannot be cleaned by mechanical means because of the sharp U-bend. Only an even number of tube passes can be accommodated, but thermal expansion is unlimited (Figure 1.11).

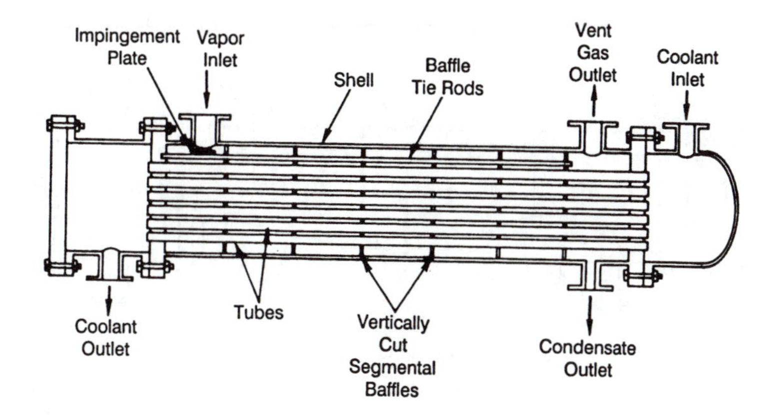

FIGURE 1.9
Shell-and-tube heat exchanger as a shell-side condenser: TEMA E-type shell with single tube side. (Adapted from Butterworth, D. [1988] In *Two-Phase Flow Heat Exchangers: Thermal–Hydraulic Fundamentals and Design*, Kluwer, Dordrecht, The Netherlands.)

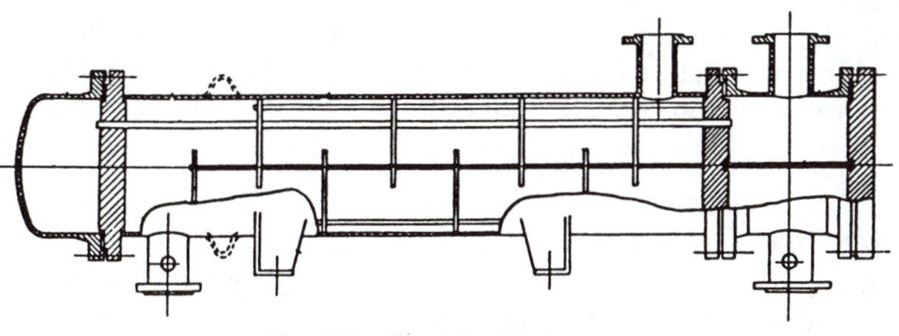

FIGURE 1.10
A two-pass tube, baffled single-pass shell, shell-and-tube heat exchanger. (From Standards of Tubular Exchangers Manufacturers Association (1988), New York. With permission.)

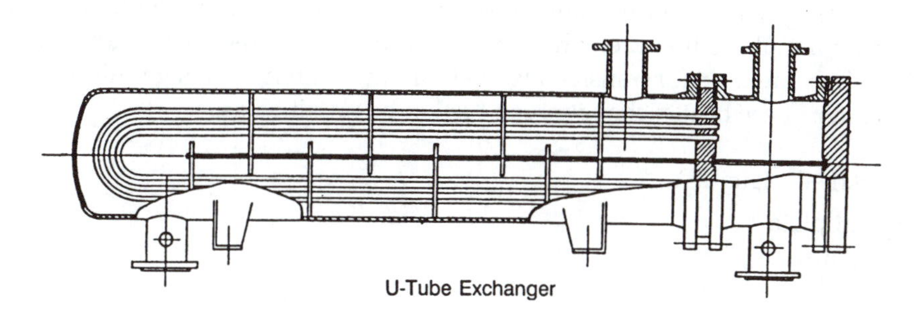

FIGURE 1.11
A U-tube, baffled single-pass shell, shell-and-tube heat exchanger. (From Standards of Tubular Exchangers Manufacturers Association (1988), New York. With permission.)

Several designs have been developed that permit the tube sheet to "float" (i.e., to move with thermal expansion). The classic type of pull-through floating head is shown, which permits tube bundle removal with minimum disassembly, as required for heavily fouling units (Figure 1.12). The cost is high.

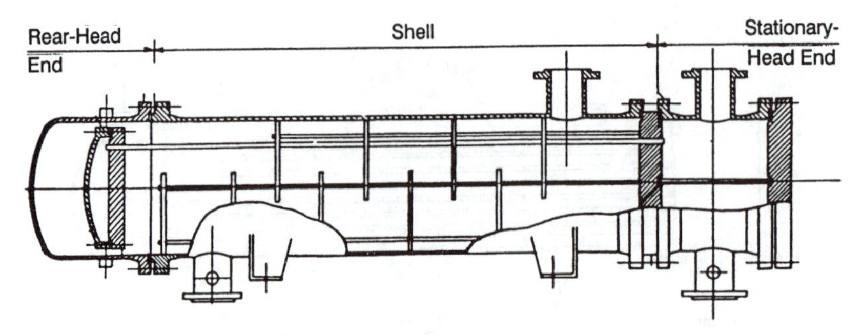

FIGURE 1.12
A heat exchanger similar to that of Figure 1.10, but with pull-through floating-head shell-and-tube exchanger. (From Standards of Tubular Exchangers Manufacturers Association (1988), New York. With permission.)

A number of shell-side and tube-side flow arrangements are used in shell-and-tube heat exchangers depending on heat duty, pressure drop, pressure level, fouling, manufacturing techniques and cost, corrosion control, and cleaning problems. The baffles are used in shell-and-tube heat exchangers to promote a better heat transfer coefficient on the shell side and to support the tubes. Shell-and-tube heat exchangers are designed on a custom basis for any capacity and operating conditions. This is contrary to many other heat exchanger types.

(3) Spiral-Tube Heat Exchangers

These consist of spirally wound coils placed in a shell, or they are designed as coaxial condensers and coaxial evaporators that are used in refrigeration systems. The heat transfer coefficient is higher in a spiral tube than that for a straight tube. These exchangers are suitable for thermal expansion and clean fluids, because cleaning them is almost impossible.

1.4.2 Plate Heat Exchangers

Plate heat exchangers are built of thin plates forming flow channels. The fluid streams are separated by flat plates that are either smooth or between which are sandwiched corrugated fins. They are used for transferring heat for any combination of gas, liquid, and two-phase streams. These heat exchangers can further be classified as:

1. Gasketed-plate
2. Spiral plate
3. Lamella

(1) Gasketed-Plate Heat Exchangers

A typical gasketed-plate heat exchanger and the flow paths are shown in Figure 1.13 and Figure 1.14.[7] A gasketed plate consists of a series of thin plates with corrugation or wavy surface that separates the fluids. The plates are provided with corner parts so arranged that the two media between which heat is to be exchanged flow through alternate interplate spaces. Appropriate design and gasketing permit a stack of plates to be held together by

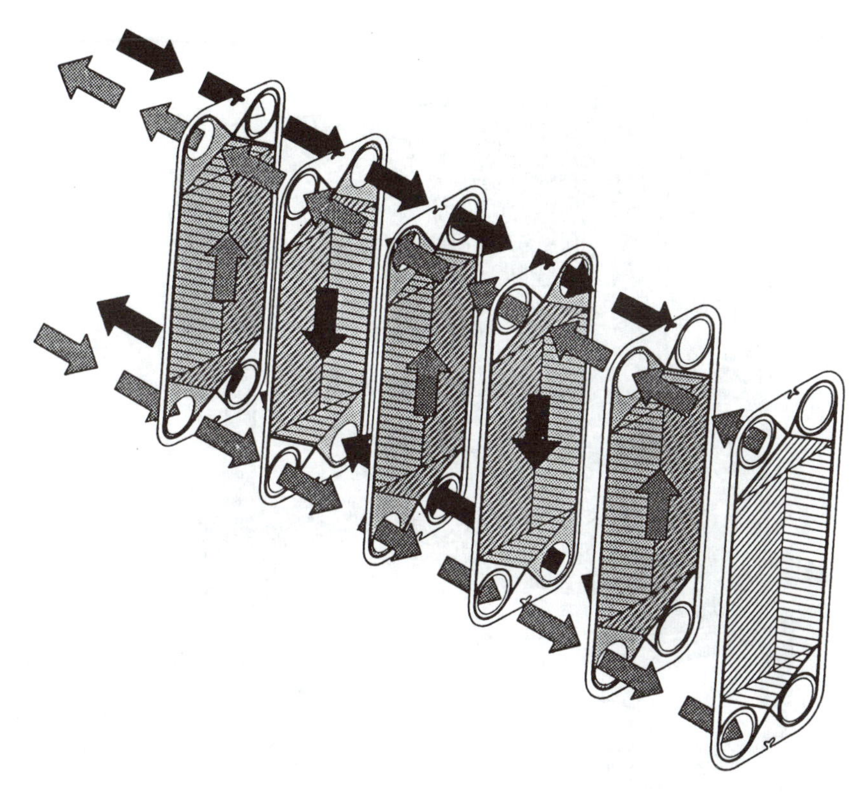

FIGURE 1.13
A diagram showing the flow paths in a gasketed-plate heat exchanger. (Courtesy of Alfa Laval Thermal AB.)

compression bolts joining end plates. Gaskets prevent the leakage to the outside and direct the fluids in the plates as desired. The flow pattern is generally chosen so that the media flow countercurrent to each other. The plate heat exchangers are usually limited to fluid stream with pressure below 25 bar and temperature below about 250°C. Because the flow passages are quite small, strong eddying gives high heat transfer coefficients, high-pressure drops, and high local shear that minimizes fouling. These exchangers provide relatively compact and lightweight heat transfer surface. They are temperature and pressure limited due to the construction details and the gasketing. They are typically used for heat exchange between two liquid streams. They are easily cleaned and sterilized because they can be completely disassembled; thus, they have a wide application in the food processing industry.

(2) Spiral Plate Heat Exchangers

Spiral heat exchangers are formed by rolling two long, parallel plates into a spiral using a mandrel and welding the edges of adjacent plates to form channels (Figure 1.15). The distance between the metal surfaces in both spiral channels is maintained by means of distance pins welded to the metal sheet. The length of the distance pins may vary between 5 and 20 mm. It is therefore possible to choose between different channel spacings according to the flow rate. This means that the ideal flow conditions and therefore the smallest possible heating surfaces are obtained.

The two spiral paths introduce a secondary flow, increasing the heat transfer and reducing fouling deposits. These heat exchangers are quite compact, but are relatively expensive

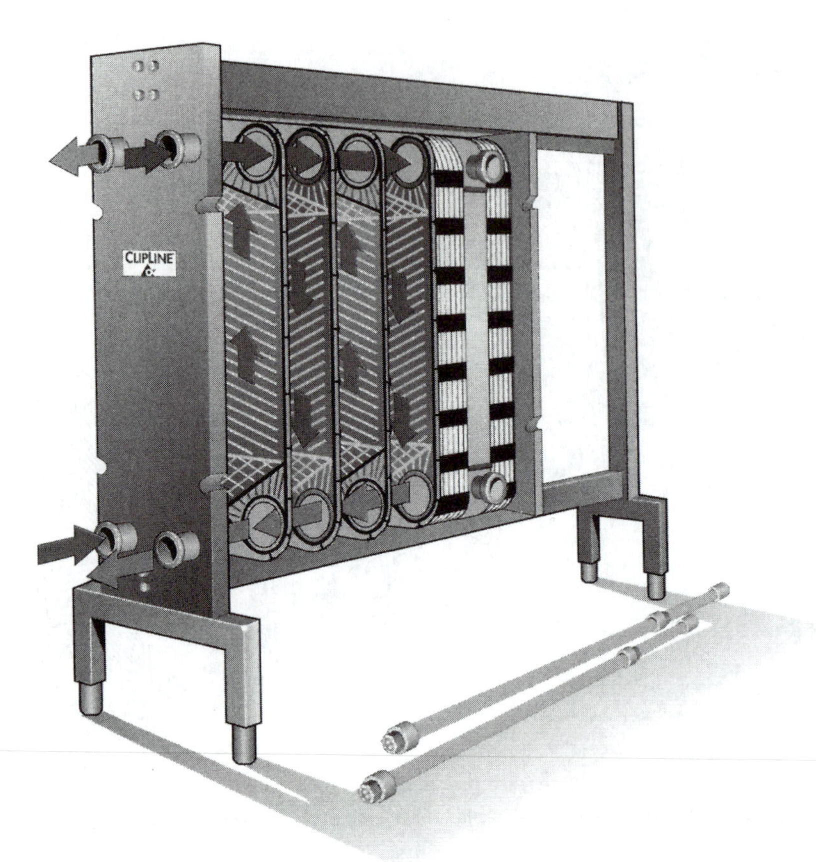

FIGURE 1.14
A gasketed plate heat exchanger. (Courtesy of Alfa Laval Thermal AB.)

due to their specialized fabrication. Sizes range from 0.5 to 500 m^2 of heat transfer surface in one single spiral body. The maximum operating pressure (up to 15 bar) and operating temperature (up to 500°C) are limited.

The spiral heat exchanger is particularly effective in handling sludges, viscous liquids, and liquids with solids in suspension including slurries.

The spiral heat exchanger is made in three main types that differ in the connections and flow arrangements.

Type I has flat covers over the spiral channels. The media flows countercurrent through the channels, via the connections in the center and at the periphery. This type is used to exchange heat between media without phase changes, such as liquid–liquid, gas–liquid, or gas–gas. One stream enters at the center of the unit and flows from inside outward. The other stream enters at the periphery and flows toward the center. Thus, true counterflow is achieved (Figure 1.15a).

Type II is designed for crossflow operation (Figure 1.15b). One channel is completely seal welded while the other is open along both sheet metal edges. Therefore, this type has one medium in spiral flow; the other, in crossflow. The passage with the medium in spiral flow

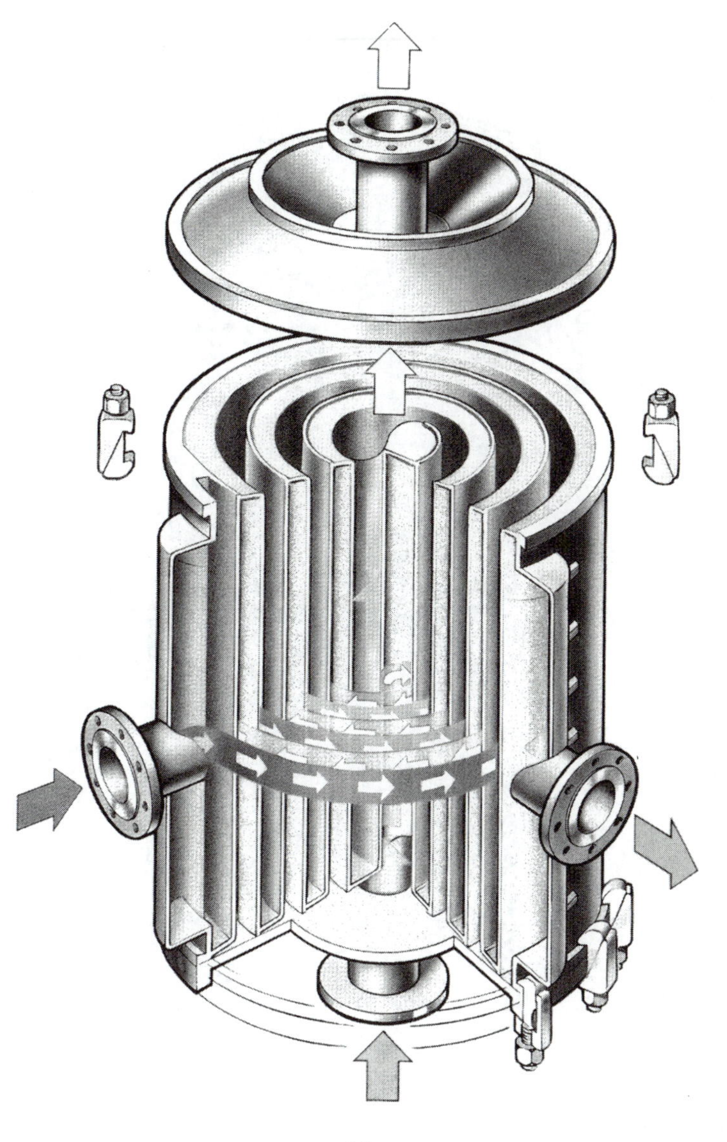

(a)

FIGURE 1.15
Spiral plate heat exchangers. (a) Type I; (b) Type II; (c) Type G. (Courtesy of Alfa Laval Thermal AB.)

is welded shut on each side, and the medium in crossflow passes through the open spiral annulus.

This type is mainly used as a surface condenser in evaporating plants. It is also highly effective as a vaporizer. Two spiral bodies are often built into the same jacket and are mounted one on top of the other.

Type III, the third standard type is in principle similar to Type I with alternately welded channels, but is provided with a specially designed top cover. This type of heat exchanger is mainly intended for condensing vapors with subcooling of condensate and noncondensable gases. The top cover therefore has a special distribution cone where the vapor is distributed to the uncovered spiral turns to maintain a constant vapor velocity along the channel opening.

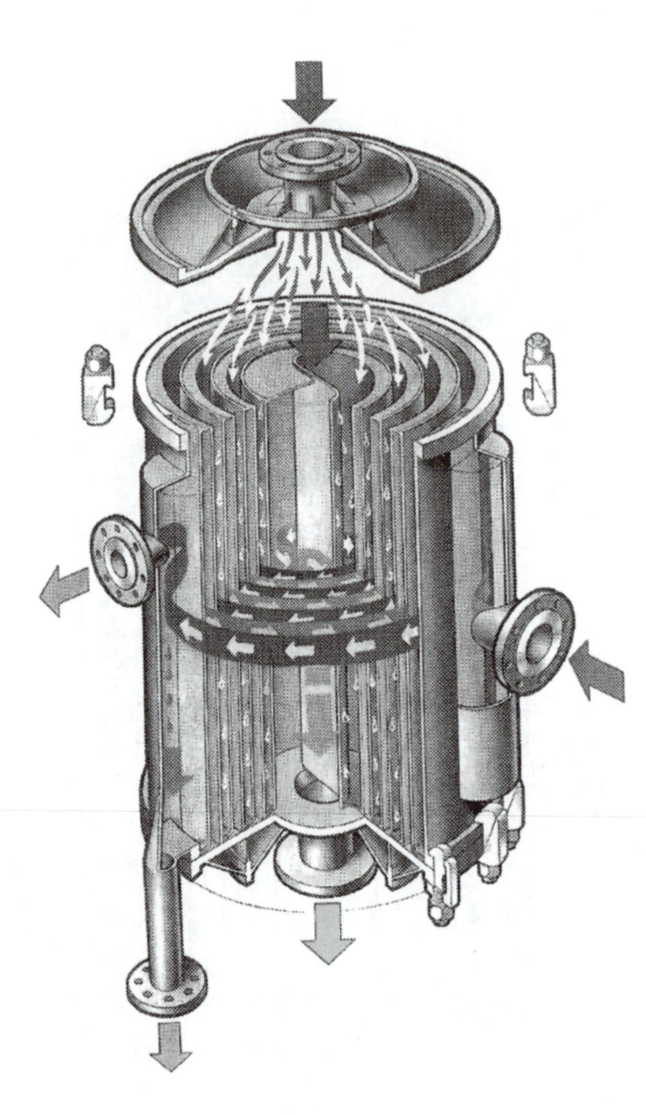

(b)

FIGURE 1.15 (CONTINUED)

For subcooling, the two to three outer turns of the vapor channel are usually covered, which means that a spiral flow path in countercurrent to the cooling medium is obtained. The vapor–gas mixture and the condensate are separated and the condensate then flows through a downward connection to the periphery box; and the gas, through an upward connection.

A spiral heat exchanger type G,[7] also shown in Figure 1.15c, is used as condenser. The vapor enters the open center tube, reverses flow direction in the upper shell extension, and is condensed in downward crossflow in the spiral element.

(3) Lamella Heat Exchangers

The lamella (Ramen) type of heat exchanger consists of a set of parallel, welded, thin plate channels or lamellae (flat tubes or rectangular channels) placed longitudinally in a shell (Figure 1.16).[7] It is a modification of the floating-head type of shell-and-tube heat

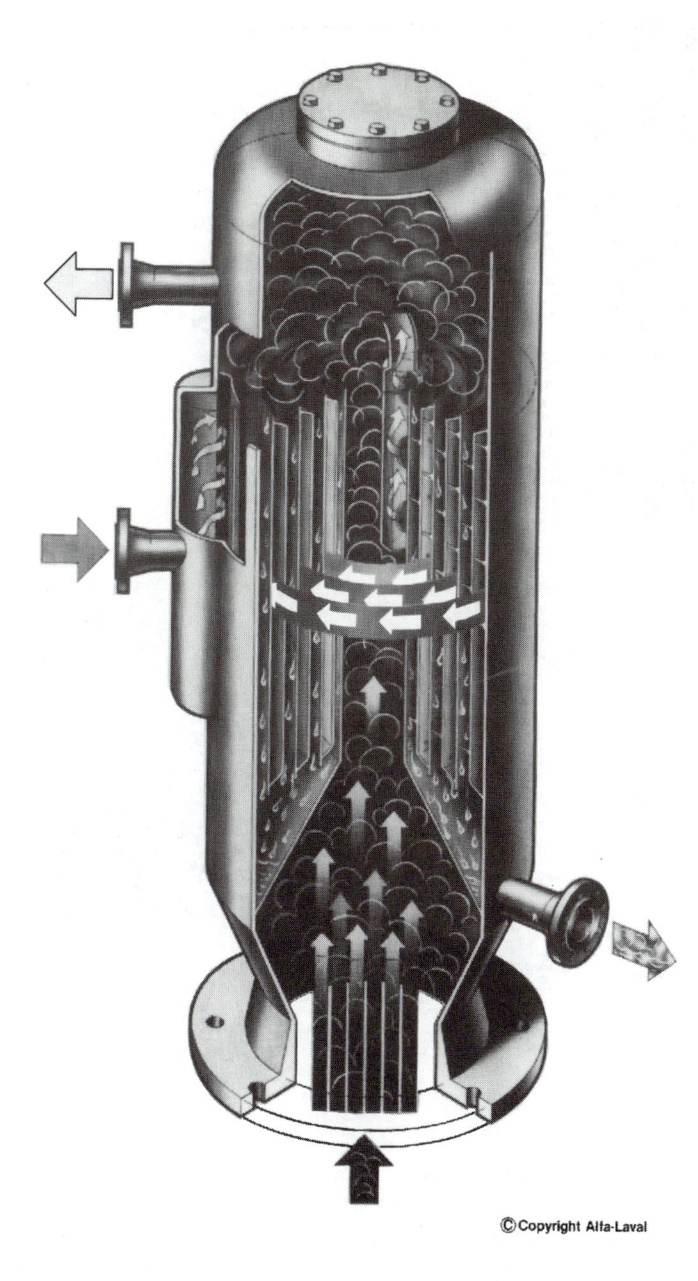

(c)

FIGURE 1.15 (CONTINUED)

exchanger. These flattened tubes, called lamellae (lamellas), are made up of two strips of plates, profiled and spot or seam welded together in a continuous operation. The forming of the strips gives the space inside the lamellas and bosses on the outside acting as spacers for the flow sections outside the lamellas on the shell side. The lamellas are welded together at both ends by joining the ends with steel bars in between, depending on the

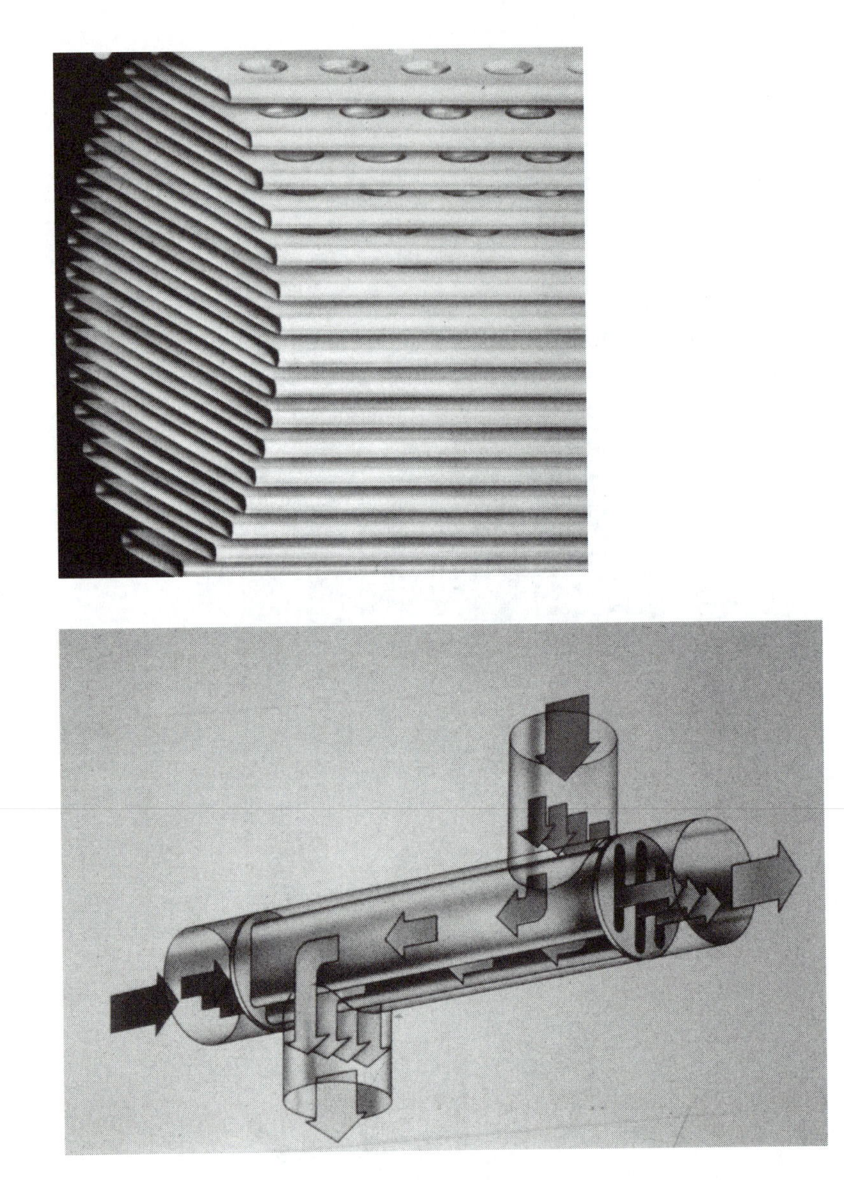

FIGURE 1.16
Lamella heat exchanger. (Courtesy of Alfa Laval Thermal AB.)

space required between lamellas. Both ends of the lamella bundle are joined by peripheral welds to the channel cover, which at the outer ends is welded to the inlet and outlet nozzle. The lamella side is thus completely sealed in by welds. At the fixed end the channel cover is equipped with an outside flange ring that is bolted to the shell flange. The flanges are of the spigot and recess type, where the spigot is an extension of the shell. The difference in expansion between the heating surface and the shell is well taken care of by a box in the floating end; this design improves the reliability and protects the lamella bundle against failure because of thermal stresses and strain from external forces. The end connection is here designed with a removable flange. By removing this flange and loosening the fixed-end shell flanges, the lamella bundle can be pulled out of the shell. The surfaces inside the lamellae are suitable for chemical cleaning. Therefore, fouling fluids should flow through the shell side. The channel walls either are plain or have dimples. The channels are welded into headers at each end of the plate bundle that is allowed to expand and contract independently of the shell by the use of a packing gland at the lower end. The shell-side flow is typically a single pass around the plates and flows longitudinally in the spaces between channels. There are no shell-side baffles, and therefore lamella heat exchangers can be arranged for true countercurrent flow. Because of the high turbulence, uniformly distributed flow, and smooth surfaces, the lamellas do not foul easily. The plate bundle can be easily removed for inspection and cleaning. This design is capable of pressure up to 35 bar, and temperature of 200°C for Teflon gaskets and 500°C for asbestos gaskets.

1.4.3 Extended Surface Heat Exchangers

Extended surface heat exchangers are devices having fins or appendages on the primary heat transfer surface (tubular or plate) with the object of increasing heat transfer area. Because it is well known that the heat transfer coefficient on the gas side is much lower than those on the liquid side, therefore, finned heat transfer surfaces are used on the gas side to increase the heat transfer area. Fins are widely used in gas-to-gas and gas–liquid heat exchangers whenever the heat transfer coefficient on one or both sides is low and there is a need for having a compact heat exchanger. The two most common types of extended surface heat exchangers are

1. Plate–fin
2. Tube–fin

(1) Plate–Fin Heat Exchangers

The plate–fin type is primarily used for gas-to-gas applications and tube–fin exchangers for liquid–air heat exchangers. In most of the applications (in trucks, cars, and airplanes), mass and volume reduction are particularly important. Because of this gain in volume and mass, compact heat exchangers are also widely used in cryogenic, energy recovery, process industry, refrigeration, and air-conditioning systems. Figure 1.17 shows the general form of a plate–fin heat exchanger. The fluid streams are separated by flat plates between which are sandwiched corrugated fins. They can be arranged into a variety of configurations with respect to the fluid streams. Figure 1.17 shows the arrangement for parallel flow or counterflow and crossflow between the streams. They are very compact units having a heat transfer area per unit volume of around 2000 m^2/m^3. Special manifold devices are provided

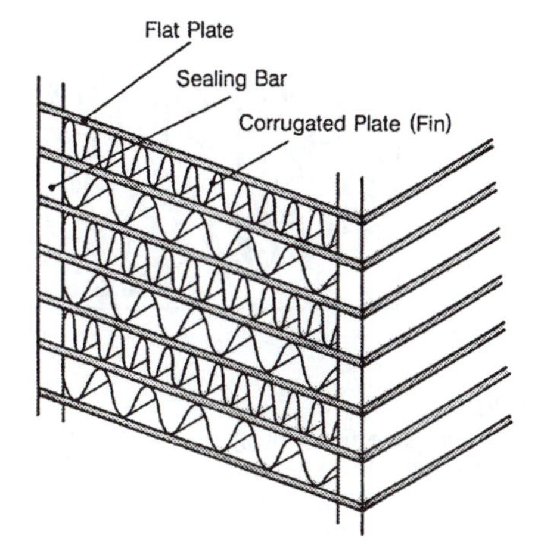

FIGURE 1.17
Basic construction of a plate–fin exchangers.

at the inlet to these exchangers to provide good flow distributions across the plates and from plate to plate. The plates are typically 0.5 to 1.0 mm thick and the fins 0.15 to 0.75 mm thick. The whole exchanger is made of an aluminum alloy, and the various components are brazed together in a salt bath or in a vacuum furnace.

The corrugated sheets that are sandwiched between the plates serve both to give extra heat transfer area and to give structural support to the flat plates. There are many different forms of corrugated sheets used in these exchangers, but the most common types are

1. Plain fin
2. Plain-perforated fin
3. Serrated fin (also called "lanced", "interrupted", "louver", or "multientry")
4. Herringbone or wavy fin

By the use of fins, discontinuous in the flow direction, the boundary layers can be completely disrupted; if the surface is wavy in the flow direction, the boundary layers are either thinned or interrupted, which results in higher heat transfer coefficients and higher pressure drop.

Figure 1.18 shows these four types.[4] The perforated type is essentially the same as the plain type except that it has been formed from a flat sheet with small holes in it. Many variations of interrupted fins have been used by industry.

The flow channels in plate–fin exchangers are small, which means that the mass velocity also has to be small (10 to 300 kg/[m^2·s]) to avoid excessive pressure drops. This may make the channel prone to fouling, which — when combined with the fact that they cannot be mechanically cleaned — means that plate–fin exchangers are restricted to clean fluids. They are frequently used for condensation duties in air liquefaction plants. Further information on these exchangers is given by Heat Transfer and Fluid Flow Service (HTFS.)[8]

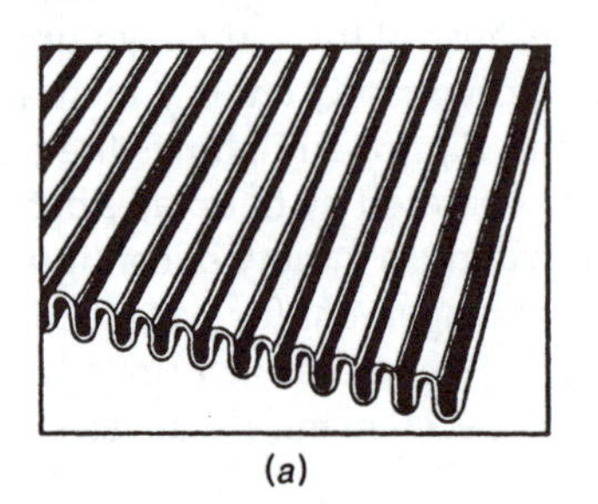

(*a*)

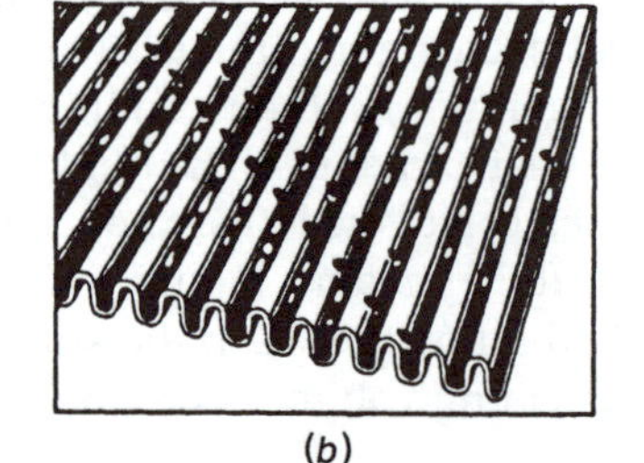

(*b*)

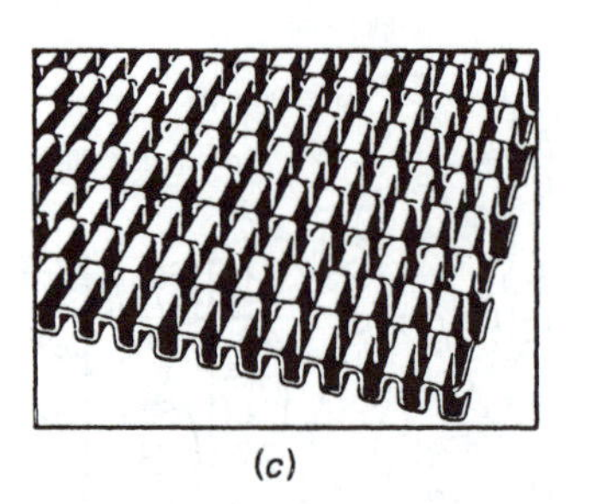

(*c*)

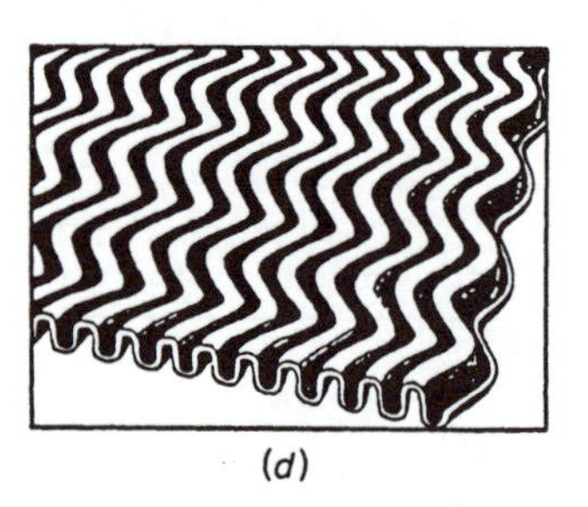

(*d*)

FIGURE 1.18
Fin types in plate–fin exchangers. (a) Plain; (b) perforated; (c) serrated; (d) herringbone. (Adapted from Butterworth, D. [1991] In *Boilers, Evaporators and Condensers*, Wiley, New York.)

Plate–fin heat exchangers have been established for use in gas turbines; conventional and nuclear power plants; propulsion engineering (airplanes, trucks, and automobiles); refrigeration, heating, ventilating, and air-conditioning; waste heat recovery systems; chemical industry; and cooling of electronic devices.

(2) Tubular-Fin Heat Exchangers

These heat exchangers are used as gas-to-liquid heat exchangers. The heat transfer coefficients on the gas side are generally much lower than those on the liquid side, and fins are required on that side. A tubular-fin heat exchanger consists of an array of tubes with fins fixed on the outside (Figures 1.19 to 1.21). The fins on the outside of the tubes may be normal on individual tubes, transverse or helical, or longitudinal (or axial) as shown in Figure 1.20. Longitudinal fins are commonly used in double-pipe or in shell-and-tube heat exchangers with no baffles. The fluids may be gases or viscous liquids (oil coolers). Alternately continuous plate–fin sheets may be fixed on the array of tubes. Examples are shown in Figure 1.19. As can be seen from Figure 1.19, in tube–fin exchangers, tubes of round, rectangular, or elliptical shapes are generally used. Fins are attached to the tubes by soldering, brazing, welding, extrusion, mechanical fit, tension wound, etc. Plate–fin–tube heat exchangers are commonly used in heating and ventilating, refrigeration, and air-conditioning systems.

Some of the extended surface heat exchangers are compact. A heat exchanger having a surface area density at least in one side of the heat transfer surface greater than 700 m^2/m^3 is quite arbitrarily referred to as a compact heat exchanger. These heat exchangers are generally used for applications where at least on one side of the heat transfer surface gas flows. These heat exchangers are generally of the plate–fin, tube–fin, and regenerative types. Extremely high heat transfer coefficients are achievable with small hydraulic diameter flow passages with gases.

Extended surfaces on the inside of the tubes are commonly used in the condensers and evaporators of refrigeration systems. Figure 1.22 shows examples of in-tube enhancement techniques for evaporating and condensing refrigerants.[9,10]

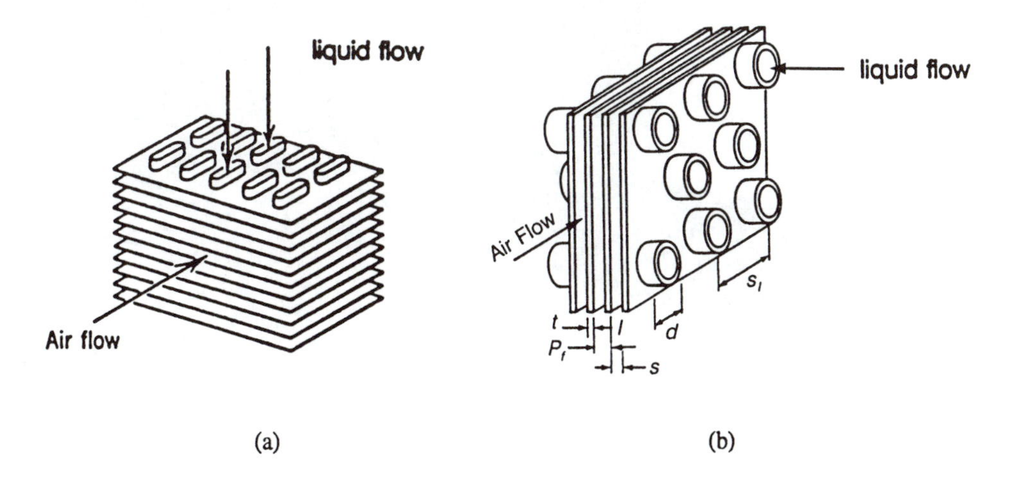

FIGURE 1.19
Tube–fin heat exchanger. (a) Flattened tube–fin; (b) round tube–fin.

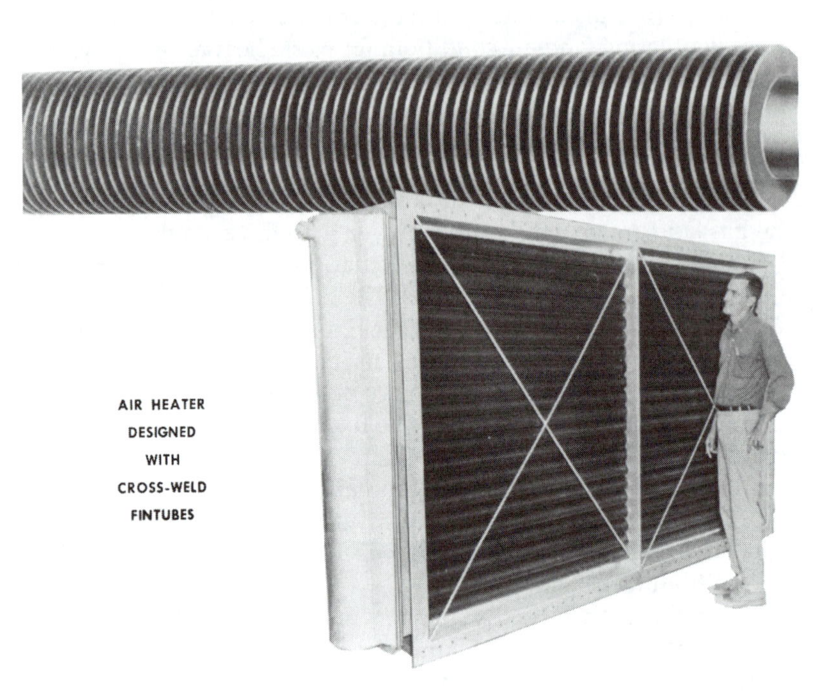

FIGURE 1.20
Fin–tube air heater. (Courtesy of Brown Fintube.)

FIGURE 1.21
Longitudinally finned tubes. (Courtesy of Brown Fintube.)

Air-cooled condensers and waste heat boilers are tube–fin exchangers that consist of a horizontal bundle of tubes with air or gas being blown across the tubes on the outside and condensation or boiling occurring inside the tubes (Figures 1.23[4] and 1.24[11]).

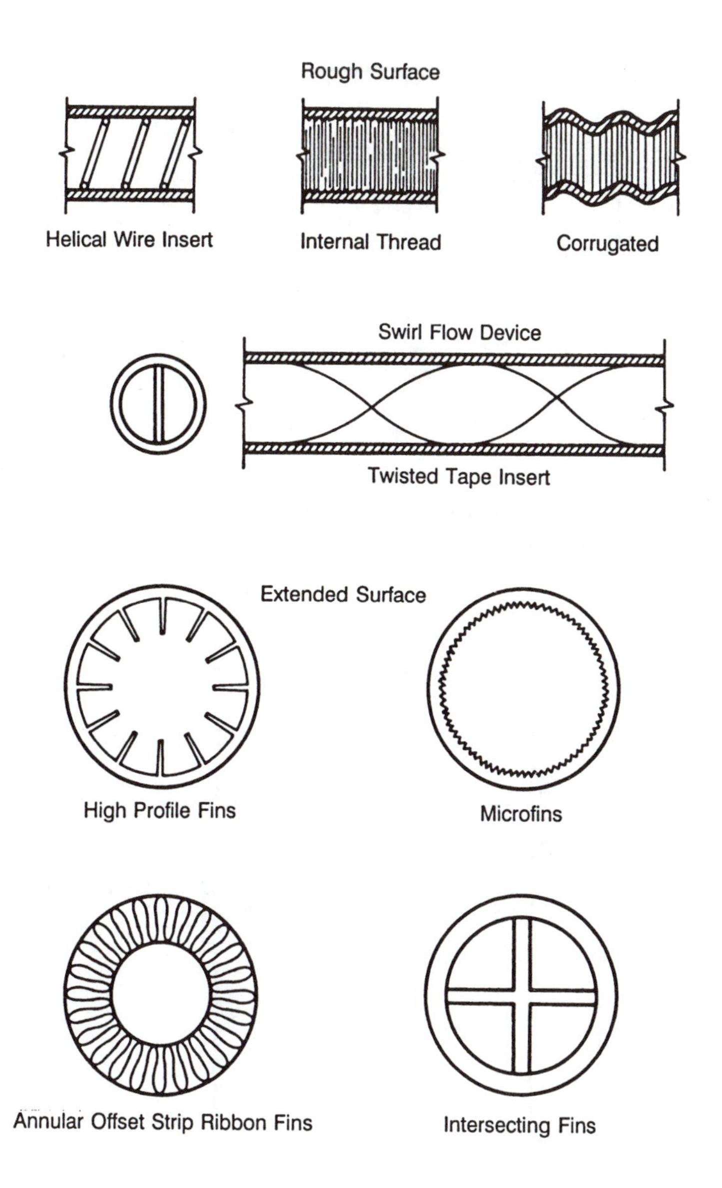

FIGURE 1.22
Examples of in-tube enhancement techniques for evaporating (and condensing) refrigerants. (From Pate, M. B. [1991] In *Boilers, Evaporators and Condensers*, Wiley, New York. With permission.)

An alternative design for air-cooled condensers is the induced-draft unit that has the fans on the top to suck the air over the tubes. The tubes are finned with transverse fins on the outside to overcome the effects of low air-side coefficients. There would normally be a few tube rows, and the process stream may take one or more passes through the unit. With multipass condensers, the problem arises with redistributing the two-phase mixture on entry to the next pass. This can be overcome in some cases by using U-tubes or by having separate passes just for subcooling or desuperheating duties. In multipass condensers, it is important to have each successive pass below the previous one to enable the condensate to continue downward. Further information about air-cooled heat exchangers is given by the American Petroleum Institute.[12]

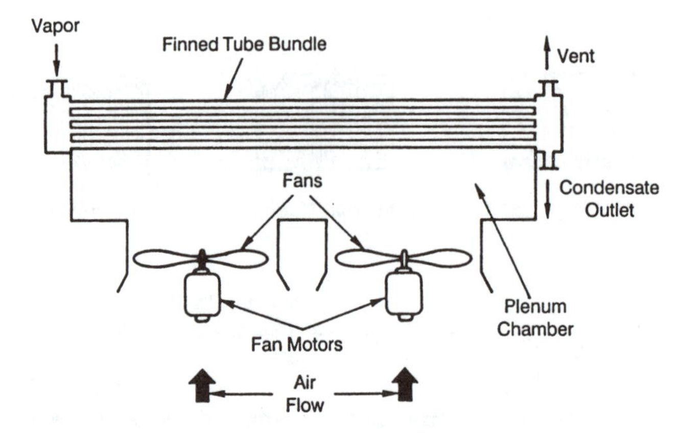

FIGURE 1.23
Forced-draft, air-cooled exchanger used as a condenser. (Adapted from Butterworth, D. [1991] In *Boilers, Evaporators and Condensers*, Wiley, New York.)

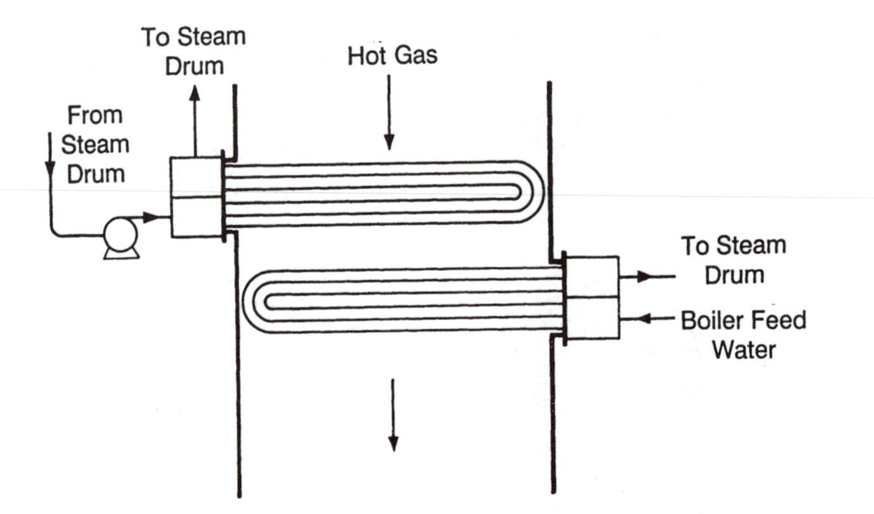

FIGURE 1.24
Horizontal U-tube waste heat boiler. (Adapted from Collier, J. G. [1988] In *Two-Phase Flow Heat Exchangers: Thermal–Hydraulic Fundamentals and Design*, Kluwer, Boston, MA.)

1.5 Heat Transfer Mechanisms

Heat exchanger equipment can also be classified according to the heat transfer mechanisms as:

1. Single-phase convection on both sides
2. Single-phase convection on one side, two-phase convection on other side
3. Two-phase convection on both sides

The principles of these types are illustrated in Figure 1.1e, f, and g. Figures 1.1f and g illustrate two possible modes of two-phase flow heat exchangers. In Figure 1.1f, fluid A is being evaporated, receiving heat from fluid B; and in Figure 1.1g, fluid A is being condensed, giving up heat to fluid B.

In heat exchangers like economizers and air heaters in boilers, compressor intercoolers, automative radiators, regenerators, oil coolers, space heaters, etc., single-phase convection occurs on both sides.

Condensers, boilers, and steam generators used in pressurized water reactors, power plants, evaporators, and radiators used in air-conditioning and space heating include the mechanisms of condensation, boiling, and radiation on one of the surfaces of the heat exchanger. Two-phase heat transfer could also occur on each side of the heat exchanger such as condensing on one side and boiling on the other side of the heat transfer surface. However, without phase change, we may also have a two-phase flow heat transfer mode as in the case of fluidized beds where a mixture of gas and solid particles transports heat to or from a heat transfer surface.

1.6 Flow Arrangements

Heat exchangers may be classified according to the fluid–flow path through the heat exchanger. The three *basic* configurations are

1. Parallel flow
2. Counter flow
3. Cross flow

In parallel flow (cocurrent) heat exchangers, the two fluid streams enter together at one end, flow through in the same direction, and leave together at the other end (Figure 1.25a). In counterflow (countercurrent) heat exchangers, two fluid streams flow in opposite directions (Figure 1.25b). In single-crossflow heat exchangers, one fluid flows through the heat transfer surface at right angles to the flow path of the other fluid. Crossflow arrangements with both fluids unmixed, and one fluid mixed and the other fluid unmixed are illustrated in Figures 1.25c and 1.25d, respectively. Multipass crossflow configurations can also be arranged by having the basic arrangements in a series. For example, in a U-baffled tube single-pass shell-and-tube heat exchanger, one fluid flows through the U-tube while the other fluid flows first downward and then upward, crossing the flow path of the other fluid stream, which is also referred to as crosscounter, cross-parallel flow arrangements (Figure 1.25c).

The multipass flow arrangements are frequently used in heat exchanger designs, especially in shell-and-tube heat exchangers with baffles (Figures 1.10 to 1.12). The main difference between the flow arrangements lies in the temperature distribution along the length of the heat exchanger, and the relative amounts of heat transfer under given temperature specifications for specified heat exchanger surfaces (i.e., as will be shown later for given flow and specified temperatures, a counterflow heat exchanger requires a minimum area, a parallel flow heat exchanger requires a maximum area, while a crossflow heat exchanger requires an area in between).

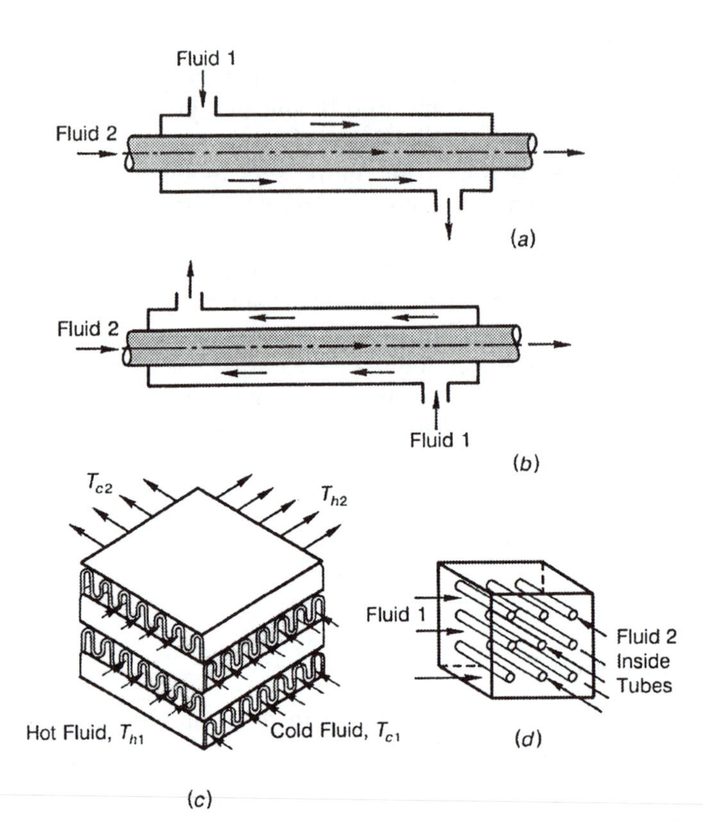

FIGURE 1.25
Heat exchanger classification according to flow arrangement. (a) Parallel-flow; (b) counterflow; (c) crossflow, both fluids unmixed; (d) crossflow, fluid 1 mixed, fluid 2 unmixed.

In the crossflow arrangement, the flow may be called *mixed* or *unmixed*, depending on the design. Figure 1.25c shows an arrangement in which both hot and cold fluids flow through individual flow channels with no fluid mixing between adjacent flow channels. Then each fluid stream is said to be unmixed.

In the flow arrangements shown in Figure 1.25d, fluid 2 flows inside the tubes, thus is not free to move in the transverse direction, and therefore is considered unmixed; on the other hand, fluid 1 is free to move in the transverse direction and mix itself, and therefore is called an *unmixed–mixed* crossflow heat exchanger.

For extended surface heat exchangers, it is also possible to have the basic crossflow system in a series to form multipass arrangements as cross-counterflow and cross-parallel flow. These usually help to increase the overall effectiveness of the heat exchanger.

In a gasketed-plate heat exchanger it is also possible to have more than one pass simply by properly gasketing around the parts in the plates.

1.7 Applications

Most common heat exchangers are two-fluid heat exchangers. Three-fluid heat exchangers are widely used in cryogenics. They are also used in various heat exchangers that are used in chemical and process industries, such as air separation systems, purification and liquefaction of hydrogen, and ammonia gas synthesis. Three- and multicomponent heat exchangers are very complex to design. They may also include multicomponent two-phase convection as in condensation of mixed vapors in the distillation of hydrocarbons.

Heat exchangers are used in a wide variety of applications as in the process, power, air-conditioning, refrigeration, cryogenics, heat recovery, and manufacturing industries.[11] In the power industry, various kinds of fossil boilers, nuclear steam generators, steam condensers, regenerators, and cooling towers are used. In the process industry, two-phase flow heat exchangers are used for vaporizing, condensing, and freezing in crystallization, and for fluidized beds with catalytic reaction. The air-conditioning and refrigeration industries need condensers and evaporators.

Energy can be saved by direct contact condensation. By direct contact condensation of a vapor in liquid of the same substance under high pressure, thermal energy can be stored in a storage tank. When the energy is needed again, the liquid is depressurized and flashing occurs that results in producing vapor. The vapor can then be used for heating or as a working fluid for an engine.

There have been abrupt developments in heat exchanger applications.[12,13] One of the main steps for the early development of boilers was the introduction of the water–tube boilers. The demand for more powerful engines created a need for boilers that operated at higher pressures, and as a result, individual boilers were built larger and larger. The boiler units used in modern power plants for steam pressures above 1200 $lb_f/in.^2$ (80 bar) consist of furnace water–wall tubes, superheaters, and such heat recovery accessories as economizers and air heaters (illustrated in Figure 1.26).[14,15] The development of modern boilers

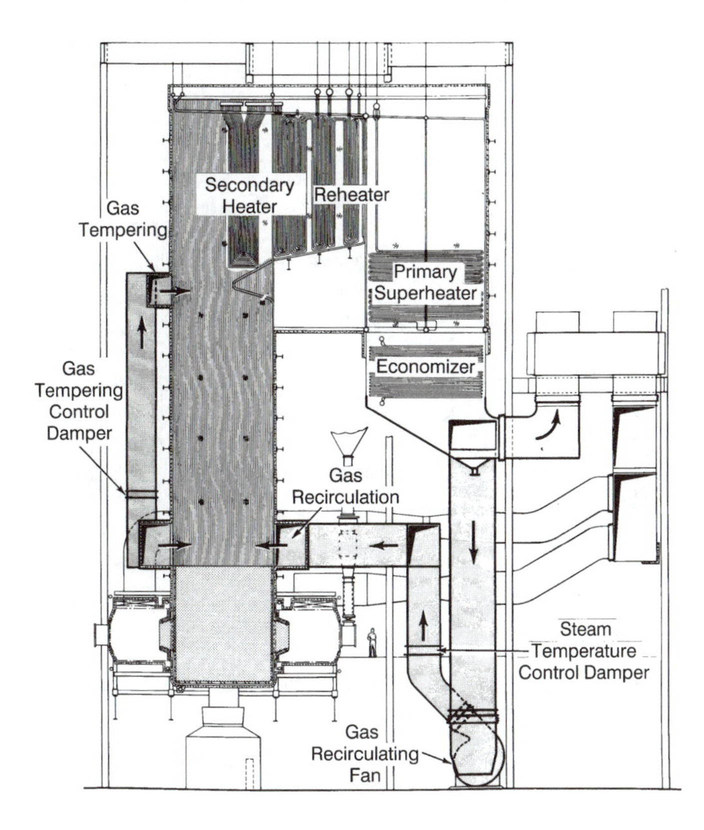

FIGURE 1.26
Boiler with water-cooled furnace with heat recovery units. (From Kitto, J. B., Jr. and Albrecht, M. J. [1991] In *Boilers, Evaporators and Condensers*, Wiley, New York.)

and of more efficient condensers for the power industry has represented a major milestone in engineering. In the pressurized water reactors (PWR), large-sized, inverted U-tube types of steam generators, each providing steam 300 to 400 MW of electrical power, are constructed (Figure 1.27).[11] In the process industry, engineers are concerned with designing equipment to vaporize a liquid. In the chemical industry, the function of an evaporator is to vaporize a liquid (vaporizer) or to concentrate a solution by vaporizing part of the solvent. Evaporizers may also be used in the crystallization process. Often, the solvent is water, but in many cases the solvent is valuable and recovered for reuse. The vaporizers used in the process chemical industry cover a wide range of sizes and applications.[9,13,16,17] The wide variety of applications of heat exchanger equipment is shown in Figure 1.28.

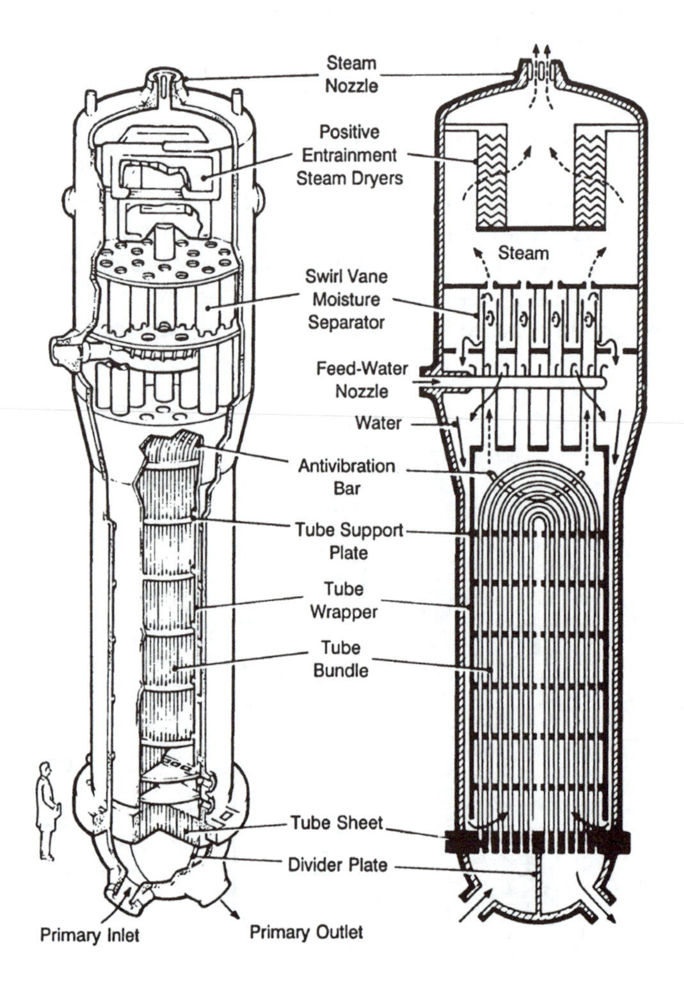

FIGURE 1.27
Inverted U-tube type of steam generator. (Adapted from Collier, J. G. [1988] In *Two-Phase Flow Heat Exchangers: Thermal–Hydraulic Fundamentals and Design*, Kluwer, Boston, MA.)

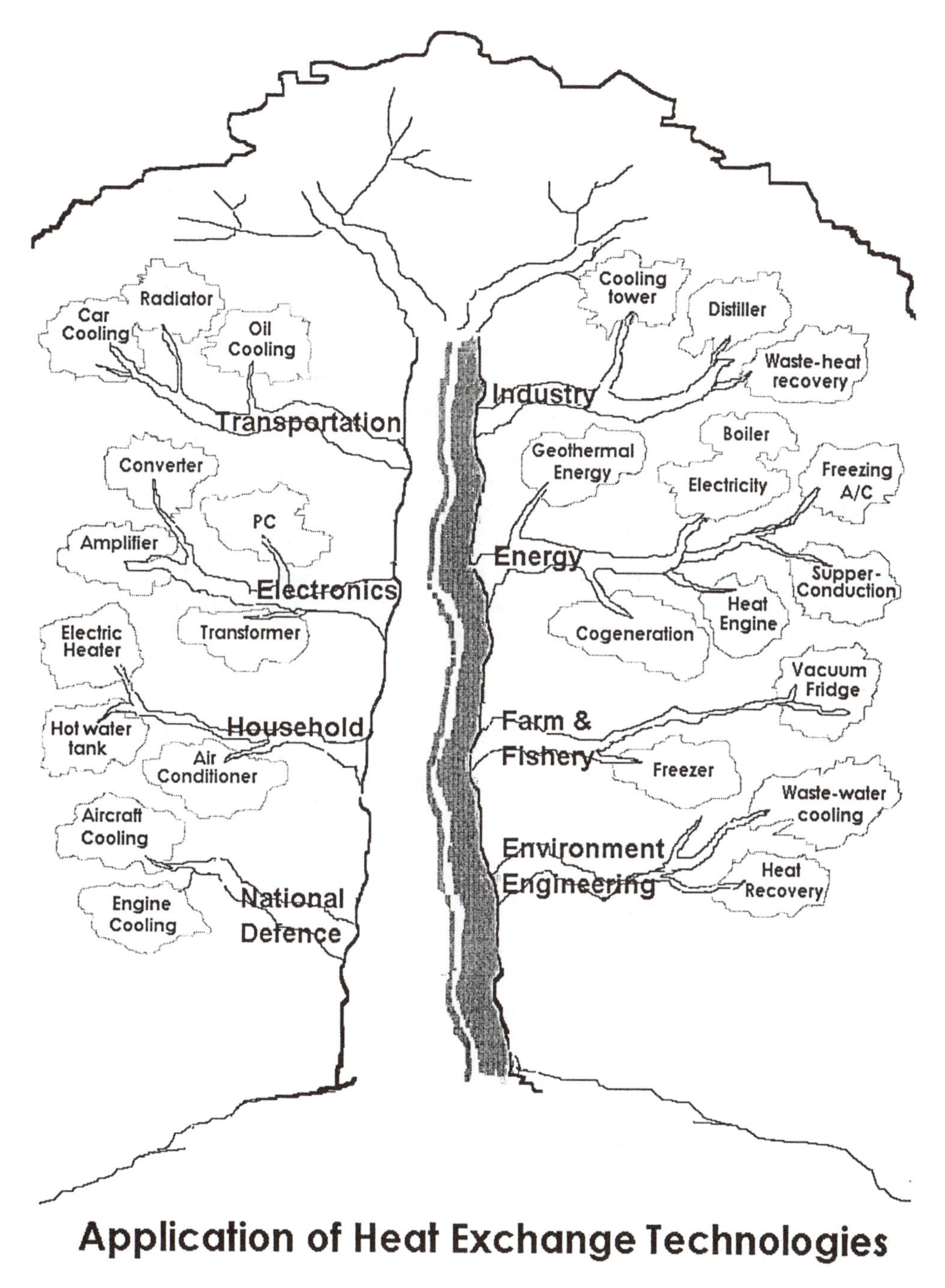

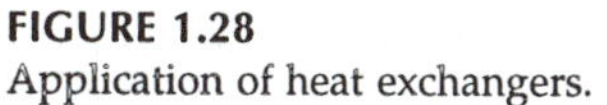

FIGURE 1.28
Application of heat exchangers.

1.8 Selection of Heat Exchangers

The basic criteria for heat exchanger selection from various types available are:[18]

- The heat exchanger must satisfy the process specifications; it must continue to the next scheduled shutdown of the plant for maintenance.
- The heat exchanger must withstand the service conditions of the plant environment. It must also resist corrosion by the process and service streams as well as the environment. The heat exchanger should also resist fouling.
- The exchanger must be maintainable. This usually implies choosing a configuration that permits cleaning; and the replacement of any components that may be especially vulnerable to corrosion, erosion, or vibration. This requirement will dictate the positioning of the exchanger and the space requirement around it.
- The heat exchanger should be cost-effective. The installed operating, maintenance costs including the loss of production due to exchanger unavailability must be calculated and the exchanger should cost as little as possible.
- There may be limitations on exchanger diameter, length, weight, and tube configurations due to site requirements, lifting and servicing, capabilities, or inventory considerations.[18]

References

1. Hewitt, G. F., Shires, G. L., and Bott, T. R. (1994) *Process Heat Transfer*, CRC Press, Boca Raton, FL.
2. Shah, R. K. (1981) Classification of heat exchangers. In *Heat Exchangers — Thermo-Hydraulic Fundamentals and Design*, S. Kakac, A. E. Bergles, and F. Mayinger (Eds.), Wiley, New York.
3. Lin, Z. H. (1991) Thermohydraulic design of fossil-fuel boiler components. In *Boilers, Evaporators and Condensers*, S. Kakac (Ed.), Wiley, New York.
4. Butterworth, D.(1991) Steam power plant and process condensers. In *Boilers, Evaporators and Condensers*, S. Kakac (Ed.), Wiley, New York.
5. Butterworth, D. (1988) Condensers and their design. In *Two-Phase Flow Heat Exchangers: Thermal–Hydraulic Fundamentals and Design*, S. Kakac, A. E. Bergles, and E. O. Fernandes (Eds.), Kluwer, Dordrecht, The Netherlands.
6. *Standards of the Tubular Exchanger Manufacturers Association* (1988), 7th ed., Tarrytown, NY.
7. Alfa Laval (1980) *Heat Exchanger Guide*, Lund, Sweden.
8. HTFS Plate–Fin Study Group (1987) *Plate–Fin Heat Exchangers — Guide to Their Specifications and Use*, M. A. Taylor (Ed.), HTFS, Oxfordshire, U.K.
9. Pate, M. B. (1991) Evaporators and condensers for refrigeration and air-conditioning systems. In *Boilers, Evaporators and Condensers*, S. Kakac (Ed.), Wiley, New York.
10. Pate, M. B. (1988) Design considerations for air-conditioning evaporators and condenser coils. In *Two-Phase Flow Heat Exchangers: Thermal–Hydraulic Fundamentals and Design*, S. Kakac, A. E. Bergles, and E. O. Fernandes (Eds.), Kluwer, Dordrecht, The Netherlands.
11. Collier, J. G. (1988) Evaporators. In *Two-Phase Flow Heat Exchangers: Thermal–Hydraulic Fundamentals and Design*, S. Kakac, A. E. Bergles, and E. O. Fernandes (Eds.), Kluwer, Boston, MA.
12. American Petroleum Institute (1986) *Air Cooled Heat Exchangers for General Refinery Service*, API Standard 661, Washington, D.C.

13. Mayinger, F. (1988) Classification and applications of two-phase flow heat exchangers. In *Two-Phase Flow Heat Exchangers: Thermal–Hydraulic Fundamentals and Design*, S. Kakac, A.E. Bergles, and E. O. Fernandes (Eds.), Kluwer, Boston, MA.
14. Kitto, J. B., Jr. and Albrecht, M. J. (1991), Fossil-fuel-fired boilers: fundamentals and elements. In *Boilers, Evaporators and Condensers*, S. Kakac (Ed.), Wiley, New York.
15. *Steam, Its Generation and Use.* (1992) Babcock & Wilcox, New York.
16. Schlünder, E. U. et al. (Eds.), (1988) *Heat Exchanger Design Handbook*, Hemisphere, New York.
17. Smith, R. A. (1986) *Vaporizers — Selection, Design and Operation*, Wiley, New York.
18. Bell, K. J. (1981) Preliminary design of shell and tube heat exchangers. In *Heat Exchangers: Thermal–Hydraulic Fundamentals and Design*, S. Kakac, A. E. Bergles, and F. Mayinger (Eds.), Hemisphere, New York.

Problems

1.1. What are the principal types of shell-and-tube construction?

1.2. What are the advantages and disadvantages of the principal types of shell-and-tube construction?

1.3. What are the different binds of tubes used in heat exchangers?

1.4. Why are baffles used in shell-and-tube heat exchangers?

1.5. Where are fins (extended surfaces) used?

1.6. What are the types of fins that are used in heat exchangers?

1.7. What is fin efficiency? What does it depend on?

1.8. What are enhanced surfaces and what are their advantages?

1.9. What characterizes a unit to be compact?

1.10. What are the types of heat exchangers used in a conventional plant? Classify them and discuss.

1.11. What are the advantages and the limitations of the gasketed-type plate heat exchangers?

1.12. What are the different kinds of spiral heat exchangers and what are their limitations?

1.13. Name the specific exchanger construction type that may be used in the following applications:

a. Milk pasteurizing

b. Power condenser

c. Automotive radiator

d. Marine oil cooler

e. Air-cooled condenser

1.14. What are the main selection criteria of a heat exchanger?

2

Basic Design Methods of Heat Exchangers

2.1 Introduction

The most common problems in heat exchanger design are rating and sizing. In this chapter, the basic design methods for two-fluid direct-transfer heat exchangers (recuperators) are reviewed.

The rating problem is concerned with the determination of the heat transfer rate and the fluid outlet temperatures for prescribed fluid flow rates, inlet temperatures, and allowable pressure drop for an existing heat exchanger; hence the heat transfer surface area and the flow passage dimensions are available.

The sizing problem, on the other hand, is concerned with the determination of the dimensions of the heat exchanger, that is, selecting an appropriate heat exchanger type and determining the size to meet the specified hot- and cold-fluid inlet and outlet temperatures, flow rates, and pressure drop requirements.

2.2 Arrangement of Flow Path in Heat Exchangers

A recuperator-type heat exchanger is classified according to the flow direction of the hot- and cold-fluid streams and the number of passes made by each fluid as it passes through the heat exchanger as discussed in Chapter 1. Therefore heat exchangers may have the following patterns of flow: (1) parallel flow with two fluids flowing in the same direction (Figure 2.1a); (2) counterflow with two fluids flowing parallel to one another but in opposite directions (Figure 2.1b); (3) crossflow with two fluids crossing each other (Figures 2.1c and d); and (4) mixed flow where both fluids are simultaneously in parallel flow, in counterflow (Figures 2.2a and b), and in multipass crossflow (Figure 2.2c). Applications include various shell-and-tube heat exchangers.[1]

2.3 Basic Equations in Design

The term *heat exchanger*, although applicable to all four categories listed previously, will be used in this chapter to designate a recuperator in which heat transfer occurs between two fluid streams that are separated by a heat transfer surface. Basic heat transfer equations will be outlined for the thermal analysis (sizing and rating calculations) of such heat

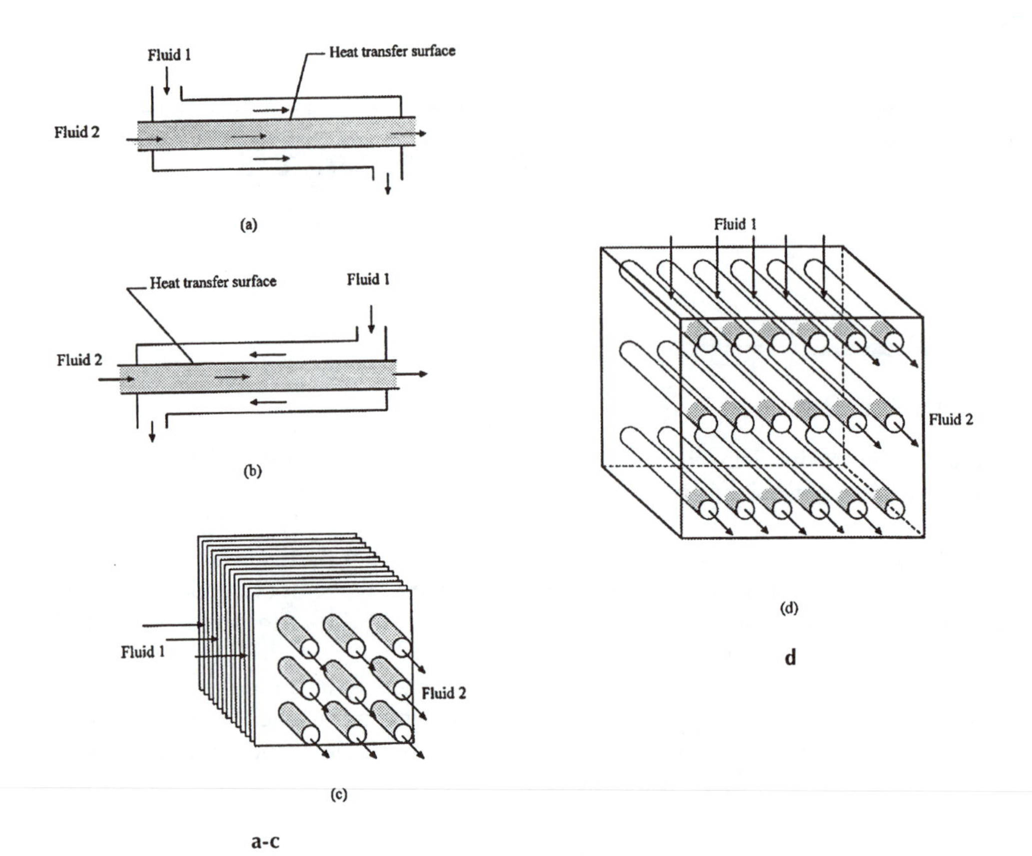

FIGURE 2.1
Heat exchanger classification according to flow arrangements.

exchangers. Although complete design of a heat exchanger requires structural and economical considerations in addition to these basic equations, the purpose of the thermal analysis given here will be to determine the heat transfer surface area of the heat exchanger (sizing problem). Performance calculations of a heat exchanger (rating problem) are conducted when the heat exchanger is available, but it is necessary to find the amount of heat transferred, pressure losses, and outlet temperatures of both fluids.

The temperature variations in usual fluid-to-fluid heat transfer processes, depending on the flow path arrangement, are shown in Figure 2.3, in which the heat transfer surface area A is plotted along the x axis and the temperature of the fluids is plotted along the y axis. Counterflow heat transfer with the two fluids flowing in opposite directions is shown in Figure 2.3a. Parallel–flow heat transfer with the two fluids flowing in the same direction is shown in Figure 2.3b. Heat transfer with the cold fluid at constant temperature (evaporator) is shown in Figure 2.3c. Heat transfer with the hot fluid at constant temperature (condenser) is shown in Figure 2.3d. The nature of the temperature profiles also depends on the heat capacity rate ($\dot{m}c_p$) of the fluids and is shown later.

From the first law of thermodynamics for an open system, under steady-state, steady flow conditions, with negligible potential and kinetic energy changes, the change of enthalpy of one of the fluid streams is (Figure 2.4):

$$\delta Q = \dot{m}di \tag{2.1}$$

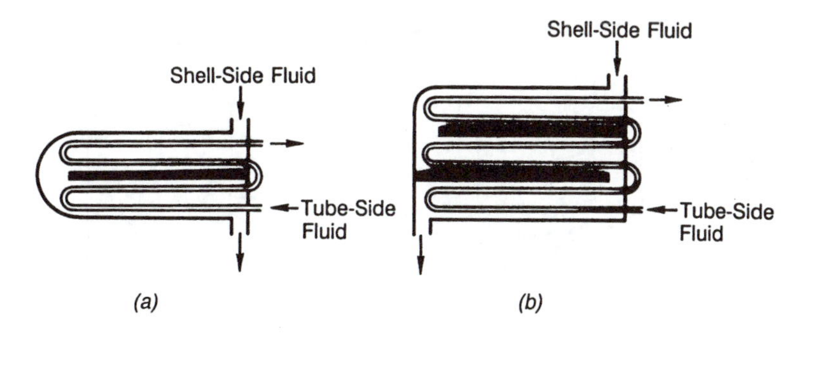

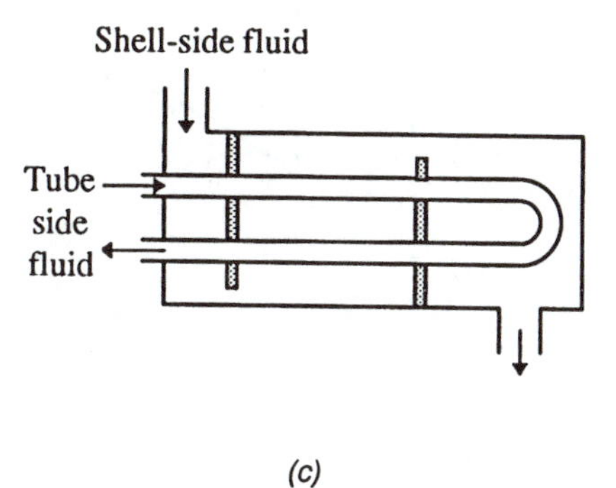

(c)

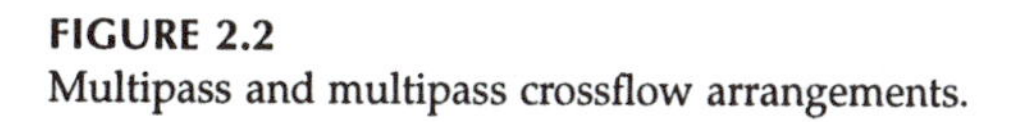

FIGURE 2.2
Multipass and multipass crossflow arrangements.

where $\dot{m}$ is the rate of mass flow, i is the specific enthalpy, and δQ is the heat transfer rate to the fluid concerned associated with the infinitesimal state change. Integration of Equation 2.1 gives:

$$Q = \dot{m}\ (i_2 - i_1) \tag{2.2}$$

where i_1 and i_2 represent the inlet and outlet enthalpies of the fluid stream. Equation (2.2) holds for all processes of Figure 2.3. Note that δQ is negative for the hot fluid. If there is negligible heat transfer between the exchanger and its surroundings (adiabatic process), integration of Equation 2.1 for hot and cold fluids gives:

$$Q = \dot{m}_h\ (i_{h1} - i_{h2}) \tag{2.3}$$

and

$$Q = \dot{m}_c\ (i_{c2} - i_{c1}) \tag{2.4}$$

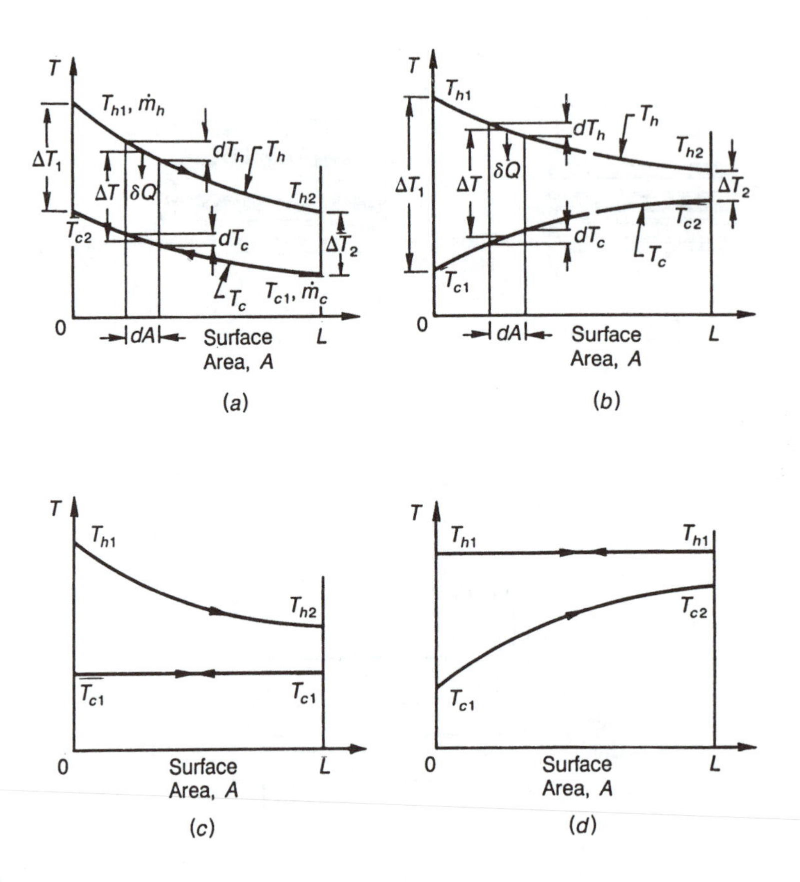

FIGURE 2.3
Fluid temperature variation in parallel-flow, counterflow, evaporator, and condenser heat exchangers. (a) Counterflow; (b) parallel flow; (c) cold fluid evaporating at constant temperature; (d) hot fluid condensing at constant temperature.

The subscripts h and c refer to the hot and cold fluids, where 1 and 2 designate the fluid inlet and outlet conditions. If the fluids do not undergo a phase change and have constant specific heats with $di = c_p dT$, then Equations (2.3) and (2.4) can be written as:

$$Q = (\dot{m}c_p)_h\,(T_{h1} - T_{h2}) \tag{2.5}$$

and

$$Q = (\dot{m}c_p)_c(T_{c2} - T_{c1}) \tag{2.6}$$

As can be seen from Figure 2.3, the temperature difference between the hot and cold fluids ($\Delta T = T_h - T_c$) varies with position in the heat exchanger. Therefore, in the heat transfer analysis of heat exchangers, it is convenient to establish an appropriate mean value of the temperature difference between the hot and cold fluids such that the total heat transfer rate Q between the fluids can be determined from the following equation:

$$Q = UA\Delta T_m \tag{2.7}$$

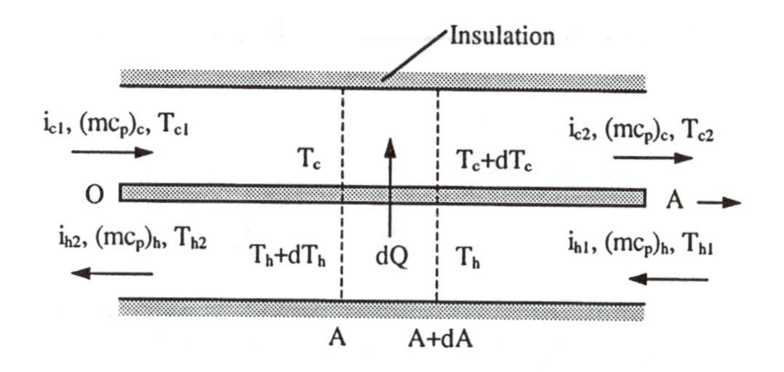

FIGURE 2.4
Overall energy balance for the hot and cold fluids of a two-fluid heat exchanger.

where A is the total hot-side or cold-side heat transfer area and U is the average overall heat transfer coefficient based on that area. ΔT_m is a function of T_{h1}, T_{h2}, T_{c1}, and T_{c2}. Therefore a specific form of ΔT_m must be obtained.

Equations (2.5) to (2.7) are the basic equations for the thermal analysis of a heat exchanger under steady-state conditions. If Q, the total heat transfer rate, is known from Equation (2.5) or (2.6), then Equation (2.7) is used to calculate the heat transfer surface area A. Therefore, it is clear that the problem of calculating the heat transfer area comes down to determining the overall heat transfer coefficient and the mean temperature difference ΔT_m.

2.4 Overall Heat Transfer Coefficient

Heat exchanger walls are usually made of a single material, although the wall may sometimes be bimetallic (steel with aluminum cladding) or coated with a plastic as a protection against corrosion. Most heat exchanger surfaces tend to acquire an additional heat transfer resistance that increases with time. This either may be a very thin layer of oxidation, or — at the other extreme — may be a thick crust deposit, such as that which results from a salt-water coolant in steam condensers. This fouling effect can be taken into consideration by introducing an additional thermal resistance, termed the fouling resistance R_s. Its value depends on the type of fluid, fluid velocity, type of surface, and length of service of the heat exchanger.[2–4] Fouling will be discussed in a separate chapter.

In addition, fins are often added to the surfaces exposed to either or both fluids; and, by increasing the surface area, they reduce the resistance to convection heat transfer. The overall heat transfer coefficient for a single smooth and clean plane wall can be calculated from

$$UA = \frac{1}{R_t} = \frac{1}{\dfrac{1}{h_i A} + \dfrac{t}{kA} + \dfrac{1}{h_o A}} \tag{2.8}$$

where R_t is the total thermal resistance to heat flow across the surface between the inside and outside flow; t is the thickness of the wall; and h_i and h_o are heat transfer coefficients for inside and outside flows, respectively.

For the unfinned and clean tubular heat exchanger, the overall heat transfer coefficient is given by:

$$U_oA_o = U_iA_i = \frac{1}{R_t} = \frac{1}{\dfrac{1}{h_iA_i} + \dfrac{\ln(r_o/r_i)}{2\pi kL} + \dfrac{1}{h_oA_o}} \tag{2.9}$$

If the heat transfer surface is fouled with the accumulation of deposits, this in turn introduces an additional thermal resistance in the path of heat flow.[2] We define a scale coefficient of heat transfer h_s in terms of the thermal resistance R_s of this scale as:

$$\frac{\Delta T_s}{Q} = R_s = \frac{1}{Ah_s} \tag{2.10}$$

where the area A is the original heat transfer area of the surface before scaling and ΔT_s is the temperature drop across the scale. $R_f = 1/h_s$ is termed as fouling factor (i.e., unit fouling resistance) that has the unit of m^2·K/W. This is discussed in detail in the following chapters, and tables are provided for the values of R_f.

We now consider heat transfer across a heat exchanger wall fouled by deposit formation on both the inside and outside surfaces. The total thermal resistance R_t can be expressed as:

$$R_t = \frac{1}{UA} = \frac{1}{U_oA_o} = \frac{1}{U_iA_i} = \frac{1}{h_iA_i} + R_w + \frac{R_{fi}}{A_i} + \frac{R_{fo}}{A_o} + \frac{1}{A_oh_o} \tag{2.11}$$

The calculation of an overall heat transfer coefficient depends on whether it is based on the cold- or hot-side surface area, because $U_o \neq U_i$ if $A_o \neq A_i$. The wall resistance R_w is obtained from the following equations:

$$R_w = t/kA \qquad \text{(for bare plane wall)} \tag{2.12a}$$

$$R_w = \frac{\ln(r_o/r_i)}{2\pi Lk} \qquad \text{(for a bare tube wall)} \tag{2.12b}$$

A separating wall may be finned differently on each side (Figure 2.5). On either side, heat transfer takes place from the fins (subscript f) as well as from the unfinned portion of the wall (subscript u). By introducing the fin efficiency η_f, the total heat transfer can be expressed as:

$$Q = (\eta_f A_f h_f + A_u h_u)\Delta T \tag{2.13}$$

where ΔT is either $(T_h - T_{w1})$ or $(T_{w2} - T_c)$. The subscripts h and c refer to the hot and cold fluids, respectively (see Figure 2.5).

By taking $h_u = h_f = h$ and rearranging the right-hand side of Equation (2.13), we get:

$$Q = hA\left[1 - \frac{A_f}{A}(1-\eta_f)\right]\Delta T \tag{2.14}$$

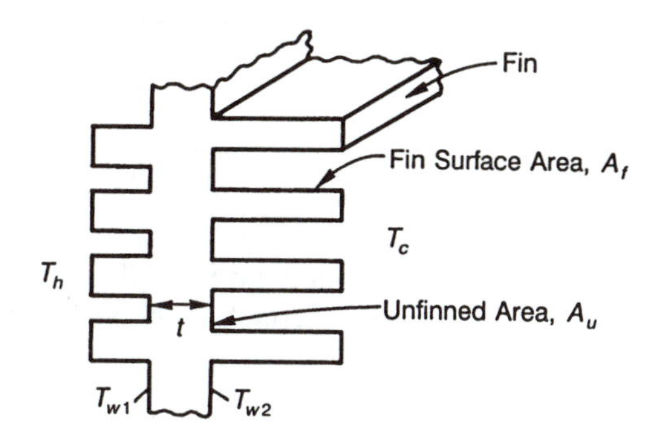

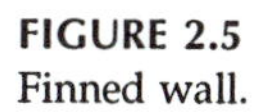
FIGURE 2.5
Finned wall.

or

$$Q = \eta h A \Delta T \tag{2.15}$$

where $\eta = [1 - (1-\eta_f)A_f/A]$ is called the overall surface efficiency, ΔT is the temperature difference between the fluid stream and the base temperature, and the total surface area on one side is $A = A_u + A_f$.

As can be seen from Equation (2.15), there will be additional thermal resistances for finned surfaces as $1/\eta h A$ on both sides of the finned wall; this is the combined effective surface resistance that accounts for parallel heat flow paths by conduction and convection in the fins and by convection from the prime surface.

Therefore, an overall thermal resistance for the entire wall is then given by:

$$R_t = \frac{1}{UA} = \frac{1}{U_o A_o} = \frac{1}{U_i A_i} = \frac{1}{\eta_i h_i A_i} + \frac{R_{fi}}{\eta_i A_i} + R_w + \frac{R_{fo}}{\eta_o A_o} + \frac{1}{\eta_o h_o A_o} \tag{2.16}$$

It should be noted that for finned surfaces, A_o and A_i represent total surface area of the outer and inner surfaces, respectively.

Contact resistance may be finite between a tube or a plate and the fin. In this case the contact resistance terms on the hot and cold sides are added to Equation (2.16).

In heat exchanger applications, the overall heat transfer coefficient is usually based on the outer area (cold side or hot side). Then Equation (2.16) can be represented in terms of the overall heat transfer coefficient based on the outside surface area of the wall as:

$$U_o = \frac{1}{\dfrac{A_o}{A_i}\dfrac{1}{\eta_i h_i} + \dfrac{A_o R_{fi}}{\eta_i A_i} + A_o R_w + \dfrac{R_{fo}}{\eta_o} + \dfrac{1}{\eta_o h_o}} \tag{2.17}$$

The expressions or magnitude of η_f for a variety of fin configurations is available in the literature.[5] If a straight or pin fin of length L and uniform cross section are used and adiabatic tip is assumed, then the fin efficiency is given by:

$$\eta_f = \frac{\tanh(mL)}{mL} \tag{2.18}$$

where

$$m = \sqrt{\frac{2h}{\delta k_f}} \tag{2.19}$$

where δ is the fin thickness and L is the fin length.

For the unfinned, tubular heat exchangers shown in Figures 2.1a and b and Figure 2.2, Equation (2.17) reduces to:

$$U_o = \frac{1}{\frac{r_o}{r_i}\frac{1}{h_i} + \frac{r_o}{r_i}R_{fi} + \frac{r_o \ln(r_o/r_i)}{k} + R_{fo} + \frac{1}{h_o}} \tag{2.20}$$

The overall heat transfer coefficient can be determined from knowledge of the inside and outside heat transfer coefficients, fouling factors, and appropriate geometric parameters.

2.4.1 Order of Magnitude of Thermal Resistance

For a plane wall of thickness t, and h_i and h_o on either side with fouling only on one side, Equation (2.17) becomes:

$$\frac{1}{U} = \frac{1}{h_i} + R_{fi} + \frac{t}{k} + \frac{1}{h_o} \tag{2.21}$$

The order of magnitude and range of h for various conditions are given in Table 2.1.

TABLE 2.1
Order of Magnitude of Heat Transfer Coefficient

Fluid	h, W/m²·K
Gases (natural convection)	3–25
Engine oil (natural convection)	30–60
Flowing superheated steam or air	30–300
Flowing liquids (nonmetal)	100–10,000
Flowing liquid metals	5,000–250,000
Boiling heat transfer	
Water, pressure <5 bar abs, ΔT = 25 K	5,000–10,000
Water, pressure 5–100 bar abs, ΔT = 20 K	4,000–15,000
Film boiling	300–400
Condensing vapor heat transfer at 1 atm	
Film condensation on horizontal tubes	9,000–25,000
Film condensation on vertical surfaces	4,000–11,000
Dropwise condensation	60,000–120,000

Example 2.1

Determine the overall heat transfer coefficient U for liquid-to-liquid heat transfer through a 0.003-m thick steel plates (k = 50 W/m·K) for the following heat transfer coefficients and fouling factor on one side:

$$h_i = 2500 \text{ W/m}^2\cdot\text{K} \quad h_o = 1800 \text{ W/m}^2\cdot\text{K}$$

$$R_{fi} = 0.0002 \text{ m}^2\cdot\text{K/W}$$

By substituting h_i, h_o, R_{fi}, t, and k into Equation (2.21), we get:

$$\frac{1}{U} = \frac{1}{2500} + 0.0002 + \frac{0.003}{50} + \frac{1}{1800}$$

$$= 0.0004 + 0.0002 + 0.00006 + 0.00056 = 0.00122 \text{ m}^2\cdot\text{K/W}$$

$$U \approx 820 \text{ W/m}^2\cdot\text{K}$$

In this case none of the resistances are negligible.

Example 2.2

In Example 2.1, replace one of the flowing liquids by a flowing gas ($h_o = 50$ W/m²·K):

$$\frac{1}{U} = \frac{1}{2500} + 0.0002 + \frac{0.003}{50} + \frac{1}{50}$$

$$= 0.0004 + 0.0002 + 0.00006 + 0.02 = 0.02066 \text{ m}^2\cdot\text{K/W}$$

$$U \approx 48 \text{ W/m}^2\cdot\text{K}$$

In this case only the gas-side resistance is significant.

Example 2.3

In Example 2.2, replace the remaining flowing liquid by another flowing gas ($h_i = 20$ W/m²·K):

$$\frac{1}{U} = \frac{1}{20} + 0.0002 + \frac{0.003}{50} + \frac{1}{50}$$

$$= 0.05 + 0.0002 + 0.00006 + 0.02 = 0.07026 \text{ m}^2\cdot\text{K/W}$$

$$U \approx 14 \text{ W/m}^2\cdot\text{K}$$

Here the wall and scale resistances are negligible.

2.5 Log Mean Temperature Difference Method for Heat Exchanger Analysis

In the heat transfer analysis of heat exchangers, the total heat transfer rate Q through the heat exchanger is the quantity of primary interest. Let us consider a simple counterflow or parallel-flow heat exchanger (Figures 2.3a and b). The form of ΔT_m in Equation (2.7) may be determined by applying an energy balance to a differential area element dA in the hot and cold fluids. The temperature of the hot fluid will drop by dT_h. The temperature of the cold fluid will also drop by dT_c over the element dA for counterflow, but it will increase by dT_c for parallel flow if the hot-fluid direction is taken as positive. Consequently, from the differential forms of Equations (2.5) and (2.6) or from Equation (2.1) for adiabatic, steady state, steady flow, the energy balance yields

$$\delta Q = -(\dot{m}c_p)_h dT_h = \pm (\dot{m}c_p)_c dT_c \tag{2.22a}$$

or

$$\delta Q = -C_h dT_h = \pm C_c dT_c \tag{2.22b}$$

where C_h and C_c are the hot- and cold-fluid heat capacity rates, respectively; and the + and – signs correspond to parallel flow and counterflow, respectively. The rate of heat transfer δQ from the hot to the cold fluid across the heat transfer area dA may also be expressed as:

$$\delta Q = U(T_h - T_c)dA \tag{2.23}$$

From Equation (2.22b) for counterflow, we get:

$$d(T_h - T_c) = dT_h - dT_c = \delta Q\left(\frac{1}{C_c} - \frac{1}{C_h}\right) \tag{2.24}$$

By substituting the value of δQ from Equation (2.23) into Equation (2.24), we obtain:

$$\frac{d(T_h - T_c)}{(T_h - T_c)} = U\left(\frac{1}{C_c} - \frac{1}{C_h}\right)dA \tag{2.25}$$

which, when integrated with constant values of U, C_h, and C_c over the entire length of the heat exchangers, results in:

$$\ln\frac{T_{h2} - T_{c1}}{T_{h1} - T_{c2}} = UA\left(\frac{1}{C_c} - \frac{1}{C_h}\right) \tag{2.26a}$$

or

$$T_{h2} - T_{c1} = (T_{h1} - T_{c2})\exp\left[UA\left(\frac{1}{C_c} - \frac{1}{C_h}\right)\right] \tag{2.26b}$$

It can be shown that for a parallel-flow heat exchanger Equation (2.26b) becomes:

$$T_{h2} - T_{c2} = (T_{h1} - T_{c1})\exp\left[-UA\left(\frac{1}{C_c} + \frac{1}{C_h}\right)\right] \tag{2.26c}$$

It is seen that from Equations (2.25) and (2.26) the temperature difference along the heat exchanger is an exponential function of A. Hence in a counterflow heat exchanger, the temperature difference $(T_h - T_c)$ increases in the direction of flow of the hot fluid, if $C_h > C_c$ (Figure 2.6a). It can be shown that both temperature curves are concave (see Problem 4).[6] If the length of heat exchanger is infinite ($A = \infty$), the cold-fluid outlet temperature becomes equal to the inlet temperature of the hot fluid. If $C_h < C_c$, both curves are convex and $(T_h - T_c)$ decreases in direction of flow of the hot fluid. If the length is infinite ($A = \infty$), the hot fluid exit temperature becomes equal to the inlet temperature of the cold fluid (Figure 2.6b). The expressions for C_c and C_h can now be obtained from Equations (2.5) and (2.6). By substituting C_c and C_h into Equation (2.26a) and solving for Q, we obtain:

$$Q = UA \frac{(T_{h1} - T_{c2}) - (T_{h2} - T_{c1})}{\ln\left(\dfrac{T_{h1} - T_{c2}}{T_{h2} - T_{c1}}\right)} \tag{2.27a}$$

or

$$Q = UA \frac{\Delta T_1 - \Delta T_2}{\ln(\Delta T_1 / \Delta T_2)} \tag{2.27b}$$

where ΔT_1 is the temperature difference between the two fluids at one end of the heat exchanger and ΔT_2 is the temperature difference of the fluids at the other end of the heat exchanger.

Comparison of the preceding expression with Equation (2.7) reveals that the appropriate average temperature difference between the hot and cold fluids, over the entire length of the heat exchanger, is given by:

$$\Delta T_{lm} = \frac{\Delta T_1 - \Delta T_2}{\ln(\Delta T_1 / \Delta T_2)} \tag{2.28}$$

which is called the LMTD. Accordingly, we may write the total heat transfer rate between the hot and cold fluids for counterflow arrangement as:

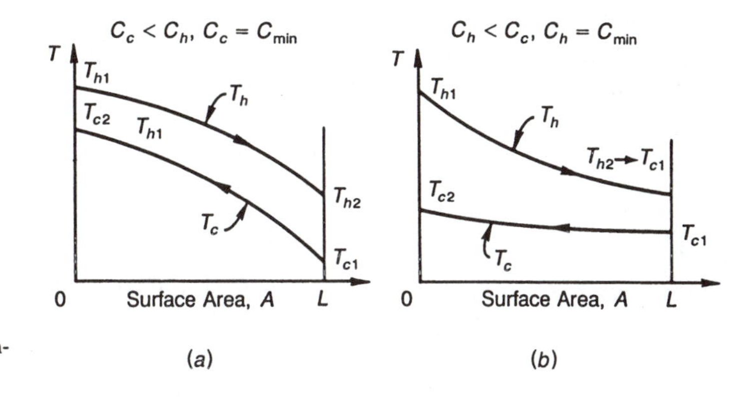

FIGURE 2.6
Temperature variation for a counterflow heat exchanger.

$$Q = AU\Delta T_{lm} \tag{2.29}$$

In the case of counterflow with $(\dot{m}c_p)_h = (\dot{m}c_p)_c$, the quantity ΔT_{lm} is indeterminate because:

$$(T_{h1} - T_{h2}) = (T_{c2} - T_{c1}) \text{ and } \Delta T_1 = \Delta T_2 \tag{2.30}$$

In this case, it can be shown using L'Hospital's rule that $\Delta T_{lm} = \Delta T_1 = \Delta T_2$ and therefore

$$Q = UA\,(T_h - T_c) \quad \text{with} \quad (T_h - T_c) = \Delta T_1 = \Delta T_2 \tag{2.31}$$

Starting with Equation (2.22) for a parallel-flow arrangement, it can be shown that Equation (2.28) is also applicable. However, for a parallel-flow heat exchanger, the end point temperature differences must now be defined as $\Delta T_1 = (T_{h1} - T_{c1})$ and $\Delta T_2 = (T_{h2} - T_{c2})$.

Moreover, the total heat transfer rate between hot and cold fluids for all single-pass flow arrangements shown in Figure 2.3 is determined from Equation (2.29).

Note that, for the same inlet and outlet temperatures, the LMTD for counterflow exceeds that for parallel flow, $\Delta T_{lm,cf} > \Delta T_{lm,pf}$; that is, LMTD represents the maximum temperature potential for heat transfer that can only be obtained in a counterflow exchanger. Hence the surface area required to affect a prescribed heat transfer rate Q is smaller for a counterflow arrangement than that for a parallel-flow arrangement, by assuming the same value of U. Also note that T_{c2} can exceed T_{h2} for counterflow but not for parallel flow.

Example 2.4

Consider a shell-and-tube heat exchanger that will be constructed from a 0.0254-m O.D. tube to cool 6.93 kg/s of a 95% ethyl alcohol solution (c_p = 3810 J/kg·K) from 66 to 42°C, using 6.30 kg/s of water available at 10°C (c_p = 4187 J/kg·K). In the heat exchanger 72 tubes will be used. Assume that the overall coefficient of heat transfer based on the outer-tube area is 568W/m²·K. Calculate the surface area and the length of the heat exchanger for each of the following arrangements:

1. Parallel-flow shell-and-tube heat exchanger
2. Counterflow shell-and-tube heat exchanger

SOLUTION

The heat transfer rate may be obtained from the overall energy balance for the hot fluid, Equation (2.5).

$$Q = (\dot{m}c_p)_h\,(T_{h1} - T_{h2})$$

$$Q = 6.93 \times 3810(66 - 42) = 633.679 \times 10^3 \text{ W}$$

By applying Equation (2.6), the water outlet temperature is

$$T_{c2} = \frac{Q}{(\dot{m}c_p)_c} + T_{c1}$$

$$T_{c2} = \frac{633.679 \times 10^3}{6.30 \times 4187} + 10 = 34°C$$

The LMTD can be obtained from Equation (2.28):

$$\Delta T_{lm} = \frac{\Delta T_1 - \Delta T_2}{\ln(\Delta T_1 / \Delta T_2)} = \frac{56 - 8}{\ln(56/8)} = 24.67°C$$

The outside surface area can be calculated from Equation (2.7):

$$Q = U_o A_o \Delta T_{lm}$$

$$A_o = \frac{Q}{U_o \Delta T_{lm}} = \frac{633.679 \times 10^3}{568 \times 24.67} = 45.22 \text{ m}^2$$

Because $A_o = \pi d_o n L$, the required heat exchanger length may now be obtained as:

$$L = \frac{A_o}{d_o \pi n} = \frac{45.22}{0.0254 \times \pi \times 72} = 7.87 \text{ m}$$

For counterflow arrangement, we have the case given by Equation (2.30):

$$\Delta T_{lm} = \Delta T_1 = \Delta T_2 = 32°C$$

$$A = \frac{Q}{U \Delta T_{lm}} = \frac{633.679 \times 10^3}{568 \times 32} = 34.86 \text{ m}^2$$

$$L = \frac{A}{d_o \pi n} = \frac{34.86}{0.0254 \times \pi \times 72} = 6.07 \text{ m}$$

Therefore the surface area required under the same condition of a prescribed heat transfer rate Q is smaller for a counterflow arrangement than that for a parallel-flow arrangement.

2.5.1 Multipass and Crossflow Heat Exchangers

The LMTD developed previously is not applicable for heat transfer analysis of crossflow and multipass flow heat exchangers. The integration of Equation (2.23) for these flow arrangements results in a form of an integrated mean temperature difference ΔT_m such that:

$$Q = UA\Delta T_m \tag{2.32}$$

where ΔT_m is the true (or effective) mean temperature difference; and it is a complex function of T_{h1}, T_{h2}, T_{c1}, and T_{c2}. Generally this function ΔT_m can be determined analytically in terms of the following quantities:[7]

$$\Delta T_{lm,cf} = \frac{(T_{h2} - T_{c1}) - (T_{h1} - T_{c2})}{\ln [(T_{h2} - T_{c1})/(T_{h1} - T_{c2})]} \tag{2.33}$$

$$P = \frac{T_{c2} - T_{c1}}{T_{h1} - T_{c1}} = \frac{\Delta T_c}{\Delta T_{max}} \tag{2.34}$$

and

$$R = \frac{C_c}{C_h} = \frac{T_{h1} - T_{h2}}{T_{c2} - T_{c1}} \tag{2.35}$$

where $\Delta T_{lm,cf}$ is the log mean temperature difference for a counterflow arrangement with the same fluid inlet and outlet temperatures. P is a measure of the ratio of the heat actually transferred to the heat transfer that would be transferred if the same cold fluid were to be raised to the hot-fluid inlet temperature; therefore P is the temperature effectiveness of the heat exchanger on the cold-fluid side. When the cold fluid has the minimum value of $(\dot{m}c_p)$, the temperature effectiveness will be equal to the temperature effectiveness P, as it is given by Equation (2.42). R is the ratio of the $(\dot{m}c_p)$ value of the cold fluid to that of the hot fluid and it is called the heat capacity rate ratio (regardless of which fluid is in the tube side or shell side in a shell-and-tube heat exchanger).

For design purposes, Equation (2.29) can also be used for multipass and crossflow heat exchangers by multiplying ΔT_{lm} that would be computed under the assumption of counterflow arrangement with a correction factor F.

$$Q = UAF\Delta T_{lm,cf} \tag{2.36}$$

F is nondimensional; it depends on the temperature effectiveness P, the heat capacity rate ratio R, and the flow arrangement.

$$F = \phi(P, R, \text{flow arrangement}) \tag{2.37}$$

The correction factors are available in chart form as prepared by Bowman et al.[8] for practical use for all common multipass shell-and-tube and crossflow heat exchangers and selected results are presented in Figures 2.7 to 2.11. In calculating P and R to determine F, it is immaterial whether the colder fluid flows through the shell or inside the tubes.

The correction factor F is less than 1 for crossflow and multipass arrangements; it is 1 for a true counterflow heat exchanger. It represents the degree of departure of the true mean temperature difference from the LMTD for a counterflow arrangement.

In a multipass or a crossflow arrangement, the fluid temperature may not be uniform at a particular location in the exchanger unless the fluid is well mixed along the path length. For example, in crossflow (Figure 2.12) the hot and cold fluids may enter at uniform temperatures, but if there are channels in the flow path (with or without corrugated spacers) to prevent mixing, the exit temperature distributions will be as shown in Figure 2.12. If such channels are not present, the fluids may be well mixed along the path length and the exit temperatures are more nearly uniform as in the flow normal to the tube bank in Figure 2.1d. A similar stratification of temperatures occurs in the shell-and-tube multipass exchanger. A series of baffles may be required if mixing of the shell fluid is to be obtained. Charts are presented for both mixed and unmixed fluids in Figures 2.13 and 2.14.

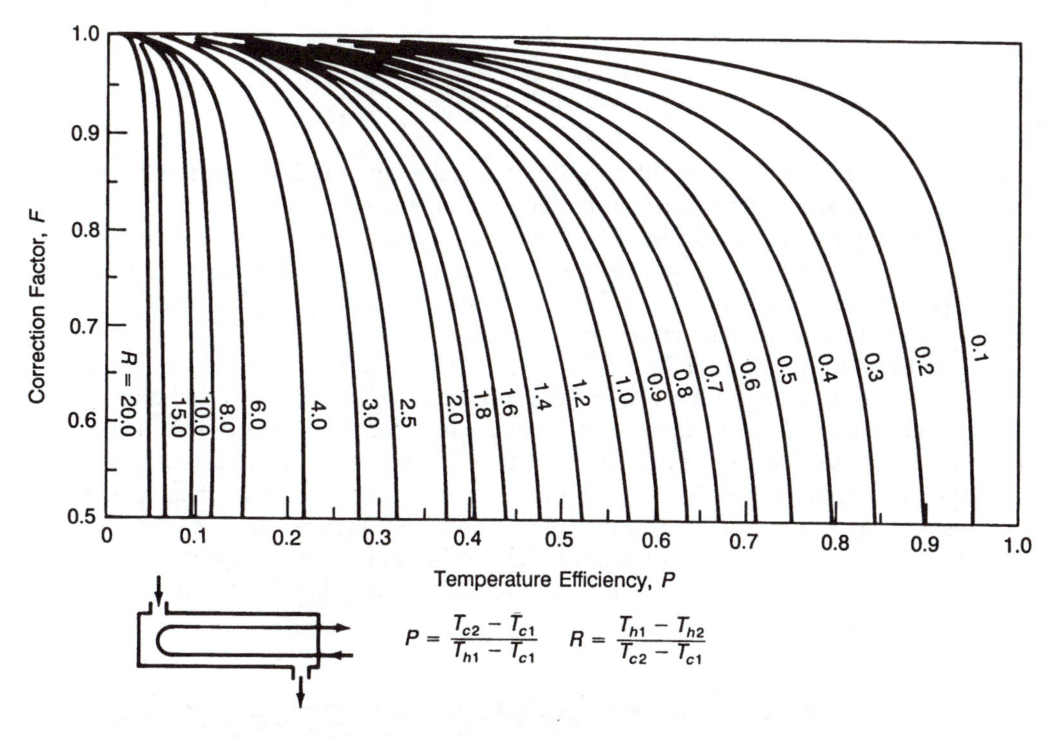

FIGURE 2.7
LMTD correction factor F for a shell-tube heat exchanger — one shell pass and two or multiples of two tube passes. (From Standards of Tubular Exchanger Manufacturers Association (1988), New York. With permission.)

Example 2.5

Repeat Example 2.4 for a shell-and-tube heat exchanger with one-shell pass and multiples of two-tube passes.

SOLUTION

For the multipass arrangement from Equation (2.36) we have:

$$Q = UAF\Delta T_{lm,cf}$$

P and R can be obtained from Equations (2.34) and (2.35):

$$P = \frac{T_{c2} - T_{c1}}{T_{h1} - T_{c1}} = \frac{24}{56} = 0.43$$

$$R = \frac{T_{h1} - T_{h2}}{T_{c2} - T_{c1}} = \frac{24}{24} = 1.0$$

From Figure 2.7 we get $F = 0.89$, then

$$A = \frac{Q}{U\Delta T_m} = \frac{633.68 \times 10^3}{568 \times 32 \times 0.89} = 39.17 \text{ m}^2$$

$$L = \frac{A}{\pi d_o N_T} = \frac{39.17}{\pi \times 0.0254 \times 72} = 6.19 \text{ m}$$

Therefore, under the same conditions, the surface area required for a multipass arrangement is between counterflow and parallel-flow arrangements.

The preceding analysis assumed U to be uniform throughout the heat exchanger. If U is not uniform, the heat exchanger calculations may be made by subdividing the heat exchanger into sections over which U is nearly uniform and by applying the previously developed relations to each subdivision (see Section 2.8).

An important implication of Figures 2.7 to 2.11, 2.13, and 2.14 is that if the temperature change of one fluid is negligible, either P or R is 0 and F is 1. Hence heat exchanger behavior is independent of specific configuration. Such will be the case if one of the fluids undergoes a phase change. We note from Figures 2.7 to 2.11, 2.13, and 2.14 that the value of temperature effectiveness P ranges from 0 to 1. The value of R ranges from 0 to ∞, with 0 corresponding to pure vapor condensation; and ∞, to evaporation. It should be noted that a value of F close to 1 does not mean a highly efficient heat exchanger; it means a close approach to the counterflow behavior for comparable operating conditions of flow rates and inlet fluid temperatures.

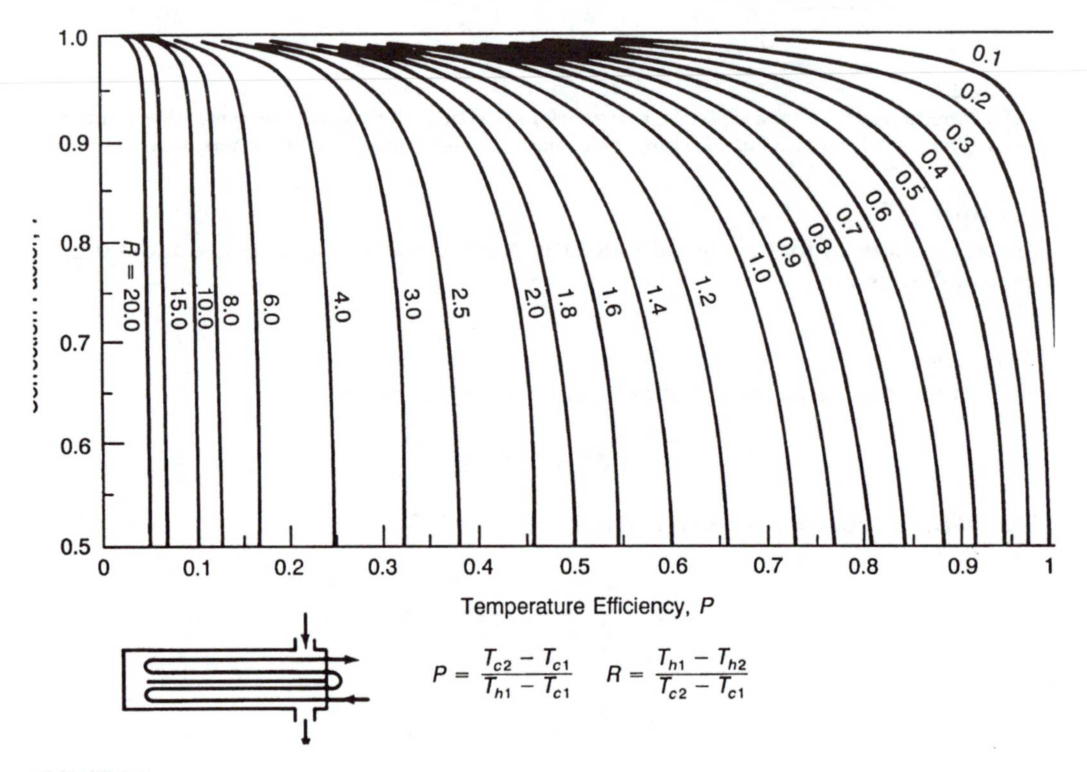

FIGURE 2.8
LMTD correction factor F for a shell-tube heat exchanger — two shell passes and four or multiples of four tube passes. (From Standards of Tubular Exchanger Manufacturers Association (1988), New York. With permission.)

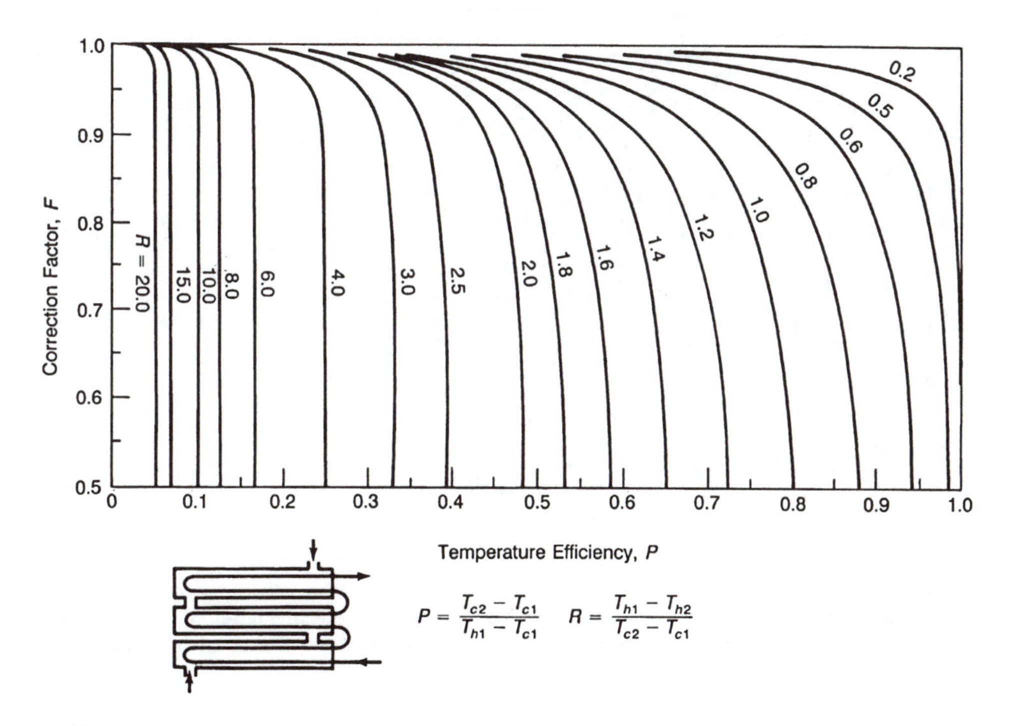

FIGURE 2.9
LMTD correction factor F for a shell-tube heat exchanger — three two-shell passes and six or more even number of tube passes. (From Standards of Tubular Exchanger Manufacturers Association (1988), New York. With permission.)

Example 2.6

A heat exchanger with two-shell passes and four-tube passes is used to heat water with hot exhaust gas. Water enters the tubes at 50°C and leaves at 125°C with a flow rate of 10 kg/s, while the hot exhaust gas enters the shell side at 300°C and leaves at 125°C. The total heat transfer surface is 800 m². Calculate the overall heat transfer coefficient.

SOLUTION

The overall heat transfer coefficient can be determined from Equation (2.36).

$$Q = UAF\Delta T_{lm,cf}$$

with

$$P = \frac{\Delta T_c}{T_{h1} - T_{c1}} = \frac{75}{300 - 50} = 0.3$$

$$R = \frac{T_{h1} - T_{h2}}{T_{c2} - T_{c1}} = \frac{300 - 125}{125 - 50} = 2.3$$

$$\Delta T_{lm,cf} = \frac{\Delta T_2 - \Delta T_1}{\ln(\Delta T_2 / \Delta T_1)} = \frac{175 - 75}{\ln(175/75)} = 118°\text{C}$$

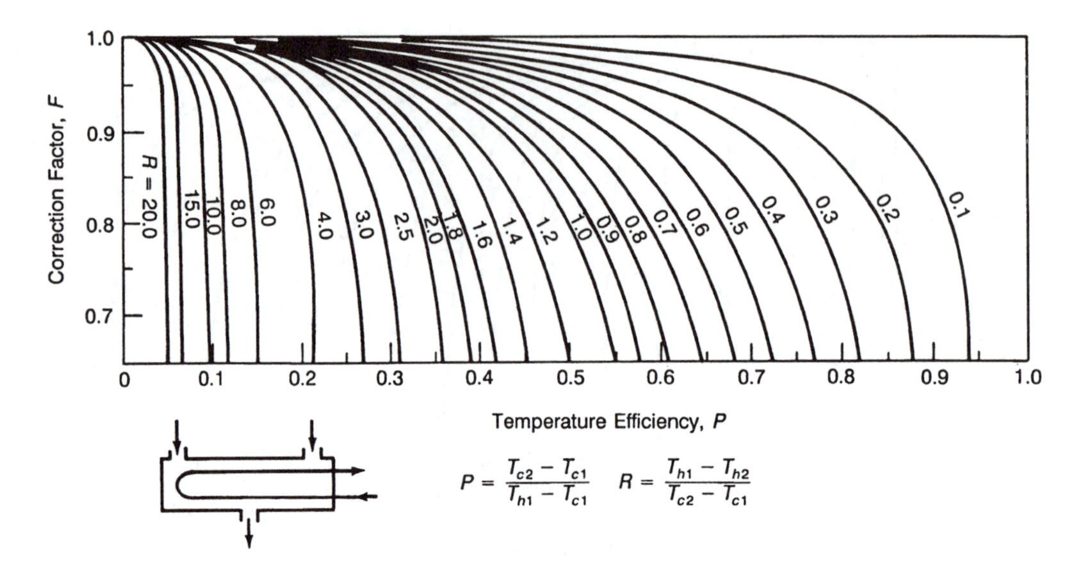

FIGURE 2.10
LMTD correction factor F for a divided-flow shell-type heat exchanger — one divided-flow shell pass and even number of tube passes. (From Standards of Tubular Exchanger Manufacturers Association (1988), New York. With permission.)

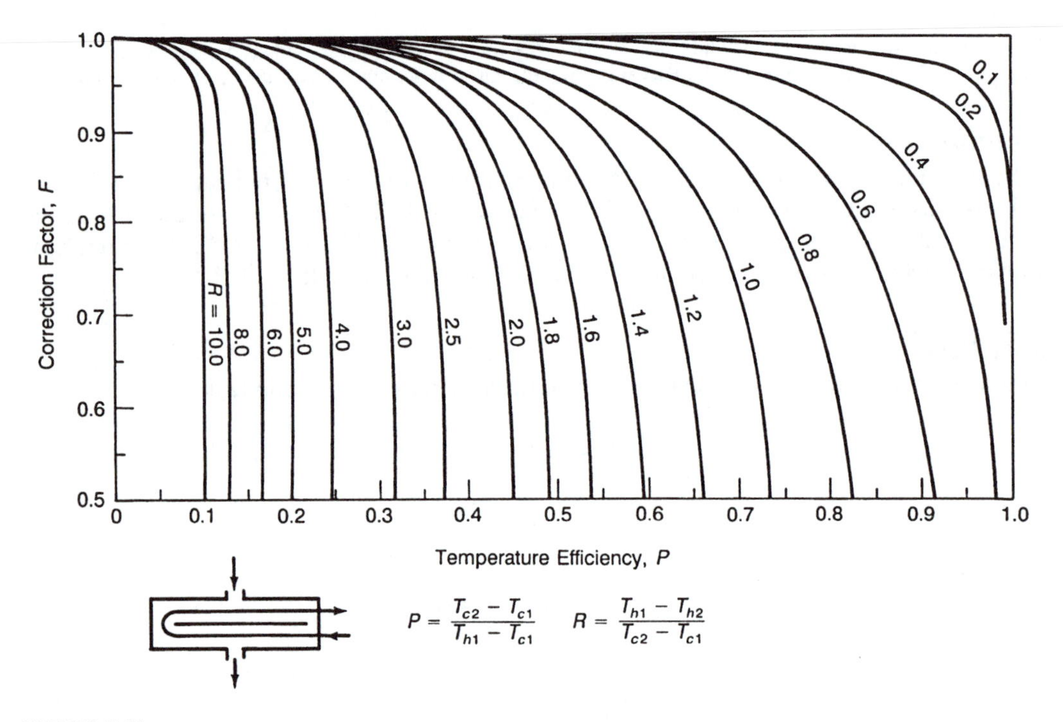

FIGURE 2.11
LMTD correction factor F for a split-flow shell-type heat exchanger — one split-flow shell pass and two-tube passes. (From Standards of Tubular Exchanger Manufacturers Association (1988), New York. With permission.)

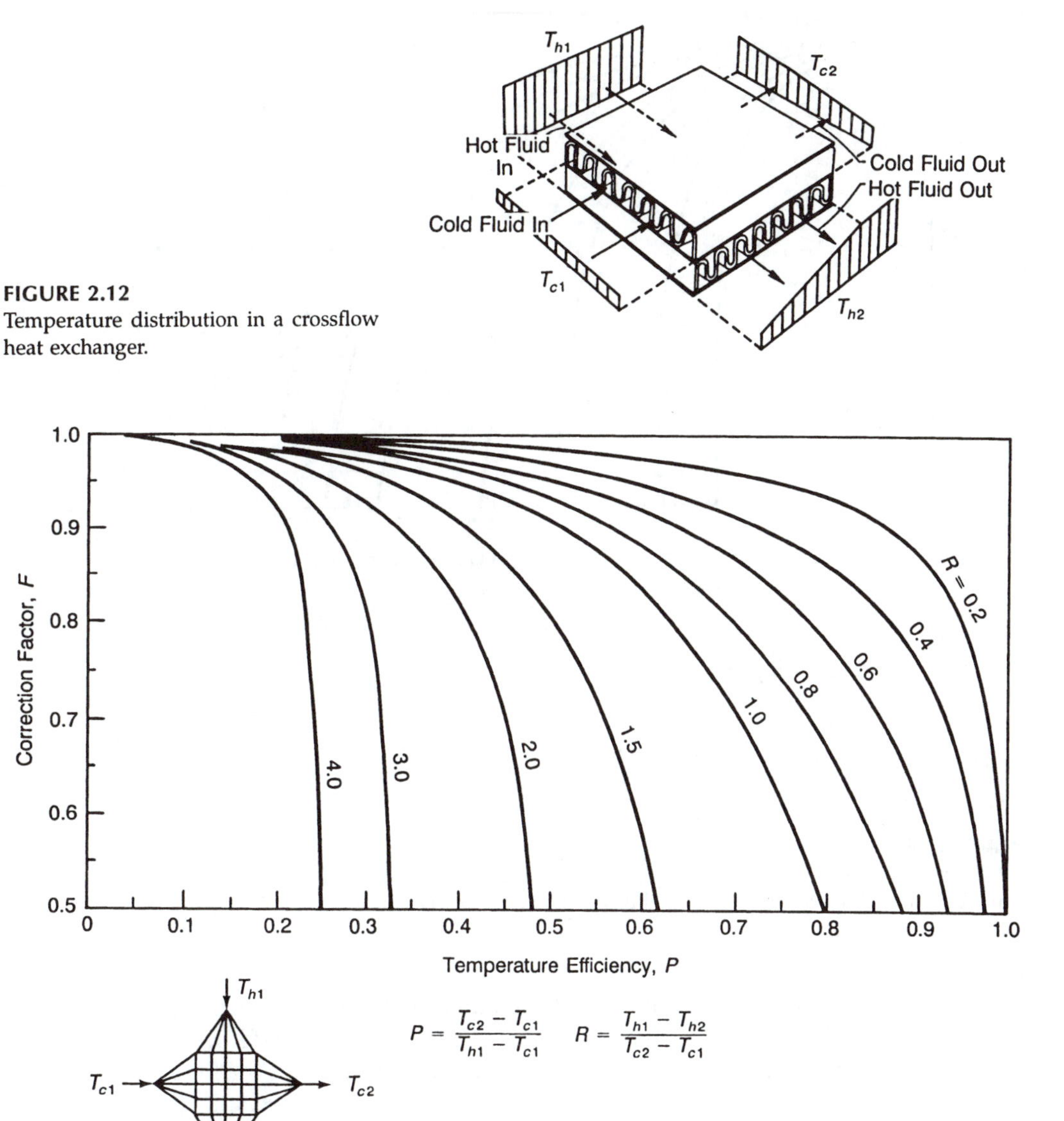

FIGURE 2.12
Temperature distribution in a crossflow heat exchanger.

FIGURE 2.13
LMTD correction factor F for a crossflow heat exchanger with both fluids unmixed. (From Bowman, R. A., Mueller, A. C., and Nagle, W. M. [1940] *Trans. ASME*, Vol. 62, 283–294.)

It follows from Figure 2.8 that $F \approx 0.96$. Water at 87.5°C, $c_p = 4203$ J/kg·K

$$U = \frac{(\dot{m}c_p)_c(T_{c2} - T_{c1})}{AF\Delta T_{m,cf}} = \frac{10 \times 4203(125 - 50)}{800 \times 0.96 \times 118} = 34.8 \text{ W/m}^2 \text{ K}$$

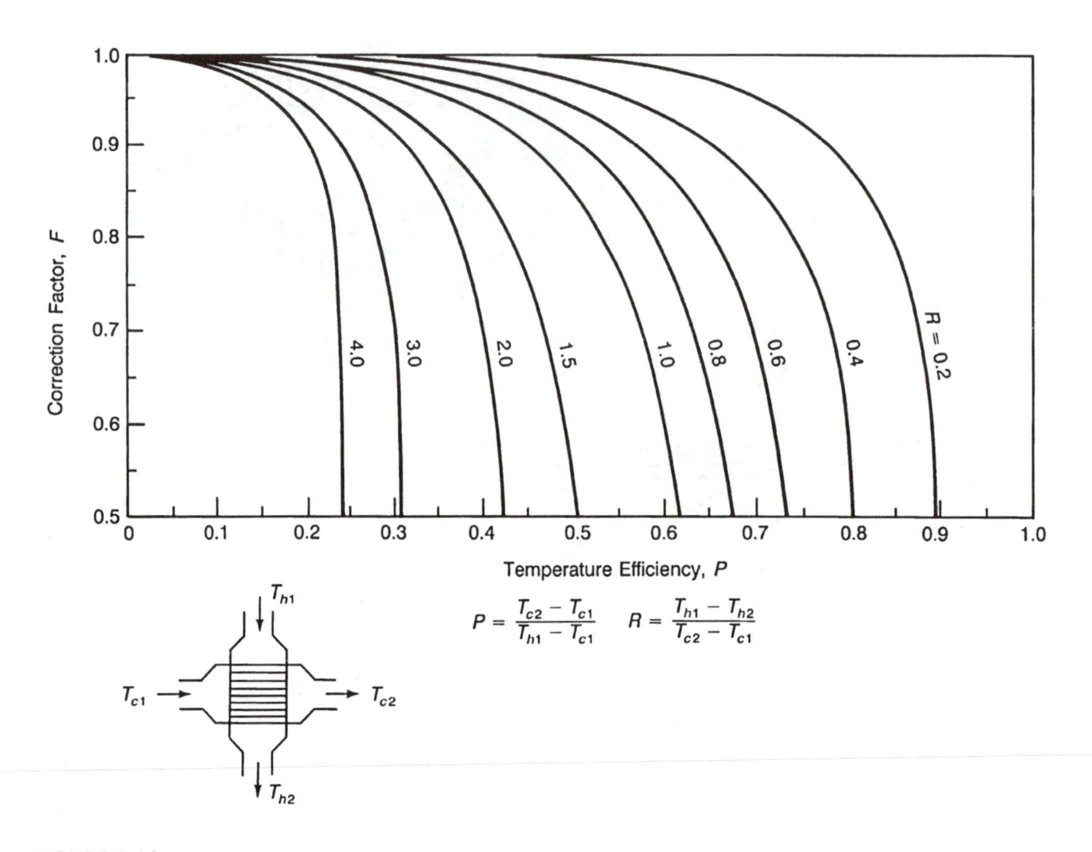

FIGURE 2.14
LMTD correction factor F for a single-pass crossflow heat exchanger with one fluid mixed and the other unmixed. (From Bowman, R. A., Mueller, A. C., and Nagle, W. M. [1940] *Trans. ASME*, Vol. 62, 283–294.)

Example 2.7

A finned-tube, single-pass, crossflow heat exchanger with both fluids unmixed as shown in Figure 2.13 is used to heat air with hot water. The total heat transfer rate is 200 kW. The water enters the tube at 85°C and leaves at 30°C, while the air enters the finned side at 15°C and leaves at 50°C. The overall heat transfer coefficient is 75 W/m²·K. Calculate the surface area required.

SOLUTION

The desired heat transfer can be determined from Equation (2.36):

$$A = \frac{Q}{UF\Delta T_{lm,cf}}$$

with

$$P = \frac{\Delta T_c}{T_{h1} - T_{c1}} = \frac{50 - 15}{85 - 15} = 0.5, \; R = \frac{T_{h1} - T_{h2}}{T_{c2} - T_{c1}} = \frac{85 - 30}{50 - 15} = 1.57$$

It follows from Figure 2.13 that $F \approx 0.8$.
From Equation (2.28) counterflow can be calculated by:

$$\Delta T_{lm,cf} = \frac{\Delta T_1 - \Delta T_2}{\ln(\Delta T_1 / \Delta T_2)} = \frac{35 - 15}{\ln(35/15)} = 23.6°\text{C}$$

$$A = \frac{200 \times 10^3}{75 \times 0.8 \times 23.6} = 141.2 \text{ m}^2$$

Example 2.8

Air flowing at a rate of 5 kg/s is to be heated in a shell-and-tube heat exchanger from 20 to 50°C with hot water entering at 90°C and exiting at 60°C. The overall heat transfer coefficient is 400 W/m²·K. The length of the heat exchanger is 2 m. Determine the surface area of the heat exchanger and the number of tubes required by using:

1. 1 to 2 shell-and-tube type
2. 2 to 4 shell-and-tube type

The tubes are schedule 40, 3/4-in. nominal diameter.

SOLUTION

For air at 35°C, c_p = 1005 J/kg·K. From the overall energy balance, Equation (2.6), the heat transfer required of the exchanger is

$$Q = (\dot{m}c_p)_c (T_{c2} - T_{c1}) = 1005 \times 5\ (50 - 20) = 150.75 \text{ kW}$$

1. The desired heat transfer area may be determined by using the LMTD method from Equation (2.36) with:

$$P = \frac{T_{c2} - T_{c1}}{T_{h1} - T_{c1}} = \frac{50 - 20}{90 - 20} - 0.43,\ R = \frac{T_{h1} - T_{h2}}{T_{c2} - T_{c1}} = \frac{90 - 60}{50 - 20} = 1$$

It follows from Figure 2.7 that $F \approx 0.90$.

From Equation (2.30):

$$\Delta T_{lm,cf} = \Delta T_1 = \Delta T_2 = 40°\text{C}$$

The heat transfer area of the heat exchanger can be obtained from Equation (2.36):

$$A = \frac{Q}{UF\Delta T_{lm,cf}} = \frac{150750}{400 \times 0.9 \times 40}$$

$$A = 10.47 \text{ m}^2$$

The heat transfer surface area of one tube (see Chapter 8, Table 8.2):

$$A_t = \pi d_o L = \pi \times 0.02667 \times 2$$

$$A_t = 0.1676 \text{ m}^2$$

The number of tubes of the heat exchanger is

$$N_T = \frac{A}{A_t} = \frac{10.47}{0.1676} = 63 \text{ tubes}$$

2. For 2 to 4 shell-and-tube heat exchanger from Figure 2.8, $F \approx 0.975$.
   The number of tubes is

$$N_T = \frac{Q}{UF\Delta T_{lm,cf}(\pi d_o L)} = \frac{150750}{400 \times 0.975 \times 40(\pi \times 0.02667 \times 2)}$$

$$N_T \cong 58 \text{ tubes}$$

2.6 The ε-NTU Method for Heat Exchanger Analysis

When the inlet or outlet temperatures of the fluid streams are not known, a trial-and-error procedure could be applied for using the LMTD method in the thermal analysis of heat exchangers. The converged value of LMTD will satisfy the requirement that the heat transferred in the heat exchanger (Equation [2.7]) be equal to the heat convected to the fluid (Equation [2.5] or [2.6]). In these cases, to avoid a trial-and-error procedure, the method of the number of transfer units (NTU) based on the concept of a heat exchanger effectiveness may be used. The method is based on the fact that the inlet or exit temperature differences of a heat exchanger are a function of UA/C_c and C_c/C_h (see Equations [2.26b] and [2.26c]).

Heat exchanger heat transfer equations such as Equations (2.5), (2.6), and (2.26) may be written in dimensionless form resulting in the groups that follow.[8]

1. Capacity rate ratio:

$$C^* = \frac{C_{min}}{C_{max}} \tag{2.38}$$

   where C_{min} and C_{max} are the smaller and larger of the two magnitudes of C_h and C_c, respectively; and $C^* \leq 1$. $C^* = 0$ corresponds to a finite C_{min} and C_{max} approaching ∞ (a condensing or evaporating fluid).
2. Exchanger heat transfer effectiveness:

$$\varepsilon = \frac{Q}{Q_{max}} \tag{2.39}$$

which is the ratio of the actual heat transfer rate in a heat exchanger to thermodynamically limited maximum possible heat transfer rate if an infinite heat transfer surface area were available in a counterflow heat exchanger.

The actual heat transfer is obtained either by the energy given off by the hot fluid or the energy received by the cold fluid, from Equations (2.5) and (2.6):

$$Q = (\dot{m}c_p)_h (T_{h1} - T_{h2}) = (\dot{m}c_p)_c (T_{c2} - T_{c1}) \tag{2.40}$$

$$\text{If } C_h > C_c, \text{ then } (T_{h1} - T_{h2}) < (T_{c2} - C_{c1})$$

$$\text{If } C_h < C_c, \text{ then } (T_{h1} - T_{h2}) > (T_{c2} - T_{c1})$$

The fluid that might undergo the maximum temperature difference is the fluid having the minimum heat capacity rate C_{min}. Therefore, the maximum possible heat transfer is expressed as:

$$Q_{max} = (\dot{m}c_p)_c (T_{h1} - T_{h2}) \text{ if } C_c < C_h \tag{2.41a}$$

or

$$Q_{max} = (\dot{m}c_p)_h (T_{h1} - T_{h2}) \text{ if } C_h < C_c \tag{2.41b}$$

which can be obtained with a counterflow heat exchanger if an infinite heat transfer surface area were available (Figure 2.6). Heat exchanger effectiveness, ε, is therefore written as:

$$\varepsilon = \frac{C_h(T_{h1} - T_{h2})}{C_{min}(T_{h1} - T_{c1})} = \frac{C_c(T_{c2} - T_{c1})}{C_{min}(T_{h1} - T_{c1})} \tag{2.42}$$

The first definition is for $C_h = C_{min}$ and the second is for $C_c = C_{min}$. Equation (2.42) is valid for all heat exchanger flow arrangements. The value of ε ranges between 0 and 1. When the cold fluid has a minimum heat capacity rate, then temperature effectiveness, P, given by Equation (2.34), will be equal to heat exchanger effectiveness, ε. For given ε and Q_{max}, the actual heat transfer rate Q from Equation (2.39) is

$$Q = \varepsilon \, (\dot{m}c_p)_{min} (T_{h1} - T_{c1}) \tag{2.43}$$

If the effectiveness ε of the exchanger is known, Equation (2.43) provides an explicit expression for the determination of Q.

3. Heat transfer area number:

$$\text{NTU} = \frac{AU}{C_{min}} = \frac{1}{C_{min}} \int_A U \, dA \tag{2.44}$$

If U is not constant, the definition of second equality applies. NTU designates the nondimensional heat transfer size of the heat exchanger.

Let us consider a single-pass heat exchanger, assuming $C_c > C_h$, so that $C_h = C_{min}$ and $C_c = C_{max}$. With Equation (2.44), Equation (2.26b) may be written as:

$$T_{h2} - T_{c1} = (T_{h1} - T_{c2}) \exp\left[-\text{NTU}\left(\pm 1 - \frac{C_{min}}{C_{max}}\right)\right] \tag{2.45}$$

where the + is for counterflow and the – is for parallel flow. With Equations (2.5), (2.6), and (2.42), T_{h2} and T_{c2} in Equation (2.45) can be eliminated and the following expression is obtained for ε for counterflow:

$$\varepsilon = \frac{1 - \exp[-\text{NTU}(1 - C_{min}/C_{max})]}{1 - (C_{min}/C_{max}) \exp[-NTU(1 - C_{min}/C_{max})]} \tag{2.46}$$

if $C_c < C_h$ ($C_c = C_{min}$, $C_h = C_{max}$), the result will be the same.

In the case of parallel flow, a similar analysis may be applied to obtain the following expression:

$$\varepsilon = \frac{1 - \exp[-\text{NTU}(1 + C_{min}/C_{max})]}{1 + (C_{min}/C_{max})} \tag{2.47}$$

Two limiting cases are of interest: C_{min}/C_{max} equal to 1 and 0. For $C_{min}/C_{max} = 1$, Equation (2.46) is indeterminate, but by applying L'Hospital's rule to Equation (2.46), the following result is obtained for counterflow:

$$\varepsilon = \frac{\text{NTU}}{1 + \text{NTU}} \tag{2.48}$$

For parallel flow, Equation (2.47) gives:

$$\varepsilon = \frac{1}{2}\left(1 - e^{-2\text{NTU}}\right) \tag{2.49}$$

For $C_{min}/C_{max} = 0$, as in boilers and condensers (Figures 2.3c and d) for parallel flow or counterflow, Equations (2.46) and (2.47) become:

$$\varepsilon = 1 - e^{-\text{NTU}} \tag{2.50}$$

It is noted from Equations (2.46) and (2.47) that:

$$\varepsilon = \phi(\text{NTU}, C^*, \text{ flow arrangement}) \tag{2.51}$$

Similar expressions have been developed for heat exchangers having other flow arrangements, such as crossflow, multipass, and so forth; and representative results are summarized in Table 2.2.[3,10,11]

TABLE 2.2
ε-NTU Expressions

Type	ε (NTU, C^*)	NTU (ε, C^*)	Particular Case	Value of ε When
Counterflow	$\varepsilon = \dfrac{1 - \exp[-(1-C^*)\text{NTU}]}{1 - C^* \exp[-(1-C^*)\text{NTU}]}$	$\text{NTU} = \dfrac{1}{1-C^*} \ln\left(\dfrac{1-\varepsilon C^*}{1-\varepsilon}\right)$	$\varepsilon = 1 - \exp(-\text{NTU})$	$\varepsilon = 1$ for all C^*
Parallel	$\varepsilon = \dfrac{1}{1+C^*}\{1 - \exp[-(1+C^*)\text{NTU}]\}$	$\text{NTU} = -\dfrac{1}{1+C^*} \ln[1+\varepsilon(1+C^*)]$	$\varepsilon = 1 - \exp(-\text{NTU})$	$\varepsilon = \dfrac{1}{1+C^*}$
Crossflow, C_{min} mixed	$\varepsilon = 1 - \exp\left[\dfrac{1-\exp(-C^*\text{NTU})}{C^*}\right]$	$\text{NTU} = -\dfrac{1}{C^*} \ln[1 + C^* \ln(1-\varepsilon)]$	$\varepsilon = 1 - \exp(-\text{NTU})$	$\varepsilon = 1 - \exp\left(-\dfrac{1}{C^*}\right)$
Crossflow, C_{max} mixed	$\varepsilon = \dfrac{1}{C^*}[1 - \exp\{-C^*[1-\exp(-\text{NTU})]\}]$	$\text{NTU} = -\ln\left[1 + \dfrac{1}{C^*}\ln(1-\varepsilon C^*)\right]$	$\varepsilon = 1 - \exp(-\text{NTU})$	$\varepsilon = \dfrac{1}{C^*}[1 - \exp(-C^*)]$
1–2 shell-and-tube heat	$\varepsilon = \dfrac{2}{1 + C^* + (1+C^{*2})^{1/2} \dfrac{1+\exp[-\text{NTU}(1+C^{*2})^{1/2}]}{1-\exp[-\text{NTU}(1+C^{*2})^{1/2}]}}$	$\text{NTU} = \dfrac{1}{(1+C^{*2})^{1/2}} \ln \dfrac{2-\varepsilon[1+C^*-(1+C^{*2})^{1/2}]}{2-\varepsilon[1+C^*+(1+C^{*2})^{1/2}]}$	$\varepsilon = 1 - \exp(-\text{NTU})$	$\varepsilon = \dfrac{2}{1+C^*+(1+C^{*2})^{1/2}}$

Some ε-NTU relations are graphically shown in Figure 2.15.[9] The following observations may be made by reviewing these figures:

1. The heat exchanger thermal effectiveness ε increases with increasing values of NTU for a specified C^*.
2. The exchanger thermal effectiveness ε increases with decreasing values of C^* for a specified NTU.
3. For $\varepsilon < 40\%$, the capacity rate ratio C^* does not have a significant influence on the exchanger effectiveness.

Because of the asymptotic nature of the ε-NTU curves, a significant increase in NTU and hence in the heat exchanger size is required for a small increase in ε at high values of ε.

The counterflow exchanger has the highest exchanger effectiveness ε for specified NTU and C^* values compared with those for all other exchanger flow arrangements. Thus, for given NTU and C^*, a maximum heat transfer performance is achieved for counterflow; alternately, the heat transfer surface is utilized most efficiently in counterflow arrangement compared with all other flow arrangements.

Table 2.3 summarizes each of the methods discussed in the preceding sections.

TABLE 2.3
Working Equations for the LMTD, and ε-NTU Methods

LMTD	ε-NTU
$Q = UAF\Delta T_{lm,cf}$	$Q = \varepsilon\,(\dot{m}c_p)_{min}(T_{h1} - T_{c1})$
$\text{LMTD} = \Delta T_{lm,cf} = \dfrac{\Delta T_1 - \Delta T_2}{\ln(\Delta T_1/\Delta T_2)}$	$\varepsilon = \dfrac{C_h(T_{h1}-T_{h2})}{C_{min}(T_{h1}-T_{c1})} = \dfrac{C_c(T_{c2}-T_{c1})}{C_{min}(T_{h1}-T_{c1})}$
$\Delta T_1 = T_{h1} - T_{c2},\ \Delta T_2 = T_{h2} - T_{c1}$	$C^* = \dfrac{C_{min}}{C_{max}} = \dfrac{(\dot{m}c_p)_{min}}{(\dot{m}c_p)_{max}}$
$P = \dfrac{T_{c2}-T_{c1}}{T_{h1}-T_{c1}},\ R = \dfrac{T_{h1}-T_{h2}}{T_{c2}-T_{c1}}$	$\text{NTU} = \dfrac{UA}{C_{min}} = \dfrac{1}{C_{min}}\int_A U\,dA$
$F = \phi\,(P, R, \text{flow arrangement})$	$\varepsilon = \phi\,(\text{NTU}, C^*, \text{flow arrangement})$

Example 2.9

A two-pass tube, baffled single-pass shell, shell-and-tube heat exchanger is used as an oil cooler. Cooling water flows through the tubes at 20°C at a flow rate of 4.082 kg/s. Engine oil enters the shell side at a flow rate of 10 kg/s. The inlet and outlet temperatures of oil are 90 and 60°C, respectively. Determine the surface area of the heat exchanger by both the LMTD and ε-NTU methods if the overall heat transfer coefficient based on the outside tube area is 262 W/m²·K. The specific heats of water and oil are 4179 J/kg·K and 2118 J/kg·K, respectively.

SOLUTION

LMTD method — We shall first calculate the heat transfer rate Q and LMTD for counterflow. Subsequently P, R, and the correction factor F will be determined. Then applying the heat transfer rate Equation (2.36), the heat transfer surface area A will be determined.

The heat capacity rates for the shell fluid (oil) and the tube fluid (water) are

$$C_h = (\dot{m}c_p)_h = 10 \text{ kg/s} \times 2118 \text{ J/kg·K} = 21180 \text{ W/·K}$$

$$C_c = (\dot{m}c_p)_c = 4.082 \text{ kg/s} \times 4179 \text{ J/kg·K} = 17058.7 \text{ W/K}$$

The heat transfer rate from the temperature drop of the oil is

$$Q = C_h \,(T_{h1} - T_{h2}) = 21180 \times (90 - 60) = 635400 \text{ W}$$

The water outlet temperature from the energy balance is

$$Q = C_c \,(T_{c2} - T_{c1}) = 635400 \text{ W}$$

$$T_{c2} = 635400/C_c + T_{c1} = 635400/17058.678 + 20 = 37.248 + 20 = 57.25°\text{C}$$

Now let us determine $\Delta T_{lm,cf}$. From the definition of $\Delta T_{lm,cf}$ in Equation (2.28) for a counter-flow arrangement:

$$\Delta T_{lm,cf} = \frac{32.75 - 40}{\ln (3.275/40)} = 36.3°\text{C}$$

Now the values of P and R from Equations (2.34) and (2.35) are

$$P = \frac{T_{c2} - T_{c1}}{T_{h1} - T_{c1}} = \frac{57.25 - 20}{90 - 20} = \frac{37.25}{70} = 0.532$$

$$R = \frac{T_{h1} - T_{h2}}{T_{c2} - T_{c1}} = \frac{C_c}{C_h} = \frac{17058.7}{21180} = 0.805$$

Therefore, from Figure 2.7 we get $F = 0.85$.
Thus the heat transfer area from Equation (2.36):

$$A = \frac{Q}{UF\Delta T_{lm}} = \frac{635400}{262 \times 0.85 \times 36.4} = 78.6 \text{ m}^2$$

The ε-NTU method — In the ε-NTU method, we will first determine ε and C^*, subsequently NTU, and then A. In this problem $C_h > C_c$, and hence $C_c = C_{min}$:

$$C^* = \frac{C_{min}}{C_{max}} = \frac{17058.7}{21180} = 0.805$$

From the given temperature, for $C_c = C_{min}$, Equation (2.42) gives:

$$\varepsilon = \frac{T_{c2} - T_{c1}}{T_{h1} - T_{c1}} = \frac{57.25 - 20}{90 - 20} = \frac{37.25}{70} = 0.532$$

Now calculate NTU either from the formula of Table 2.2 or from Figure (2.15) with a proper interpretation for ε, NTU, and C^*. From Table 2.2 for 1 to 2 shell-and-tube heat exchanger, we have

$$\mathrm{NTU} = -\frac{1}{(1+C^{*2})^{1/2}} \ln\left[\frac{2-\varepsilon[1+C^*-(1+c^{*2})^{1/2}]}{2-\varepsilon[1+C^*+(1+C^{*2})^{1/2}]}\right]$$

$$= \frac{1}{(1+0.805^2)^{-1/2}} \ln\left[\frac{2-0.532[1+0.805-(1+0.805^2)^{1/2}]}{2-0.532[1+0.805+(1+0.805^2)^{1/2}]}\right]$$

$$= 0.778965 \ln\left[\frac{1.7227}{0.35679}\right] = 1.226$$

Hence

$$A = \frac{C_{min}}{U}\,\mathrm{NTU} = \frac{17058.7}{262} \times 1.226 = 79.8\ \mathrm{m}^2$$

In obtaining expressions for the basic design methods that we discussed in the preceding sections, Equations (2.22) and (2.23) are integrated across the surface area under the following assumptions:

1. The heat exchanger operates under steady-state, steady-flow conditions.
2. Heat transfer to the surroundings is negligible.
3. There is no heat generation in the heat exchanger.
4. In counterflow and parallel-flow heat exchangers, the temperature of each fluid is uniform over every flow cross section; in crossflow heat exchangers each fluid is considered mixed or unmixed at every cross section depending on the specifications.
5. If there is a phase change in one of the fluid streams flowing through the heat exchanger, phase change occurs at a constant temperature for a single-component fluid at constant pressure.
6. The specific heat at constant pressure is constant for each fluid.
7. Longitudinal heat conduction in the fluid and in the wall is negligible.
8. The overall heat transfer coefficient between the fluids is constant throughout the heat exchanger including the case of phase change.

Assumption 5 is an idealization of a two-phase-flow heat exchanger. Especially, for two-phase flows on both sides, many of the foregoing assumptions are not valid. The design theory of these types of heat exchangers is discussed and practical results are presented in the following chapters.

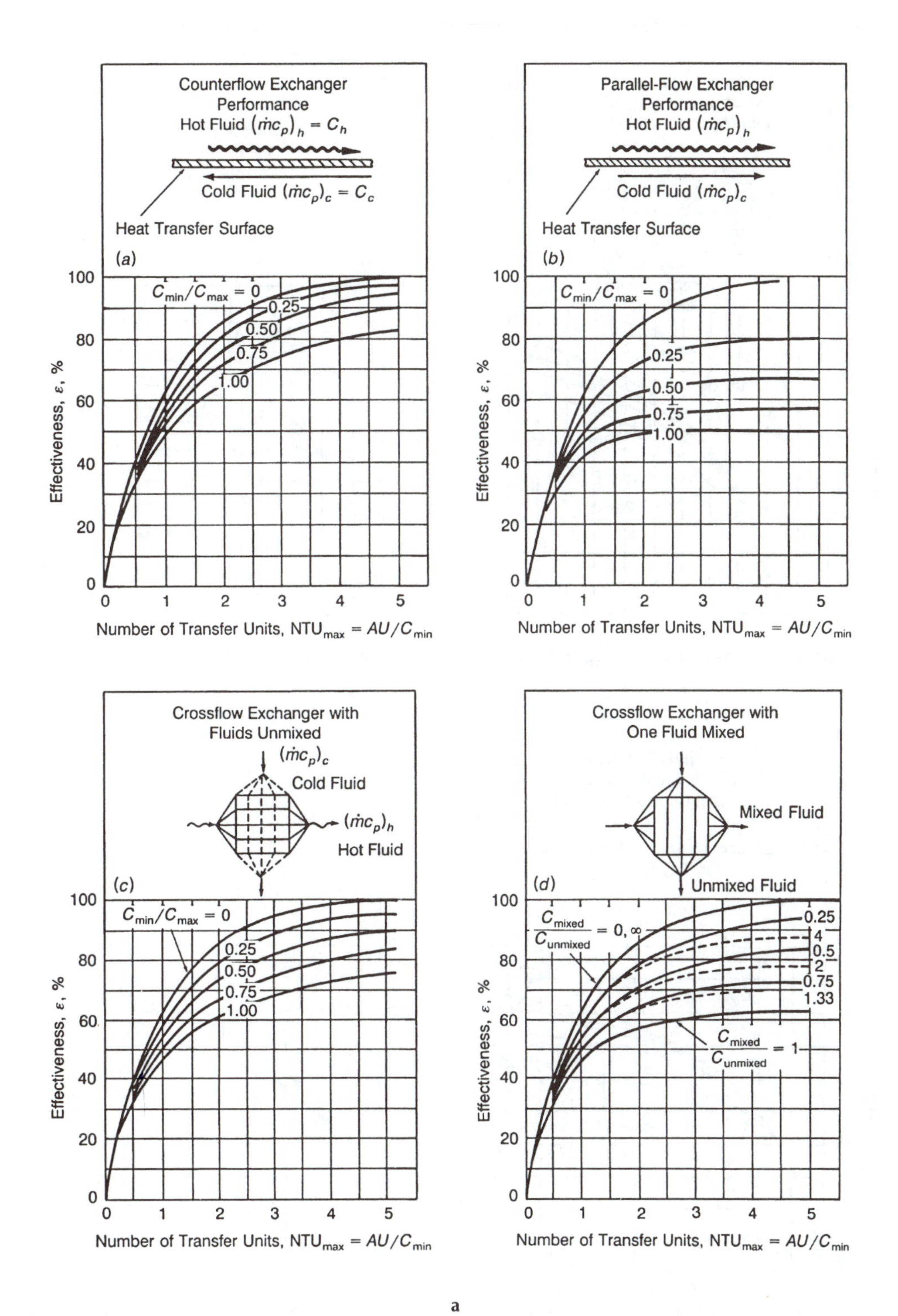

FIGURE 2.15
Effectiveness vs. NTU for various types of heat exchangers. (From Kays, W. M. and London, A. L. [1984] *Compact Heat Exchangers*, 3rd ed., McGraw-Hill, New York. With permission.)

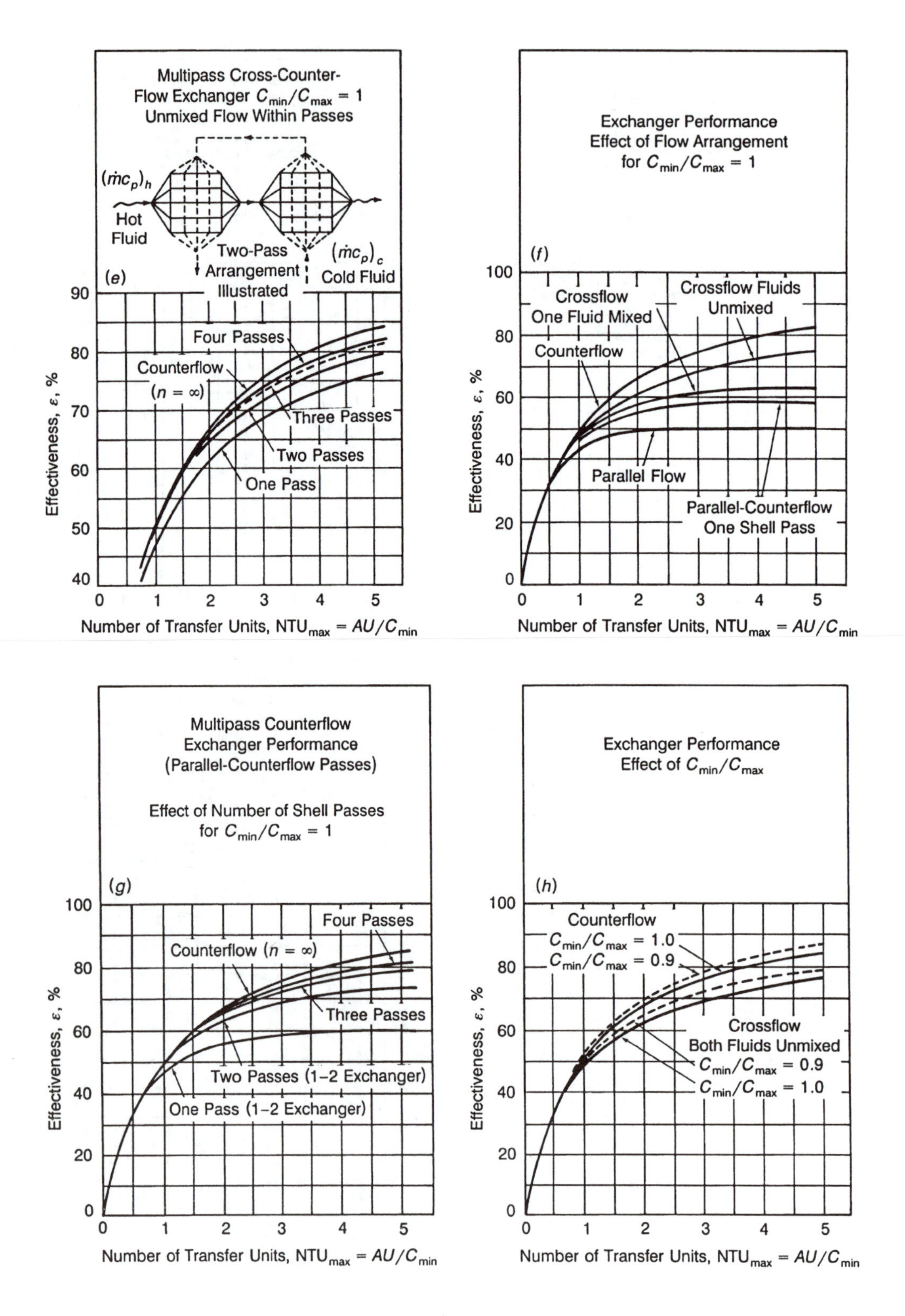

FIGURE 2.15 (CONTINUED)

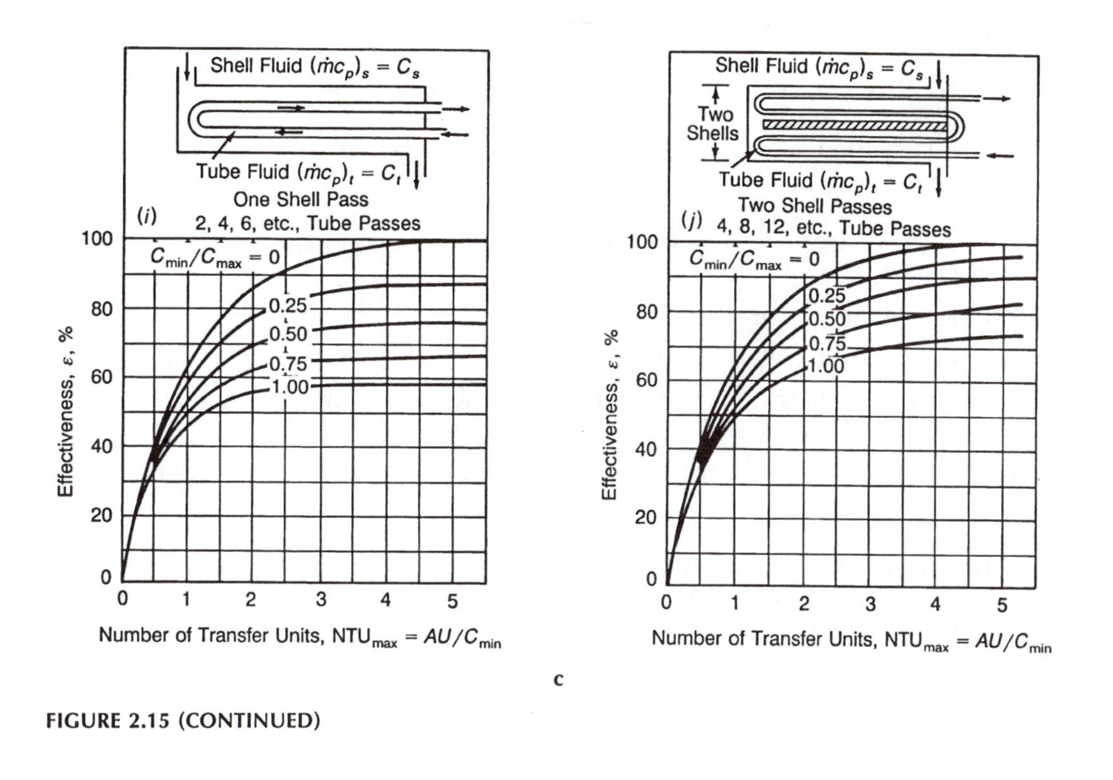

c

FIGURE 2.15 (CONTINUED)

2.7 Heat Exchanger Design Calculation

We have discussed two methods for performing a heat exchanger thermal analysis (Table 2.3). The rating and sizing of heat exchangers are two important problems encountered in the thermal analysis of heat exchangers.

For example, if inlet temperatures, one of the fluid outlet temperatures, and mass flow rates are known, then the unknown outlet temperature can be calculated from heat balances and the LMTD method can be used to solve this sizing problem with the following steps:

1. Calculate Q and the unknown outlet temperature from Equations (2.5) and (2.6).
2. Calculate ΔT_{lm} from Equation (2.28) and obtain the correction factor F if necessary.
3. Calculate the overall heat transfer coefficient U.
4. Determine A from Equation (2.36).

The LMTD method may also be used for rating problems (performance analysis) for an available heat exchanger, but computation would be tedious, requiring iteration because the outlet temperatures are not known. In such situations the analysis can be simplified by using the ε-NTU method. The rating analysis with the ε-NTU method is as follows:

1. Calculate the capacity rate ratio $C^* = C_{min}/C_{max}$ and $NTU = UA/C_{min}$ from the input data.
2. Determine the effectiveness ε from the appropriate charts or ε-NTU equations for the given heat exchanger and specified flow arrangement.

3. When ε is known, calculate the total heat transfer rate from Equation (2.43).
4. Calculate the outlet temperatures from Equations (2.5) and (2.6).

The ε-NTU method may also be used for the sizing problem, and the procedure is as follows:

1. When the outlet and inlet temperatures are known, calculate ε from Equation (2.42).
2. Calculate the capacity rate ratio $C^* = C_{min}/C_{max}$.
3. Calculate the overall heat transfer coefficient U.
4. When ε, C^*, and the flow arrangement are known, determine NTU from charts or from ε-NTU relations.
5. When NTU is known, calculate the heat transfer surface area A from Equation (2.44).

The use of the ε-NTU method is generally preferred in the design of compact heat exchangers for automotive, aircraft, air-conditioning, and other industrial applications where the inlet temperatures of the hot and cold fluids are specified and the heat transfer rates are to be determined. The LMTD method is traditionally used in the process, power, and petrochemical industries.

2.8 Variable Overall Heat Transfer Coefficient

In practical applications, the overall heat transfer coefficient varies along the heat exchanger and it is strongly dependent on the flow Reynolds number, heat transfer surface geometry, and fluid physical properties. Methods to account for specific variations in U are given for counterflow, crossflow, and multipass shell-and-tube heat exchangers.

Figure 2.16[12] shows typical situations in which the variation of U within a heat exchanger might be substantially large. The case in which both fluids are changing phase is shown in Figure 2.16a, where there is no sensible heating and cooling; the temperatures simply remain constant throughout. The condenser shown in Figure 2.16b is perhaps more common than the condenser of Figure 2.3d. In the former, the fluid vapor enters at a temperature greater than the saturation temperature and the condensed liquid becomes subcooled before leaving the condenser. A corresponding situation — where the cold fluid enters as a subcooled liquid is heated, evaporated, and then superheated — is shown in Figure 2.16c. When the hot fluid consists of both condensable vapor and noncondensable gases, the temperature distribution is more complex as represented in a general way in Figure 2.16d. The difficulty that one faces in designing such a heat exchanger is the continuous variation of U with position within the heat exchanger. If the three parts of the heat exchanger (Figures 2.16b and c) had constant values of U, then the heat exchanger could be treated as three different heat exchangers in series. For arbitrary variation of U through the heat exchanger, the exchanger is divided into many segments and a different value of U is then assigned to each segment. The analysis is best performed by a numerical or finite difference method.

Let us consider a parallel-flow double-pipe heat exchanger (Figure 2.1a). The heat exchanger is divided into increments of surface area ΔA_i. For any incremental surface area, the hot- and cold-fluid temperatures are T_{hi} and T_{ci}, respectively; and it will be assumed that

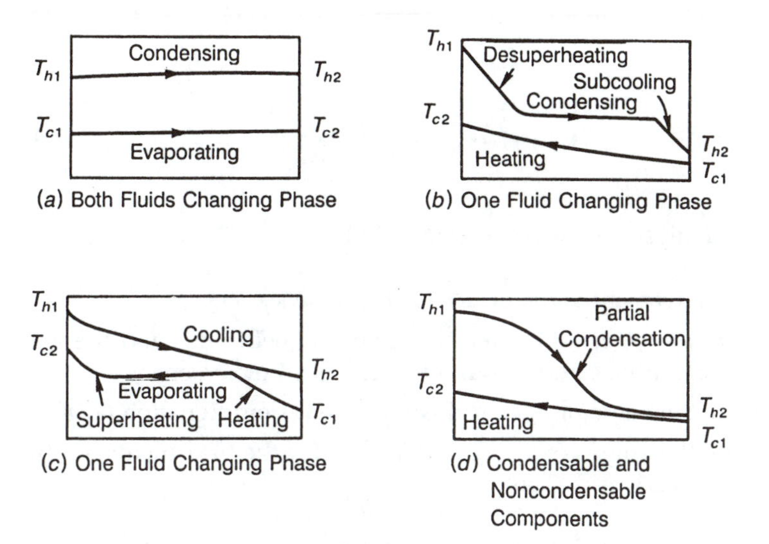

FIGURE 2.16
Typical cases of a heat exchanger with variable U. (From Walker, G. [1990] *Industrial Heat Exchanger. A Basic Guide*, Hemisphere, New York.)

the overall heat transfer coefficient can be expressed as a function of these temperatures. Thus

$$U_i = U_i(T_{hi}, T_{ci}) \tag{2.52}$$

The incremental heat transfer in ΔA_i can be calculated from Equation (2.22):

$$\Delta Q_i = -\ (\dot{m}c_p)_{hi}\ (T_{h(i+1)} - T_{hi}) = \ (\dot{m}c_p)_{ci}\ (T_{c(i+1)} - T_{ci}) \tag{2.53}$$

From Equation (2.23), δQ_i is also given by:

$$\delta Q_i = U_i \Delta A_i (T_h - T_c)_i \tag{2.54}$$

where ΔA_i must be small enough for the accuracy of the solution.

Equation (2.25) can be written for parallel-flow arrangement as:

$$\frac{d(T_h - T_c)}{(T_h - T_c)} = -U\left(\frac{1}{C_h} + \frac{1}{C_c}\right)dA \tag{2.55}$$

Equation (2.55) can be written in the finite-difference form as:

$$\frac{(T_h - T_c)_{i+1} - (T_h - T_c)_i}{(T_h - T_c)_i} = -U_i\left(\frac{1}{C_{ci}} + \frac{1}{C_{hi}}\right)\Delta A_i \tag{2.56}$$

which can be solved for $(T_h - T_c)_{i+1}$:

$$(T_h - T_c)_{i\ +1} = (T_h - T_c)_i\ (1 - M_i \Delta A_i) \tag{2.57}$$

where

$$M_i = U_i\left(\frac{1}{C_{ci}} + \frac{1}{C_{hi}}\right) \tag{2.58}$$

The numerical analysis can be conducted as follows:

1. Select a convenient value of ΔA_i for the analysis.
2. Calculate the inner and outer heat transfer coefficients and the value of U for the inlet conditions and through the initial ΔA increment.
3. Calculate the value of ΔQ_i for this increment from Equation (2.54).
4. Calculate the values of T_h, T_c, and $T_h - T_c$ for the next increment by the use of Equations (2.53) and (2.56).

The total heat transfer rate is then calculated from:

$$Q_i = \sum_{i=1}^{n} \Delta Q_i \tag{2.59}$$

where n is the number of the increments in ΔA_i.

The treatment of counterflow shell-and-tube and the exchanger with one shell-side pass and two-tube passes for variable overall heat transfer coefficient is given in Reference 13.

For the overall heat transfer coefficient U and ΔT varying linearly with Q, Colburn recommended the following expression to calculate Q:[14]

$$Q = U_m A \Delta T_{lm} = \frac{A(U_2\Delta T_1 - U_1\Delta T_2)}{\ln[(U_2\Delta T_1)/(U_1\Delta T_2)]} \tag{2.60}$$

where U_1 and U_2 are the values of the overall heat transfer coefficients on the ends of the exchanger having temperature differences of ΔT_1 and ΔT_2, respectively.

When both $1/U$ and ΔT vary linearly with Q, Butterworth[13] has shown that:

$$Q = U_m A \Delta T_{lm} \tag{2.61}$$

where

$$\frac{1}{U_m} = \frac{1}{U_1}\left[\frac{\Delta T_{lm} - \Delta T_2}{\Delta T_1 - \Delta T_2}\right] = \frac{1}{U_2}\left[\frac{\Delta T_1 - \Delta T_{lm}}{\Delta T_1 - \Delta T_2}\right] \tag{2.62}$$

Equations (2.60) and (2.61) will not usually be valid over the whole of the heat exchanger but may apply to a small portion of it.

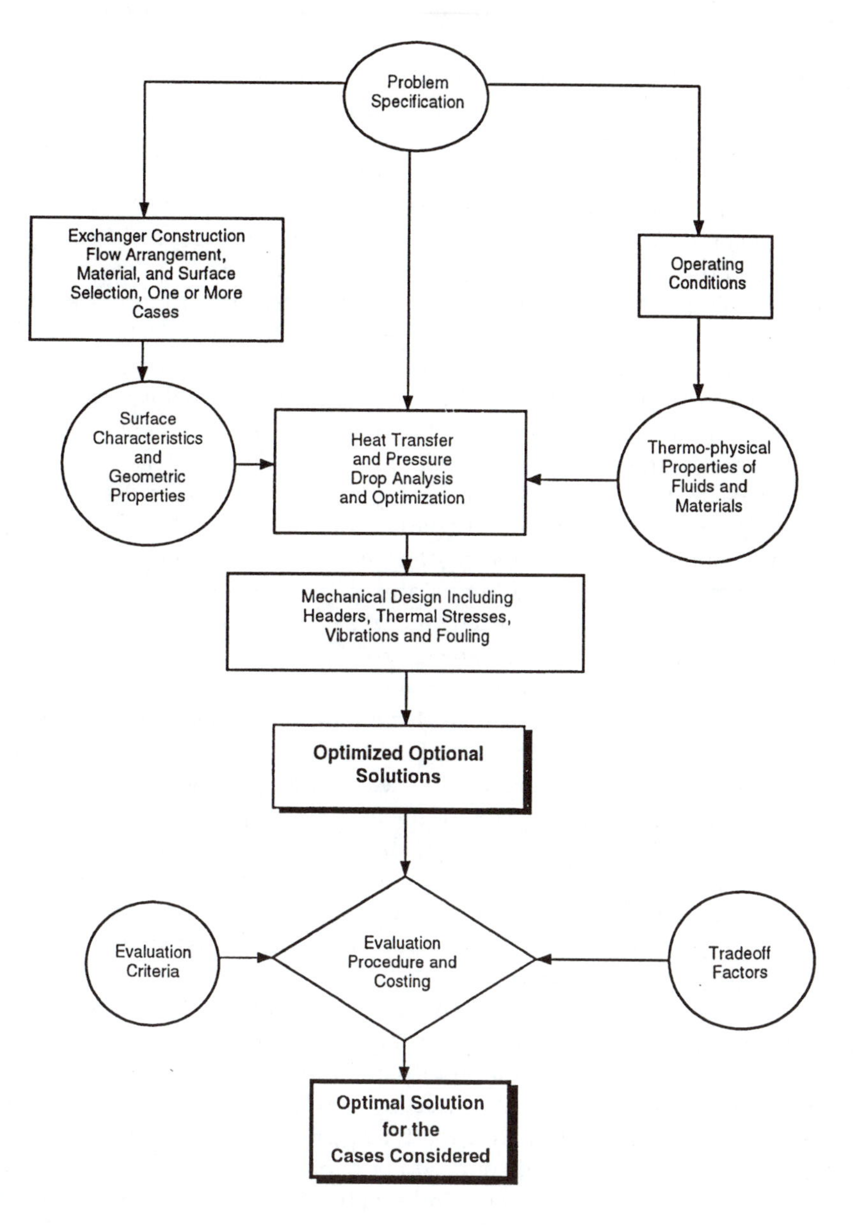

FIGURE 2.17
Heat exchanger design methodology. (From Shah, R. K. [1981] In *Heat Exchangers–Thermal–Hydraulic Fundamentals and Design*, pp. 455–459. Hemisphere, New York; Bell, K. J. [1988] In *Heat Transfer Equipment Design*, pp. 131–144. Hemisphere, New York.)

2.9 Heat Exchanger Design Methodology

The flowchart of heat exchanger design methodology is given in Figure 2.17.[15] The first criterion that a heat exchanger should satisfy is the fulfillment of the process requirements.

The design specifications may contain all the necessary detailed information on flow rates of streams; operating pressures; pressure drop limitations for both streams; temperatures; size; length; and other design constraints such as cost, type of materials, heat exchanger type, and arrangements. The heat exchanger design provides missing information based on experiences, judgment, and the requirements of the customer.

Based on the problem specifications, the exchanger construction type, flow arrangement, surface or core geometry, and materials must be selected. In the selection of the type of heat exchanger, the operating pressure and temperature levels, maintenance requirements, reliability, safety, availability and manufacturability of surfaces, and cost must be considered.

As discussed in previous sections, heat exchanger thermal design may be classified as *sizing* (design problem) or *rating* (performance analysis). In the sizing problem, the surface area and heat exchanger dimensions are to be determined; inputs to the sizing problem are flow rates, inlet temperatures and one outlet temperature, surface geometries, pressure drop limitations, and thermophysical properties of streams and materials.

In the rating problem, the heat exchanger already exists or the heat exchanger configuration is selected by approximate sizing. Therefore, inputs to the rating problem are heat exchanger surface geometry and dimensions, fluid flow rates, inlet temperatures, and pressure drop limitations. The fluid outlet temperatures, total heat transferred, and pressure drop for both streams through the heat exchanger are to be calculated in the rating (performance) analysis. If the rating gives acceptable thermal performance with pressure drops in both streams below the maximum allowable ones, this heat exchanger configuration may be considered a solution to the problem. It is often possible to find a number of variant configurations that will meet these requirements; then the choice must be made on other criteria, usually the cost of the heat exchanger. If the heat exchanger selected for rating is found to be unsatisfactory, a new design must be chosen for the next rating. The process by which one moves from one design to a more satisfying design can be called design modification.[15] Rating is the computational process (hand method or computer method) by which one determines the thermal performance and pressure drops for two streams in a completely defined heat exchanger. The rating and sizing problems are discussed and worked examples are given in the various chapters of this book.

The selection criterion is that the heat exchanger must withstand the service conditions of the plant environment. Therefore after thermal design analysis, the *mechanical design* is conducted, which includes the calculation of plate, tube, shell, header thicknesses, and arrangements. The heat exchanger must resist corrosion by the service and process streams and by the environment; this is mostly a matter of proper material selection. A proper design of inlet and outlet nozzles and connections, supporting materials, location of pressure and temperature measuring devices, and manifolds are to be made. Thermal stress calculations must be performed under steady-state and transit operating conditions. The additional important factors that should be considered and checked in the design are flow vibrations, and the level of velocities to eliminate or minimize fouling and erosion.

Another criterion is the *maintance* requirement. The configuration and placement of the heat exchanger must be properly chosen to permit cleaning as required and replacement of tubes, gaskets, and any other components that are especially vulnerable to corrosion, erosion, vibration, or aging. *Operating* problems under severe climatic conditions (freezing) and the transportation of the unit should also be considered.

There may be limitations on exchanger diameter, length, weight, or tube, matrix core specifications due to size requirements, lifting and servicing capabilities, and availability of replacement tubes and gaskets.

After the mechanical design is completed, the final cost analysis to arrive at an optimum solution must be considered. An overall optimum design, in general, is the one that meets the performance requirements at a minimum cost, which includes capital cost (the costs of materials, manufacturing, testing, shipment, and installation) and operating and maintenance costs (the costs of fluid pumping powers, repair, and cleaning). There are many interdependent factors that must be considered while designing and optimizing a heat exchanger.

The problem of heat exchanger design is intricate. Only a part of the total design process consists of quantitative analytic evaluation. Because of a large number of qualitative judgments, trade-offs, and compromises, heat exchanger design is more of an art than a science at this stage. In general, no two engineers will come up with the same heat exchanger design for a given application.[16]

Nomenclature

A	total heat transfer area on one side of a recuperator, m^2
A_c	cross-sectional area of a heat exchanger passage, m^2
A_f	fin surface area on one side of a heat exchanger, m^2
A_u	unfinned surface area on one side of a heat exchanger, m^2
C	flow stream heat capacity rate, $\dot{m}c_p$, W/K
C_{max}	maximum of C_c and C_h, W/K
C_{min}	minimum of C_c and C_h, W/K
C^*	heat capacity rate ratio, C_{min}/C_{max}
c_p	specific heat at constant pressure, J/kg·K
d_i	tube I.D., m
d_o	tube O.D., m
e	tube surface roughness, m
F	LMTD correction factor
i	specific enthalpy, J/kg
k	thermal conductivity, W/m·K
L	length of the heat exchanger, m
m	fluid mass flow rate, kg/s
N_T	number of tube
NTU	number of heat transfer units based on C_{min} , UA/C_{min}
P	temperature effectiveness, defined by Equation (2.34)
Q	heat transfer rate, W
R	heat capacity rate ratio, defined by Equation (2.35)
R	thermal resistance, m^2·K/W
R_f	fouling factor, m^2·K/W
r	tube radius, m
T	temperature,°C, K
T_c	cold-fluid temperature,°C, K
T_h	hot-fluid temperature,°C, K
ΔT	local temperature difference between two fluids,°C, K

ΔT_{lm} log mean temperature difference, defined by Equation (2.28),°C, K
ΔT_m true mean temperature difference, defined by Equation (2.7),°C, K
t wall thickness, m
U overall heat transfer coefficient, W/m^2·K
u_m fluid mean velocity, m/s

Greek Symbols

Δ difference
δ fin thickness, m
ε heat exchanger effectiveness, defined by Equation (2.39)
η_f fin efficiency
η extended surface efficiency, defined by Equation (2.15)
μ dynamic viscosity, Pa·s
ν kinematic viscosity, m^2/s
ρ fluid density, kg/m^3
ϕ parameter, function of

Subscripts

c cold fluid
cf counterflow
f fin, finned, friction
h hot fluid, heat transfer
i inner, inside
m mean
max maximum
min minimum
o outer, outside, overall
u unfinned
w wall
x local
1 inlet
2 outlet

References

1. Shah, R. K. (1981) Classification of heat exchanger. In *Heat Exchangers: Thermal–Hydraulic Fundamentals and Design*, S. Kakaç, A. E. Bergles, and F. Mayinger, (Eds.), pp. 9–46. Hemisphere, New York.

2. Chenoweth, J. M. and Impagliazzo, M. (Eds.) (1981) *Fouling in Hot Exchange Equipment*, ASME Symposium, Volume HTD-17. ASME, New York.
3. Kakaç, S., Shah, R. K., and Bergles, A. E. (Eds.) (1981) *Low Reynolds Number Flow Heat Exchangers*, pp. 21–72. Hemisphere, New York.
4. Kakaç, S., Shah, R. K., and Aung, W. (Eds.) (1987) *Handbook of Single Phase Convective Heat Transfer*, Chapters 4 and 18. Wiley, New York.
5. Kern, D. Q. and Kraus, A. D. (1972) *Extended Surface Heat Transfer*, McGraw-Hill, New York.
6. Padet, J. (1994) *Echangeurs Thermiques*, Masson, Paris, France.
7. *Standards of the Tubular Exchange Manufacturers Association* (1978) 6th ed., TEMA, New York.
8. Bowman, R. A., Mueller, A. C., and Nagle, W. M. (1940) Mean temperature difference in design. *Trans. ASME*, Vol. 62, 283–294.
9. Kays, W. M. and London, A. L. (1984) *Compact Heat Exchangers*, 3rd ed., McGraw-Hill, New York.
10. Shah, R. K. and Mueller, A. C. (1985) Heat exchangers. In *Handbook of Heat Transfer Applications*, W. M. Rohsenow, J. P. Hartnett, and E. N. Ganič (Eds.), Chapter 4. McGraw-Hill, New York.
11. Kays, W. M., London, A. L., and Johnson, K. W. (1951) *Gas Turbine Plant Heat Exchangers*, ASME, New York.
12. Walker, G. (1990) *Industrial Heat Exchanger. A Basic Guide*, Hemisphere, New York.
13. Butterworth, D. (1981) Condensers: thermohydraulic design. In *Heat Exchangers: Thermal–Hydraulic Fundamentals and Design*, S. Kakaç, A. E. Bergles, and F. Mayinger (Eds.), pp. 647–679. Hemisphere, New York.
14. Colburn, A. P. (1933) Mean temperature difference and heat transfer coefficient in liquid heat exchangers. *Ind. Eng. Chem.*, Vol. 25, 873–877.
15. Shah, R. K. (1981) Heat exchanger design methodology — An Overview. In *Heat Exchangers–Thermal–Hydraulic Fundamentals and Design*, S. Kakaç, A. E. Bergles, and F. Mayinger (Eds.), pp. 455–459. Hemisphere, New York.
16. Bell, K. J. (1988) Overall design methodology for shell-and-tube exchangers. In *Heat Transfer Equipment Design*, R. K. Shah, E. C. Subbarao, and R. A. Mashelkar(Eds.),pp. 131–144. Hemisphere, New York.

Problems

2.1. Starting from Equation 2.22, show that for a parallel flow heat exchanger, Equation 2.26a becomes

$$\frac{T_{h_2} - T_{c_2}}{T_{h_1} - T_{c_1}} = \exp\left[-\left(\frac{1}{C_h} + \frac{1}{C_c}\right)UA\right]$$

2.2. Show that for a parallel flow heat exchanger the variation of the hot fluid temperature along the heat exchanger is given by

$$\frac{T_h - T_{h_1}}{T_{h_1} - T_{c_1}} = \frac{C_c}{C_h + C_c}\left\{1 - \exp\left[-\left(\frac{1}{C_h} + \frac{1}{C_c}\right)UA\right]\right\}.$$

Obtain a similar expression for the variation of the cold fluid temperatures along the heat exchanger. Also show that for $A \to \infty$, the temperature will be equal to the mixing cup temperature of the fluids which is given by

$$T_\infty = \frac{C_h T_{h_1} + C_c T_{c_1}}{C_c + C_h}$$

2.3. Show that the variations of the hot and cold fluid temperatures along a counterflow heat exchanger are given by

$$\frac{T_h - T_{h_1}}{T_{h_1} - T_{c_2}} = \frac{C_c}{C_c - C_h}\left\{\exp\left[\left(\frac{1}{C_c} - \frac{1}{C_h}\right)UA\right] - 1\right\}$$

and

$$\frac{T_c - T_{c_2}}{T_{h_1} - T_{c_2}} = \frac{C_h}{C_c - C_h}\left\{\exp\left[\left(\frac{1}{C_c} - \frac{1}{C_h}\right)UA\right] - 1\right\}$$

2.4. From problem 2.3, show that for the case $C_h < C_c$,

$$\frac{d^2T_h}{dA^2} > 0 \text{ and } \frac{d^2T_c}{dA^2} > 0,$$

and therefore temperature curves are convex and for the case $C_h > C_c$,

$$\frac{d^2T_h}{dA^2} < 0, \text{ and } \frac{d^2T_c}{dA^2} < 0,$$

therefore the temperature curves are concave (see Figure 2.6).

2.5. Show that when the heat capacity rates of hot and cold fluids are equal ($C_c = C_h = C$), the variation of the hot and cold fluid temperatures along the heat exchanger are linear with the surface area as:

$$\frac{T_c - T_{c_2}}{T_{h_1} - T_{c_2}} = \frac{T_h - T_{h_1}}{T_{h_1} - T_{c_2}} = -\frac{UA}{C}$$

2.6. Assume that in a condenser, there will be no subcooling and condensate leaves the condenser at saturation temperature, T_h. Show that variation of the coolant temperature along the condenser is given by

$$\frac{T_c - T_{c_1}}{T_h - T_{c_1}} = 1 - \exp\left(-\frac{UA}{C_c}\right)$$

2.7. In a boiler (evaporator), the temperature of hot gases decreases from T_{h_1} to T_{h_2}, while boiling occurs at a constant temperature T_c. Obtain an expression, as in Problem 2.6, for the variation of hot fluid temperature with the surface area.

2.8. Show that Equation 2.46 is also applicable for $C_h > C_c$, that is $C^* = C_c/C_h$.

2.9. Obtain the expression for exchanger heat transfer effectiveness, ε, for parallel flow given by Equation 2.47.

2.10. 5000 kg/hr of water will be heated from 20°C to 35°C by hot water at 140°C. A 15°C hot water temperature drop is allowed. A number of double-pipe heat exchangers with annuli and pipes each connected in series will be used. Hot water flows through the inner tube. The thermal conductivity of the material is 50 W/m·K.

Fouling factors:	R_i = 0.000176 m²·K/W,
	R_o = 0.000352 m²·K/W.
Inner tube diameters:	ID = 0.0525 m, OD = 0.0603 m.
Annulus diameters:	ID = 0.0779 m, OD = 0.0889 m.

The heat transfer coefficients in the inner tube and in the annulus are 4620 W/m²·K and 1600 W/m²·K, respectively. Calculate the overall heat transfer coefficient and the surface area of the heat exchanger.

2.11. Water at a rate of 45,500 kg/hr is heated from 80°C to 150°C in a shell-and-tube heat-exchanger having two shell passes and eight tube passes with a total outside heat transfer surface area of 925 m². Hot exhaust gases having approximately the same thermophysical properties as air enter at 350°C and exit at 175°C. Determine the overall heat transfer coefficient based on the outside surface area.

2.12. A shell-and-tube heat exchanger given in Problem 2.11 is used to heat 62,000 kg/hr of water from 20°C to about 50°C. Hot water at 100°C is available. Determine how the heat transfer rate and the water outlet temperature vary with the hot water mass flow rate. Calculate the heat transfer rates and the outlet temperatures for hot water flow rates:

a. 80,000 kg/hr;

b. 40,000 kg/hr.

2.13. Water at a flow rate of 5000 kg/hr is heated from 10°C to 35°C in an oil cooler (c_p = 4182 J/kg·K) by engine oil with an inlet temperature of 65°C (c_p = 2072 J/kg·K) and a flow rate of 6000 kg/hr. Take the overall heat transfer coefficient to be 3500 W/m²·K. What are the areas required for:

a. parallel flow;

b. counterflow?

2.14. In order to cool a mass flow rate of 9.4 kg/hr of air from 616°C to 232°C, it is passed through the inner tube of a double-pipe heat exchanger with counterflow, which is 1.5 m long and the outer diameter of the inner tube is 2 cm.

a. Calculate the heat transfer rate. For air c_{ph} = 1060 J/kg·K.

b. The cooling water enters the annular side at 16°C with a mass flow rate of 0.3 l/min. Calculate the exit temperature of the water. For water, c_{pc} = 4180 J/kg·K.

c. Determine the effectiveness of this heat exchanger, then NTU. The overall heat transfer coefficient based on the outside heat transfer surface area is 38.5 W/m²·K. Calculate the surface area of the heat exchanger and the number of hairpins.

2.15. A shell-and-tube heat exchanger is designed to heat water from 40°C to 60°C with a mass flow rate of 20,000 kg/hr. Water at 180°C flows through tubes with a mass flow rate of 10,000 kg/hr. The tubes have an inner diameter of d_i = 20 mm, the Reynolds number is $Re = 10{,}000$. The overall heat transfer coefficient-based outside heat transfer surface area is estimated to be $U = 450$ W/m²·K.

1. Calculate the heat transfer rate Q of the heat exchanger and the exit temperature of the hot fluid.
2. If the heat exchanger is counterflow with one tube and one shell pass; determine (by the use of LMTD and ε-NTU methods):
   a. the outer heat transfer area;
   b. the velocity of the fluid through the tubes;
   c. the cross-sectional area of the tubes;
   d. the number of the tubes and the length of the heat exchanger.

2.16. An oil cooler is used to cool lubrication oil from 70°C to 30°C. The cooling water enters the exchanger at 15°C and leaves at 25°C. The specific heat capacities of the oil and water are 2 and 4.2 kJ/kg·K, respectively, and the oil flow rate is 4 kg/s.

a. Calculate the water flow rate required.
b. Calculate the true mean temperature difference for two-shell-pass-and-four-tube passes, one-shell-pass-and-two-tube passes shell-and-tube heat exchanger, and an unmixed-unmixed cross-flow configuration.
c. Find the effectiveness of the heat exchangers.

2.17. For the oil cooler described in Problem 16, calculate the surface area required for the shell-and-tube and unmixed-unmixed cross-flow exchangers, assuming the overall heat transfer coefficient $U = 90$ W/m²·K. For the shell-and-tube exchanger, calculate the stream outlet temperatures and compare it with the given values.

2.18. In an oil cooler, oil flows through the heat exchanger with a mass flow rate of 8 kg/s and inlet temperature of 70°C. Specific heat of oil is 2 kJ/kg·K. The cooling stream is treated cooling water that has specific heat capacity of 4.2 kJ/kg·K, flow rate of 20 kg/s, and inlet temperature of 15°C. Assuming a total heat exchanger surface area of 150 m² and an overall heat transfer coefficient of 150 W/m²·K, calculate the outlet temperature for two-pass shell-and-tube and unmixed-unmixed cross-flow units, respectively. Estimate the respective F-corrections factor and examine.

3

Forced Convection Correlations for Single-Phase Side of Heat Exchangers

3.1 Introduction

In this chapter recommended correlations for the single-phase side (or sides) of heat exchangers are given. In many two-phase-flow heat exchangers — such as boilers, nuclear steam generators, power condensers, and air-conditioning and refrigeration evaporators and condensers — one side has single-phase flow while the other side has two-phase flow. Generally, the single-phase side represents higher thermal resistance, particularly with gas or oil flow. In this chapter a comprehensive review of the available correlations for laminar and turbulent flow of single-phase Newtonian fluid through circular and noncircular ducts with and without the effect of property variations is made. A large number of experimental and analytic correlations are available for heat transfer coefficients and flow friction factors for laminar and turbulent flows through ducts and across tube banks.

Laminar and turbulent forced convection correlations for single-phase fluids represent an important class of heat transfer solution for heat exchanger applications. When a viscous fluid enters a duct, a boundary layer will form along the wall. The boundary layer gradually fills the entire duct cross section and the flow then is said to be fully developed. The distance at which the velocity becomes fully developed is called the hydrodynamic or velocity entrance length (L_{he}). Although theoretically, the approach to the fully developed velocity profile is asymptotic and it is therefore impossible to describe a definite location where the boundary layer completely fills the duct. Nevertheless, for all practical purposes, the hydrodynamic entrance length is finite.

If the walls of the duct are heated or cooled, then a thermal boundary layer will also develop along the duct wall. At a certain point downstream, one can talk about the fully developed temperature profile where the thickness of the thermal boundary layer is approximately equal to half the distance across the cross section. The distance at which the temperature profile becomes fully developed is called the thermal entrance length (L_{te}).

If heating or cooling starts from the inlet of the duct, then both the velocity and the temperature profiles develop simultaneously. The associated heat transfer problem is referred to as the combined hydrodynamic and thermal entry length problem, or simultaneously developing region problem. Therefore, there are four regimes in duct flows with heating or cooling, namely, hydrodynamically and thermally fully developed, hydrodynamically fully developed but thermally developing, thermally developed but hydrodynamically developing, and simultaneously developing; and the design correlations should be selected accordingly.

The relative rates of development of the velocity and temperature profiles in the combined entrance region depend on the fluid Prandtl number ($Pr = \nu/\alpha$). For high Prandtl number fluids, such as oils, even though the velocity and temperature profiles are uniform at the tube entrance, the velocity profile is established much more rapidly than the temperature profile. In contrast, for very low Prandtl number fluids, such as liquid metals, the temperature profile is established much more rapidly than the velocity profile. However, for Prandtl numbers about 1, as for gases, the temperature and velocity profiles develop at a similar rate simultaneously along the duct, starting from uniform temperature and uniform velocity at the duct entrance.

For the limiting case of $Pr \to \infty$, the velocity profile is developed before the temperature profile starts developing. For the other limiting case of $Pr \to 0$, the velocity profile never develops and remains uniform while the temperature profile is developing. The idealized $Pr \to \infty$ and $Pr \to 0$ cases are good approximations for highly viscous fluids and liquid metals (high thermal conductivity), respectively.

When fluids flow at very low velocities, the fluid particles move in definite paths called streamlines. This type of flow is called laminar flow. There is no component of fluid velocity normal to the duct axis in the fully developed region. Depending on the roughness of the circular duct inlet and inside surface, fully developed laminar flow will be obtained up to $Re_d \leq 2300$ if the duct length L is longer than the hydrodynamic entry length L_{he}. However, if $L < L_{he}$, developing laminar flow would exist over the entire duct length. The hydrodynamic and thermal entrance lengths for laminar flow inside conduits have been given in References 1 and 2.

If the velocity of the fluid is gradually increased, there will be a point where the laminar flow becomes unstable in the presence of small disturbances and the fluid no longer flows along smooth lines (streamlines), but by a series of eddies that result in a complete mixing of the entire flow field. This type of flow is called turbulent flow. The Reynolds number at which the flow ceases to be purely laminar is referred to as the critical (value of) Reynolds number. The critical Reynolds number in circular ducts is between 2100 and 2300. Although the value of the critical Reynolds number depends on the duct cross-sectional geometry and surface roughness, for particular applications it can be assumed that the transition from laminar to turbulent flow in noncircular ducts will also take place between $Re_{cr} = 2100 - 2300$ when the hydraulic diameter of the duct, which is defined as four times the cross-sectional (flow) area A_c divided by the wetted perimeter P of the duct, is used in calculation of the Reynolds number.

At a Reynolds number $Re > 10^4$, the flow is completely turbulent. Between the lower and upper limits lies the transition zone from laminar to turbulent flow. Therefore, fully turbulent flow in a duct occurs at a Reynolds number $Re \geq 10^4$.

The heat flux between the duct wall and a fluid flowing inside the duct can be calculated at any position along the duct by:

$$\frac{\delta Q}{dA} = h_x (T_w - T_b)_x \tag{3.1}$$

where h_x is called the local heat transfer coefficient or film coefficient and is defined based on the inner surface of the duct wall by using the convective boundary condition:

$$h_x = \frac{-k(\partial T/\partial y)_w}{(T_w - T_b)_x} \tag{3.2}$$

where k is the thermal conductivity of the fluid, T is the temperature distribution in the fluid, and T_w and T_b are the local wall and the fluid bulk temperatures, respectively. The local Nusselt number is calculated from:

$$Nu_x = \frac{h_x d}{k} = \frac{-d(\partial T / \partial y)_w}{(T_w - T_b)_x} \tag{3.3}$$

The local fluid bulk temperature T_{bx}, also referred to as the "mixing cup" or average fluid temperature, is defined for incompressible flow as:

$$T_b = \frac{1}{A_c u_m} \int_{A_c} uTdA_c \tag{3.4}$$

where u_m is the mean velocity of the fluid, A_c is the flow cross section area, and u and T are the velocity and temperature profiles of the flow at position x along the duct, respectively.

In design problems, it is necessary to calculate the total heat transfer rate over the total (entire) length of a duct using a mean value of the heat transfer coefficient based on the mean value of the Nusselt number defined as:

$$Nu - \frac{hL}{k} = \frac{1}{L} \int_0^L Nu_x dx \tag{3.5}$$

3.2 Laminar Forced Convection

Laminar duct flow is generally encountered in compact heat exchangers, cryogenic cooling systems, heating or cooling of heavy (highly viscous) fluids such as oils, and many other applications. Different investigators performed extensive experimental and theoretical studies with various fluids for numerous duct geometries and under different wall and entrance conditions. As a result, they formulated relations for the Nusselt number vs. the Reynolds and Prandtl numbers for a wide range of these dimensionless groups. Shah and London[1] and Shah and Bhatti[2] have compiled the laminar flow results.

Laminar flow can be obtained for a specified mass velocity $G = \rho u_m$ for (1) small hydraulic diameter D_h of the flow passage or (2) high fluid viscosity μ. Flow passages with small hydraulic diameter are encountered in compact heat exchangers because they result in large surface area per unit volume of the exchanger. The internal flow of oils and other liquids with high viscosity in noncompact heat exchangers is generally of a laminar nature too.

3.2.1 Hydrodynamically Developed and Thermally Developing Laminar Flow in Smooth Circular Ducts

The well-known Nusselt–Graetz problem for heat transfer to an incompressible fluid with constant properties flowing through a circular duct with a constant wall temperature boundary condition (subscript T) and fully developed laminar velocity profile was solved numerically by several investigators.[1,2] The asymptotes of the mean Nusselt number for a circular duct of length L are

$$Nu_T = 1.61\left(\frac{Pe_b d}{L}\right)^{1/3} \quad \text{for} \quad \frac{Pe_b d}{L} > 10^3 \tag{3.6}$$

and

$$Nu_T = 3.66 \quad \text{for} \quad \frac{Pe_b d}{L} < 10^2 \tag{3.7}$$

Properties are evaluated at mean bulk temperature.

The superposition of two asymptotes for the mean Nusselt number derived by Schlünder[3] gives sufficiently good results for most of the practical cases:

$$Nu_T = \left[3.66^3 + 1.61^3 \left(\frac{Pe_b d}{L} \right) \right]^{1/3} \tag{3.8}$$

An empirical correlation has also been developed by Hausen[4] for laminar flow in the thermal entrance region of the circular duct at constant wall temperature and is given as:

$$Nu_T = 3.66 + \frac{0.19 (Pe_b d / L)^{0.8}}{1 + 0.117 (Pe_b d / L)^{0.467}} \tag{3.9}$$

The results of Equations (3.8) and (3.9) are comparable to each other. These equations may be used for the laminar flow of gases and liquids in the range of $0.1 < Pe_b d/L < 10^4$. Axial conduction effects must be considered if $Pe_b d/L < 0.1$. All physical properties are evaluated at the fluid mean bulk temperature of T_b, defined as:

$$T_b = \frac{T_i + T_o}{2} \tag{3.10}$$

where T_i and T_o are the bulk temperatures of the fluid at the inlet and outlet of the duct, respectively.

The asymptotic mean Nusselt numbers in circular ducts with a constant wall heat flux boundary condition (subscript H) are[1,2]

$$Nu_H = 1.953 \left(\frac{Pe_b d}{L} \right)^{1/3} \quad \text{for} \quad \frac{Pe_b d}{L} > 10^2 \tag{3.11}$$

and

$$Nu_H = 4.36 \quad \text{for} \quad \frac{Pe_b d}{L} < 10 \tag{3.12}$$

The fluid properties are evaluated at the mean bulk temperature T_b as defined by Equation (3.10).

The results given by Equations (3.7) and (3.12) represent the dimensionless heat transfer coefficients for laminar forced convection inside a circular duct in the hydrodynamically and thermally developed region (fully developed conditions) under constant wall temperature and constant wall heat flux boundary conditions, respectively.

3.2.2 Simultaneously Developing Laminar Flow in Smooth Ducts

When heat transfer starts as soon as the fluid enters a duct, the velocity and temperature profiles start developing simultaneously. The analysis of the temperature distribution in the flow, and hence of the heat transfer between the fluid and the duct wall, for such situations is more complex because the velocity distribution varies in the axial direction as well as normal to it. Heat transfer problems involving simultaneously developing flow have been mostly solved by numerical methods for various duct cross sections. A comprehensive review of such solutions are given by Shah and Bhatti[2] and Kakaç.[5]

Shah and London[1] and Shah and Bhatti[2] presented the numerical values of the mean Nusselt number for this region. In the case of a short duct length, *Nu* values are represented by the asymptotic equation of Pohlhausen[6] for simultaneously developing flow over a flat plate; for a circular duct, this equation becomes:

$$Nu_T = 0.664\left(\frac{Pe_b d}{L}\right)^{1/2} Pr_b^{-1/6} \tag{3.13}$$

The range of validity is $0.5 < Pr_b < 500$ and $Pe_b d/L > 10^3$.

For most engineering applications with short circular ducts ($d/L > 0.1$), it is recommended that whichever of Equations (3.8), (3.9), and (3.13) gives the highest Nusselt number be used.

3.2.3 Laminar Flow Through Concentric Annular Smooth Ducts

Correlations for concentric annular ducts are very important in heat exchanger applications. The simplest form of a two-fluid heat exchanger is a double-pipe heat exchanger made up of two concentric circular tubes (Figure 3.1). One fluid flows inside the inner tube while the other flows through the annular passage. Heat is usually transferred through the wall of the inner tube while the outer wall of the annular duct is insulated. The heat transfer coefficient in the annular duct depends on the ratio of the diameters D_i/d_o, because of the shape of the velocity profile.

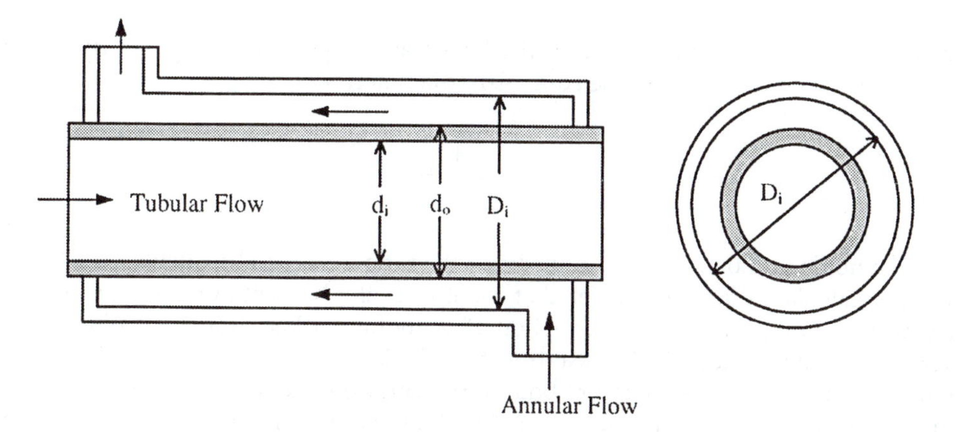

FIGURE 3.1
Concentric tube annulus.

The hydraulic (equivalent) diameter approach is the simplest method to calculate the heat transfer and the pressure drop in the annulus. In this approach, the hydraulic diameter of annulus D_h is substituted instead of the tube diameter in internal flow correlations:

$$D_h = 4 \frac{\text{net free - flow area}}{\text{wetted (or heat transfer)perimeter}} \quad (3.14)$$

This approximation is acceptable for heat transfer and pressure drop calculations. The validity of the hydraulic diameter approach has been substantiated by the results of experiments performed with finned annuli.[7] The wetted perimeter for pressure drop calculation in the annulus is defined as:

$$P_w = \pi (D_i + d_o) \quad (3.15)$$

and the heat transfer perimeter of the annulus can be calculated by:

$$P_h = \pi d_o \quad (3.16)$$

The only difference between P_w and P_h is D_i, which is the I.D. of the shell (outer tube) of the annulus. This difference is due to the fluid friction on the inner surface of the shell; however, this is not the case for the heat transfer perimeter because the heat transfer takes place only through the walls of the inner tube. The net free-flow area of the annulus is given by:

$$A_c = \frac{\pi (D_i^2 - d_o^2)}{4} \quad (3.17)$$

The hydraulic diameter based on the total wetted perimeter for pressure drop calculation is

$$D_h = \frac{4A_c}{P_w} \quad (3.18)$$

and the hydraulic diameter based on the heat transfer perimeter is given by Equation (3.19), which is hereafter called the equivalent diameter:

$$D_e = \frac{4A_c}{P_h} \quad (3.19)$$

The Reynolds number, Graetz number, and the ratio d/L are to be calculated with D_h. D_e is used to calculate the heat transfer coefficient from the Nusselt number and to evaluate the Grashof number in natural convection. Slightly higher heat transfer coefficients arise when D_h is used instead of D_e for heat transfer calculations.

For the constant wall temperature boundary condition, Stephan[8] has developed a heat transfer correlation based on Equation (3.9). The Nusselt number for hydrodynamically developed laminar flow in the thermal entrance region of an isothermal annulus, the outer wall of which is insulated, may be calculated by the following correlation:

$$Nu_T = Nu_\infty + \left[1 + 0.14\left(\frac{d_o}{D_i}\right)^{-1/2}\right] \frac{0.19(P_e D_h / L)^{0.8}}{1 + 0.117(Pe_b D_h / L)^{0.467}} \tag{3.20a}$$

where Nu_∞ is the Nusselt number for fully developed flow, which is given by:

$$Nu_\infty = 3.66 + 1.2\left(\frac{d_o}{D_i}\right)^{-1/2} \tag{3.20b}$$

where the outer wall of the annulus is insulated.[3]

A detailed review of laminar convective heat transfer in ducts for various hydrodynamic and thermal boundary conditions is given in Reference 2.

3.3 The Effect of Variable Physical Properties

When the previously mentioned correlations are applied to practical heat transfer problems with large temperature differences between the wall and the fluid mean bulk temperatures, the constant–property assumption could cause significant errors. The transport properties of most fluids vary with temperature, which influence the variation of velocity and temperature through the boundary layer or over the flow cross section of a duct.

For practical applications, a reliable and appropriate correlation based on the constant–property assumption can be modified and/or corrected so that it may be used when the variable–property effect becomes important.

Two correction methods for constant–property correlations for the variable–property effect have been employed, namely, the reference temperature method and the property ratio method. In the former, a characteristic temperature is chosen at which the properties appearing in nondimensional groups are evaluated so that the constant–property results at that temperature may be used to account for the variable–property behavior; in the latter, all properties are taken at the bulk temperature and then all variable–property effects are lumped into a function of the ratio of one property evaluated at the wall (surface) temperature to that property evaluated at the average bulk temperature. Some correlations may involve a modification or combination of these two methods.

For liquids, the variation of viscosity with temperature is responsible for most of the property effects. Therefore, the variable–property Nusselt numbers and friction factors in the property ratio method for liquids are correlated by:

$$\frac{Nu}{Nu_{cp}} = \left(\frac{\mu_b}{\mu_w}\right)^n \tag{3.21a}$$

$$\frac{f}{f_{cp}} = \left(\frac{\mu_b}{\mu_w}\right)^m \tag{3.21b}$$

where μ_b and k in Nu are the viscosity and conductivity evaluated at the bulk mean temperature, μ_w is the viscosity evaluated at the wall temperature, and cp refers to the constant–property solution. The friction coefficient usually employed is the so-called Fanning friction factor based on the wall shear rather than the pressure drop.

For gases, the viscosity, thermal conductivity, and density vary with the absolute temperature. Therefore, in the property ratio method, temperature corrections of the following forms are found to be adequate in practical applications for the temperature-dependent property effects in gases:

$$\frac{Nu}{Nu_{cp}} = \left(\frac{T_w}{T_b}\right)^n \tag{3.22a}$$

$$\frac{f}{f_{cp}} = \left(\frac{T_w}{T_b}\right)^m \tag{3.22b}$$

where T_b and T_w are the absolute bulk mean and wall temperatures, respectively.

It must be noted that the constant–property portion of the specific correlation is evaluated in terms of the parameters and conditions defined by its author(s).

Extensive theoretical and experimental investigations on convective heat transfer of fluids with variable properties have been reported in the literature to obtain the values of the exponents n and m, which will be cited in the following sections of this chapter.

3.3.1 Laminar Flow of Liquids

Deissler[9] conducted a numerical analysis, as described previously, for laminar flow through a circular duct at a constant heat flux boundary condition for liquid viscosity variation with temperature given by:

$$\frac{\mu}{\mu_w} = \left(\frac{T}{T_w}\right)^{-1.6} \tag{3.23}$$

and obtained $n = 0.14$ to be used with Equation (3.21a). This has been used widely to correlate experimental data for laminar flow for $Pr > 0.6$.

Deissler[9] also obtained $m = -0.58$ for heating and $m = -0.50$ for cooling of liquids to be used with Equation (3.21b).

A simple empirical correlation has been proposed by Seider and Tate[11] to predict the mean Nusselt number for laminar flow in a circular duct for the combined entry length with constant wall temperature as:

$$Nu_T = 1.86\left(\frac{Pe_b d_i}{L}\right)^{1/3}\left(\frac{\mu_b}{\mu_w}\right)^{0.14} \tag{3.24}$$

which is valid for smooth tubes for $0.48 < Pr_b < 16{,}700$ and $0.0044 < (\mu_b / \mu_w) < 9.75$. This correlation has been recommended by Whitaker[12] for values of:

$$\left(\frac{Pe_b d_i}{L}\right)^{1/3}\left(\frac{\mu_b}{\mu_w}\right)^{0.14} \geq 2 \tag{3.25}$$

Below this limit, fully developed conditions will be established and Equation (3.7) may be used to a good approximation. All physical properties are evaluated at the fluid bulk mean temperature except μ_w, which is evaluated at the wall temperature.

Example 3.1

Determine the total heat transfer coefficient at 30 cm from the inlet of a heat exchanger where engine oil flows through tubes with a diameter of 0.5 in. Oil flows with a velocity of 0.5 m/s and at a local bulk temperature of 30°C, while the local tube wall temperature is 60°C.

SOLUTION

From Appendix B (Table B.12) the properties of engine oil at T_b = 30°C are

ρ = 882.3 kg/m³ $\qquad$ c_p = 1922 J/kg·K

μ = 0.416 N·s/m² $\qquad$ k = 0.144 W/m·K

Pr = 5550 $\qquad$ μ_w = 0.074 N·s/m² (at T_b = 60°C)

To determine the heat transfer coefficient, we need to find the Reynolds number first:

$$Re_b = \frac{\rho U_m d_i}{\mu} = \frac{882.3 \times 0.5 \times 0.0127}{0.416} = 13.47$$

Because $Re < 2300$, the flow inside the tube is laminar. We can calculate the heat transfer coefficient from the Sieder and Tate correlation, Equation (3.24), as long as the following conditions are satisfied:

$$\left(Re_b Pr_b \frac{d}{L}\right)^{1/3}\left(\frac{\mu_b}{\mu_w}\right)^{0.14} = \left[\frac{13.47 \times 5550 \times 0.0127}{0.3}\right]^{1/3}\left(\frac{0.416}{0.074}\right)^{0.14} = 18.7 > 2$$

$$\left(\frac{\mu_b}{\mu_w}\right) = \left(\frac{0.416}{0.074}\right) = 5.62 < 9.75$$

Therefore

$$Nu_T = 1.86(RePr)_b^{1/3}\left(\frac{d_i}{L}\right)^{1/3}\left(\frac{\mu_b}{\mu_w}\right)^{0.14}$$

and

$$Nu_T = 1.86 \times 18.7 = 34.8$$

$$h = \frac{Nu_T k}{d_i} = \frac{34.8 \times 0.144}{0.0127} = 394.6\ \text{W/m}^2 \cdot \text{K}$$

The Nusselt-Graetz correlation given by Equation (3.11), which is applicable with constant heat flux boundary conditions, can also be used, because:

$$\frac{Pe_b d_i}{L} = \frac{Re_b Pr_b d_i}{L} = \frac{13.47 \times 5550 \times 0.0127}{0.3} = 3164.8 > 100$$

$$Nu_H = 1.953\left(Pe_b \frac{d}{L}\right)^{1/3} = 1.953 \times \left(5550 \times 13.47 \times \frac{0.0127}{0.3}\right)^{1/3} = 28.67$$

$$h = \frac{Nu_h k}{d_i} = \frac{28.67 \times 0.144}{0.0127} = 325\ \text{W/m}^2 \cdot \text{K}$$

Equation (3.24) takes into account the effect that higher fluid temperature due to heating increases the heat transfer coefficient. Although the constant heat flux boundary condition gives a higher heat transfer coefficient than the constant wall temperature boundary condition, here because of the heating of the fluid, the Nusselt–Graetz correlation gives a more conservative answer.

It is not surprising that alternative correlations have been proposed for specific fluids. Oskay and Kakaç[13] performed experimental studies with mineral oil in laminar flow through a circular duct under constant wall heat flux boundary conditions in the range of $0.8 \times 10^3 < Re_b < 1.8 \times 10^3$ and $1 < (T_w/T_b) < 3$, and suggested that the viscosity ratio exponent for Nu in Equation (3.24) should be increased to 0.152 for mineral oil.

Kuznetsova[14] conducted experiments with transformer oil and fuel oil in the range of $400 < Re_b < 1900$ and $170 < Pr_b < 640$ and recommended:

$$Nu_b = 1.23\left(\frac{Pe_b d}{L}\right)^{0.4}\left(\frac{\mu_b}{\mu_w}\right)^{1/6} \tag{3.26}$$

Test[15] conducted an analytic and experimental study on the heat transfer and fluid friction of laminar flow in a circular duct for liquids with temperature-dependent viscosity. The analytic approach is a numerical solution of the continuity, momentum, and energy equations. The experimental approach involves the use of a hot-wire technique for determination of the velocity profiles. Test[15] obtained the following correlation for the local Nusselt number:

$$Nu_b = 1.4\left(\frac{Pe_b d}{L}\right)^{1/3}\left(\frac{\mu_b}{\mu_w}\right)^n \tag{3.27}$$

where

$$n = 0.05 \text{ for heating liquids}$$
$$n = 1/3 \text{ for cooling liquids}$$

Test[15] also obtained the friction factor as:

$$f = \frac{16}{Re}\frac{1}{0.89}\left(\frac{\mu_b}{\mu_w}\right)^{0.2} \tag{3.28}$$

Equations (3.24) and (3.27) should not be applied to extremely long ducts.

3.3.2 Laminar Flow of Gases

The first reasonably complete solution for laminar heat transfer of a gas flowing in a tube with temperature-dependent properties was developed by Worsøe-Schmidt,[16] who solved the governing equations with a finite difference technique for fully developed and developing gas flow through a circular duct. Heating and cooling with a constant surface temperature and heating with a constant heat flux are considered. In the entrance region solution, the radial velocity was included. Worsøe-Schmidt concluded that near the entrance, and also well downstream (thermally developed), the results could be satisfactorily correlated for heating $1 < (T_w/T_b) < 3$ by $n = 0$, $m = 1.00$; and for cooling $0.5 < (T_w/T_b) < 1$ by $n = 0$, $m = 0.81$ in Equations (3.22a) and (3.22b).

Laminar forced convection and fluid flow in ducts have been studied extensively, and numerous results are available for circular and noncircular ducts under various boundary conditions. These results have been compiled by Shah and London[1] and Shah and Bhatti.[2] The laminar forced convection correlations discussed in previous sections are summarized in Table 3.1. The constant–property correlations can be corrected for the variable physical properties by the use of Table 3.2, in which the exponents m and n are summarized. For fully developed laminar flow, $n = 0.14$ is generally recommended for heating liquids.

3.4 Turbulent Forced Convection

Extensive experimental and theoretical efforts have been made to obtain the solutions for turbulent forced convection heat transfer and flow friction problems in ducts because of their frequent occurrence and application in heat transfer engineering. A compilation of such solutions and correlations for circular and noncircular ducts has been summarized by Bhatti and Shah.[17] There are a large number of correlations available in the literature for the fully developed (hydrodynamically and thermally) turbulent flow of single-phase

TABLE 3.1

Laminar Forced Convection Correlations in Smooth Straight Circular Ducts

Number	Correlation	Limitations and Remarks	Ref.
1	$Nu_T = 1.61(Pe_b d/L)^{1/3}$	$Pe_b d/L > 10^3$, constant wall temperature	1, 2
	$Nu_T = 3.66$	$Pe_b d/L > 10^3$, fully developed flow in a circular duct, constant wall temperature	
2	$Nu_T = [(3.66)^3 + (1.61)^3 Pe_b d/L)^{1/3}$	Superposition of two asymptoics given in case 1 for the mean Nusselt number, $0.1 < Pe_b d/L < 10^4$	3
3	$Nu_T = 3.66 + \dfrac{0.19(Pe_b d/L)^{0.8}}{1+0.117(Pe_b d/L)^{0.467}}$	Thermal entrance region, constant wall temperature, $0.1 < Pe_b d/L < 10^4$	4
4	$Nu_H = 1.953(Pe_b d/L)^{1/3}$	$Pe_b d/L > 10^2$, constant heat flux	1, 2
	$Nu_H = 4.36$	$Pe_b d/L < 10$, fully developed flow in a circular duct, constant heat flux	1, 2
5	$Nu_T = 0.664 \dfrac{1}{(Pr)^{1/6}} \left(Pe_b \dfrac{d}{L}\right)^{1/2}$	$Pe_b d/L > 10^4$, $0.5 < Pr < 500$, simultaneously developing flow	6
6	$Nu_T = Nu + \phi\left(\dfrac{d_o}{D_i}\right) \dfrac{0.19(PeD_{h/L})^{0.8}}{1+0.117(PeD_{h/L})^{0.467}}$	Circular annular duct, constant wall temperature, thermal entrance region	8
	$\phi(d_o/D_i) = 1 + 0.14(d_o/D_i)^{-1/2}$	Outer wall is insulated, heat transfer through the inner wall	
	$\phi(d_o/D_i) = 1 + 0.14(d_o/D_i)^{0.1}$	Heat transfer through outer and inner wall	
7	$Nu_T = 1.86(Re_b Pr_b d/L)^{1/3}(\mu_b/\mu_w)^{0.14}$	Thermal entrance region, constant wall temperature, $0.48 < Pr_b < 16{,}700$, $4.4 \times 10^{-3} < (\mu_b/\mu_w) < 9.75$, $(Re_b Pr_b d/L)^{1/3}(\mu_b/\mu_w)^{0.14} > 2$	11
8	$Nu_H = 1.86(Re_b Pr_b d/L)^{1/3}(\mu_b/\mu_w)^{0.152}$	Thermal entrance region, constant wall heat flux, for oils $0.8 \times 10^3 < Re_b < 1.8 \times 10^3$, $1 < (T_w/T_b) < 3$	13
9	$Nu_H = 1.23(Re_b Pr_b d/L)^{0.4}(\mu_b/\mu_w)^{1/6}$	Thermal entrance region, constant heat flux, $400 < Re_b < 1900$, $170 < Pr_b < 640$, for oils	14
10	$Nu_b = 1.4(Re_b Pr_b d/L)^{1/3}(\mu_b/\mu_w)^n$	Thermal entrance region, $n = 0.05$ for heating liquids, $n = {}^1/_3$ for cooling liquids	15

Note: Unless otherwise stated, fluid properties are evaluated at the bulk mean fluid temperature, $T_b = (T_i + T_o)/2$.

TABLE 3.2

Exponents n and m Associated with Equations (3.21) and (3.22) for Laminar Forced Convection Through Circular Ducts, $Pr > 0.5$

Number	Fluid	Condition	n	m^a	Limitations	Ref.
1	Liquid	Laminar, heating	0.14	−0.58	Fully developed flow,	
	Liquid	Laminar, cooling	0.14	−0.50	q''_w = constant $Pr > 0.6$, $\mu/\mu_w = (T/T_w)^{-1.6}$	9
2	Liquid	Laminar, heating	0.11	—	Developing and fully developed regions of a circular duct, T_w = constant q''_w = constant	10
3	Gas	Laminar, heating	0	1.00	Developing and fully developed regions, q''_w = constant, T_w = constant, $1 < (T_w/T_b) < 3$	16
	Gas	Laminar, cooling	0	0.81	T_w = constant, $0.5 < (T_w/T_b) < 1$	

[a] Fanning friction factor f is defined as $f = 2\tau_w/(\rho u_m^2)$ and for hydrodynamically developed isothermal laminar flow as $f = 16/Re$.

Newtonian fluids in smooth, straight circular ducts with constant and temperature-dependent physical properties. The objective of this section is to highlight some of the existing correlations to be used in the design of heat exchange equipment and to emphasize the conditions or limitations imposed on the applicability of these correlations.

3.4.1 Turbulent Flow in Circular Ducts with Constant Properties

Extensive efforts have been made to obtain empirical correlations that either represent a best-fit curve to experimental data or to adjust coefficients in the theoretical equations to best fit the experimental data. An example of the latter is the correlation given by Petukhov and Popov.[18] Their theoretical calculations for the case of fully developed turbulent flow with constant properties in a circular tube with constant heat flux boundary conditions yielded a correlation, which was based on the three-layer turbulent boundary layer model with constants adjusted to match the experimental data. A simplified form of this correlation has also been given by Petukhov and Kirillov as reported in Reference 19:

$$Nu_b = \frac{(f/2)Re_b Pr_b}{1.07 + 12.7(f/2)^{1/2}(Pr_b^{2/3} - 1)} \tag{3.29}$$

where $f = (1.58 \ln Re_b - 3.28)^{-2}$.

Equation (3.29) predicts the results in the range $10^4 < Re_b < 5 \times 10^6$ and $0.5 < Pr_b < 200$ with 5 to 6% error, and in the range $0.5 < Pr_b < 2000$ with 10% error.

Webb[20] has examined a range of data for turbulent flow under fully developed conditions in smooth tubes; he concluded that the relation developed by Petukhov and Popov, given previously, provided the best agreements with the measurements (Table 3.3). Sleicher and Rouse[21] correlated analytic and experimental results for the range $0.1 < Pr_b < 10^4$ and $10^4 < Re_b < 10^6$, and obtained:

$$Nu_b = 5 + 0.015 Re_b^m Pr_b^n \tag{3.30}$$

TABLE 3.3

Correlations for Fully Developed Turbulent Forced Convection Through a Circular Duct with Constant Properties

Number	Correlation[a]	Remarks and Limitations	Ref.
1	$Nu_b = \dfrac{(f/2)Re_b Pr_b}{1+8.7(f/2)^{1/2}(Pr_b - 1)}$	Based on three-layer turbulent boundary layer model, $Pr > 0.5$	23, 24
2	$Nu_b = 0.021\ Re_b^{0.8} Pr_b^{0.4}$	Based on data for common gases; recommended for Prandtl numbers ≈ 0.7	25
3	$Nu_b = \dfrac{(f/2)Re_b Pr_b}{1.07+12.7(f/2)^{1/2}(Pr_b^{2/3} - 1)}$	Based on three-layer model with constants adjusted to match experimental data $0.5 < Pr_b < 2000$, $10^4 < Re_b < 5 \times 10^6$	19
4	$Nu_b = \dfrac{(f/2)Re_b Pr_b}{1.07+9(f/2)^{1/2}(Pr_b - 1)Pr_b^{-1/4}}$	Theoretically based; Webb found case 3 better at high Pr and this one the same at other Pr	20
	$f = (1.58 \ln Re_b - 3.28)^{-2}$		
5	$Nu_b = 5 + 0.015\ Re_b^{m}\ Pr_b^{n}$ $m = 0.88 - 0.24/(4 + Pr_b)$	Based on numerical results obtained for $0.1 < Pr_b < 10^4$, $10^4 < Re_b < 10^6$	21
	$n = 1/3 + 0.5 \exp(-0.6\ Pr_b)$	Within 10% of case 6 for $Re_b > 10^4$	
	$Nu_b = 5 + 0.012\ Re_b^{0.87}\ (Pr_b + 0.29)$	Simplified correlation for gases, $0.6 < Pr_b < 0.9$	
6	$Nu_b = \dfrac{(f/2)(Re_b - 1000)Pr_b}{1+12.7(f/2)^{1/2}(Pr_b^{2/3} - 1)}$	Modification of case 3 to fit experimental data at low Re ($2300 < Re_b < 10^4$)	22
	$f = (1.58 \ln Re_b - 3.28)^{-2}$	Valid for $2300\ Re_b < 5 \times 10^6$ and $0.5 < Pr_b < 2000$	
	$Nu_b = 0.0214\ (Re_b^{0.8} - 100)Pr_b^{0.4}$	Simplified correlation for $0.5 < Pr < 1.5$; agrees with case 4 within −6% and +4%	
	$Nu_b = 0.012\ (Re_b^{0.87} - 280)Pr_b^{0.4}$	Simplified correlation for $1.5 < Pr < 500$; agrees with case 4 within −10% and +0% for $3 \times 10^3 < Re_b < 10^6$	
7	$Nu_b = 0.022\ Re_b^{0.8} Pr_b^{0.5}$	Modified Dittus–Boelter correlation for gases ($Pr \approx 0.5 - 1.0$); agrees with case 6 within 0 to 4% for $Re_b \geq 5000$	23

[a] Properties are evaluated at bulk temperatures.

with

$$m = 0.88 - \frac{0.24}{4 + Pr_b}$$

$$n = 1/3 + 0.5 \exp(-0.6\ Pr_b)$$

Equations (3.29) and (3.30) give the average Nusselt numbers and are not applicable in the transition region. Gnielinski[22] further modified the Petukhov–Kirillov correlation by comparing it with the experimental data so that the correlation covers a lower Reynolds number range. Gnielinski recommended the following correlation for the average Nusselt number that is also applicable in the transition region where the Reynolds numbers are between 2300 and 10^4:

$$Nu_b = \frac{(f/2)(Re_b - 1000)Pr_b}{1 + 12.7(f/2)^{1/2}(Pr_b^{2/3} - 1)} \tag{3.31}$$

where

$$f = (1.58 \ln Re_b - 3.28)^{-2} \tag{3.32}$$

The effect of thermal boundary conditions is almost negligible in turbulent forced convection;[23] therefore the empirical correlations given in Table 3.3 can be used for both constant wall temperature and constant wall heat flux boundary conditions.

Example 3.2

A two-tube pass, baffled single-pass shell-and-tube heat exchanger is used as an oil cooler. Cooling water enters the tubes at 25°C at a total flow rate of 8.154 kg/s and leaves at 32°C. The inlet and the outlet temperatures of the engine oil are 65 and 55°C, respectively. The heat exchanger has 12.25-in. I.D. shell, 1813 wire gauge (WG), 0.75-in. O.D. tubes. A total of 160 tubes are laid out on 15/16 in. equilateral triangular pitch. $R_{fo} = 1.76 \times 10^{-4}$ m²·K/W, $A_oR_W = 1.084 \times 10^{-5}$ m²·K/W, $h_o = 686$ W/m²·K, $A_o/A_i = 1.1476$, and $R_{fi} = 0.00008$ m²·K/W; find:

1. The heat transfer coefficient inside the tube
2. The total surface area of the heat exchanger by the LMTD method

SOLUTION

The properties of water at 28.5°C from Appendix B (Table B.2) are c_p= 4.18 kJ/kg·K, k = 0.611 W/m·K, Pr = 5.64, $\mu = 8.24 \times 10^{-4}$ Pa·s. The proper heat transfer correlations can be selected from the knowledge of the Reynolds number:

$$Re_b = \frac{\rho U_m d_i}{\mu}$$

or

$$Re_b = \frac{4\dot{m}_c}{\left(\frac{N_T}{2}\right)\pi\mu d_i}$$

where N_T is the total number of tubes and I.D. = 0.652 in. = 0.0165 m.

$$Re_b = \frac{4 \times 8.154}{80 \times \pi \times 8.24 \times 10^{-4} \times 0.0165} = 9510$$

Because $Re \gg 2300$, the flow inside the tube is turbulent.

By working with Gnielinski's correlation, Equation (3.31):

$$Nu_b = \frac{(f/2)(Re_b - 1000)Pr_b}{1 + 12.7(f/2)^{1/2}(Pr_b^{2/3} - 1)}$$

where

$$f = (1.58 \ln Re_b - 3.28)^{-2} = (1.58 \ln 9510 - 3.28)^{-2} = 7.982 \times 10^{-3}$$

Thus

$$Nu_b = \frac{(7.982 \times 10^{-3})/2(9510 - 1000)5.64}{1 + 12.7\left(\frac{7.982 \times 10^{-3}}{2}\right)^{1/2}(5.64^{2/3} - 1)}$$

Hence

$$h_i = \frac{Nu_b k}{d_i} = 70 \times \frac{0.611}{0.0165} = 2583 \text{ W/m}^2 \cdot \text{K}$$

The required heat transfer surface area can be obtained from Equation (2.36):

$$Q = U_o A_o F \Delta T_{lm,cf}$$

By using Equation (2.11), U_o can be calculated:

$$U_o = \left[\frac{A_o}{A_i} \cdot \frac{1}{h_i} + A_o R_w + R_{fo} + \frac{A_o}{A_i} R_{fi} + \frac{1}{h_o}\right]^{-1}$$

$$U_o = \left[1.1476 \times \frac{1}{2583} + 1.084 \times 10^{-5} + 1.76 \times 10^{-4} + (1.1476)(0.00008) + \frac{1}{686}\right]^{-1}$$

$$= 458.6 \text{ W/m}^2 \cdot \text{K}$$

Correction factor F may be obtained from Figure 2.7 where:

$$P = \frac{T_{c2} - T_{c1}}{T_{h1} - T_{c1}} = \frac{32 - 25}{65 - 25} = 0.175$$

$$R = \frac{T_{h1} - T_{h2}}{T_{c2} - T_{c1}} = \frac{65 - 55}{32 - 25} = 1.43$$

Hence $F \approx 0.98$.

From Equation (2.33) it follows that:

$$\Delta T_{lm,cf} = \frac{\Delta T_1 - \Delta T_2}{\ln \frac{\Delta T_1}{\Delta T_2}} = \frac{(65 - 32) - (55 - 25)}{\ln \frac{65 - 32}{55 - 25}} = 31.48°\text{C}$$

From heat balance:

$$Q = \dot{m}_c c_{pc}(T_{C2} - T_{C1}) = 8.154 \times 4.18 \times (32 - 25) = 238586 \text{ W}$$

and

$$A_o = \pi d_o L N_T$$

$$A_o = \frac{Q}{U_o F \Delta T_{lm,cf}} = \frac{238586}{(458.6)(0.98)(31.5)} = 16.85 \text{ m}^2$$

The length of the heat exchanger is calculated by:

$$L = \frac{A_o}{\pi d_o N_T} = \frac{16.85}{\pi \times 0.75 \times 2.54 \times 10^{-2} \times 160} = 1.76 \text{ m}$$

3.5 Turbulent Flow in Smooth Straight Noncircular Ducts

The heat transfer and friction coefficients for turbulent flow in noncircular ducts are compiled in Reference 17. A common practice is to employ the hydraulic diameter in the circular duct correlations to predict *Nu* and *f* for turbulent flow in noncircular ducts. For most of the noncircular smooth ducts, the accurate constant–property experimental friction factors are within ±10% of those predicted using the smooth circular duct correlation with hydraulic (equivalent) diameter D_h in the place of circular duct diameter *d*. The constant–property experimental Nusselt numbers are also within ±10 to ±15% except for some sharp-cornered and narrow channels. This order of accuracy is adequate for the overall heat transfer coefficient and the pressure drop calculations in most of the practical design problems.

Many attempts have been reported in the literature to arrive at a universal characteristic dimension for internal turbulent flows that would correlate the constant–property friction factors and Nusselt numbers for all noncircular ducts.[25–30] It must be emphasized that any improvement made by these attempts is only a few percentage, and therefore the circular duct correlations may be adequate for many engineering applications.

The correlations given in Table 3.3 do not account for entrance effects occurring in short ducts. Gnielinski[3] recommends the entrance correction factor derived by Hausen[27] in obtaining the Nusselt number for short ducts from the following correlation:

$$Nu_L = Nu_\infty \left[1 + \left(\frac{d}{L}\right)^{2/3}\right] \tag{3.33}$$

where Nu_∞ represents the fully developed Nusselt number calculated from the correlations given in Table 3.3. It should be noted that the entrance length depends on the Reynolds and Prandtl numbers and the thermal boundary condition. Thus Equation (3.33) should be used cautiously.

Example 3.3

Water flowing at 5000 kg/h will be heated from 20 to 35°C by hot water at 140°C. A 15°C hot water temperature drop is allowed. A number of 15-ft (4.5-m) hairpins (Figure 6.4) of 3-in. (I.D. = 3.068 in., O.D. = 3.5 in.) by 2-in. (I.D. = 2.067 in., O.D. = 2.357 in.) double-pipe heat exchangers with annuli and pipes each connected in series will be used. Hot water flows through the inner tube and the outside of the annulus is insulated against heat loss. Calculate (1) the heat transfer coefficient in the inner tube and (2) the heat transfer coefficient inside the annulus.

SOLUTION

1. We first calculate the Reynolds number to determine whether the flow is laminar or turbulent, and then select the proper correlation to calculate the heat transfer coefficient. From Appendix B (Table B.2), the properties of hot water at T_b =132.5°C are

$$\rho = 932.4 \text{ kg/m}^3 \qquad c_p = 4271 \text{ J/kg}\cdot\text{K}$$

$$k = 0.688 \text{ W/m}\cdot\text{K} \qquad \mu = 0.213\times10^{-3} \text{N}\cdot\text{s/m}^2$$

$$Pr = 1.325$$

We now make an energy balance to calculate the hot-water mass flow rate:

$$(\dot{m}c_p)_h\Delta T_h = (\dot{m}c_p)_c\Delta T_c$$

$$\dot{m}_h = \frac{\dot{m}_c c_{pc}}{c_{ph}} = \frac{(5000/3600)(4179)}{4271} = 1.36 \text{ kg/s}$$

where

$$c_{pc} = 4179 \text{ J/kg}\cdot\text{K} \quad \text{at} \quad T_b = 27.5°\text{C}$$

$$Re_b = \frac{\rho U_m d_i}{\mu} = \frac{4\dot{m}}{\pi\mu d_i} = \frac{4\times1.36}{\pi\times0.0525\times0.213\times10^{-3}} = 154850$$

Therefore, the flow is turbulent and we can select a correlation from Table 3.3. The Petukhov–Kirillov correlation is used here:

$$Nu_b = \frac{(f/2)Re_b Pr_b}{1.07+12.7(f/2)^{0.5}(Pr^{2/3}-1)}$$

where f is Flonenko's friction factor (Table 3.4), calculated as

$$f = (1.58 \ln Re_b - 3.28)^{-2} = [1.58 \ln (154850) - 3.28]^{-2} = 0.0041$$

$$Nu_b = \frac{(4.1\times10^{-3}/2)(154850\times1.325)}{1.07+12.7(4.1\times10^{-3}/2)^{0.5}(1.325^{2/3}-1)} = 353.9$$

$$h_i = \frac{Nu_b k}{d_i} = \frac{353.9\times0.688}{2.067\times2.54\times10^{-2}} = 4637.6 \text{ W/m}^2\cdot\text{K}$$

The effect of property variations can be found from Equation (3.21a) with n = 0.25 for cooling of a liquid in turbulent flow (see also Table 3.5), and the heat transfer coefficient can be corrected.

2. Calculate the heat transfer coefficient in the annulus. From Appendix B (Table B.2), the properties of cold water at T_b = 27.5°C are

$$\rho = 996.3 \text{ kg/m}^3 \qquad c_p = 4179 \text{ J/kg}\cdot\text{K}$$

$$k = 0.61 \text{ W/m}\cdot\text{K} \qquad \mu = 0.848\times 10^{-3}\text{N}\cdot\text{s/m}^2$$

$$Pr = 5.825$$

The hydraulic diameter of the annulus from Equation (3.17) is

$$D_i - d_o = (3.068 - 2.375)\times 2.54\times 10^{-2} = 0.0176 \text{ m}$$

$$Re_b = \frac{4D_h\dot{m}_c}{\pi\mu(D_i^2 - d_o^2)} = \frac{4(0.0176)(5000/3600)}{\pi\times(848\times 10^{-6})(0.002434)} = 15{,}079$$

Therefore, the flow inside the annulus is turbulent. One of the correlations can be selected from Table 3.3. The Gnielinski correlation is used here. It should be noted that for the annulus, the Nusselt number should be based on the hydraulic diameter (or equivalent diameter) calculated from Equation (3.19):

$$D_e = \frac{4A_c}{P_h} = \frac{4[\pi/4(D_i^2 - d_o^2)]}{\pi d_o} = \frac{[0.0779^2 - 0.0603^2}{0.0603} = 0.0403 \text{ m}$$

$$Nu_b = \frac{(f/2)(Re_b - 1000)Pr_b}{1.07 + 12.7(f/2)^{1/2}(Pr_b^{2/3} - 1)}$$

$$f = (1.58 \ln Re_b - 3.28)^{-2} = [1.58 \ln (15079) - 3.28]^{-2} = 0.007037$$

$$Nu_b = \frac{(0.007037/2)(15079 - 1000)(5.825)}{1.07 + 12.7(0.007037/2)^{1/2}(5.825^{2/3} - 1)} = 104.8$$

$$h_o = \frac{Nu_b k}{D_e} = \frac{104.8\times 0.614}{0.0403} = 1586 \text{ W/m}^2\cdot\text{K}$$

3.6 Effect of Variable Physical Properties in Turbulent Forced Convection

When there is a large difference between the duct wall and fluid bulk temperatures, heating and cooling influence heat transfer and the fluid friction in turbulent duct flow because of the distortion of turbulent transport mechanisms, in addition to the variation of fluid properties with temperature as for laminar flow.

3.6.1 Turbulent Liquid Flow in Ducts

Petukhov[19] reviewed the status of heat transfer and friction coefficient in fully developed turbulent pipe flow with both constant and variable physical properties.

To choose the correct value of n in Equation (3.21a), the heat transfer experimental data corresponding to heating and cooling for several liquids over a wide range of values (μ_w/μ_b) were collected by Petukhov,[19] who found that the data were well correlated by:

$$\frac{\mu_w}{\mu_b} < 1, \quad n = 0.11 \quad \text{for heating liquids} \tag{3.34}$$

$$\frac{\mu_w}{\mu_b} > 1, \quad n = 0.25 \quad \text{for cooling liquids} \tag{3.35}$$

which are applicable for fully developed turbulent flow in the range $10^4 < Re_b < 5 \times 10^6$, $2 < Pr_b < 140$, and $0.08 < (\mu_w/\mu_b) < 40$. The value of Nu_{cp} in Equation (3.21a) is calculated from Equation (3.30) or (3.31). The value of Nu_{cp} can also be calculated from the correlations listed in Table 3.3.

Petukhov[19] collected data from various investigators for the variable viscosity influence on friction in water for both heating and cooling, and suggested the following correlations for the friction factor:

$$\frac{\mu_w}{\mu_b} < 1, \quad \frac{f}{f_{cp}} = \frac{1}{6}\left(7 - \frac{\mu_b}{\mu_w}\right) \quad \text{for heating liquids} \tag{3.36}$$

$$\frac{\mu_w}{\mu_b} > 1, \quad \frac{f}{f_{cp}} = \left(\frac{\mu_w}{\mu_b}\right)^{0.24} \quad \text{for cooling liquids} \tag{3.37}$$

The friction factor for an isothermal (constant–property) flow f_{cp} can be calculated by the use of Table 3.4 or directly from Equation (3.32) for the range $0.35 < (\mu_w/\mu_b) < 2$, $10^4 < Re_b < 23 \times 10^4$, and $1.3 < Pr_b < 10$.

3.6.2 Turbulent Gas Flow in Ducts

The heat transfer and friction coefficients for turbulent fully developed gas flow in a circular duct were obtained theoretically by Petukhov and Popov[18] by assuming physical properties ρ, c_p, k, and μ as given functions of temperature. This analysis is valid only for small subsonic velocities, because the variations of density with pressure and heat dissipation in the flow were neglected. The eddy diffusivity of momentum was extended to the case of variable properties. The turbulent Prandtl number was taken to be 1 (i.e., $\epsilon_H = \epsilon_M$). The analyses were conducted for hydrogen and air for the following range of parameters: $0.37 < (T_w/T_b) < 3.1$ and $10^4 < Re_b < 4.3 \times 10^6$ for air, and $0.37 < (T_w/T_b) < 3.7$ and $10^4 < Re_b < 5.8 \times 10^6$ for hydrogen. The analytic results are correlated by Equation (3.22a), where Nu_{cp} is given by Equation (3.30) or (3.31), and the following values for n are obtained:

TABLE 3.4

Turbulent Flow Isothermal Fanning Friction Factor Correlations for Smooth Circular Ducts

Number	Correlation[a]	Remarks and Limitations	Ref.[b]
1	$f = \tau_w/1/2\ \rho u_m^2 = 0.0791\ Re^{-1/4}$	This approximate explicit equation agrees with case 3 within ±2.5%, $4 \times 10^3 < Re < 10^5$	Blasius
2	$f = 0.00140 + 0.125\ Re^{-0.32}$	This correlation agrees with case 3 within −0.5% and +3%, $4 \times 10^3 < Re < 5 \times 10^6$	Drew, Koo, and McAdams
3	$1/\sqrt{f} = 1.737 \ln\left(Re\sqrt{f}\right) - 0.4$	von Kármán's theoretical equation with the constants adjusted to best fit Nikuradse's experimental data, also referred to as the Prandtl correlation, should be valid for very high values of Re. $4 \times 10^3 < Re < 3 \times 10^6$	von Kármán and Nikuradse
	or $1/\sqrt{f} = 4 \log\left(Re\sqrt{f}\right) - 0.4$ approximated as $f = (3.64 \log Re - 3.28)^{-2}$		
	$f = 0.046\ Re^{-0.25}$	This approximate explicit equation agrees with the preceding within −0.4 and +2.2% for $3 \times 10^4 < Re < 10^6$	
4	$f = 1/(1.58 \ln Re - 3.28)^2$	Agrees with case 3 within ±0.5% for $3 \times 10^4 < Re < 10^7$ and within ±1.8% at $Re = 10^4$. $10^4 < Re < 5 \times 10^5$	Flonenko
5	$1/f = \left(1.7372 \ln \dfrac{Re}{1.964 \ln Re - 3.8215}\right)^2$	An explicit form of case 3; agrees with it within ±0.1%, $10^4 < Re < 2.5 \times 10^8$	Techo, Tickner, and James

[a] Cited in References 17, 23, 24, 26.
[b] Properties are evaluated at bulk temperatures.

$$\frac{T_w}{T_b} < 1 \quad n = -0.36 \quad \text{for cooling gases} \tag{3.38}$$

$$\frac{T_w}{T_b} > 1 \quad n = -\left[0.3 \log\left(\frac{T_w}{T_b}\right) + 0.36\right] \quad \text{for heating gases} \tag{3.39}$$

With these values for n, Equation (3.22a) describes the solution for air and hydrogen within an accuracy of ±4%. For simplicity, one can take n to be constant for heating as $n = -0.47$ (Table 3.5); then Equation (3.22a) describes the solution for air and hydrogen within ±6%. These results have also been confirmed experimentally and can be used for practical calculations when $1 < (T_w/T_b) < 4$.

TABLE 3.5

Exponents n and m Associated with Equations (3.21) and (3.22) for Turbulent Forced Convection Through Circular Ducts

Number	Fluid	Condition	n	m	Limitations	Ref.
1	Liquid	Turbulent heating	0.11	—	$10^4 < Re_b < 1.25 \times 10^5$, $2 < Pr_b < 140$, $0.08 < \mu_w/\mu_b < 1$	19
	Liquid	Turbulent cooling	0.25	—	$1 < \mu_w/\mu_b < 40$	
	Liquid	Turbulent heating	—	Eq. (3.39) or −0.25	10^4 $Re_b < 23 \times 10^4$, $1.3 < Pr_b < 10^4$, $0.35 < \mu_w/\mu_b < 1$	
	Liquid	Turbulent cooling	—	−0.24	$1 < \mu_w/\mu_b < 2$	
2	Gas	Turbulent heating	−0.47	—	$10^4 < Re_b < 4.3 \times 10^6$, $1 < T_w/T_b < 3.1$	18
	Gas	Turbulent cooling	−0.36	—	$0.37 < T_w/T_b < 1$	
	Gas	Turbulent heating	—	−0.52	$14 \times 10^4 < Re_w \leq 10^6$, $1 < T_w/T_b < 3.7$	
	Gas	Turbulent cooling	—	−0.38	$0.37 < T_w/T_b < 1$	
3	Gas	Turbulent heating	—	−0.264	$1 \leq T_w/T_b \leq 4$	31
4	Gas	Turbulent heating	—	−0.1	$1 < T_w/T_b < 2.4$	32

Example 3.4

Air at a mean bulk temperature of 40°C flows through a heated pipe section with a velocity of 6 m/s. The length and diameter of the pipe are 300 and 2.54 cm, respectively. The average pipe wall temperature is 300°C. Determine the average heat transfer coefficient.

SOLUTION

Because the wall temperature is so much greater than the initial air temperature, variable–property flow must be considered. From Appendix B (Table B.1), the properties of air at T_b = 40°C are

$$\rho = 1.128 \text{ kg/m}^3 \qquad c_p = 1005.3 \text{ J/kg}\cdot\text{K}$$
$$k = 0.0267 \text{ W/m}\cdot\text{K} \qquad \mu = 1.912\times10^{-5} \text{ N}\cdot\text{s/m}^2$$
$$Pr = 0.719$$

The inside heat transfer coefficient can be obtained from knowledge of the flow regime, that is, the Reynolds number:

$$Re_b = \frac{\rho U_m d_i}{\mu} = \frac{1.128\times6\times0.0254}{1.912\times10^{-5}} = 8991$$

Hence the flow in the tube can be assumed turbulent. On the other hand, $L/d = 3/0.0254 = 118 > 60$; then fully developed conditions can be assumed. Because $Pr > 0.6$, we can use one of the correlations given in Table 3.3. Gnielinski's correlation, Equation (3.31), with constant properties:

$$Nu_b = \frac{(f/2)(Re_b - 1000)Pr_b}{1 + 12.7(f/2)(Pr_b^{2/3} - 1)}$$

is used here to determine the Nusselt number:

$$f = (1.58 \ln Re_b - 3.28)^{-2} - [1.58 \ln (8991) - 3.28]^{-2} = 0.00811$$

$$Nu_b = \frac{hd}{k} = \frac{(0.00811/2)(8991-1000)(0.719)}{1+12.7(0.00811/2)^{0.5}(0.719^{2/3}-1)} = 27.712$$

$$h = Nu_b\frac{k}{d} = \frac{27.72\times0.0267}{0.0254} = 29.14 \text{ W/m}^2\cdot\text{K}$$

The heat transfer coefficient with variable properties can be calculated from Equation (3.22a), where n is given in Table 3.5 as $n = -0.47$:

$$Nu_b = Nu_{cp}\left(\frac{T_w}{T_b}\right)^{-0.47} = 27.712\left(\frac{573}{313}\right)^{-0.47} = 20.856$$

Then

$$h = \frac{Nu_b k}{d} = \frac{(20.856)\times(0.0267)}{0.0254} = 21.92 \text{ W/m}^2\cdot\text{K}$$

As can be seen in the case of a gas with temperature-dependent properties, heating a gas decreases the heat transfer coefficient.

A large number of experimental studies are available in the literature for heat transfer between tube walls and gas flows with large temperature differences and temperature-dependent physical properties. The majority of the work deals with gas heating at constant wall temperature in a circular duct; experimental studies on gas cooling are limited.

The results of heat transfer measurements at large temperature differences between the wall and the gas flow are usually presented as:

$$Nu_b = CRe_b^{0.8}Pr_b^{0.4}\left(\frac{T_w}{T_b}\right)^n \tag{3.40}$$

For fully developed temperature and velocity profiles (i.e., $L/d > 30$), C becomes constant and becomes independent of L/d (Table 3.7).

TABLE 3.6

Turbulent Forced Convection Correlations in Circular Ducts for Liquids with Variable Properties

Number	Correlation	Comments and Limitations	Ref.
1	$St_b Pr_f^{2/3} = 0.023\ Re_f^{-0.2}$	$L/d > 60$, $Pr_b > 0.6$, $T_f = (T_b + T_w)/2$; inadequate for large $(T_w - T_b)$	33
2	$Nu_b = 0.023\ Re_b^{0.8} Pr_b^{1/3}\left(\frac{\mu_b}{\mu_w}\right)^{0.14}$	$L/d > 60$, $Pr_b > 0.6$, for moderate $(T_w - T_b)$	11
3	$Nu_b = \frac{(f/8)Re_b Pr_b}{1.07 + 12.7\sqrt{f/8}(Pr_b^{2/3} - 1)}\left(\frac{\mu_b}{\mu_w}\right)^n$	$L/d > 60$, $0.08 < \mu_w/\mu_b < 40$, $10^4 < Re_b < 5 \times 10^6$, $2 < Pr_b < 140$, $f = (1.82 \log Re_b - 1.64)^{-2}$, $n = 0.11$ (heating), $n = 0.25$ (cooling)	19
4	$Nu_b = \frac{(f/8)Re_b Pr_b}{1.07 + 12.7\sqrt{f/8}(Pr_b^{2/3} - 1)}\left(\frac{Pr_b}{Pr_w}\right)^{0.11}$	Water, $2 \times 10^4 < Re_b < 6.4 \times 10^5$, $2 < Pr_b < 5.5$, $f = (1.82 \log Re_b - 1.64)^{-2}$, $0.1 < Pr_b/Pr_w < 10$	34
5	$Nu_b = 0.0277\ Re_b^{0.8} Pr_b^{0.36}\left(\frac{Pr_b}{Pr_w}\right)^{0.11}$	Fully developed conditions, the use of the Prandtl group was first suggested by the author in 1960	35
6	$Nu_b = 0.023\ Re_b^{0.8} Pr_b^{0.4}\left(\frac{\mu_b}{\mu_w}\right)^{0.262}$	Water, $L/d > 10$, $1.2 \times 10^4 < Re_b < 4 \times 10^4$	13
	$Nu_b = 0.023\ Re_b^{0.8} Pr_b^{0.4}\left(\frac{\mu_b}{\mu_w}\right)^{0.487}$	30% glycerine–water mixture $L/d > 10$, $0.89 \times 10^4 < Re_b < 2.0 \times 10^4$	
7	$Nu_b = 0.0235(Re_b^{0.8} - 230 \times 1.8\, Pr_b^{0.3} - 0.8) \times \left[1 + \left(\frac{d}{L}\right)^{2/3}\right]\left(\frac{\mu}{\mu_w}\right)^{0.14}$	Altered form of equation presented in 1959[4]	36
8	$Nu_b = 5 + 0.015\ Re_f^m Pr_w^n$		
	$m = 0.88 - 0.24/(4 + Pr_w)$	$L/d > 60$, $0.1 < Pr_b < 10^4$, $10^4 < Re_b < 10^6$	21
	$n = 1/3 + 0.5e^{-0.6\,Pr_w}$		
	$Nu_b = 0.015Re_f^{0.88} Pr_w^{1/3}$	$Pr_b > 50$	
	$Nu_b = 4.8 + 0.015Re_f^{0.85} Pr_w^{0.93}$	$Pr_b < 0.1$, uniform wall temperature	
	$Nu_b = 6.3 + 0.0167Re_f^{0.85} Pr_w^{0.93}$	$Pr_b < 0.1$, uniform wall heat flux	

TABLE 3.7

Turbulent Forced Convection Correlations in Circular Ducts for Gases with Variable Properties

Number	Correlation	Gas	Comments and Limitations	Ref.
1	$Nu_b = 0.023\, Re_b^{0.8} Pr_b^{0.4} \left(\frac{T_w}{T_b}\right)^n$ $T_w/T_b < 1$, $n = 0$ (cooling) $T_w/T_b > 1$, $n = -0.55$ (heating)	Air	$30 < L/d < 120$, $7 \times 10^3 < Re_b < 3 \times 10^5$, $0.46 < T_w/T_b < 3.5$	37
2	$Nu_b = 0.022\, Re_b^{0.8} Pr_b^{0.4} \left(\frac{T_w}{T_b}\right)^{-0.5}$	Air	$29 < L/d < 72$, $1.24 \times 10^5 < Re_b < 4.35 \times 10^5$, $1.1 < T_w/T_b < 1.73$	19
3	$Nu_b = 0.023\, Re_b^{0.8} Pr_b^{0.4} \left(\frac{T_w}{T_b}\right)^n$ $n = -0.4$ for air, $n = -0.185$ for helium, $n = -0.27$ for carbon dioxide	Air, helium, carbon dioxide	$1.2 < T_w/T_b < 2.2$, $4 \times 10^3 < Re_b < 6 \times 10^4$, $L/d > 60$	38
4	$Nu_b = 0.021\, Re_b^{0.8} Pr_b^{0.4} \left(\frac{T_w}{T_b}\right)^{-0.5}$ $Nu_b = 0.021\, Re_b^{0.8} Pr_b^{0.4} \left(\frac{T_w}{T_b}\right)^{-0.5} \times \left[1 + \left(\frac{L}{d}\right)^{-0.7}\right]$	Air, helium, nitrogen	$L/d > 30$, $1 < T_w/T_b < 2.5$, $1.5 \times 10^4 < Re_{ib} < 2.33 \times 10^5$, $L/d > 5$, local values	32
5	$Nu_b = 0.024\, Re_b^{0.8} Pr_b^{0.4} \left(\frac{T_w}{T_b}\right)^{-0.7}$	Nitrogen	$L/d > 40$, $1.24 < T_w/T_b < 7.54$, $18.3 \times 10^3 < Re_{ib} < 2.8 \times 10^5$	31
	$Nu_b = 0.023\, Re_w^{0.8} Pr_w^{0.4}$		Properties evaluated at wall temperature, $L/d > 24$	
	$Nu_b = 0.024\, Re_b^{0.8} Pr_b^{0.4} \left(\frac{T_w}{T_b}\right)^{-0.7} \times \left[1 + \left(\frac{L}{d}\right)^{-0.7} \left(\frac{T_w}{T_b}\right)^{0.7}\right]$		$1.2 \le L/d \le 144$	
6	$Nu_b = 0.021\, Re_b^{0.8} Pr_b^{0.4} \left(\frac{T_w}{T_b}\right)^n$ $n = -\left(0.9 \log \frac{T_w}{T_b} + 0.205\right)$	Nitrogen	$80 < L/d < 100$, $13 \times 10^3 < Re_b < 3 \times 10^5$, $1 < T_w/T_b < 6$	19
7	$Nu_b = 5 + 0.012\, Re_f^{0.83} (Pr_w + 0.29)$		For gases $0.6 < Pr_b < 0.9$	21
8	$Nu_b = 0.0214\, (Re_b^{0.8} - 100)\, Pr_b^{0.4} \left(\frac{T_b}{T_w}\right)^{0.45} \times \left[1 + \left(\frac{d}{L}\right)^{2/3}\right]$ $Nu_b = 0.012\, (Re_b^{0.87} - 280)\, Pr^{0.4} \left(\frac{T_b}{T_w}\right)^{0.4} \times \left[1 + \left(\frac{d}{L}\right)^{2/3}\right]$	Air, helium, carbon dioxide	$0.5 < Pr_b < 1.5$, for heating of gases; the author collected the data from the literature; second for $1.5 < Pr_b < 500$	3
9	$Nu_b = 0.022\, Re_b^{0.8}\, Pr_b^{0.4} \left(\frac{T_w}{T_b}\right)^{-10.29+0.0019L/d}$	Air, helium	$10^4 < Re_b < 10^5$, $18 < L/d < 316$	39

A number of heat transfer correlations have been developed for variable–property fully developed turbulent liquid and gas flows in a circular duct, some of which are also summarized in Tables 3.5 to 3.7.

Comprehensive information and correlations for convective heat transfer and friction factor in noncircular rectangular straight and curved ducts and coils, in crossflow arrangements, over rod bundles, and in various fittings and liquid metals are given in Reference 40. The comparison of the important correlations for forced convection in ducts is also given in Reference 41.

3.7 Summary Of Forced Convection In Straight Ducts

Important and reliable correlations for Newtonian fluids in single-phase laminar and turbulent flows through ducts have been summarized, and can be used in the design of heat transfer equipment.

The tables presented in this chapter cover the recommended specific correlations for laminar and turbulent forced convection heat transfer through ducts with constant and variable fluid properties. Tables 3.2 and 3.5 provide exponents m and n associated with Equations (3.21) and (3.22) for laminar and turbulent forced convection in circular ducts, respectively. By the use of these tables, the effect of variable properties are incorporated by the property ratio method. The correlations for turbulent flow of liquids and gases with temperature-dependent properties are also summarized in Tables 3.6 and 3.7.

Turbulent forced convection heat transfer correlations for fully developed flow (hydrodynamically and thermally) through a circular duct with constant properties are summarized in Table 3.3. The correlations by Gnielinski, Petukhov and Kirillov, Webb, Sleicher, and Rouse are recommended for constant–property Nusselt number evaluation for gases and liquids, and the entrance correction factor is given by Equation (3.33). Recommended Fanning friction factor correlations for isothermal turbulent flow in smooth circular ducts are listed in Table 3.4. The correlations given in Tables 3.3, 3.4, 3.6 and 3.7 can also be utilized for turbulent flow in smooth straight noncircular ducts for engineering applications by the use of the hydraulic diameter concept for heat transfer and pressure drop calculations as discussed in Section 3.2.3. Except for sharp-cornered or very irregular duct cross sections, the fully developed turbulent Nusselt number and friction factor vary from their actual values within ±15 and ±10%, respectively, when the hydraulic diameter is used in circular duct correlations; however, for laminar flows, the use of circular tube correlations are less accurate with the use of hydraulic diameter concept, particularly with cross sections characteristized by sharp corners. For such cases, the Nusselt number for fully developed conditions may be obtained from Table 3.8, using the following nomenclatures:

Nu_T = average Nusselt number for uniform wall temperature

Nu_{H1} = average Nusselt number for uniform heat flux in flow direction and uniform wall temperature at any cross section

Nu_{H2} = average Nusselt number for uniform heat flux both axially and circumferentially

The calculation of heat transfer coefficient of plain and finned tubes for various arrangements in external flow is important in heat exchanger design. Various correlations for these can be found in Reference 3.

TABLE 3.8

Nusselt Number and Friction Factor for Hydrodynamically and Thermally Developed Laminar Flow in Ducts of Various Cross Sections

Geometry ($L/D_h > 100$)	Nu_T	Nu_{H1}	Nu_{H2}	$f\,Re$
Circle	3.657	4.364	4.364	64.00
Hexagon	3.34	4.002	3.862	60.22
Triangle (2b, 2a), 60° $\frac{2b}{2a}=\frac{\sqrt{3}}{2}$	2.47	3.111	1.892	53.33
Rectangle (2b, 2a) $\frac{2b}{2a}=1$	2.976	3.608	3.091	56.91
Rectangle (2b, 2a) $\frac{2b}{2a}=\frac{1}{2}$	3.391	4.123	3.017	62.20
Rectangle (2b, 2a) $\frac{2b}{2a}=\frac{1}{4}$	3.66	5.099	4.35	74.8
Rectangle (2b, 2a) $\frac{2b}{2a}=\frac{1}{8}$	5.597	6.490	2.904	82.34
Parallel plates (2b) $\frac{2b}{2a}=0$	7.541	8.235	8.235	96.00
Parallel plates, one insulated $\frac{b}{a}=0$ (Insulated)	4.861	5.385	—	96.00

From Shah, R. K. and Bhatti, M. S. (1987) In *Handbook of Single-Phase Convective Heat Transfer*, pp. 3.1–3.137. Wiley, New York. With permission.

3.8 Heat Transfer From Smooth-Tube Bundles

A circular tube array is one of the most complicated and common heat transfer surfaces, particularly in shell-and-tube heat exchangers. The most common tube arrays are staggered and in line as shown in Figure 3.2. Other arrangements are also possible (see Chapter 8). The bundle is characterized by the cylinder diameter, d_o; the longitudinal spacing of the consecutive rows, X_l; and the transversal spacing of two consecutive cylinders, X_t.

A considerable amount of work has been published on heat transfer in tube bundles. The most up-to-date review of this work was presented by Zukauskas.[42]

The average heat transfer from bundles of smooth tubes is generally determined from the following equation:

$$Nu_b = c\,Re_b^m Pr_b^n \left(\frac{Pr_b}{Pr_w}\right)^p \tag{3.41}$$

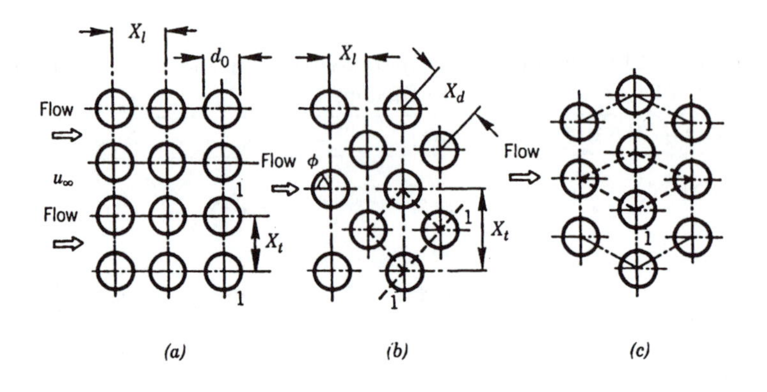

FIGURE 3.2
Tube bundle arrangements. (a) In-line array; (b) - (c) staggered array. Minimum intertube spacing at Section 1-1 between two tubes. (From Zukauskas, A. A. [1987] In *Handbook of Single-Phase Convective Heat Transfer*, pp. 6.1–6.45. Wiley, New York. With permission.)

where Pr_w stands for Prandtl number estimated at the wall temperature. Fluid properties are evaluated at the bulk mean temperature.

The variation of c in Equation (3.41) for staggered arrangements may be represented by a geometric parameter X_t^*/X_l^* with an exponent of 0.2 for $X_t^*/X_l^* < 2$. For $X_t^*/X_l^* > 2$, a constant value of $c = 0.40$ may be assumed. For in-line bundles, the effect of change in either longitudinal or transverse pitch is not so evident, and $c = 0.27$ may be assumed for the whole subcritical regime ($10^3 < Re < 2 \times 10^5$).[42]

For in-line tube bundles in crossflow, the array averaged Nusselt numbers for $n < 16$[42] are given by:

$$\overline{Nu}_b = 0.9\ c_n\ Re_b^{0.4} Pr_b^{0.36} \left(\frac{Pr_b}{Pr_w} \right)^{0.25} \quad \text{for} \quad Re_b = 1 - 10^2 \tag{3.42a}$$

$$\overline{Nu}_b = 0.52\ c_n\ Re_b^{0.5} Pr_b^{0.36} \left(\frac{Pr_b}{Pr_w} \right)^{0.25} \quad \text{for} \quad Re_b = 10^2 - 10^3 \tag{3.42b}$$

$$\overline{Nu}_b = 0.27\ c_n\ Re_b^{0.63} Pr_b^{0.36} \left(\frac{Pr_b}{Pr_w} \right)^{0.25} \quad \text{for} \quad Re_b = 10^3 - 2 \times 10^5 \tag{3.42c}$$

$$\overline{Nu}_b = 0.033\ c_n\ Re_b^{0.8} Pr_b^{0.4} \left(\frac{Pr_b}{Pr_w} \right)^{0.25} \quad \text{for} \quad Re_b = 2 \times 10^5 - 2 \times 10^6 \tag{3.42d}$$

where c_n is a correction factor for the number of tube rows because of the shorter bundle; the effect of the number of tubes becomes negligible only when $n > 16$, as shown in Figure 3.3.

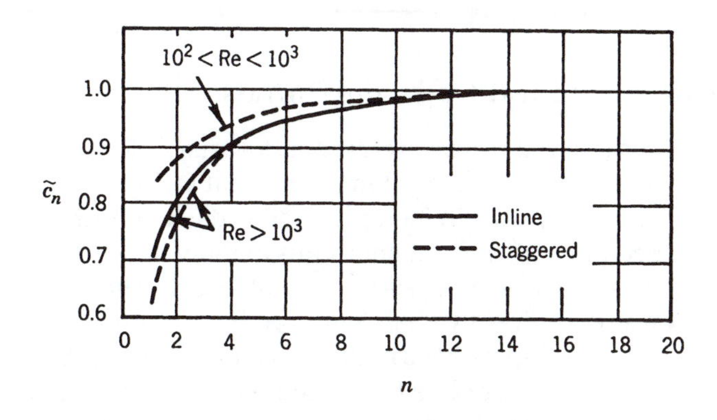

FIGURE 3.3
Correlation factor for the number of rows for the average heat transfer from tube bundles. (From Zukauskas, A. A. [1987] In *Handbook of Single-Phase Convective Heat Transfer*, pp. 6.1–6.45. Wiley, New York. With permission.)

The average heat transfer from staggered bundles in crossflow is presented in Reference 42 as:

$$\overline{Nu}_b = 1.04\ c_n\ Re_b^{0.4} Pr_b^{0.36} \left(\frac{Pr_b}{Pr_w}\right)^{0.25} \quad \text{for} \quad Re_b = 1 - 500 \tag{3.43a}$$

$$\overline{Nu}_b = 0.71\ c_n\ Re_b^{0.5} Pr_b^{0.36} \left(\frac{Pr_b}{Pr_w}\right)^{0.25} \quad \text{for} \quad Re_b = 500 - 10^3 \tag{3.43b}$$

$$\overline{Nu}_b = 0.35\ c_n\ Re_b^{0.6} Pr_b^{0.36} \left(\frac{Pr_b}{Pr_w}\right)^{0.25} \left(\frac{X_t}{X_l}\right)^{0.2} \quad \text{for} \quad Re_b = 10^3 - 2\times 10^5 \tag{3.43c}$$

$$\overline{Nu}_b = 0.031\ c_n\ Re_b^{0.8} Pr_b^{0.4} \left(\frac{Pr_b}{Pr_w}\right)^{0.25} \left(\frac{X_t}{X_l}\right)^{0.2} \quad \text{for} \quad Re_b = 2\times 10^5 - 2\times 10^6 \tag{3.43d}$$

All physical properties except Pr_w are evaluated at the bulk mean temperature of the fluid that flows around the cylinders in the bundle.

The uncertainty of the results given by Equations (3.42) and (3.43) is within ±15%.

The Reynolds number, *Re*, is based on the average velocity through the narrowest cross section formed by the array, that is, the maximum average velocity, U_o:

$$Re_b = \frac{U_o d_o \rho}{\mu} \tag{3.44}$$

Heat transfer over a tube in a bundle, heat transfer for rough-tube bundles, and finned-tube bundles are also given in Reference 42; pressure drop analysis of bundles is given in Chapter 4.

3.9 Heat Transfer in Helical Coils and Spirals

Helical coils and spirals (curved tubes) are used in a variety of heat exchangers such as chemical reactors, agitated vessels, storage tanks, and heat recovery systems. Curved tube heat exchangers are extensively used in food processing, dairy, refrigeration and air-conditioning, and hydrocarbon processing. An enormous amount of work has been published on heat transfer and pressure drop analysis on this subject; the most up-to-date review of the work in this area was presented by Shah and Joshi.[43]

Nomenclature for helical and spiral coils is illustrated in Figure 3.4. A horizontal coil with tube diameter 2a, coiled curvature radius R, and coil pitch b are shown in Figure 3.4a. It is called a horizontal coil because the tube in each turn is approximately horizontal. If the coil in Figure 3.4a were turned 90°, it would be referred to as a vertical coil. A spiral coil is shown in Figure 3.4b. A simple spiral coil of circular cross section is characterized by the tube diameter 2a, the constant pitch b, and the minimum and maximum radii of curvature (R_{min} and R_{max}) of the beginning and the end of the spiral.

3.9.1 Nusselt Numbers of Helical Coils: Laminar Flow

Several theoretical and experimental studies have reported Nusselt numbers for Newtonian fluids through a helical coil subjected to constant temperature boundary conditions.[44,45]

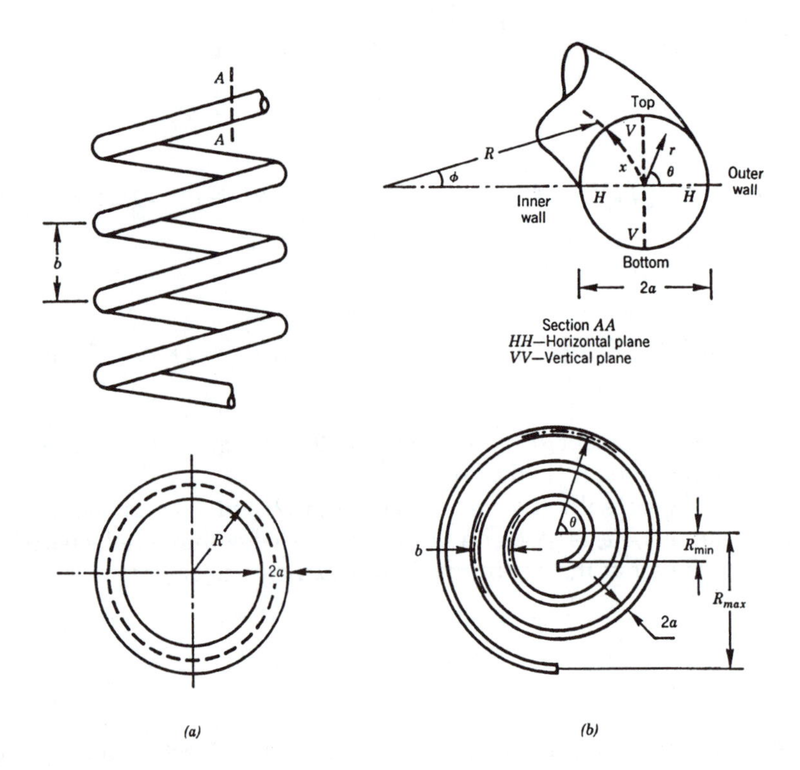

FIGURE 3.4
(a) A schematic of a helical coil; (b) a schematic of a spiral. (From Shah, R. K. and Joshi, S. D. [1987] In *Handbook of Single-Phase Convective Heat Transfer*, pp. 5.1–5.46. Wiley, New York. With permission.)

Experimental and theoretical results are compared with the following Manlapaz–Churchill correlation[46] based on a regression analysis of the available data and acceptable agreement is obtained.

$$Nu_T = \left[\left(3.657 + \frac{4.343}{x_1}\right)^3 + 1.158\left(\frac{De}{x_2}\right)^{3/2}\right]^{1/3} \tag{3.45}$$

where

$$x_1 = \left(1.0 + \frac{957}{De^2 Pr}\right)^2, \quad x_2 = 1.0 + \frac{0.477}{Pr} \tag{3.46}$$

and De is the Dean number, defined as:

$$De = Re\left(\frac{a}{R}\right)^{1/2} \tag{3.47}$$

and the characteristic lengths in the Nusselt and Reynolds numbers is the I.D. of the coil.

Manlapaz and Churchill[46] derived the following correlation under constant heat flux boundary conditions by comparing theoretical prediction with available experimental Nusselt number data:

$$Nu_H = \left[\left(4.364 + \frac{4.636}{x_3}\right)^3 + 1.816\left(\frac{De}{x_4}\right)^{3/2}\right]^{1/3} \tag{3.48}$$

where

$$x_3 = \left(1.0 + \frac{1342}{De^2 Pr}\right)^2, \quad x_4 = 1.0 + \frac{1.15}{Pr} \tag{3.49}$$

In the preceding correlations, the properties are evaluated at the average bulk temperature or at an arithmetic mean temperature of the inlet and outlet temperatures.

3.9.2 Nusselt Numbers for Spiral Coils: Laminar Flow

Kubair and Kuloor[47,48] obtained the Nusselt number for two spirals that were enclosed in a steam chamber using glycerol solutions. They recommended the following correlation:

$$Nu_T = \left(1.98 + \frac{1.8}{R_{ave}}\right) Gz^{0.7} \tag{3.50}$$

for $9 \le Gz < 1000$, $80 < Re < 6000$, and $20 < Pr < 100$. The properties in Equation (3.50) should be evaluated at an arithmetic mean of the inlet and outlet temperatures.

3.9.3 Nusselt Numbers for Helical Coils: Turbulent Flow

A large number of experimental and theoretical correlations are available to predict Nusselt numbers for a turbulent flow through a helical coil, but experimental data for spiral coils is very limited.

Most of the turbulent fluid flow and heat transfer analysis are limited to fully developed flow. Limited data on turbulent developing flow indicate flow becoming fully developed within the first half turn of the coil.[43] Entrance length for developing turbulent flow is usually shorter than that for laminar flow. Various correlations are available.

Schmidt's correlation[43] has the largest application range:

$$\frac{Nu_C}{Nu_s} = 1.0 + 3.6\left[1 - \left(\frac{a}{R}\right)\right]\left(\frac{a}{R}\right)^{0.8} \tag{3.51}$$

where Nu_c is the Nusselt number for the curved coil and Nu_s is the straight-tube Nusselt number. It is applicable for $2 \times 10^4 < Re < 1.5 \times 10^5$ and $5 < R/a < 84$. This correlation was developed using air and water in coils under constant wall temperature boundary conditions. The properties were evaluated at fluid mean bulk temperature. For low Reynolds number, Pratt's correlation[49] is recommended:

$$\frac{Nu_C}{Nu_s} = 1.0 + 3.4\left(\frac{a}{R}\right) \quad \text{for } 1.5\times10^3 < Re < 2\times10^4 \tag{3.52}$$

For the influence of temperature-dependent properties, especially viscosity for liquid, Orlov and Tselishchev[50] recommended the following correlation by Mikheev:

$$\frac{Nu_C}{Nu_s} = \left(1.0 + 3.4\frac{a}{R}\right)\left(\frac{Pr_b}{Pr_w}\right)^{0.25} \quad \text{for } 1.5\times10^3 < Re < 2\times10^4 \tag{3.53}$$

The only experimental data for spiral coils are reported by Orlov and Tselishchev.[50] They indicate that Mikheev's correlation, Equation (3.53), represented their data within ±15% deviation when an average radius of curvature of a spiral was used in the correlation. This indicates that most helical coil correlations can be used for spiral coils if the average radius of curvature of the spiral, R_{ave} is used in the correlations.

3.10 Heat Transfer in Bends

Bends are used in pipelines, in circuit lines of heat exchangers, and in tubular heat exchangers. In some cases bends are heated; in other applications (e.g., inner tube returns in hairpin-type [double-pipe] exchangers), bend is not heated but tubes leading to and from a bend are heated. There are various investigations on heat transfer and pressure drop analysis available in the open literature; an excellent review of heat transfer in bends and fittings is given by Shah and Joshi.[51]

Bends are characterized by a bend angle ϕ in degrees and a bend curvature ratio R/a. Bends with circular and rectangular cross sections are shown in Figure 3.5. In this section,

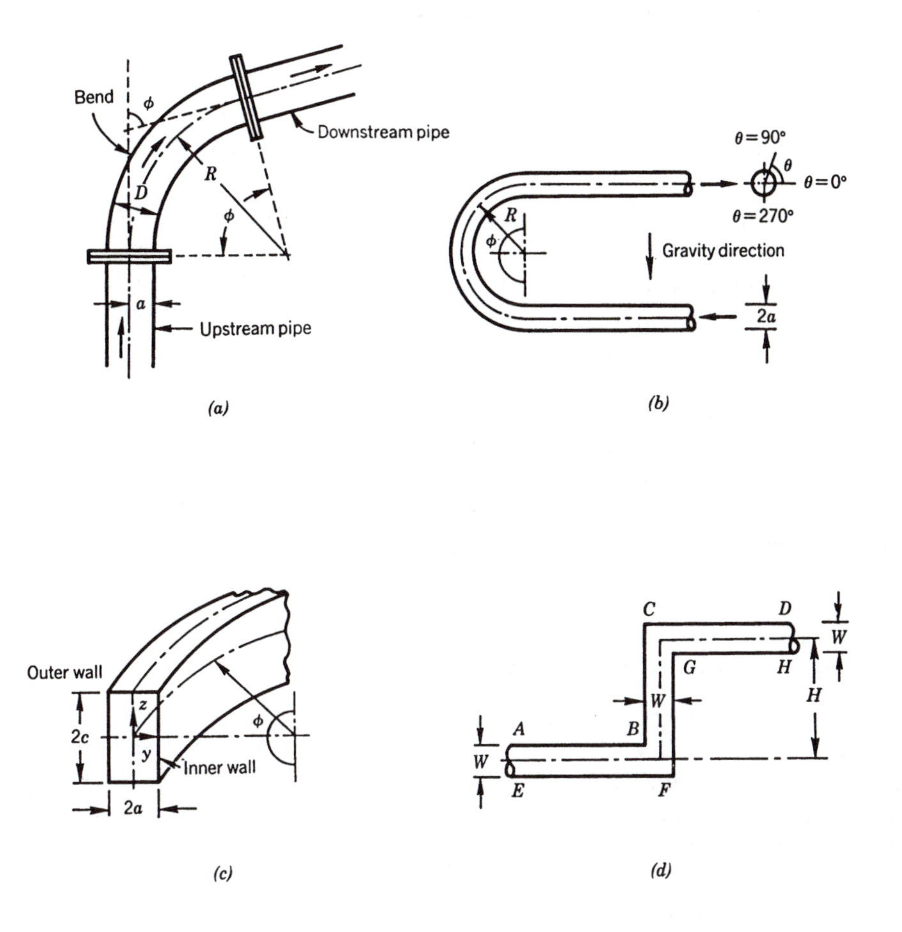

FIGURE 3.5
Schematic diagrams of bends. (a) A bend with an angle $\phi < 90°$; (b) 180° bend; (c) rectangular cross-section bend; (d) two 90° bends in series. (From Shah, R. K. and Joshi, S. D. [1987] In *Handbook of Single-Phase Convective Heat Transfer*, pp. 5.1–5.46. Wiley, New York. With permission.)

heat transfer in 90° and 180° smooth bends for Newtonian fluids will be briefly outlined. As an engineering approximation, the heat transfer results for bends may be used for geometrically similar smooth fittings.

3.10.1 Heat Transfer in 90° Bends

No experimental data are available to calculate the heat transfer coefficient in laminar flow through a 90° bend. As an engineering application, it is suggested that helical coil thermal entrance length Nusselt number correlations that are given in Reference 43 can be used to calculate the bent heat transfer coefficient.

Experimental studies indicate that a bend has a negligible effect on the heat transfer in a tube upstream of a bend, but it does have a significant influence downstream.[52,53]

Ede[52] and Tailby and Staddon[53] measured Nusselt number for turbulent flow in a 90° bend, at a constant wall temperature; they reported 20 to 30% increase above the straight-tube Nusselt numbers.

Tailby and Staddon[53] proposed the following correlation for the local heat transfer coefficient in turbulent air cooling in a 90° bend:

$$Nu_x = 0.0336\ Re^{0.81}\ Pr^{0.4}\left(\frac{R}{a}\right)^{-0.06}\left(\frac{x}{d_i}\right)^{-0.06} \tag{3.54}$$

$$2.3 \le R/a \le 14.7, \quad 7 \le x/D \le 30, \quad 10^4 \le Re \le 5\times10^4$$

where x is measured along the axis of a bend starting from the bend inlet.

For a turbulent flow fluid cooling, the average heat transfer in a bend is only about 20 to 30% above the straight-tube values. It is important to note that the preceding correlation is valid only for fluid cooling. For practical applications, the fully developed turbulent flow fluid-heating heat transfer correlation given by Equation (3.51) may be used.[51]

For turbulent flow through a 90° bend, the overall effect of incremental heat transfer in a bend and its downstream pipe is equivalent to increasing the length of a heated pipe by about eight diameters for an elbow, and three diameters for bends with $R/a = 8$ and 4.[52,53]

Experimental results show that a bend affects heat transfer in a downstream straight pipe a distance of about 10 diameters.[46] In turbulent flow, heat transfer in the downstream pipe was found to be independent of upstream heating.[52,53]

3.10.2 Heat Transfer in 180° Bends

For laminar flow through 180° bends, the most extensive results are reported by Moshfeghian and Bell[54] for $R/a = 4.84, 7.66, 12.32$, and 25.62; however, they could not reach a satisfying correlation for laminar flow. Because of the lack of information, helical coil entrance region correlations given in Reference 48 may be used to estimate laminar flow Nusselt numbers in a 180° bend.

Moshfeghian and Bell[54] measured turbulent flow heat transfer coefficients in a heated bend using different fluids and presented the following correlation for turbulent flow local Nusselt numbers:

$$Nu_x = 0.0285\ Re^{0.81}\ Pr^{0.4}\left(\frac{x}{d_i}\right)^{0.046}\left(\frac{R}{a}\right)^{-0.133}\left(\frac{\mu_b}{\mu_w}\right)^{0.14} \tag{3.55}$$

$$\text{for}\quad 4.8 \le R/a \le 26, \quad 10^4 \le Re \le 3\times10^4, \quad 0 < x/d_i \le \pi R/(2a)$$

The properties are evaluated at local mean temperature. It shows less than 8% deviation from the experimental data.

Pressure drops in helical coils, spiral coils, and bends are discussed in Chapter 4.

Nomenclature

a	coil tube inside radius, m
A	constant
A_c	net free-flow cross-sectional area, m^2
b	coil pitch, m

c_p specific heat at constant pressure, J/kg·K
c_v specific heat at constant volume, J/kg·K
De Dean number, $Re(a/R)^{1/2}$
D_e equivalent diameter for heat transfer, $4\ A_c/P_h$, m
D_i I.D. of a circular annulus, m
D_h hydraulic diameter for pressure drop, $4\ A_c/P_w$, m
d circular duct diameter, or distance between parallel plates, m
d_i I.D. of circular or curved duct, m
d_o O.D. of the inner tube of an annular duct, m
Eu n-row average Euler number = $2\Delta p/\rho U_o^2 n$
f Fanning friction factor, $\tau_w/\rho u_m^2\ 1/2$
G fluid mass velocity, ρu_m, kg/m²·s
Gz Graetz number, $(P/4x)\cdot RePr$
h average heat transfer coefficient, W/m²·K
h_x local heat transfer coefficient, W/m²·K
k thermal conductivity of fluid, W/m·K
L distance along the duct, or tube length, m
L_{he} hydrodynamic entrance length, m
L_{te} thermal entrance length, m
m exponent, Equations (3.21b) and (3.22b)
$\dot{m}$ mass flow rate, kg/s
n number of tube rows in the flow direction
n exponent, Equations (3.21a) and (3.22a)
Nu Nusselt number, hd/k
P duct wetted perimeter, m
Pe Péclet number, $RePr$
Pr Prandtl number, $c_p\mu/k = \nu/\alpha$
q'' heat flux, W/m²
R radius of curvature (see Figures 3.4 and 3.5), m
R_{ave} mean radius of curvature for a spiral, m
Re Reynolds number, $\rho u_m d/\mu$, $\rho u_m D_h/\mu$
T temperature, °C, K
T_f film temperature, $(T_w + T_b)/2$, °C, K
u velocity component in axial direction, m/s
u_m mean axial velocity, m/s
U_o mean velocity in the minimum free-flow area or intertube spacing, m/s
x Cartesian coordinate, axial distance, or axial distance along the axis of the coil, m
X_l tube longitudinal pitch, m
X_t tube transverse pitch, m
X_l^* ratio of longitudinal pitch to tube diameter for crossflow to a tube bundle, X_l/d_o
X_t^* ratio of transverse pitch to tube diameter for crossflow to a tube bundle, X_t/d_o
y Cartesian coordinate, distance normal to the surface, m

Greek Symbols

α thermal diffusivity of fluid, m^2/s
μ dynamic viscosity of fluid, Pa·s
υ kinematic viscosity of fluid, m^2/s
ρ density of fluid, kg/m^3
τ_w shear stress at the wall, Pa
ϕ bend angle, or angle measured along the tube perimeter from the front stagnation point, rad or degree

Subscripts

a arithmetic mean
b bulk fluid condition or properties evaluated at bulk mean temperature
c curved coil or duct
cp constant property
e equivalent
f film fluid condition or properties evaluated at film temperature
H constant heat flux boundary condition
l laminar
i inlet condition or inner
o outlet condition or outer
r reference condition
s straight duct
T constant temperature boundary condition
t turbulent
w wall condition or wetted
x local value at distance *x*
∞ fully developed condition

References

1. Shah, R. K. and London, A. L. (1978) *Laminar Forced Convection in Ducts*. Academic, New York.
2. Shah, R. K. and Bhatti, M. S. (1987) Laminar convective heat transfer in ducts, Chapter 3, In *Handbook of Single-Phase Convective Heat Transfer*, S. Kakaç, R. K. Shah, and W. Aung (Eds.), pp. 3.1–3.137. Wiley, New York.
3. Gnielinski, V. (1983) Forced convection ducts, in *Heat Exchanger Design Handbook*, Schlünder, E. U. (Ed.), pp. 2.5.1–2.5.3. Hemisphere, New York.
4. Hausen, H. (1959) Neue Gleichungen für die Warmeübertragung bei freier oder erzqungener Strömung. *Allg. Wärmetech.*, Vol. 9, 75–79.
5. Kakaç, S. (1985) Laminar forced convection in the combined entrance region of ducts, in *Natural Convection: Fundamentals and Applications*, S. Kakaç, W. Aung, and R. Viskanta (Eds.), pp. 165–204. Hemisphere, New York.

6. Pohlhausen, E. (1921) Der Warmeaustausch Zwischen festen Körpern und Flüssigkeiten mit Kleiner Reibung und Kleiner Warmeleitung. *Z. Angew. Math. Mech.*, Vol. 1, 115–121.
7. Delorenzo, B. and Anderson, E. D. (1945) Heat transfer and pressure drop of liquids in double pipe fintube exchangers. *Trans. ASME*, Vol. 67, 697.
8. Stephan, K. (1959) Warmeübergang und Druckabfall beinichtausgebildeter Laminar Stömung in Rohren und evenen Spalten. *Chem. Ing. Tech.*, Vol. 31, 773–778.
9. Deissler, R. G. (1951) Analytical investigation of fully developed laminar flow in tubes with heat transfer with fluid properties variable along the radius. NACA Report TN 2410.
10. Yang, K. T. (1962) Laminar forced convection of liquids in tubes with variable viscosity. *J. Heat Transfer*, Vol. 84, 353–362.
11. Sieder, E. N. and Tate, G. E. (1936) Heat transfer and pressure drop if liquids in tubes. *Ind. Eng. Chem.*, Vol. 28, 1429–1453.
12. Whitaker, S. (1972) Forced convection heat-transfer correlations for flow in pipes, past flat plates, single cylinders, single spheres, and flow in packed beds and tube bundles. *AIChE J.*, Vol. 18, 361–371.
13. Oskay, R. and Kakaç, S. (1973) Effect of viscosity variations on turbulent and laminar forced convection in pipes. *METU J. Pure Appl. Sci.*, Vol. 6, 211–230.
14. Kuznetsova, V.V. (1972) Convective heat transfer with flow of a viscous liquid in a horizontal tube (in Russian). *Teploenergetika*, Vol. 19, no. 5, 84.
15. Test, F. L. (1968) Laminar flow heat transfer and fluid flow for liquids with a temperature dependent viscosity. *J. Heat Transfer*, Vol. 90, 385–393.
16. Worsøe-Schmidt, P. M. (1966) Heat transfer and friction for laminar flow of helium and carbon dioxide in a circular tube at high heating rate. *Int. J. Heat Mass Transfer*, Vol. 9, 1291–1295.
17. Bhatti, M. S. and Shah, R. K. (1987) Turbulent forced convection in ducts. In *Handbook of Single-Phase Convective Heat Transfer*, S. Kakaç, R. K. Shah, and W. Aung (Eds.), pp. 4.1–4.166. Wiley, New York.
18. Petukhov, B. S. and Popov, V. N. (1963) Theoretical calculation of heat exchanger and frictional resistance in turbulent flow in tubes of incompressible fluid with variable physical properties. *High Temp.*, Vol. 1, no. 1, 69–83.
19. Petukhov, B. S. (1970) Heat transfer and friction in turbulent pipe flow with variable physical properties. In *Advances in Heat Transfer*, J. P. Hartnett and T. V. Irvine (Eds.), Vol. 6, pp. 504–564. Academic, New York.
20. Webb, R. I. (1971) A critical evaluation of analytical solutions and Reynolds analogy equations for heat and mass transfer in smooth tubes. *Wärme Staffübertragung*, Vol. 4, 197–204.
21. Sleicher, C. A. and Rouse, M. W. (1975) A convenient correlation for heat transfer to constant and variable property fluids in turbulent pipe flow. *Int. J. Heat Mass Transfer*, Vol.18, 677–683.
22. Gnielinski, V. (1976) New equations for heat and mass transfer in turbulent pipe and channel flow. *Int. Chem. Eng.*, Vol. 16, 359–368.
23. Kays, W. M. and Crawford, M. E. (1981) *Convective Heat and Mass Transfer*, 2nd ed., McGraw-Hill, New York.
24. Kakaç, S. and Yener, Y. (1994) *Convective Heat Transfer*, 2nd ed., CRC Press, Boca Raton, FL.
25. McAdams, W. H. (1954) *Heat Transmission*, 3rd ed., McGraw-Hill, New York.
26. Kakaç, S. (1987) The effects of temperature-dependent fluid properties on convective heat transfer, Chapter 18. In *Handbook of Single-Phase Convective Heat Transfer*, S. Kakaç, R. K. Shah, and W. Aung (Eds.), pp. 18.1–18.92. Wiley, New York.
27. Hausen, H. (1943) Darstellung des Warmeüberganges in Rohren durch verallgeineinerte Potenzbeziebungen. *Z. Ver. Dtsch. Ing. Beigeft Verfahrenstech.*, Vol. 4, 91–134.
28. Rehme, K. (1973) A simple method of predicting friction factors of turbulent flow in noncircular channels. *Int. J. Heat Mass Transfer*, Vol. 16, 933–950.
29. Malak, J., Hejna, J., and Schmid, J. (1975) Pressure losses and heat transfer in noncircular channels with hydraulically smooth walls. *Int. J. Heat Mass Transfer*, Vol. 18, 139–149.
30. Brundrett, E. (1979) Modified hydraulic diameter. In *Turbulent Forced Convection in Channels and Bundles*, S. Kakaç and D. B. Spalding (Eds.), Vol. 1, pp. 361–367. Hemisphere, New York.
31. Perkins, H. C. and Worsøe-Schmidt, P. (1965) Turbulent heat and momentum transfer for gases in a circular tube at wall to bulk temperature ratios to seven. *Int. J. Heat Mass Transfer*, Vol. 8, 1011–1031.

32. McElligot, D. M., Magee, P. M., and Leppert, G. (1965) Effect of large temperature gradients on convective heat transfer: the downstream region. *J. Heat Transfer*, Vol. 87, 67–76.
33. Colburn, A. P. (1933) A method of correlating forced convection heat transfer data and comparison with fluid friction. *Trans. AIChE*, Vol. 29, 174–210.
34. Hufschmidt, W., Burck, E., and Riebold, W. (1966) Die Bestimmung örlicher und Warmeübergangs-Zahlen in Rohren bei Hohen Warmestromdichten. *Int. J. Heat Mass Transfer*, Vol. 9, 539–565.
35. Rogers, D. G. (1980) Forced convection heat transfer in single phase flow of a newtonian fluid in a circular pipe. CSIR Report CENG 322, Pretoria, South Africa.
36. Hausen, H. (1974) Extended equation for heat transfer in tubes at turbulent flow. *Wärme Stoübertragung*, Vol. 7, 222–225.
37. Humble, L. V., Lowdermilk, W. H., and Desmon, L. G. (1951) Measurement of average heat transfer and friction coefficients for subsonic flow if air in smooth tubes at high surface and fluid temperature. NACA Report 1020.
38. Barnes, J. F. and Jackson, J. D. (1961) Heat transfer to air, carbon dioxide and helium flowing through smooth circular tubes under conditions of large surface / gas temperature ratio. *J. Mech. Eng. Sci.*, Vol. 3, no. 4, 303–314.
39. Dalle-Donne, M. and Bowditch, P. W. (1963) Experimental local heat transfer and friction coefficients for subsonic laminar transitional and turbulent flow of air or helium in a tube at high temperatures. Dragon Project Report 184, Winfirth, Dorchester, Dorset, U.K.
40. Kakaç, S., Shah, R. K., and Aung, W. (Eds.) (1987) *Handbook of Single-Phase Convective Heat Transfer*. Wiley, New York.
41. Kakaç, S., Bergles, A. E., and Fernandes, E. O. (Eds.) (1988) *Two-Phase Flow Heat Exchangers*, pp. 123–158. Kluwer, Dordrecht, The Netherlands.
42. Zukauskas, A. A. (1987) Convective heat transfer in cross flow, Chapter 6, In *Handbook of Single-Phase Convective Heat Transfer*, S. Kakaç, R. K. Shah, and W. Aung (Eds.), pp.6.1–6.45. Wiley, New York.
43. Shah, R. K. and Joshi, S. D. (1987) Convective heat transfer in curved pipes, Chapter 5, In *Handbook of Single-Phase Convective Heat Transfer*, S. Kakaç, R. K. Shah, and W. Aung (Eds.), 5.1–5.46. Wiley, New York.
44. Mori, Y. and Nakayoma, W. (1967) Study on forced convective heat transfer in curved pipes (3rd Report, Theoretical Analysis under the Condition of Uniform Wall Temperature and Practical Formulae), *Int. J. Heat Mass Transfer*, Vol. 15, 1426–1431.
45. Akiyama, M. and Cheng, K. C. (1972) Laminar forced convection heat transfer in curved pipes with uniform wall temperature. *Int. J. Heat Mass Transfer*, Vol. 15, 1426–1431.
46. Manlapaz, R.L. and Churchill, S. W. (1981) Fully developed laminar convection from a helical coil, *Chem. Eng. Commun.*, Vol. 9, 185–200.
47. Kubair, V. and Kuloor, N. R. (1966) Heat transfer to Newtonian fluids in coiled pipes in laminar flow, *Int. J. Heat Mass Transfer*, Vol. 9, pp. 63–75.
48. Kubair, V. and Kuloor, N. R. (1965) Heat transfer to Newtonian fluids in spiral coils at constant tube wall temperature in laminar flow, *Indian. J. Technol.*, Vol. 3, 144–146.
49. Pratt, N. H. (1947) The heat transfer in a reaction tank cooled by means of a coil, *Trans. Inst. Chem. Eng.*, Vol. 25, 163–180.
50. Orlov, V. K. and Tselishchev, P. A. (1964) Heat exchange in a spiral coil with turbulent flow of water, *Thermal Eng.*, (translated from Teploenergetika), Vol. 11, no. 12, 97–99.
51. Shah, R. K. and Joshi, S. D. (1987) Convective heat transfer in bends and fittings, Chapter 10 in *Handbook of Single-Phase Convective Heat Transfer*, S. Kakaç, R. K. Shah, and W. Aung (Eds.), 10.1–10.32. Wiley, New York.
52. Ede, A. J. (1962) The effect of a right-angled bend on heat transfer in a pipe. In *Int. Dev. Heat Transfer*, 634–642. ASME, New York.
53. Tailby, S. R. and Staddon, P. W. (1970) The influence of 90° and 180° pipe bends on heat transfer from an internally flowing gas stream, *Heat Transfer*, 1970, Vol. 2, Paper no. FC4.5.
54. Moshfeghian, M. and Bell, K. J. (1979) Local heat transfer measurements in and downstream from a U-bend. ASME Paper no. 79-HT-82.

Problems

3.1. A fluid flows steadily with a velocity of 6 m/s through a commercial iron rectangular duct whose sides are 1 by 2 in. and the length of the duct is 6 m. The average temperature of the fluid is 60°C. The fluid completely fills the duct. Calculate the surface heat transfer coefficient if the fluid is

a. Water

b. Air at atmospheric pressure

c. engine oil (ρ = 864 kg/m^3, c_p = 2047 J/(kg·K), ν = 0.0839 × 10^{-3}m^2/s, Pr = 1050, k = 0.140 W/m ·K).

3.2. Air at 1.5 atm and 40°C flows through a 10-m long rectangular duct of 40 by 25 cm with a velocity of 5 m/s. The surface temperature of the duct is maintained at 120°C and the average air temperature at exit is 80°C. Calculate the total heat transfer.

3.3. Calculate the heat transfer coefficient for water flowing though a 2-cm diameter tube with a velocity of 1 m/s. The average temperature of water is 60°C and the surface temperature is

a. Slightly 60°C

b. 120°C

3.4. An oil with k = 0.120 W/m·K, c_p = 2000 J/kg·K, ρ = 895 kg/m^3, μ = 0.0041 kg/m·s flows through a 2-cm diameter tube that is 2-m long. The oil is cooled from 60 to 30°C. The mean flow velocity is 0.4 m/s, and the tube wall temperature is maintained at 24°C (μ_w = 0.021 kg/m·s). Calculate the heat transfer rate.

3.5. A shell-and-tube type of condenser is to be made with 3/4-in. O.D. (0.654 in. I.D.) brass tubes, and the length of the tubes between tube plates is 3 m. Under the worst conditions, cooling water is available at 21°C and the outlet temperature is to be 31°C. Water velocity inside the tubes is to be approximately 2 m/s. The vapor-side film coefficient can be taken as 10,000 W/m^2·K. Calculate the overall heat transfer coefficient for this heat exchanger.

3.6. Water at 1.15 bar and 30°C is heated as it flows through a 1-in. I.D. tube at a velocity of 3 m/s. The pipe surface temperature is kept constant by condensing steam outside the tube. If water outlet temperature is 80°C, calculate the surface temperature of the tube by assuming the inner surface of the tube to:

a. Be smooth

b. Have surface with a relative roughness of 0.001

3.7. A counterflow double-pipe heat exchanger is used to cool the lubricating oil for a large industrial gas turbine engine. The flow rate of cooling water through the inner tube is m_c = 0.2 kg/s, while the flow rate of oil through the outer annulus is m_h = 0.4 kg/s. The oil and water enter at temperatures of 60 and 30°C, respectively. The heat transfer coefficient in the annulus is calculated to be 8 W/m·K. The I.D. of the tube of the annulus is 25 and 45 mm, respectively. The outlet temperature of the oil is 40°C. Take c_p = 4178 J/kg·K for water and c_p = 2006 J/kg·K for oil.

a. Calculate the heat transfer coefficients in the inner tube and in the annulus.

b. Neglect the wall resistance and the curvature of the tube wall (assume a flat plate surface); calculate the overall heat transfer coefficient assuming water used is the city water with a fouling factor of 0.000176 $m^2 \cdot K/W$ inside the tube. The oil side is clean.

3.8. City water flowing at 0.5 kg/s will be heated from 20 to 35°C by hot treated boiler water at 140°C. A 15°C hot water temperature drop is allowed. A number of 4.50-m hairpins of 3 by 2 in. (schedule 40) double-pipe heat exchangers with annuli and tubes each connected in series will be used. Hot water flows through the inner tube. Calculate the heat transfer coefficient in the tube and in the annulus.

3.9. It is proposed that the stationary diesel exhaust gases (CO_2) be used for heating air by the use of a double-pipe heat exchanger. From the measurements of the velocity, it is calculated that the mass flow rate of gases from the engine is 100 kg/h. The exhaust gas temperature is 600 K. The air is available at 20°C and it will be heated to 80°C at a mass flow rate of 90 kg/h. Standard (schedule 40) tube size 4- × 3-in. double-pipe heat exchanger (hairpins) in 2-m length (copper tubes) will be used. The air-side heat transfer coefficient in tube is 25 $W/m^2 \cdot K$. By neglecting the tube wall resistance, and assuming clean and smooth surfaces, calculate the overall heat transfer coefficient.

3.10. Consider the laminar flow of an oil inside a duct with a Reynolds number of 1000. The length of the duct is 2.5 m and the diameter is 2 cm. The duct is heated electrically by the use of its walls as an electrical resistance. Properties of the oil at the average oil temperature are ρ = 870 kg/m^3, μ = 0.004 $N \cdot s/m^2$, c_p = 1959 kJ/kg·K, and k = 0.128 W/m·K. Obtain the local Nusselt number at the end of the duct.

3.11. In a crossflow heat exchanger, hot air at atmosphere pressure with an average velocity of 3 m/s flows across a bank of tubes in an array with $X_l = X_t$ = 5 cm (see Figure 3.2). The tube diameter is 2.5 cm. The array has 20 rows in the direction of flow. The tube wall temperature is 30°C, and the average air temperature in the bundle is assumed to be 300°C. Calculate the average heat transfer coefficient and repeat the calculation if the array has six rows in the flow direction.

3.12. Repeat Problem 3.11, if the heat exchanger employs a bank of staggered bare tubes with a longitudinal pitch of 4 cm and transverse pitch of 5 cm.

3.13. A shell-and-tube heat exchanger is to be used to cool 20 kg/s of water from 40 to 20°C. The exchanger has one shell-side pass and two tube-side passes. The hot water flows through the tubes and the cooling water flows through the shell. The cooling water enters at 15°C and leaves at 25°C. The maximum permissible pressure drop is 10 kPa. The tube wall thickness is 1.25 mm and is selected as 18 Birmingham wire gauge (BWG) copper. The length of the heat exchanger is 5 m. Assume that the pressure losses at the inlet and outlet are equal to two of a velocity head ($\rho u_m^2/2$). Find the number of tubes and the proper tube diameter to expand the available pressure drop. (Hint: assume a tube diameter and average velocity inside the tubes.)

3.14. Repeat Problem 3.13, assuming that an overall heat transfer coefficient is given or estimated as 2000 $W/m^2 \cdot K$.

3.15. Calculate the average heat transfer coefficient for water flowing at 1.5 m/s with an average temperature of 20°C in a long, 2.5-cm I.D. tube by four different correlations from Table 3.6 (No. 2, 3, 4, and 6) considering the effect of temperature-dependent properties. The inside wall temperature of the tube is 70°C.

3.16. A double-pipe heat exchanger is used to condense steam at 40°C saturation temperature. Water at an average bulk temperature of 20°C flows at 2 m/s through the inner tube (copper, 2.54-cm I.D., 3.05-cm O.D.). Steam at its saturation temperature flows in the annulus formed between the outer surface of the inner tube and outer tube of 6-cm I.D. The average heat transfer coefficient of the condensing steam is 6000 $W/m^2 \cdot K$, and the thermal resistance of a surface scale on the outer surface of the copper pipe is 0.000176 $m^2 \cdot K/W$.

a. Determine the overall heat transfer coefficient between the steam and the water basedon the outer area of the copper tube.
b. Evaluate the temperature at the inner surface of the tube.
c. Estimate the length required to condense 0.5 kg/s of steam.

3.17. Carbon dioxide gas at 1 atm pressure is to be heated from 30 to 75°C by pumping it through a tube bank at a velocity of 4 m/s. The tubes are heated by steam condensing within them at 200°C. The tubes have a 2.5-cm O.D., are in in-line arrangement, and have a longitudinal spacing of 4 cm and a transverse spacing of 4.5 cm. If 15 tube rows are required, what is the average heat transfer coefficient?

4

Heat Exchanger Pressure Drop and Pumping Power

4.1 Introduction

The thermal design of heat exchanger is directed to calculating an adequate surface area to handle the thermal duty for the given specifications. Fluid friction effects in the heat exchanger are equally important because they determine the pressure drop of the fluids flowing in the system, and consequently the pumping power or fan work input necessary to maintain the flow. Provision of pumps or fans adds to the capital cost and is a major part of the operating cost of the exchanger. Savings in exchanger capital cost achieved by designing a compact unit with high fluid velocities may soon be lost by increased operating costs. The final design and selection of a unit will, therefore, be influenced just as much by effective use of the permissible pressure drop and the cost of pump or fan power as they are influenced by the temperature distribution and provision of adequate area for heat transfer.

4.2 Tube-Side Pressure Drop

In fully developed flow in a tube, the following functional relationship can be written for the frictional pressure drop for either laminar or turbulent flow:

$$\frac{\Delta p}{L} = \phi(u_m, d_i, \rho, \mu, e) \tag{4.1}$$

where the quantity e is a statistical measure of the surface roughness of the tube and has the dimension of length. It is assumed that Δp is proportional to the length L of the tube. With mass M, length L, and time θ as the fundamental dimensions; and u_m, d_i, and ρ as the set of maximum number of quantities that in themselves cannot form a dimensionless group, the π theorem leads to:

$$\frac{\Delta p}{4(L/d_i)(\rho u_m^2/2)} = \phi\left(\frac{u_m d_i \rho}{\mu}, \frac{e}{d_i}\right) \tag{4.2}$$

where the dimensionless numerical constants 4 and 2 are added for convenience.

The preceding dimensionless group involving Δp has been defined as the Fanning friction factor f:

$$f = \frac{\Delta p}{4(L/d_i)(\rho u_m^2/2)} \tag{4.3}$$

Equation 4.2 becomes:

$$f = \phi(Re, e/d_i) \tag{4.4}$$

Figure 4.1 shows this relationship as deduced by Moody[1] from experimental data for fully developed flow. In the laminar region, existing empirical data on the frictional pressure drop within circular tubes can be correlated by a simple relationship between f and Re, independent of the surface roughness:

$$f = \frac{16}{Re} \tag{4.5}$$

The transition from laminar to turbulent flow is somewhere in the neighborhood of Reynolds number from 2300 to 4000.

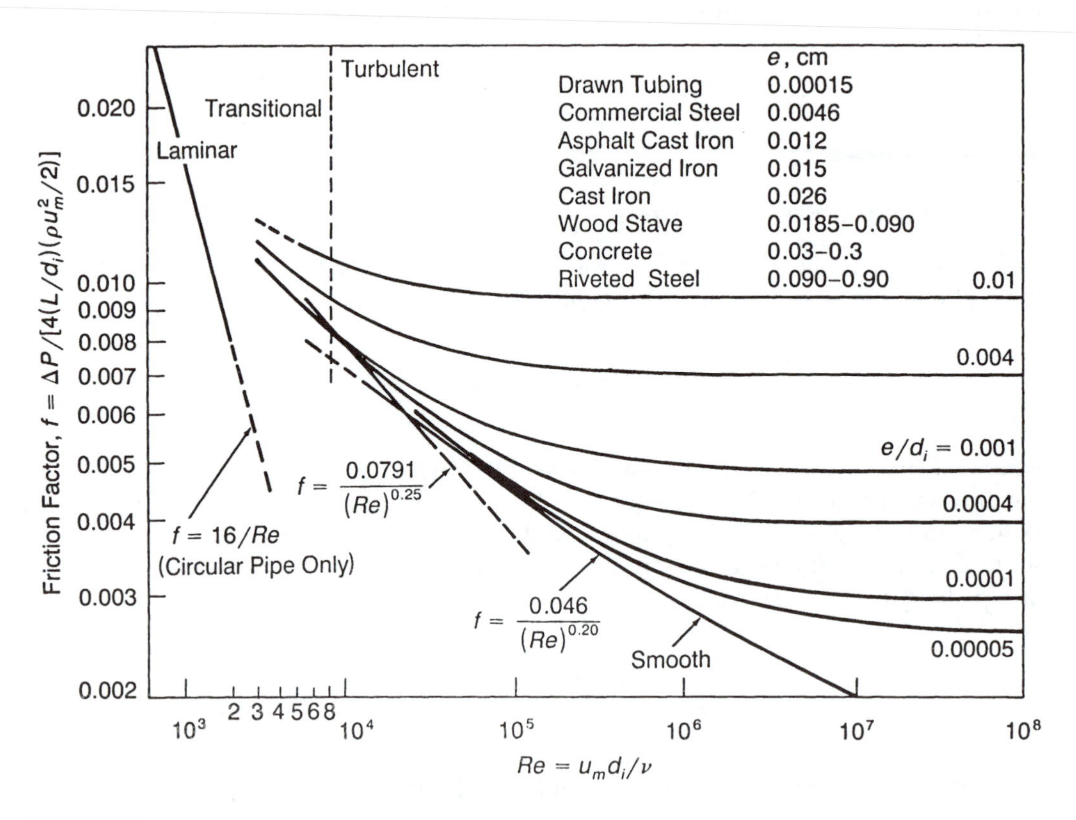

FIGURE 4.1
Friction factor for fully developed flow inside a circular duct. (From Moody, L. F. [1994] *Trans. ASME*, Vol. 66, 671–684. With permission.)

The f vs. Re relation for smooth tubes in turbulent flow has a slight curvature on a log–log plot. A few recommended correlations for turbulent flow in smooth pipes are given in Table 4.1.[1] The two linear approximations shown by dotted lines in Figure 4.1 for turbulent flow are:

$$f = 0.046\, Re^{-0.2} \quad \text{for} \quad 3 \times 10^4 < Re < 10^6 \tag{4.6}$$

and

$$f = 0.079\, Re^{-0.25} \quad \text{for} \quad 4 \times 10^3 < Re < 10^5 \tag{4.7}$$

f can be read from the graph, but the correlations for f are useful for computer analysis of heat exchanger, and they also show the functional relationship of various quantities.

For fully developed flow in a tube, a simple force balance yields (Figure 4.2):

$$\Delta p \frac{\pi}{4} d_i^2 = \tau_w (\pi d_i L) \tag{4.8}$$

TABLE 4.1

Turbulent Flow Isothermal Fanning Friction Factor Correlations for Smooth Circular Ducts[3]

Source	Correlation[a]	Limitations
Blasius	$f = \dfrac{\tau_w}{\rho u_m^2/2} = 0.0791\, Re^{-0.25}$	$4 \times 10^3 < Re < 10^5$
Drew, Koo, and McAdams	$f = 0.00140 + 0.125\, Re^{-0.32}$	$4 \times 10^3 < Re < 5 \times 10^6$
Karman–Nikuradse	$\dfrac{1}{\sqrt{f}} = 1.737 \ln (Re\sqrt{f}) - 0.4$	$4 \times 10^3 < Re < 3 \times 10^6$
	or	
	$\dfrac{1}{\sqrt{f}} = 4 \log_{10} (Re\sqrt{f}) - 0.4$	
	approximated as	
	$f = 0.046\, Re^{-0.2}$	$3 \times 10^4 < Re < 10^6$
Filonenko	$f = (3.64 \log_{10} Re - 3.28)^{-2}$	

[a] Properties are evaluated at the bulk temperature.

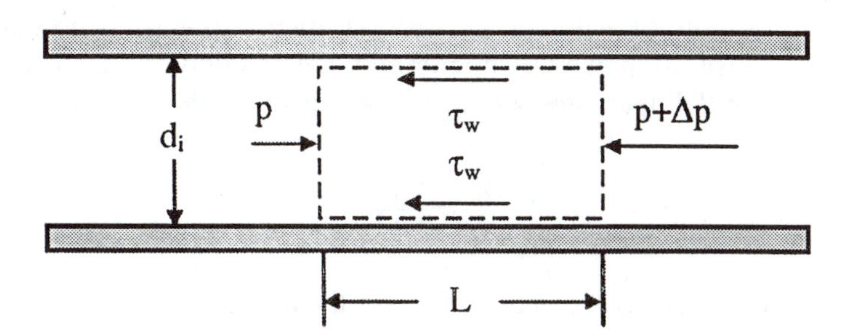

FIGURE 4.2
Force balance of a cylindrical fluid element within a pipe. (From Rohsenow, W. M. [1981] In *Heat Exchangers: Thermal–Hydraulic Fundamentals and Design*, Hemisphere, New York. With permission.)

which may be combined with Equation (4.3) to get an equivalent form for the friction factor:

$$f = \frac{\tau_w}{\rho u_m^2 / 2} \tag{4.9}$$

Equation (4.9) and (4.3) are valid for either laminar or turbulent flow, provided that the flow is fully developed within the length L.

4.2.1 Noncircular Cross-Sectional Ducts

A duct of noncircular cross section is not geometrically similar to a circular duct; hence, dimensional analysis does not relate the performance of these two geometric shapes. However, in turbulent flow, f for noncircular cross sections (annular spaces, rectangular and triangular ducts, etc.) may be evaluated from the data for circular ducts if d_i is replaced by the hydraulic diameter, D_h, defined by Equation (3.14):

$$D_h = 4\frac{A_c}{p} = \frac{4\ (\text{net free flow area})}{\text{wetted perimeter}} \tag{4.10}$$

Using the hydraulic diameter in turbulent flow gives f values within about ±8% of the measured values.[2]

The hydraulic diameter of an annulus of inner and outer diameters D_i and d_o, respectively is

$$D_h = \frac{4(\pi/4)(D_i^2 - d_o^2)}{\pi(D_i + d_o)} = (D_i - d_o) \tag{4.11}$$

For a circular duct, Equation (4.10) reduces to $D_h = d_i$.

In using the correlations in Table 4.1, the effect of property variations has been neglected. When these correlations are applied to the design of heat exchange equipment with large temperature differences between the heat transfer surface and fluid, the friction coefficient must be corrected according to Equations (3.21b) and (3.22b).

The transition Reynolds number for noncircular ducts $(\rho \mu_m D_h / \mu)$ is also found to be approximately 2300, as for circular ducts.

For laminar flows, however, the results for noncircular cross sections are not universally correlated. In a thin annulus, the flow has a parabolic distribution perpendicular to the wall and has this same distribution at every circumferential position. If this flow is treated as a flow between two parallel flat plates separated by a distance $2b$, one obtains:

$$\frac{\Delta p}{\Delta x} = \frac{12\mu u_m}{b^2} \tag{4.12}$$

Here $D_h = 2b$ and Equation (4.12) can be written in the form:

$$f = \frac{24}{Re} \tag{4.13}$$

with D_h replacing d_i in the definitions of f and Re. This equation is different from Equation (4.5), which applies to laminar flow in circular ducts.

Flow in a rectangular duct (dimensions $a \times b$) in which $b \ll a$ is similar to this annular flow. For rectangular ducts of other aspect ratios (b/a):

$$f = \frac{16}{\phi Re} \tag{4.14}$$

where

$$D_h = \frac{4ab}{2(a+b)} \tag{4.15}$$

and ϕ is given in Figure 4.3.[3,4]

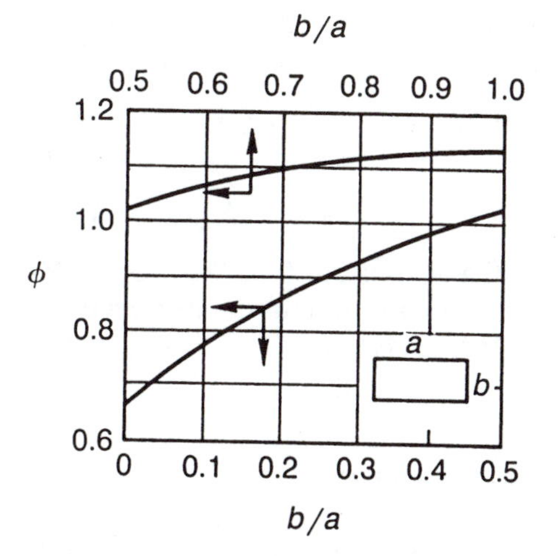

FIGURE 4.3
Values of ϕ for rectangular ducts.

For laminar flow in ducts of triangular and trapezoidal cross section, Nikuradse[4] showed that f is approximated by $16/Re$ with D_h given by Equation (4.10) and transition occurs approximately at $Re = 2300$.

The frictional pressure drop for flow through a duct of length L is generally expressed as:

$$\Delta p = 4f \frac{L}{D_h} \frac{\rho u_m^2}{2} \tag{4.16a}$$

or

$$\Delta p = 4f \frac{L}{D_h} \frac{G^2}{2\rho} \tag{4.16b}$$

where $G = u_m \rho$ is referred to as the mass velocity.

The exact form of Equation (4.16) depends on the flow situation. The pressure drop experienced by a single-phase fluid in the tubes of a shell-and-tube heat exchanger must include the effects of all the tubes. The pressure drop for all the tubes can be calculated by:

$$\Delta p_t = 4f \frac{LN_p}{D_h} \frac{G^2}{2\rho} \tag{4.17}$$

where N_p is the number of tube passes and $D_h = d_i$. The fluid will experience an additional loss due to sudden expansions and contractions that the tube undergoes during a return. Experiments show that the return pressure loss is given by:[5]

$$\Delta p_r = 4N_p \frac{\rho u_m^2}{2} \tag{4.18}$$

where N_p is the number of tube passes, which is 2 in a 1 to 2 heat exchanger; $4N_p$ is the return pressure loss coefficient; and $\rho u_m^2/2$ is called dynamic head.

The pumping power expenditure that is required to circulate each stream to flow through its heat exchange passage is equally important as the heat transfer characteristics of the heat exchanger. The pumping power expenditure is proportional to the total pressure drop, Δp, that is experienced by the stream across the passage. For an incompressible stream with a mass flow rate $\dot{m}$, the power required by an adiabatic pump is:[6]

$$W_p = \frac{1}{\eta_p} \frac{\dot{m}}{\rho} \Delta p \tag{4.19}$$

In the expression, Δp is the pressure rise through the pump, ρ is the liquid density, and η_p is the isentropic efficiency of the pump. The group $\dot{m}\Delta p/\rho$ is the isentropic (minimum) power requirement.

4.3 Pressure Drop in Tube Bundles in Crossflow

A circular tube bundle is one of the most common heat transfer surfaces, especially in shell-and-tube heat exchangers. The two most common tube arrangements are staggered and in-line, which are shown in Figure 3.2. In-line tubes form rectangulars with the centers of their cross sections, where staggered tubes form isosceles triangles.

Bundles are characterized by the cylinder diameter d_o; and the ratios of pitches to the tube diameter in the transverse $X_t^* = X_t/d_o$, longitudinal $X_l^* = X_l/d_o$, or diagonal $X_d^* = X_d/d_o$ directions. An enormous amount of work has been published on pressure drop across the tube bundles and on heat transfer performance. The experimental results obtained by Zukauskas[7,8] with air, water, and several oils will be presented here.

The pressure drop of a multirow bundle is given by:

$$\Delta p = \left(\frac{Eu}{\chi}\right) \chi \frac{1}{2} \rho u_o^2 \cdot n \tag{4.20}$$

where the Euler number, $Eu = 2\Delta p/(\rho u_o^2 \cdot n)$, is defined per tube row; the velocity, u_o, is the mean velocity in the minimum intertube spacing; and n is the number of tube rows counted in the flow direction.

The charts of average Euler number per tube row for multirow in-line bundles are presented in Figure 4.4 as a function of $Re = u_o d_o \rho / \mu$ and X_l^* (= X_t^*). For other tube pitches ($X_l^* \neq X_t^*$), a correction factor χ is determined from the inset of Figure 4.4, and then Eu/χ from the main Figure 4.4 to find Eu for specified in-line tube bundle.

The charts of average Euler number per tube row of multirow staggered tube bundles with 30° tube layout ($X_t = X_d$) and $X_l = (\sqrt{3}/2)X_t$ are presented in Figure 4.5 as a function of Re and X_t^*. For other tube pitches [$X_l \neq (\sqrt{3}/2)X_t$], the correction factor x is obtained from the upper right inset of Figure 4.5. The Reynolds number, Re, is based on the average velocity through the narrow cross section formed by the array, that is, the maximum average velocity u_o:

$$Re = \frac{u_o d_o \rho}{\mu} \tag{4.21}$$

In the case of in-line tubes, the conservation of mass gives:

$$u_\infty X_t = u_b (X_t - d_o) \tag{4.22}$$

For staggered arrangement, the minimum cross-sectional area may be either between two tubes in a row or between one tube and a neighboring tube in the succeeding row (see Problem 4.9).

Equations (4.20) and Figures (4.4) and (4.5) are valid for $n > 9$, and these two charts refer to isothermal conditions. They also apply to nonisothermal flows if the fluid physical properties are evaluated at the bulk mean temperature and a correction is applied to account for variable fluid properties for liquids as follows:

$$Eu_b = Eu \left(\frac{\mu_w}{\mu_b} \right)^p \tag{4.23}$$

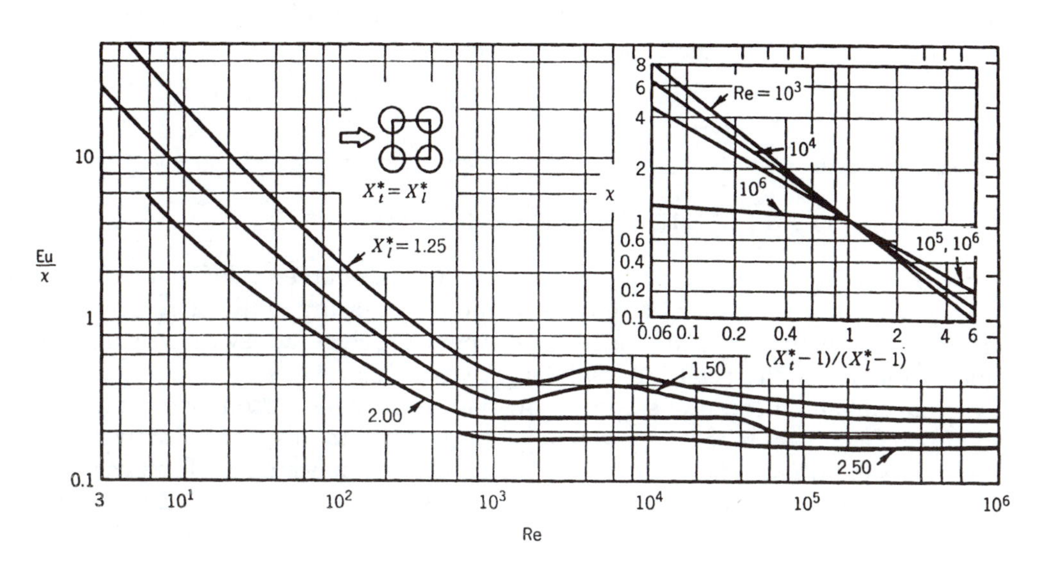

FIGURE 4.4
The hydraulic drag coefficient for in-line bundles for $n > 9$. (From Zukauskas, A. [1987] Chapter 6. In *Handbook of Convective Heat Transfer.* Wiley, New York. With permission.)

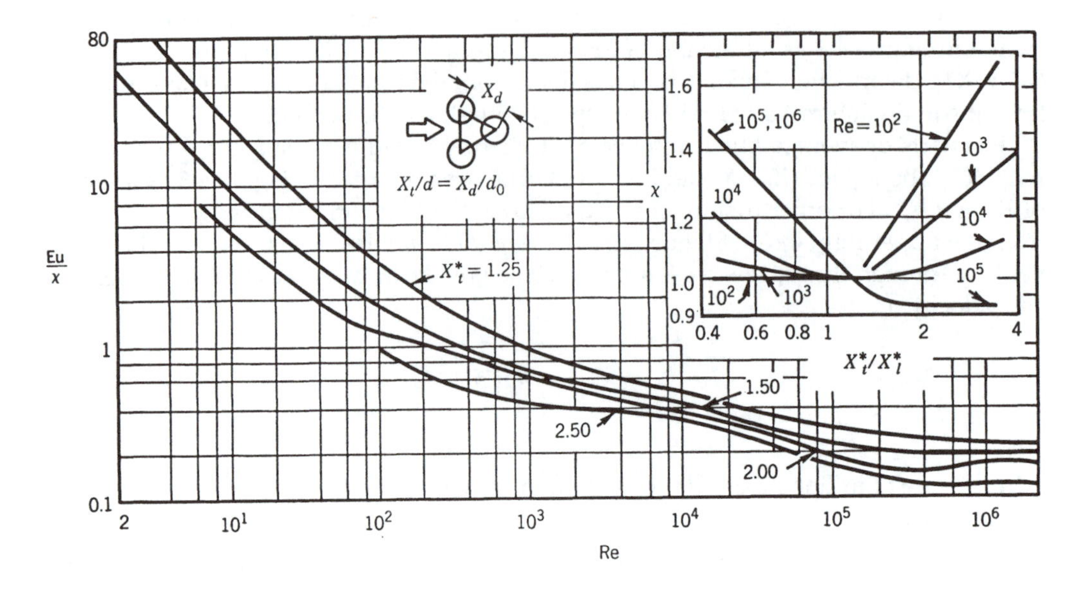

FIGURE 4.5
The hydraulic drag coefficient for staggered bundles for $n > 9$. (From Zukauskas, A. [1987] Chapter 6, In *Handbook of Convective Heat Transfer*, Wiley, New York. With permission.)

Here Eu_b is the Euler number for both heating and cooling, Eu is the Euler number for isothermal conditions, and μ_w and μ_b are the liquid dynamic viscosities at the wall temperature and at the bulk mean temperature, respectively. For $Re > 10^3$, $p \approx 0$. For $Re \leq 10^3$:

$$p = -0.0018\,Re + 0.28 \quad \text{for liquid heating} \tag{4.24}$$

$$p = -0.0026\,Re + 0.43 \quad \text{for liquid cooling} \tag{4.25}$$

Gaddis and Gnielinski[9] also suggested curves and formulas for the pressure drop for in-line and staggered tube bundles.

4.4 Pressure Drop in Helical and Spiral Coils

Helical and spiral coils are curved pipes that are used as curved tube heat exchangers in various applications such as dairy and food processing, and refrigeration and air-conditioning industries, and in medical equipment such as kidney dialysis machines. Nomenclature for a helical coil and a spiral is given in Figure 3.4.

Experimental and theoretical studies show that coiled tube friction factors are higher than those in a straight tube for a given Reynolds number. In this section, friction factors for laminar and turbulent flows in heat exchanger design application will be given. There are various studies, and an up-to-date review of this subject was presented by Shah and Jashi.[10]

4.4.1 Helical Coils: Laminar Flow

Srinivasan et al.[11] proposed the following correlation for laminar flow in terms of Dean number for their experimental data for several coils ($7 < R/a < 104$):

$$\frac{f_c}{f_s} = \begin{cases} 1 & \text{for } De < 30 \\ 0.419\, De^{0.275} & \text{for } 30 < De < 300 \\ 0.1125\, De^{0.5} & \text{for } De > 30 \end{cases} \tag{4.26}$$

where $De = Re\sqrt{(a/R)}$ and c and s stand for curved duct and straight duct, respectively.

Some correlations include R/a as a separate term in addition. Manlapaz and Churchill[12] reviewed available theoretical and experimental studies on laminar flow and recommended the following correlation:

$$\frac{f_c}{f_s} = \left[\left(1.0 + \frac{0.18}{[(1+35/De)^2]^{0.5}}\right)^m + \left(1.0 + \frac{a/R}{3}\right)^2 \left(\frac{De}{88.33}\right)\right]^{0.5} \tag{4.27}$$

where $m = 2$ for $De < 20$, $m = 1$ for $20 < De < 40$, and $m = 0$ for $De > 40$.

The preceding two correlations are in excellent agreement for $R/a > 7$. If $R/a < 3$, the correlation (4.27) is recommended.[10]

Kubair and Kuloor[13] proposed the following correlation to account for the effect of temperature-dependent fluid properties (for liquid viscosity):

$$\frac{f_c}{f_{cp}} = 0.91\left(\frac{\mu_w}{\mu_b}\right)^{0.25} \tag{4.28}$$

4.4.2 Spiral Coils: Laminar Flow

Kubair and Kuloor[14–16] and Srinivasan et al.[11] have measured friction factors for different spirals. For design purposes the following correlation[10,11] is recommended:

$$f_c = \frac{0.63(n_2^{0.7} - n_1^{0.7})^2}{Re^{0.6}(b/a)^{0.3}} \tag{4.29}$$

$$\text{for } 500 < Re(b/a) < 20{,}000 \text{ and } 7.3 < b/a < 15.5$$

where n_1 and n_2 are the number of turns from the origin to the start and the end of a spiral, respectively.

4.4.3 Helical Coils: Turbulent Flow

Several experimental studies have proposed correlations to calculate friction factor for turbulent flow in a helical coil. Ito[17] obtained the following correlation:

$$f_c\left(\frac{R}{a}\right)^{0.5} = 0.00725 + 0.076\left[Re\left(\frac{R}{a}\right)^{-2}\right]^{-0.25} \tag{4.30}$$

$$\text{for } 0.034 < Re\left(\frac{R}{a}\right)^{-2} < 300$$

As a result of extensive friction factor data Srinivasan et al.[11] proposed the following correlation:

$$f_c\left(\frac{R}{a}\right)^{0.5} = 0.0084\left[Re\left(\frac{R}{a}\right)^{-2}\right]^{-0.2} \tag{4.31}$$

$$\text{for } Re\left(\frac{R}{a}\right)^{-2} < 700 \text{ and } 7 < \frac{R}{a} < 10$$

These two correlations for smooth coils are in good agreement with each other and also show good agreement with experimental data for air and water (within ±10%).[18,19]

For temperature-dependent fluid properties, Rogers and Mayhew[19] proposed the following relation:

$$\frac{f_c}{f_{cp}} = \left(\frac{Pr_m}{Pr_w}\right)^{-0.33} \tag{4.32}$$

where f_c is calculated using Ito's correlation, Equation (4.30).

4.4.4 Spiral Coils: Turbulent Flow

Kubair and Kuloor[14] and Srinivasan et al.[11] measured friction factors in five spirals for water and fuel oil, and their experimental correlation is

$$f_c = \frac{0.0074\,(n_2^{0.9} - n_1^{0.9})^{1.5}}{[Re\,(b/a)^{0.5}]^{0.2}} \tag{4.33}$$

4.5 Pressure Drop in Bends and Fittings

4.5.1 Pressure Drop in Bends

Bends are used in heat exchanger piping circuits and in turbulent heat exchangers. The schematic diagram of bend geometry is shown in Figure 3.5.

Friction factor correlations in smooth bends with circular cross section for laminar flow of Newtonian fluids are given in this section.

Total pressure drop in a bend is the sum of the frictional head loss due to the length of the bend, head loss due to curvature, and head loss due to excess pressure drop in the downstream pipe because of the velocity profile distortion.

Ito[20] defined a total loss coefficient K as:

$$\Delta p = K \frac{\rho u_m^2}{2} \tag{4.34}$$

where

$$K = \frac{4 f_c L}{D_h} = \frac{4 f L}{D_h} + K^* \tag{4.35}$$

and f_c is the bend friction factor, f is the friction factor for a straight pipe at the Reynolds number in the bend, and K^* represents a combined loss coefficient other than friction loss.

Here f is given by:

$$f = 0.0791\, Re^{-0.25} \quad \text{for} \;\; 10^4 < Re < 10^5 \tag{4.36}$$

and

$$f = 0.0008 + 0.005525\, Re^{-0.237} \quad \text{for} \;\; 10^5 < Re < 10^7 \tag{4.37}$$

Idelchik[21] reported the following correlation to calculate laminar friction factors in smooth bends of any degree $\phi < 360°$:

$$f_c = 5\, Re^{-6.5} (R/a)^{-0.175} \quad \text{for} \;\; 50 < De \le 600 \tag{4.38}$$

$$f_c = 2.6\, Re^{-0.55} (R/a)^{-0.225} \quad \text{for} \;\; 600 < De \le 1400 \tag{4.39}$$

$$f_c = 1.25\, Re^{-0.45} (R/a)^{-0.275} \quad \text{for} \;\; 1400 < De \le 5000 \tag{4.40}$$

For turbulent flow, Ito[20] obtained extensive experimental data for 45°, 90°, and 180° bends for different values of R/a; he proposed the following correlation for $2 \times 10^4 < \mathrm{Re} < 4 \times 10^5$:

$$K = \begin{cases} 0.00873\, B\, \phi f_c (R/a) & \text{for} \;\; Re(R/a)^{-2} < 91 \\ 0.00241\, B\, \phi\, Re^{-0.17} (R/a)^{0.84} & \text{for} \;\; Re(R/a)^{-2} \ge 91 \end{cases} \tag{4.41}$$

where ϕ is a bend angle in degrees and f_c is the curved tube friction factor obtained from:

$$f_c (R/a)^{0.5} = 0.00725 + 0.076 [Re\, (a/R)^2]^{-0.25} \tag{4.42}$$

for $0.034 < Re(a/R)^2 < 300$ and B is a numerical constant that is determined from the following relations:

For $\phi = 45°$

$$B = 1 + 14.2(R/a)^{-1.47} \tag{4.43}$$

For $\phi = 90°$

$$B = \begin{cases} 0.95 + 17.2(R/a) & \text{for } R/a < 19.7 \\ 1 & \text{for } R/a > 19.7 \end{cases} \tag{4.44}$$

For $\phi = 180°$

$$B = 1 + 116(R/a)^{-4.52} \tag{4.45}$$

For turbulent flow, Powle[22] also recommended the use of the following loss coefficient of K^* in Equation (4.35):

$$K^* = B\phi[0.051 + 0.38(R/a)^{-1}] \tag{4.46}$$

where

$$B(\phi) = \begin{matrix} 1 & \text{for } \phi = 90° \\ 0.9 \sin\phi & \text{for } \phi < 70° \\ 0.7 + 0.35 \sin(\phi/90) & \text{for } \phi \geq 70° \end{matrix} \tag{4.47}$$

It is recommended that Powle's correlation may be used for other values of ϕ rather than $\phi = 45°$, 90°, and 180°.

Friction factors for bends with rectangular cross sections and non-Newtonian fluids are also given in Reference 10.

4.5.2 Pressure Drop in Fittings

In the absence of irreversibility, the energy balance per unit mass for flow requires that $p/\rho + u^2/2 + z$ must be conserved. In real fluid flows some of the mechanical energy is converted into heat due to friction in the fluid itself. This loss of mechanical energy can be expressed as a loss in the total pressure as:

$$\left(p_1 + \frac{1}{2}\rho u_1^2 + \rho z_1\right) - \left(p_2 + \frac{1}{2}\rho u_2^2 + \rho z_2\right) = K\frac{1}{2}\rho u_m^2 \tag{4.48a}$$

The pressure drop in fittings is usually given as the equivalent length in pipe diameters of straight pipe, L_e/d_i, which gives the same pressure drop as the piece of fittings in question (Table 4.2); and the pressure drop is calculated by the use of Equation (4.3). The pressure drop can also be calculated, as a total loss coefficient K by which the dynamic head:

$$\frac{\rho u_m^2}{2} \tag{4.48b}$$

has to be multiplied to give the pressure drop (Table 4.3).[23]

TABLE 4.2

Equivalent Length in Pipe Diameters (L_e/d_i) of Various Valves and Fittings

	Description of Product		Equivalent Length in Pipe Diameters, (L_e/d_i)
Angle valves	Conventional with no obstruction in flat-, bevel-, or plug-type seat	Fully open	145
	Conventional with with wing or pin-guided disk	Fully open	200
Gate valves	Conventional wedge	Fully open	13
	Disk, double disk	Three quarters open	35
	Plug disk	One half open	160
Check valves	Conventional swing	0.5... Fully open	135
	Clearway swing	0.5... Fully open	50
Fittings	90° Standard elbow		30
	45° Standard elbow		16
	90° Long radius elbow		20
	90° Street elbow		50
	45° Street elbow		26
	Square corner elbow		57
	Standard tee with flow through run		20
	Standard tee with flow through branch		60
	Close pattern return bend		50

Note: Minimum calculated pressure drop (lb±/in.^{2}) across valve to provide sufficient flow to lift disk fully. (Courtesy of the Crane Co.)

From Shah, R. K. and Joshi, S. D. (1987) In *Handbook of Convective Heat Transfer,* Wiley, New York. With permission.

Hooper [24] also compiled available laminar and turbulent flow pressure drop data for Standard and long radius elbows (45°, 90°, 180°), miter bends, tees, and valves. The compiled data were presented as the following single correlation:

$$K^* = K_l Re^{-1} + K_\infty (1 + 0.5a^*) \tag{4.49}$$

where a^* is the fitting radius in inches. The values of K_l and K_∞ for various fittings are given in Reference 10.

Example 4.1

Consider a piping circuit of a heat exchanger. In the circuit there are four 90° standard elbows, three close pattern return bends, two check valves (clearway swing type), two angle valves (with no obstruction in a flat-type seat), and three gate valves (conventional wedge type); the valves are fully open. The straight part of the circuit pipe is 150 m, and water at 50°C flows with a velocity of 4 m/s. The pressure drop through the heat exchanger is 12 kPa. Nominal pipe size is 2 in. O.D. (I.D. = 0.052 m).

TABLE 4.3

Pressure Drop Δp Due to Friction in Fittings, Valves, Bends, Contraction, and Enlargement

Sudden Contraction

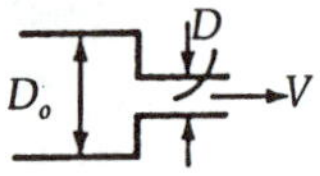

D/D_o	0.0	0.5	0.75
L_e/D	25	20	14
k	0.5	0.4	0.3

Sudden Enlargement

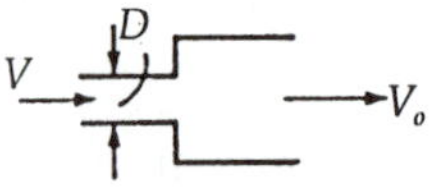

$$\Delta p_c r = \rho \frac{V^2 - V_o^2}{2}$$

Standard Tee

L_e/D	20	60	70	46
k	0.4	1.3	1.5	1.0

Standard Elbow

L_e/D	15	32
k	0.3	0.74

Sharp Elbow

L_e/D	60
k	1.3

90° Bend

R/D	0.5	1.0	2.0	4.0	8.0
L_e/D	36	16.5	10	10	14.5

R/D	0.5	1.0	2.0	4.0	8.0
L_e/D	50	23	20	26	35

180° Bend

	Small radius	Large radius
L_e/D	75	50
k	1.7	1.2

Gate Valve

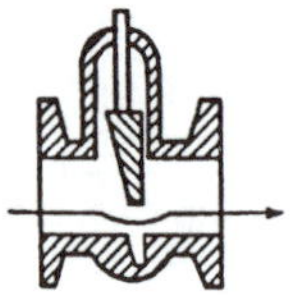

Opening	L_e/D	K
Full	7	0.13
3/4	40	0.8
1/2	200	3.8
1/4	800	15

Globe Valve

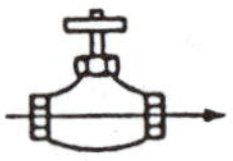

Opening	L_e/D	K
Full	330	6
Half	470	8.5

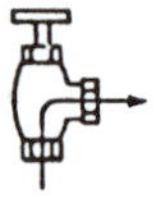

Fully open $L_e/D = 170$, $k = 3$

Diaphragm Valve

Opening	L_e/D	K
Full	125	2.3
3/4	140	2.6
1/2	235	4.3
1/4	1,140	21

Plug Cock

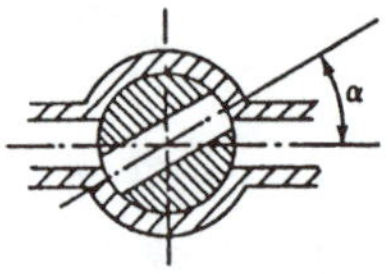

α	L_e/D	K
5	2.7	0.05
10	16	0.29
20	85	1.56
40	950	17.3
60	11,200	206

Check valve fully open

	L_e/D	K
Hinged	110	2
Disk	500	10
Ball	3,500	65

Water meter

	L_e/D	K
Wheel	300	6
Disk	400	8
Piston	600	12

Note: Given as length of straight pipe, L_e, with the same pressure drop, or as factor K in the Equation $\Delta p = K\rho V^2/2$. For further details the reader is referred to the literature.[23]

a. Calculate the total pressure drop in the system.
b. Calculate the mass flow rate and the pumping power (kW), by assuming the isentropic efficiency of the pump to be 0.8.

SOLUTION

The pressure drop can be calculated by the use of Equation (4.3):

$$\Delta p = 4f \frac{L}{d_i} \rho \frac{u_m^2}{2}$$

From Table 4.2, the following equivalent length in pipe diameters can be obtained:

	L_e/d_i
90° Standard elbows	30
Close patten return bends	50
Check valve clearway swing	50
Angles valves	145
Gate valves	13

Properties of water at 50°C from Appendix B (Table B.2) are

$c_p = 4181.2 \text{ J/(kg·K)}$ $\quad$ $k = 0.640 \text{ W/(m·K)}$

$\rho = 987.9 \text{ kg/m}^3$ $\quad$ $Pr = 3.62$

$\mu = 5.531 \times 10^{-4} \text{ Pa·s}$

Reynolds number:

$$Re = \frac{U_m d_i \rho}{\mu} = \frac{4 \times 0.052 \times 987.9}{5.531 \times 10^{-4}} = 371512$$

Therefore the flow is turbulent. The friction coefficient can be calculated using one of the correlations given in Table 4.1. Equation (4.6) is used here:

$$f = 0.046\ (Re)^{-0.2}$$

$$f = 0.046\ (371512)^{-0.2} = 0.00354$$

$$\Delta p_1 = 4f \frac{L}{d_i} \rho \frac{u_m^2}{2} = 4 \times 0.00354 \times \frac{L}{d_i} \times 987.9 \times \frac{16^2}{2}$$

$$\Delta p_1 = 111.91 \Sigma \left(\frac{L_e}{d_i} \right)$$

$$\Delta p_1 = 111.91[4 \times 30 + 3 \times 50 + 2 \times 50 + 2 \times 145 + 3 \times 13] = 78225 \text{ Pa}$$

For the straight part of the pipe:

$$\Delta p_2 = 4f\left(\frac{L}{d_i}\right)\rho\frac{u_m^2}{2} = 111.91(150/0.052) = 322817 \text{ Pa}$$

Then the total pressure drop is

$$\Delta p_t = \Delta p_1 + \Delta p_2 + \Delta p_{HE} = 78.225 + 322.817 + 12 = 413.042 \text{ kPa}$$

The pumping power can be calculated from Equation (4.19):

$$\dot{m} = \frac{\pi}{4} d_i^2 u_m \rho = \frac{\pi}{4} \times (0.052)^2 \times 4 \times 987.9 = 8.4 \text{ kg/s}$$

$$p = \frac{\dot{m}\Delta p_t}{\rho \eta_p} = \frac{8.4 \times 413.042}{987.9 \times 0.80} = 4.36 \text{ kW}$$

4.6 Pressure Drop for Abrupt Contraction, Expansion, and Momentum Change

In heat exchangers, the streams can experience a sudden contraction followed by a sudden enlargement, when flowing in and out of a heat exchanger core, such as the cases in the tube side of shell-and-tube heat exchangers and compact tube–fin and plate–fin heat exchangers. Pressure drop is expressed in terms of the contracting loss coefficient, K_c, and the enlargement loss coefficient, K_e, and the dynamic head; the values of K_c and K_e are given in Reference 26. Therefore, the total pressure drop of the stream is

$$\Delta p_i = K_c \frac{1}{2}\rho u^2 + \Delta p_s + K_e \frac{1}{2}\rho u^2 \tag{4.50}$$

where Δp_s is the straight-duct pressure drop. The velocity represents the mean velocity in the narrowest portion of the flow channel. It is assumed that the density is constant and the contraction ratio equals the enlargement ratio. It should be noted that the pressure rises during enlargement.

For an incompressible flow, Δp_s is given by Equation (4.16). The incompressibility assumption is a good approximation for such liquids as water and oil, but it is not adequate for gases. As a result of change in density in heating, the momentum increases from $\dot{m}u_i$ to $\dot{m}u_o$. Therefore, the change in longitudinal momentum must be balanced by the pressure difference applied due to the acceleration of the stream between inlet and outlet of the heat exchanger passage:

$$\dot{m}(u_0 - u_i) = \Delta p_a A_c \tag{4.51}$$

or

$$\Delta p_a = G^2\left(\frac{1}{\rho_o} - \frac{1}{\rho_i}\right) \tag{4.52}$$

The total pressure drop in the straight section, Δp_s, is then:

$$\Delta p_s = 4f\frac{L}{D_n}\rho\frac{u_m^2}{2} + G^2\left(\frac{1}{\rho_o} - \frac{1}{\rho_i}\right) \tag{4.53}$$

The density in the wall frictional effect is the average density between inlet and outlet.

It should be noted that when the gas is cooled, Δp_a is negative, which is the case of the deceleration of the stream that decreases the overall pressure drop, Δp, across the passage.

4.7 Heat Transfer and Pumping Power Relationship

The fluid pumping power, p, is proportional to the pressure drop in the fluid across a heat exchanger. It is given by:

$$P = \frac{\dot{m}\Delta p}{\rho\eta_p} \tag{4.54}$$

where η_p is the pump or fan efficiency ($\eta_p = 0.80 - 0.85$).

Frequently, the cost in terms of increased fluid friction requires an input of pumping work greater than the realized benefit of increased heat transfer.

In the design of heat exchangers involving high-density fluids, the fluid pumping power requirement is usually quite small relative to the heat transfer rate, and thus the friction power expenditure (i.e., pressure drop) has hardly any influence on the design. However, for gases and low-density fluids and also for very high viscosity fluids, pressure drop is always of equal importance to the heat transfer rate and it has a strong influence on the design of heat exchangers.

Let us consider single-phase side passage (a duct) of a heat exchanger where the flow is turbulent and the surface is smooth.

Forced convection correlation and the friction coefficient can be expressed as:[24,25]

$$St \cdot Pr^{2/3} = \phi_h(Re) \tag{4.55}$$

and

$$f = \phi_f(Re) \tag{4.56}$$

Equation (4.55) gives the heat transfer coefficient:

$$h = (\mu c_p) Pr^{-2/3} \frac{Re}{D_h} \phi_h \tag{4.57}$$

h can be expressed in W/m^2·K and it can be interpreted as the heat transfer power per unit surface area.

If the pressure drop through the passage is Δp and the associated heat transfer surface area is A, the pumping power per unit heat transfer area (W/m^2) is given by:

$$\frac{P}{A} = \frac{\Delta p\, \dot{m}}{\rho \eta_p} \frac{1}{A} \tag{4.58}$$

By substituting Δp from Equation (4.16) into Equation (4.58) and noting that $D_h = 4A_c/P_w$:

$$\frac{P}{A} = 8\left(\frac{\mu^3}{\eta_p \rho^2}\right)\left(\frac{1}{D_h}\right)^3 Re^3 \phi_f \tag{4.59}$$

If it is assumed for simplicity that the friction coefficient is given by Equation (4.5) and the Reynolds analogy is applicable, then:

$$\phi_h = 0.023\, Re^{-0.2} \tag{4.60}$$

$$\phi_f = 0.046\, Re^{-0.2} \tag{4.61}$$

Equations (4.60) and (4.61) approximate the typical characteristics of fully developed turbulent flow in smooth tubes. Substituting these relations into Equations (4.57) and (4.59) and combining them to eliminate the Reynolds number, the pumping power per unit heat transfer area (W/m^2) is obtained as:

$$\frac{P}{A} = \frac{C h^{3.5} \mu^{1.83} (D_h)^{0.5}}{k^{2.33} c_p^{\,1.17} \rho^2 \eta_p} \tag{4.62}$$

where $C = 1.2465 \times 10^4$.

As can be seen from Equation (4.62), the pumping power depends strongly on fluid properties, as well as on the hydraulic diameter of the flow passage. Some important conclusions can be drawn from Equation (4.62) (see Table 4.4):

1. With a high-density fluid such as a liquid, the heat exchanger surface can be operated at large values of h without excessive pumping power requirements.
2. A gas with its very low density results in high values of pumping power for even very moderate values of heat transfer coefficient.
3. A large value of viscosity causes the friction power to be large even though density may be high. Thus, heat exchangers using oils must be designed for relatively low values of h to hold the pumping power within acceptable limits.

4. The thermal conductivity, k, also has a very strong influence; and therefore for liquid metals with very large values of thermal conductivity, the pumping power is seldom of significance.
5. Small values of hydraulic diameter, D_h, tend to minimize the pumping power.

TABLE 4.4.

Pumping Power Expenditure for Various Fluid Conditions

Fluid Conditions	Power Expenditure (W/m^2)
Water at 300 K	
$h = 3850\ W/m^2{\cdot}K$	3.85
Ammonia at 500 K, atm pressure	
$h = 100\ W/m^2{\cdot}K$	29.1
$h = 248\ W/m^2{\cdot}K$	697
Engine oil at 300 K	
$h = 250\ W/m^2{\cdot}K$	0.270×10^4
$h = 500\ W/m^2{\cdot}K$	3.06×10^4
$h = 1200\ W/m^2{\cdot}K$	65.5×10^4

Note: $\eta_p = 80\%$, $D_e = 0.0241$ m.

Nomenclature

A	area, m^2
A_c	heat exchanger passage area, m^2
a	dimension of rectangular duct, or tube inside radius, m
b	dimension of rectangular duct, m
c_p	specific heat at constant pressure, J/(kg·K)
De	Dean number, $Re\,(0.5\ D_h/R)^{1/2}$
D_e	equivalent diameter, m
D_h	hydraulic diameter of the duct, m
d	diameter, m
Eu	average Euler number, $2\Delta P/(\rho u_o^2 \cdot n)$
e	tube surface roughness, m
f	Fanning friction factor, $\tau_w/(\rho u_m^2/2)$
G	mass velocity, $kg/m^2{\cdot}s$
h	heat transfer coefficient, $W/(m^2{\cdot}K)$
k	thermal conductivity, W/(m·K)
K	modified pressure loss coefficient, $4f_cL/D_h$
K^*	pressure loss coefficient, defined in Equation (4.35)
L	tube length or bend length, m
m	mass flow rate, kg/s

N_p number of tube passes
p pressure, Pa
P pumping power, W
Pr Prandtl number, $c_p \mu/k = \upsilon/\alpha$
P_w wetted perimeter, m
P_T pitch, m
R radius of curvature, m
Re Reynolds number, $\rho u_m d/\mu$
r radius, m
St Stanton number, $h/\rho c_p d$
T temperature,°C, K
ΔT_m mean temperature difference,°C, K
t wall thickness, m
u_m mean velocity, m/s
x axial distance along the axis of a bend measured from the bend inlet, m

Greek Symbols

η_p pump efficiency
μ dynamic viscosity, kg/ms
υ kinematic viscosity, m^2/s
ρ density, kg/m^3
τ shear stress, Pa
ϕ parameter, function of or bend angle, degree

Subscripts

a acceleration
c curved coil or duct
i inner, inside
m mean
o outer, outside, overall
p pump
r return
s straight tube
t tube, thermal, total
w wall, wetted
x axial value
1 inlet
2 outlet

References

1. Moody, L. F. (1994) Friction factor for pipe flow. *Trans. ASME*, Vol. 66, 671–684.
2. Brundrett, E. (1979) Modified hydraulic diameter for turbulent flow. In *Turbulent Forced Convection in Channels and Bundles*, S. Kakaç and D. B. Spalding, (Eds.), Vol. 1, pp. 361–367. Hemisphere, New York.
3. Rohsenow, W. M. (1981) Heat exchangers — basic methods. In *Heat Exchangers: Thermal–Hydraulic Fundamentals and Design*, S. Kakaç, A. E. Bergles, and F. Mayinger (Eds.), Hemisphere, New York.
4. Kakaç, S., Shah, R. K., and Aung, W. (Eds.), (1987) In *Handbook of Single Phase-Convective Heat Transfer*, Chapters 4 and 8. Wiley, New York.
5. Kern, D. Q. (1950) *Process Heat Transfer*, McGraw-Hill, New York.
6. Moran, J. M. and Shapiro, N. H. (1995) *Fundamentals of Engineering Thermodynamics*, 3d ed., Wiley, New York.
7. Zukauskas, A. (1987) Convective heat transfer in cross flow. In *Handbook of Convective Heat Transfer*, S. Kakaç, R. K. Shah, and W. Aung (Eds.), Chapter 6. Wiley, New York.
8. Zukauskas, A. and Uliniskas, R. (1983) Banks of plane and finned tubes. *Heat Exchanger Design Handbook*, Vol. 2, Section 2.2.4. Hemisphere, New York.
9. Gaddis, E. S. and Gnielinski, V. (1985) Pressure drop in cross flow across tube bundles. *Int. J. Eng.*, Vol. 25, no. 1, 1–15.
10. Shah, R. K. and Joshi, S. D. (1987) Convective heat transfer in curved ducts. Bends and fittings, In *Handbook of Convective Heat Transfer*, S. Kakaç, R. K. Shah, and W. Aung (Eds.), Chapters 5 and 10. Wiley, New York.
11. Srinivasan, P. S., Nandapurkar, S. S., and Holland, F. A. (1970) Friction for coils. *Trans. Inst. Chem. Eng.*, Vol. 48, T156–T161, 1970.
12. Manlapaz, R. L. and Churchill, S. W. (1980) Fully developed laminar flow in a helically coiled tube of finite pitch. *Chem. Eng. Commun.*, Vol. 7, 557–578.
13. Kubair, V. and Kuloor, N. R. (1965) Non-isothermal pressure drop data for liquid flow in helical coils. *Indian J. Technol.*, Vol. 3, 5–7.
14. Kubair V. and Kuloor, N. R.(1966) Flow of Newtonian fluids in Archimedean spiral tube coils, *Indian J. Technol.*, Vol. 4, 3–8.
15. Kubair V. and Kuloor, N. R. (1965) Non-isothermal pressure drop data for spiral tube coils, *Indian J. Technol.*, Vol. 3, 382–383.
16. Kubair V. and Kuloor, N. R. (1966) Heat transfer for Newtonian fluids in coiled pipes in laminar flow, *Int. J. Heat Transfer*, Vol. 9, 63–75.
17. Ito, H. (1959) Friction factors for turbulent flow in curved pipes. *J. Basic Eng.*, Vol. 81, 123–134.
18. Boyce, B. E., Collier, J. G., and Levy, J. (1969) Hole-up and pressure drop measurements in the two-phase flow of air-water mixtures in helical coils. *Co-current Gas Liquid Flow*, pp. 203–231. Plenum Press, New York.
19. Rogers, G. F. C. and Mayhew, Y. R. (1964) Heat transfer and pressure loss in helically coiled tubes with turbulent flow. *Int. J. Heat Mass Transfer*, Vol. 7, 1207–1216.
20. Ito, H. (1960) Pressure losses in smooth bends. *J. Basic Eng.*, Vol. 82, 131–143.
21. Idelchik, I. E. (1986) *Handbook of Hydraulic Resistance*, 2nd ed., Hemisphere, New York.
22. Powle, U. S. (1981) Energy losses in smooth bends. *Mech. Eng. Bull. (India)*, Vol. 12, no. 4, 104–109.
23. Perry, R. N. and Chilton, C. H. (1973) *Chemical Engineers' Handbook*, 5th ed., McGraw-Hill, New York.
24. Hooper, W. B. (1981) The two-k method predicts head losses in pipe fittings, *Chem. Eng.*, pp. 96–100, 24 Aug.
25. Kakac, S. (Ed.), (1991) *Boilers, Evaporators and Condensers*, Wiley, New York.
26. Kays, W. M. and London, A. L. (1984) *Compact Heat Excahngers*, McGraw-Hill, New York.

Problems

4.1. Consider the laminar flow of 20°C water through a circular duct with an I.D. of 2.54 cm. The average velocity of the water is 4 m/s. Calculate the pressure drop per unit length ($\Delta p/L$).

4.2. Water at 5°C flows through a parallel–plate channel in a flat-plate heat exchanger. The spacing between the plates is 2 cm, and the mean velocity of water is 3.5 m/s. Calculate the pressure drop per unit length in the hydrodynamically fully developed region.

4.3. It is proposed that the stationary diesel exhaust gases (CO_2) be used for heating air by the use of a double-pipe heat exchanger. CO_2 flows through the annulus. From the measurements of the velocity, it is calculated that the mass flow rate of gases from the engine is 100 kg/h. The exhaust gas temperature is 600 K. The air is available at 20°C and it will be heated to 80°C at a mass flow rate of 90 kg/h. One standard (schedule 40) tube size of 4- × 3-in. double-pipe exchanger (hairpin) of 2-m length (copper tubes) will be used.

a. Calculate the frictional pressure drop in the inner tube.
b. Calculate the pressure drop (Pa) in the annulus.
c. Calculate the pumping power for both streams.

Assume the average temperature of CO_2 at 500 K and the following properties: ρ = 1.0732 kg/m^3, c_p = 1013 J/kg·K, k = 0.03352 W/m·K, Pr = 0.702, and ν = 21.67 × 10^{-6}m^2/s. For air: c_p = 1007 J/kg·K.

4.4. Calculate the pumping power for both streams given in Problem 3.8.

4.5. Consider a shell-and-tube heat exchanger. Air with a flow rate of 1.5 kg/s at 500°C and at atmospheric pressure flows through 200 parallel tubes. Each tube has an internal diameter of 2 cm and length of 4 m. Assume that the inlet tube plate cross-sectional area (frontal) is 40% larger than the total cross-sectional area of the tubes. The outlet temperature of the cooled air is 100°C. Calculate the pressure drop for:

a. Abrupt contraction
b. Friction
c. Acceleration
d. Enlargement

and compare the total pressure drop with the frictional pressure drop.

4.6. The core of a shell-and-tube heat exchanger contains 60 tubes with an I.D. of 2.5 cm and length of 2 m. The shell I.D. is 35 cm. The air, which flows through the tubes, is to be heated from 100 to 300°C .

a. Calculate the total pressure drop as the sum of the frictional pressure drop in the tube; and the pressure drop due to acceleration, abrupt contraction, and abrupt enlargement. Is the total pressure drop smaller than the frictional pressure drop in the tube? Why?
b. Calculate the fan power needed.

4.7. Consider a tube bundle arrangement in crossflow heat exchanger. Tubes are staggered bare tubes with a longitudinal pitch of 20 mm, transverse pitch of 24.5 mm, and diameter of 1/2 in. The length of the heat exchanger is 0.80 m. The frontal area seen by the air stream is 0.6 × 0.6 m square. The air flows at 2 atm pressure with a mass flow rate of 1500 kg/h. Assume that the mean air temperature is 200°C . Calculate the air frictional pressure drop across the core of the heat exchanger.

4.8. Assume that the tube bundle arrangement given in Problem 3.11 is the core of a heat exchanger. Calculate the pressure drop caused by the tubes for in-line and staggered arrangements, and compare the two pressure drops and the two pumping powers.

4.9. In a crossflow heat exchanger with in-line tubes, air flows across a bundle of tubes at 5°C and is heated to 32°C. The inlet velocity of air is 15 m/s. Dimensions of tubes are $d_o = 25$ mm, $X_t = X_l = 50$ mm. There are 20 rows in the flow direction and 20 columns counted in the heat exchanger. The air mass flow rate is 0.5 kg/s. Air properties may be evaluated at 20°C and 1 atm. Calculate the frictional pressure drop and the pumping power.

4.10. In a heat exchanger for two different fluids, power expenditure per heating surface area is to be calculated. Fluids flow through the channels of the heat exchanger. Hydraulic diameter of the channel is 0.0241 m.

   a. For air at an average temperature of 30°C and 2 atm pressure, the heat transfer coefficient is 55 W/m^2·K.

   b. For water at an average temperature of 30°C, the heat transfer coefficient is 3850 W/m^2·K.

   By comparing the results of these two fluids, outline your conclusions.

4.11. City water will be cooled in a heat exchanger by seawater entering at 15°C . The outlet temperature of the seawater is 20°C. City water will be recirculated to reduce water consumption. The suction line of the pump has an I.D. of 154 mm, is 22 m long, and has two 90° bends and a hinged check valve. The pipe from the pump to the heat exchanger has an I.D. of 127 mm, is 140 m long, and has six 90° bends. The 90° bends are all made of steel with a radius equal to the I.D. of the pipe, $R/d = 1.0$. The heat exchanger has 62 tubes in parallel, each tube 6 m long. The I.D. of the tubes is 18 mm. All pipes are made of drawn mild steel ($\epsilon = 0.0445$ mm). The seawater flow rate is 120 m^3/h. Assume that there is one velocity head loss at the inlet and 0.5 velocity head loss at the outlet of the heat exchanger. The elevation difference is 10.5 m.

   a. Calculate the total pressure drop in the system (kPa and m liquid head = H m).

   b. Calculate the power of the seawater pump (pump efficiency $\eta = 60\%$).

   c. Plot the pumping power as a function of the seawater flow rate (for three flow rate values: 2000 l/min, 2500 l/min, and 3000 l/min).

4.12. For Problem 4.11:

   a. Plot the total pressure head vs. flow rate (m liquid head vs. l/min).

   b. Change the pipe diameter before and after the pump and repeat a, b, and c in Problem 4.11.

   c. Obtain the optimum pipe diameter by considering the cost of pumping power (operational cost) plus the fixed charges based on capital investment for the pipe installed.

4.13. A shell-and-tube heat exchanger is to be used to cool 25.3 kg/s of water from 38 to 32°C. The exchanger has one shell-side pass and two tube-side passes. The hot water flows through the tubes, and the cooling water flows through the shell. The cooling water enters at 20°C and leaves at 30°C. The shell-side (outside) heat transfer coefficient is estimated to be 5678 W/m²·K. Design specifications require that the pressure drop through the tubes be as close to 13.8 kPa as possible, that the tubes be 18 Birmingham wire gauge (BWG) copper tubing (1.24-mm wall thickness), and that each pass be 4.9 m long. Assume that the pressure losses at the inlet and outlet are equal to one and one half of a velocity head ($\rho u_m^2/2$), respectively. For these specifications, what is the tube diameter and how many tubes are needed?

4.14. Water is to be heated from 10 to 30°C at the rate of 300 kg/s by atmospheric pressure steam in a single-pass shell-and-tube heat exchanger consisting of 1-in. schedule 40 steel pipe. The surface coefficient on the steam side is estimated to be 11,350 W/m²·K. An available pump can deliver the desired quantity of water provided that the pressure drop through the pipes does not exceed 15 $lb_f/in.^2$. Calculate the number of tubes in parallel and the length of each tube necessary to operate the heat exchanger with the available pump.

5

Fouling of Heat Exchangers

5.1 Introduction

Fouling can be defined as the accumulation of undesirable substances on a surface. In general, it refers to the collection and growth of unwanted material resulting in inferior performance of the surface. Fouling occurs in natural as well as in synthetic systems. Arteriosclerosis serves as an example of fouling in the human body where the deposit of cholesterol and the proliferation of connective tissues in an artery wall form plaque that grows inward. The resulting blockage or narrowing of arteries places increased demand on the heart.

In the present context the term *fouling* is used specifically to refer to undesirable deposits on the heat exchanger surface. A heat exchanger must affect a desired change in the thermal conditions of the process streams within allowable pressure drops and continue to do so for a specified time period. During operation, the heat transfer surface fouls resulting in increased thermal resistance and often an increase in the pressure drop and pumping power as well. Both of these effects complement each other in degrading the performance of the heat exchanger. The heat exchanger may deteriorate to the extent that it must be withdrawn from service for replacement or cleaning.

Fouling may significantly influence the overall design of a heat exchanger and may determine the amount of material employed for construction and performance between cleaning schedules. Consequently, fouling causes an enormous economic loss because it directly impacts the initial cost, operating cost, and heat exchanger performance. An up-to-date review of the design of heat exchangers with fouling is given in References 1 and 2.

Fouling reduces the effectiveness of a heat exchanger by reducing the heat transfer and by affecting the pressure drop Figure (5.1).

5.2 Basic Considerations

Thermal analysis of a heat exchanger is governed by the conservation of energy in that the heat release by the hot fluid stream equals the heat gain by the cold fluid. The heat transfer rate, Q, is related to the exchanger geometry and flow parameters as:

$$Q = UA_o \Delta T_m \tag{5.1}$$

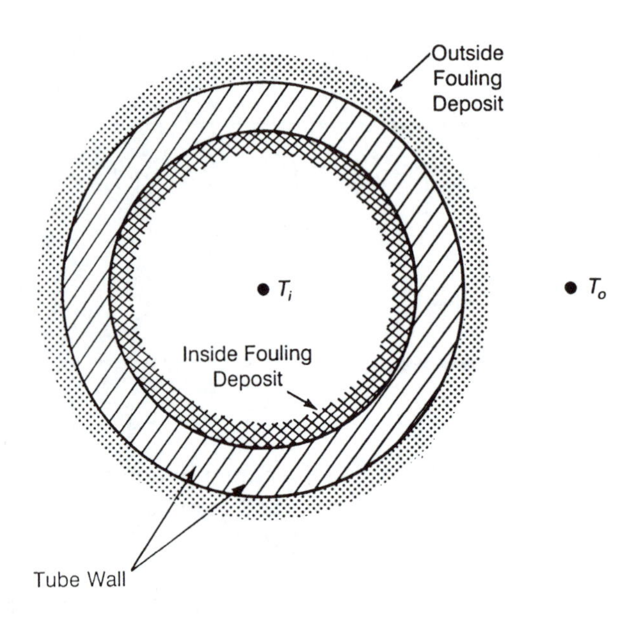

FIGURE 5.1
Fouled tube.

where U is based on the outside heat transfer surface area of the exchanger. We have to distinguish the overall heat transfer coefficient for clean (U_c), and fouled surfaces (U_f).

U_f can be related to the clean surface overall heat transfer coefficient, U_c, as:

$$\frac{1}{U_f} = \frac{1}{U_c} + R_{ft} \tag{5.2}$$

where R_{ft} is the total fouling resistance given as:

$$R_{ft} = \frac{A_o R_{fi}}{A_i} + R_{fo} \tag{5.3}$$

The heat transfer rate under fouled conditions, Q_f, can be expressed as:

$$Q_f = U_f A_f \Delta T_{mf} \tag{5.4}$$

where the subscript f refers to the fouled conditions. Process conditions usually set the heat duty and fluid temperatures at specified values, i.e., $Q_f = Q_c$ and $\Delta T_{mf} = \Delta T_{mc}$. Under these conditions, it can be shown from Eqations (5.1), (5.2), and (5.4) that:

$$\frac{A_f}{A_c} = 1 + U_c R_{ft} \tag{5.5}$$

where A_c is the required surface area if the heat exchanger remains clean. $U_c R_{ft}$ in Equation (5.5) represents the additional surface area required because of fouling of the heat exchanger. For a range of fouling resistances, Figure 5.2 shows the percentage of increase in the heat transfer surface area due to fouling. Obviously, the added surface is small if U_c is low (1 to 10 W/m²·K) even though the total fouling resistance may be high (R_{ft} = 50.0 m²·K/kW). However, for high U_c (1000 to 10,000 W/m²·K) even a small fouling resistance

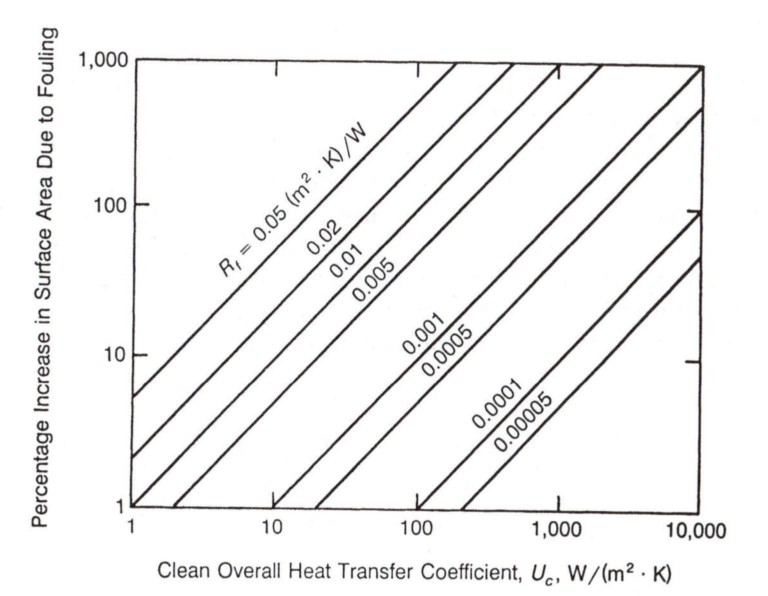

FIGURE 5.2
Effects of fouling on surface area. (From Chenoweth, J. M. [1987] In *Heat Transfer in High Technology and Power Engineering*, pp. 406–419. Hemisphere, New York. With permission.)

(R_{ft} = 0.5 m²·K/kW) results in a substantial increase in the required heat transfer surface area. A 100% increase in the surface area due to fouling alone is not uncommon.

5.3 Effects of Fouling

Lower heat transfer and increased pressure drop resulting from fouling decrease the effectiveness of a heat exchanger. These effects and the basic thermohydraulic aspects of heat exchanger design are discussed in this section.

5.3.1 Effect of Fouling on Heat Transfer

A simple visualization of fouling, shown in Figure 5.1, depicts fouling buildup on the inside and outside of a circular tube. It is evident that fouling adds an insulating layer to the heat transfer surface. For a plain tubular heat exchanger the overall heat transfer coefficient based on the outer surface area under fouled conditions, U_f, can be obtained by adding the inside and outside thermal resistances in Equation (5.2):

$$U_f = \frac{1}{\dfrac{A_o}{A_i h_i} + \dfrac{A_o}{A_i} R_{fi} + \dfrac{A_o \ln (d_o / d_i)}{2\pi k L} + R_{fo} + \dfrac{1}{h_o}} \tag{5.6}$$

The overall heat transfer coefficient for a finned tube (based on the outside surface area) is given by Equation (2.17). Fouling resistances R_{fi} or R_{fo}, also defined in Chapter 2, are sometimes referred to as *fouling factors*. The heat transfer in the unwanted fouling material takes place by conduction.

Average total fouling resistance values specified in the design of some 750 shell-and-tube heat exchangers, by five different manufacturers, are shown in Table 5.1.[3] By using the typical values of heat transfer coefficients — for boiling, condensation, and gas flow — and an average total fouling resistance given in Table 5.1, Table 5.2 can be prepared; the latter shows the percentage increase in the heat transfer surface area for shell-and-tube boiler, and evaporator and condenser applications with single-phase flow in the tube side. It should be noted that if liquids are used for sensible heating and cooling, fouling may substantially increase the required surface area by a factor of 2 or even 3. For high heat transfer applications, fouling may even dictate the design of the heat exchange device.

TABLE 5.1

Specified Total Fouling Resistance, $R_{ft} \times 10^4$, m²·K/W

	Shell Side		
Tube Side	Vapor	Liquid	Two Phase
Vapor	3.7	5.1	4.8
Liquid	6.0	7.9	6.5
Two Phase	5.1	6.7	5.1

From Garrett-Price, B. A. et al. (1985) *Fouling of Heat Exchangers: Characteristics, Costs, Prevention, Control, and Removal*, Noyes, Park Ridge, NJ. With permission.

TABLE 5.2

Added Surface Area for Typical Fluid Combinations

Shell Side (Boiling or Condensation)	h_o	$R_{ft} \times 10^4$	U_c	U_f	Increase in Area (%)
Tube Side: Gas at Very Low Pressure, h_i = 50 W/m²·K					
Medium organics	1,000	4.8	47.6	46.5	2.3
Water, low pressure	5,000	4.8	49.5	48.4	2.4
Water, high pressure	10,000	4.8	49.8	48.6	2.4
Tube Side: Gas at High Pressure, h_i = 500 W/m²·K					
Light organics	1,000	4.8	333.3	287.3	16.0
Medium organics	5,000	4.8	454.5	373.1	21.8
Steam	10,000	4.8	476.2	387.6	22.9
Tube Side: Medium organic liquids, h_i = 1,000 W/m²·K					
Medium organics	1,000	6.5	500.0	377.4	32.5
Water, low pressure	5,000	6.5	833.3	540.5	54.2
Water, high pressure	10,000	6.5	909.1	571.4	59.1
Tube Side: Water, h_i = 5,000 W/m²·K					
Light organics	1,000	6.5	833.3	287.3	54.2
Medium organics	5,000	6.5	2500.0	373.1	162.5
Water	10,000	6.5	3333.3	387.6	216.7

5.3.2 Effect of Fouling on Pressure Drop

Interestingly, more heat exchangers must be removed from service for cleaning due to excessive pressure drop than for an inability to meet the heat transfer requirements. As shown in Figure 5.1, fouling always has a finite, although sometimes small thickness. The change in the flow geometry due to fouling affects the flow field and the pressure drop (or pumping power). For example, in a tubular heat exchanger, fouling layer roughens the surface, decreases the I.D., and increases the O.D. of the tubes.

The frictional pressure drop in the tube for a single-phase flow can be calculated by Equation (4.3):

$$\Delta P = 4f\left(\frac{L}{d_i}\right)\frac{\rho u_m^2}{2} \tag{5.7}$$

where f is the Fanning friction factor. Various graphs and correlations to determine the friction factor for single-phase flow are available in the literature.[1,2]

The fouling layer decreases the I.D. and roughens the surface, thus causing an increase in the pressure drop given by Equation (5.7). Pressure drop under fouled and clean conditions can be related as:

$$\frac{\Delta P_f}{\Delta P_c} = \frac{f_f d_c}{f_c d_f}\left(\frac{u_f}{u_c}\right)^2 \tag{5.8}$$

By assuming that the mass flow rate ($m = u_m \rho A$) under clean and fouled conditions is the same, Equation (5.8) can be modified as:

$$\frac{\Delta P_f}{\Delta P_c} = \frac{f_f}{f_c}\left(\frac{d_c}{d_f}\right)^2 \tag{5.9}$$

The fouling factor can be related to the fouling thermal conductivity k_f and the fouling thickness t_f as:

$$R_f = \frac{t_f}{k_f} \quad \text{(for a plane wall)} \tag{5.10a}$$

$$R_f = \frac{d_c \ln(d_c / d_f)}{2\pi k_f} \quad \text{(for a cylindrical tube wall)} \tag{5.10b}$$

I.D. under fouled conditions, d_f, can be obtained by rearranging Equation (5.10b):

$$d_f = d_c \exp\left(-\frac{2\pi k_f R_f}{d_c}\right) \tag{5.11}$$

and the fouling thickness, t_f, is expressed as:

$$t_f = 0.5d_c\left[1 - \exp\left(-\frac{2\pi k_f R_f}{d_c}\right)\right] \tag{5.12}$$

In general, the fouling layer consists of several materials. Thermal conductivity may also be nonuniform. Approximate thermal conductivities of pure materials that can constitute fouling deposits are given in the second column of Table 5.3.[3] These values have been used to estimate fouling thickness in a 25.4-mm O.D., 16 BWG (22.1-mm I.D.) tube with a tube-side fouling resistance of 0.0004 m²·K/W. It is assumed that the fouling layer is composed solely of one material. The percentage of remaining flow area and percentage increase in the pressure drop are given in columns 3 and 4 of Table 5.3, respectively. As can be seen, the pressure drop increases by about 70% in some instances for the assumed fouling. In these calculations, it is assumed that the fouling does not affect the friction factor (i.e., $f_f = f_c$). Moreover, an increase in the pressure drop due to excess surface area (required for heat transfer) has not been taken into account. Aging of the deposit starts soon after it is laid on the surface. Mechanical properties of the deposit can change during this phase, for example, due to the changes in crystal or chemical structure. Slow poisoning of micro-organisms due to corrosion at the surface may weaken the biofouling layer. A chemical reaction taking place at the deposit surface may alter the chemical composition of the deposit and thereby change its mechanical strength.

5.3.3 Cost of Fouling

Fouling of heat transfer equipment introduces an additional cost to the industrial sector. The added cost is in the form of (1) increased capital expenditure, (2) increased maintenance cost, (3) loss of production, and (4) energy losses.

To compensate for fouling, the heat transfer area of a heat exchanger is increased. Pumps and fans are oversized to compensate for oversurfacing and increased pressure drop resulting from a reduction in the flow area. Duplicate heat exchangers may have to be installed to ensure continuous operation while a fouled heat exchanger is cleaned. High-cost

TABLE 5.3

Added Pressure Drop for Typical Fouling Materials

Material	Thermal Conductivity (W/m·K)	Fouling[a] Thickness, t (mm)	Area Remaining (%)	Increase in Pressure Drop (%)
Hematite	0.6055	0.24	95.7	11.6
Biofilm	0.7093	0.28	95.0	13.7
Calcite	0.9342	0.37	93.5	18.4
Serpentine	1.0380	0.41	92.8	20.7
Gypsum	1.3148	0.51	90.9	26.9
Magnesium phosphate	2.1625	0.83	85.5	47.9
Calcium sulphate	2.3355	0.90	84.4	52.6
Calcium phosphate	2.595	0.99	82.9	59.9
Magnetic iron oxide	2.8718	1.09	81.2	68.2
Calcium carbonate	2.941	1.12	80.8	70.3

[a] Assume fouling resistance of 4×10^{-4} m²·K/W; O.D. = 25.4-mm and I.D. = 22.1-mm tube.

materials such as titanium, stainless steel, or graphite may be required for certain fouling situations. Cleaning equipment may be required for on-line cleaning. All these items contribute to increasing the capital expenditure.

On-line and off-line cleaning add to the maintenance cost. Fouling increases the normally scheduled time incurred in maintaining and repairing equipment. Loss of production because of operation at reduced capacity or downtime can be costly. Finally, energy losses due to reduction in heat transfer and increase in pumping–power requirements can be a major contributor to the cost of fouling.

The annual costs of fouling and corrosion in U.S. industries, excluding electric utilities, were placed between $3 and $10 billion in 1982 dollars.[4] It is clear that the deleterious effects of fouling are extremely costly.

5.4 Aspects of Fouling

A landmark article by Taborek et al.[5] cited fouling as the major unresolved problem in heat transfer. Since then the great financial burden imposed by fouling on the industrial sector has been recognized. This has resulted in a significant increase in the literature on fouling, and various aspects of the problem have been resolved. The major unresolved problem in 1972 is now still the major unsolved problem. This is because the large amount of fouling research has not brought about a significant solution to the prediction and mitigation of fouling.

In the next section some fundamental aspects that help in understanding the types and mechanisms of fouling are discussed. The commonly used methods that aid in developing models to predict fouling are also outlined.

5.4.1 Categories of Fouling

Fouling can be classified in a number of different ways. These may include the type of heat transfer service (boiling, condensation), the type of fluid stream (liquid, gas), or the kind of application (refrigeration, power generation). Because of the diversity of process conditions, most fouling situations are virtually unique. However, to develop a scientific understanding, it is best to classify fouling according to the principal process that results in it. Such a classification, developed by Epstein,[6] has received wide acceptance. Accordingly, fouling is classified into the following categories: particulate, crystallization, corrosion, biofouling, and chemical reaction.

Particulate Fouling

The accumulation of solid particles suspended in the process stream onto the heat transfer surface results in particulate fouling. In boilers this may occur when unburned fuel or ashes are carried over by the combustion gases. Air-cooled condensers are often fouled because of dust deposition. Particles are virtually present in any condenser cooling water. The matter involved may cover a wide range of materials (organic, inorganic), and sizes and shapes (from the submicron to a few millimeters in diameter). Heavy particles settle on a horizontal surface because of gravity. However, other mechanisms may be involved for fine particles to settle onto a heat transfer surface at an inclination.

Crystallization Fouling

A common way in which heat exchangers become fouled is through the process of crystallization. Crystallization arises primarily from the presence of dissolved inorganic salts in the process stream that exhibits supersaturation during heating or cooling. Cooling–water systems are often prone to crystal deposition because of the presence of salts such as calcium and magnesium carbonates, silicates, and phosphates. These are inverse solubility salts that precipitate as the cooling water passes through the condenser (i.e., as the water temperature increases). The problem becomes serious if the salt concentration is high. Such a situation may arise, for example, because of accumulation in cooling–water systems with an evaporative cooling tower. The deposits may result in a dense, well-bonded layer referred to as scale, or a porous, soft layer described as a soft-scale, sludge, or powdery deposit.

Corrosion Fouling

A heat transfer surface exposed to a corrosive fluid may react producing corrosion products. These corrosion products can foul the surface provided the pH value of the fluid is not such that it dissolves the corrosion products as they are formed. For example, impurities in fuel like alkali metals, sulfur, and vanadium can cause corrosion in oil-fired boilers. Corrosion is particularly serious on the liquid side. Corrosion products may also be swept away from the surface where they are produced and transported to other parts of the system.

Biofouling

Deposition or growth of material of a biological origin on a heat transfer surface results in biofouling. Such material may include microorganisms (e.g., bacteria, algae, and molds) and their products, and the resulting fouling is known as microbial fouling. In other instances organisms such as seaweed, water weeds, and barnacles form deposits known as macrobial fouling. Both types of biofouling may occur simultaneously. Marine or power plant condensers using seawater are prone to biofouling.

Chemical Reaction Fouling

Fouling deposits are formed as a result of chemical reaction(s) within the process stream. Unlike corrosion fouling, the heat transfer surface does not participate in the reaction although it may act as a catalyst. Polymerization, cracking, and coking of hydrocarbons are prime examples.

It must be recognized that most fouling situations involve a number of different types of fouling. Moreover, some of the fouling processes may complement each other. For example, corrosion of a heat transfer surface promotes particulate fouling. Significant details about each type of fouling are available in the literature. Somerscales and Knudsen[7] and Melo et al.[8] are excellent sources of such information.

5.4.2 Fundamental Processes of Fouling

Even without complications arising from the interaction of two or more categories, fouling is an extremely complex phenomenon. This is primarily due to the large number of variables that affect fouling. To organize our thought about the topic, it will be extremely useful to approach fouling from a fundamental point of view. Accordingly, the fouling mechanisms (referred to as sequential events by Epstein,[9]) — initiation, transport, attachment, removal, and aging — are described in the following discussion.

Initiation

During initiation the surface is conditioned for the fouling that will take place later. Surface temperature, material, finish, roughness, and coatings strongly influence the initial delay, induction, or incubation period. For example, in crystallization fouling the induction period tends to decrease as the degree of supersaturation increases with respect to the heat transfer surface temperature. For chemical reaction fouling the delay period decreases with increasing temperature because of the acceleration of induction reactions. Surface roughness tends to decrease the delay period.[10] Roughness projections provide additional sites for crystal nucleation thereby promoting crystallization, while grooves provide regions for particulate deposition.

Transport

During this phase fouling substances from the bulk fluid are transported to the heat transfer surface. Transport is accomplished by a number of phenomena including diffusion, sedimentation, and thermophoresis. A great deal of information available for each of these phenomena has been applied to study the transport mechanism for various fouling categories.[7,8]

The difference between fouling species, oxygen, or reactant concentration in the bulk fluid (C_b), and that in the fluid adjacent to the heat transfer surface (C_s), results in transport by diffusion. The local deposition flux (m_d), can be written as:

$$m_d = h_D(C_b - C_s) \tag{5.13}$$

where h_D is the convective mass transfer coefficient. h_D is obtained from the Sherwood number ($Sh = h_D d/D$) which, in turn, depends on the flow and geometric parameters.

Because of gravity, particulate matter in a fluid is transported to the inclined or horizontal surface. This phenomenon, known as sedimentation, is important in applications where particles are heavy and fluid velocities are low.

Thermophoresis is the movement of small particles in a fluid stream when a temperature gradient is present. Cold walls attract colloidal particles while hot walls repel these particles. Thermophoresis is important for particles below 5 μm in diameter and becomes dominant at about 0.1 μm.

A number of other processes such as electrophoresis, inertial impaction, and turbulent downsweeps may be present. Theoretical models to study these processes are available in the literature.[11–13] However, application of these models for fouling prediction is often limited by the fact that several of the preceding processes may be involved simultaneously in a particular fouling situation.

Attachment

Part of the fouling material transported attaches to the surface. Considerable uncertainty about this process exists. Probabilistic techniques are often used to determine the degree of adherence. Forces acting on the particles as they approach the surface are important in determining attachment. Additionally, properties of the material such as density, size, and surface conditions are important.

Removal

Some material is removed from the surface immediately after deposition and some is removed later. In general, shear forces at the interface between the fluid and deposited fouling layer are considered responsible for removal. Shear forces, in turn, depend on the velocity gradients at the surface, viscosity of the fluid, and surface roughness. Dissolution, erosion, and spalling have been proposed as plausible mechanisms for removal. In dissolution the material exits in ionic form. Erosion, whereby the material exits in particulate form, is affected by fluid velocity, particle size, surface roughness, and bonding of the material. In spalling the material exits as a large mass. Spalling is affected by thermal stress setup in the deposit by the heat transfer process.

Aging

Once deposits are laid on the surface, aging begins. The mechanical properties of the deposit can change during this phase because of changes in the crystal or chemical structure, for example. Slow poisoning of microorganisms due to corrosion at the surface may weaken the biofouling layer. A chemical reaction taking place at the deposit surface may alter the chemical composition of the deposit and thereby change its mechanical strength.

5.4.3 Prediction of Fouling

The overall result of the processes listed previously is the net deposition of material on the heat transfer surface. Clearly the deposit thickness is time dependent. In the design of heat exchangers, a constant value of fouling resistance, R_f, interpreted as the value reached in a time period after which the heat exchanger will be cleaned, is used. To determine the cleaning cycle, one should be able to predict how fouling progresses with time. Such information is also required for proper operation of the heat exchanger. Variation of fouling with time can be expressed as the difference between a deposit rate, ϕ_d, and a removal rate, ϕ_r, functions.[14]

$$\frac{dR_f}{dt} = \phi_d - \phi_r \tag{5.14}$$

The behavior of ϕ_d and ϕ_r depends on a large number of parameters. Resulting fouling behavior can be represented as a fouling factor–time curve as shown in Figure 5.3. The shape of the curve relates to the phenomena occurring during the fouling process.

If either deposition rate is constant and removal rate is negligible or the difference between deposition and removal rates is constant, the fouling–time curve will assume a straight line function as shown by curve A in Figure 5.3. Linear fouling is generally represented by tough, hard, adherent deposits. Fouling in such cases will continue to increase unless some type of cleaning is employed. If the deposit rate is constant and the removal rate is ignored, Equation (5.14) can be integrated to yield:

$$R_f = \phi_f t \tag{5.15}$$

which was first developed by McCabe and Robinson.[15]

Asymptotic fouling curve C is obtained if the deposit rate is constant and the removal rate is proportional to the fouling layer thickness, suggesting that the shear strength of the layer is decreasing or that other mechanisms deteriorating the stability of the layer are taking place. Such a situation will generally occur if the deposits are soft because they can flake off easily. Fouling in such cases reaches an asymptotic value.

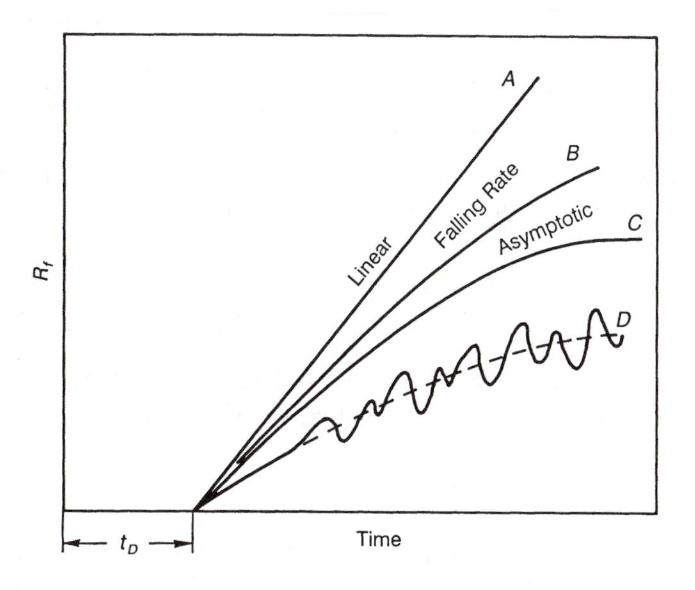

FIGURE 5.3
Typical fouling factor-time curve. (From Chenoweth, J. M. [1987] In *Heat Transfer in High Technology and Power Engineering*, pp. 406–419. Hemisphere, New York. With permission.)

Falling rate fouling, shown by curve B, lies between the linear and asymptotic fouling curves. Such behavior may result if the deposition rate is inversely proportional to the fouling thickness. A periodic change in operating conditions results in the sawtooth configuration shown in curve D. This situation is typical of commercial cooling tower water. By assuming the removal function to be proportional to the fouling resistance and the deposition to be constant, Kern and Seaton[16] obtained the classical relation:

$$R_f = R_f^*(1 - e^{-t/\theta}) \tag{5.16}$$

where θ is the time constant that indicates how quickly asymptotic fouling conditions are approached and R_f^* is the asymptotic fouling factor.

A number of semiempirical models have been developed over the years to predict the nature of the fouling curve in a given application. The general applicability of these models is limited because the various constants or coefficients involved are site dependent and would usually be unknown. Much of the current fouling research is directed toward establishing predictive models. Epstein[6] has tabulated a number of deposition and removal models developed over the past several years. The qualitative effects of increasing certain parameters on the deposition and removal rates and the asymptotic fouling factor are summarized in Table 5.4.[17] Velocity is the only parameter for which an increase causes a reduction in the asymptotic fouling factor even though there may be some exceptions.

5.5 Design of Heat Exchangers Subject to Fouling

Although fouling is time dependent, only a fixed value can be prescribed during the design stage. Therefore, the operating characteristics and cleaning schedules of the heat exchanger

TABLE 5.4

Effect of Parameters on Fouling

Parameter Increased	Deposition Rate	Removal Rate	Asymptotic Fouling
Stickness	Increases	Decreases	Increases
Surface temperature	Increases	Questionable	Increases
Toughness	Questionable	Decreases	Increases
Roughness	Increases (?)	Increases	Questionable
In situ corrosion	Increases	Questionable	Increases
Ex situ corrosion	Increases	Questionable	Increases
Velocity	Decreases	Increases	Decreases

depend on the design fouling factor. Many heat exchangers operate for long periods without being cleaned while others might require frequent cleaning.

5.5.1 Providing a Fouling Allowance

If fouling is anticipated, provisions should be made during the design stage. A number of different approaches are used to provide an allowance for fouling, all of which result in an excess surface area for heat transfer. Current methods include specifying the fouling resistances, the cleanliness factor, or the percentage over surface.

Fouling Resistance

A common practice is to prescribe a fouling resistance or fouling factor (as in Equation [5.6]) on each side of the surface where fouling is anticipated. The result is a lower overall heat transfer coefficient. Consequently, excess surface area is provided to achieve the specified heat transfer. The heat exchanger will perform satisfactorily until the specified value of the fouling resistance is reached, after which it must be cleaned. The cleaning interval is expected to coincide with the regular maintenance schedule of the plant so that additional shutdowns can be avoided.

It is extremely difficult to predict a specific fouling behavior for most cases because a large number of variables can materially alter the type of fouling and its rate of buildup. Sources of fouling resistances in the literature are rather limited, in part, because of the relatively recent interest in fouling research. Tables found in the standards of the Tubular Exchanger Manufacturers Association (TEMA),[19] reproduced here as Tables 5.5 to 5.11, are probably the most referenced source of fouling factors used in the design of heat exchangers. Unfortunately, the TEMA tables do not cover the large variety of possible process fluids, flow conditions, and heat exchanger configurations. These values allow the exchanger to perform satisfactorily in a designated service for a "reasonable time" between cleaning. The interval between cleaning is not known *a priori* because it depends on the performance of the heat exchanger while it is in service. Quite often sufficient excess area is provided for the exchanger to perform satisfactorily under fouled conditions. Proprietary research data, plant data, and personal or company experience are other sources of information on fouling resistances.

Cleanliness Factor

Another approach for providing an allowance for fouling is by specifying cleanliness factor (*CF*), a term developed for the steam power industry. *CF* relates the overall heat transfer coefficient when the heat exchanger is fouled to when it is clean:

TABLE 5.5

TEMA Design Fouling Resistances for Industrial Fluids

Industrial Fluids	$m^2 \cdot K/W$
Oils	
Fuel oil no. 2	0.000352
Fuel oil no. 6	0.000881
Transformer oil	0.000176
Engine lube oil	0.000176
Quench oil	0.000705
Gases and Vapors	
Manufactured gas	0.001761
Engine exhaust gas	0.001761
Steam (nonoil bearing)	0.000088
Exhaust stream (oil bearing)	0.000264–0.000352
Refrigerant vapors (oil bearing)	0.000352
Compressed air	0.000176
Ammonia vapor	0.000176
CO_2 vapor	0.000176
Chlorine vapor	0.000352
Coal flue gas	0.001761
Natural gas flue gas	0.000881
Liquids	
Molten heat transfer salts	0.000088
Refrigerant liquids	0.000176
Hydraulic fluid	0.000176
Industrial organic heat transfer media	0.000352
Ammonia liquid	0.000176
Ammonia liquid (oil bearing)	0.000528
Calcium chloride solutions	0.000528
Sodium chloride solutions	0.000528
CO_2 liquid	0.000176
Chlorine liquid	0.000352
Methanol solutions	0.000352
Ethanol solutions	0.000352
Ethylene glycol solutions	0.000352

From *Standards of the Tubular Exchanger Manufacturers Association* (1988) 7th ed., Tubular Exchanger Manufacturers Association, New York. With permission.

$$CF = \frac{U_f}{U_c} \tag{5.17}$$

It is apparent from Equation (5.17) that this approach provides a fouling allowance that varies directly with the clean surface overall heat transfer coefficient. From Equations (5.2) and (5.17), the cleanliness factor can be related to the total fouling resistance:

$$R_{ft} = \frac{1 - CF}{U_c CF} \tag{5.18}$$

TABLE 5.6

Fouling Resistances for Chemical Processing Streams

Streams	$m^2 \cdot K/W$
Gases and Vapors	
Acid gases	0.000352–0.000528
Solvent vapors	0.000176
Stable overhead products	0.000176
Liquids	
MEA and DEA solutions	0.000352
DEG and TEG solutions	0.000352
Stable side draw and bottom product	0.000176–0.000352
Caustic solutions	0.000352
Vegetable oils	0.000528

From *Standards of the Tubular Exchanger Manufacturers Association* (1988) 7th ed., Tubular Exchanger Manufacturers Association, New York. With permission.

TABLE 5.7

Fouling Resistances for Natural Gas–Gasoline Processing Streams

Streams	$m^2 \cdot K/W$
Gases and Vapors	
Natural gas	0.000176–0.000352
Overhead products	0.000176–0.000352
Liquids	
Lean oil	0.000352
Rich oil	0.000176–0.000352
Natural gasoline and liquified petroleum gases	0.000176–0.000352

From *Standards of the Tubular Exchanger Manufacturers Association* (1988) 7th ed., Tubular Exchanger Manufacturers Association, New York. With permission.

or

$$CF = \frac{1}{1 + R_{ft} U_c} \tag{5.19}$$

Equation (5.18) is used to obtain Figure 5.4 that shows fouling resistance vs. clean surface overall heat transfer coefficient curves for various cleanliness factors. Figure 5.4 illustrates that a given CF will respond to higher fouling resistance, R_{ft}, if the overall heat transfer coefficient, U_c, is low. Such a trend is desirable for the design of steam condensers, where U_c is proportional to the velocity. As shown in Table 5.4, lower velocities (hence low U_c) result in increased fouling. Although the cleanliness factor results in favorable trends, the

TABLE 5.8

Fouling Resistances for Oil Refinery Streams

Streams	m²·K/W		
Crude and Vacuum Unit Gases and Vapors			
Atmospheric tower overhead vapors	0.000176		
Light naphthas	0.000176		
Vacuum overhead vapors	0.000352		
Crude and Vacuum Liquids			
Crude Oil			
	–30 to 120°C Velocity m/s		
	<0.6	**0.6–1.2**	**>1.2**
Dry	0.000528	0.000352	0.000352
Salt[a]	0.000528	0.000352	0.000352
	120 to 175°C Velocity m/s		
	<0.6	**0.6–1.2**	**>1.2**
Dry	0.000528	0.000352	0.000352
Salt[a]	0.000881	0.000705	0.000705
	175 to 230°C Velocity m/s		
	<1.5	**0.6–1.2**	**>1.2**
Dry	0.000705	0.000528	0.000528
Salt[a]	0.001057	0.000881	0.000881
	230°C and over Velocity m/s		
	<1.5	**0.6–1.2**	**>1.2**
Dry	0.000881	0.000705	0.000705
Salt[a]	0.001233	0.001057	0.001057
Gasoline	0.000352		
Naphtha and light distillates	0.000352–0.000528		
Kerosene	0.000352–0.000528		
Light gas oil	0.000352–0.000528		
Heavy gas oil	0.000528–0.000881		
Heavy fuel oils	0.000881–0.001233		
Asphalt and Residuum			
Vacuum tower bottoms	0.001761		
Atmosphere tower bottoms	0.001233		
Cracking and Coking Unit Streams			
Overhead vapors	0.000352		
Light cycle oil	0.000352–0.000528		
Heavy cycle oil	0.000528–0.000705		
Light coker gas oil	0.000528–0.000705		
Heavy coker gas oil	0.000705–0.000881		
Bottoms slurry oil (1.4 m/s minimum)	0.000528		
Light liquid products	0.000176		

[a] Assumes desalting at approximately 120°C.

From *Standards of the Tubular Exchanger Manufacturers Association* (1988) 7th ed., Tubular Exchanger Manufacturers Association, New York. With permission.

TABLE 5.9

Fouling Resistances for Oil Refinery Streams

Streams	m²·K/W
Catalytic Reforming, Hydrocracking, and Hydrodesulfurization	
Reformer charge	0.000264
Reformer effluent	0.000264
Hydrocracker charge and effluent[a]	0.000352
Recycle gas	0.000176
Hydrodesulfurization charge and effluent[a]	0.000352
Overhead vapors	0.000176
Liquid product over 50° A.P.I.	0.000176
Liquid product 30–50° A.P.I.	0.000352
Light Ends Processing	
Overhead vapors and gases	0.000176
Liquid products	0.000176
Absorption oils	0.000352–0.000528
Alkylation trace acid streams	0.000352
Reboiler streams	0.000352–0.000528
Lube Oil Processing	
Feed stock	0.000352
Solvent feed mix	0.000352
Solvent	0.000176
Extract[b]	0.000528
Raffinate	0.000176
Asphalt	0.000881
Wax slurries[b]	0.000528
Refined lube oil	0.000176
Visbreaker	
Overhead vapor	0.000528
Visbreaker bottoms	0.001761
Naphtha Hydrotreater	
Feed	0.000528
Effluent	0.000352
Naphthas	0.000352
Overhead vapors	0.000264

[a] Depending on charge, characteristics, and storage history, charge resistance may be many times this value.
[b] Precautions must be taken to prevent wax deposition on cold tube walls.
From *Standards of the Tubular Exchanger Manufacturers Association* (1988) 7th ed., Tubular Exchanger Manufacturers Association, New York. With permission.

designer is still left with the problem of selecting the appropriate *CF* for his application. Typical designs are based on a cleanliness factor of 0.85.

Percentage Over Surface

In this approach the designer simply adds a certain percentage of clean surface area to account for fouling. The added surface implicitly fixes the total fouling resistance

TABLE 5.10

Fouling Resistances for Oil Refinery Streams

Streams	$m^2 \cdot K/W$
Catalytic Hydrodesulfurizer	
Charge	0.000705–0.000881
Effluent	0.000352
H.T. separator overhead	0.000352
Stripper charge	0.000528
Liquid products	0.000352
HF Alky Unit	
Alkylate, deprop. bottoms, main fract. overhead main fract. feed	0.000528
All Other Process Streams	0.000352

From *Standards of the Tubular Exchanger Manufacturers Association* (1988) 7th ed., Tubular Exchanger Manufacturers Association, New York. With permission.

TABLE 5.11

Fouling Resistances for Water

Temperature of Heating Medium	Up to 115°C		115 to 205°C	
Temperature of Water	50°C		Over 50°C	
Water Velocity (m/s)	0.9 and Less	Over 0.9	0.9 and Less	Over 0.9
Seawater	0.000088	0.000088	0.000176	0.000176
Brackish water	0.000352	0.000176	0.000528	0.000352
Cooling tower and artificial spray pond				
Treated make up	0.000176	0.000176	0.000352	0.000352
Untreated	0.000528	0.000528	0.000881	0.000705
City or well water	0.000176	0.000176	0.000352	0.000352
River water				
Minimum	0.000352	0.000176	0.000528	0.000352
Average	0.000528	0.000352	0.000705	0.000528
Muddy or silty	0.000528	0.000352	0.000705	0.000528
Hard (over 15 grains/gal)	0.000528	0.000528	0.000881	0.000881
Engine jacket	0.000176	0.000176	0.000176	0.000176
Distilled or closed cycle				
Condensate	0.000088	0.000088	0.000088	0.000088
Treated boiler feedwater	0.000176	0.000088	0.000176	0.000176
Boiler blowdown	0.000352	0.000352	0.000352	0.000352

Note: $Rf = m^2 \cdot K/W$.

From *Standards of the Tubular Exchanger Manufacturers Association* (1988) 7th ed., Tubular Exchanger Manufacturers Association, New York. With permission.

depending on the clean surface overall heat transfer coefficient. If heat transfer rate, and fluid temperatures under clean and fouled conditions are the same ($Q_f = Q_c$ and $\Delta T_{mf} = \Delta T_{mc}$), the percentage over surface (OS) can be obtained from Equation (5.5) as:

$$OS = 100\left(\frac{A_f}{A_c} - 1\right) = 100\, U_c R_{ft} \tag{5.20}$$

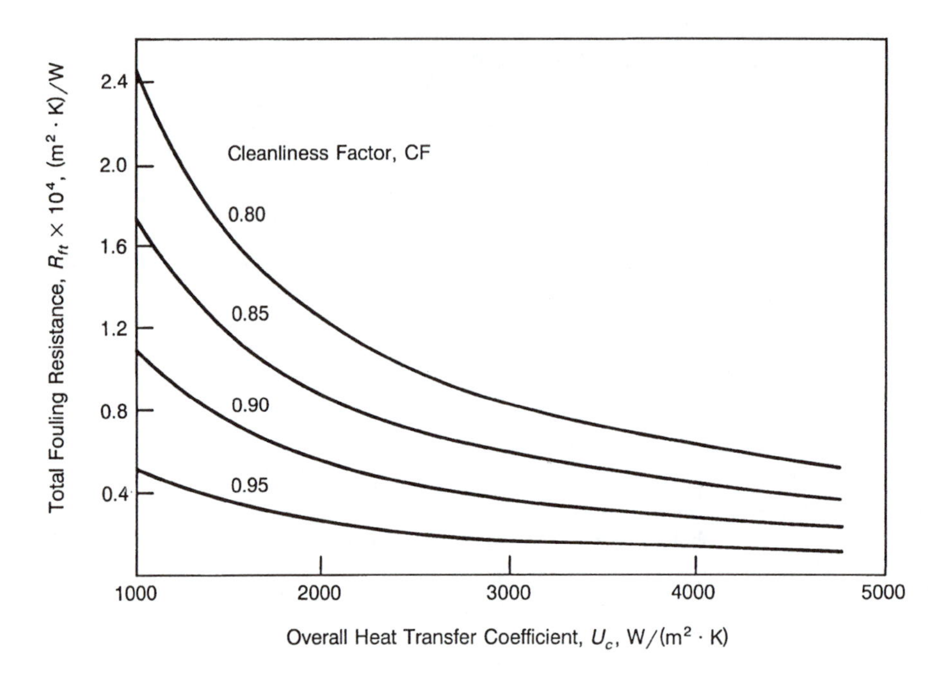

FIGURE 5.4
Calculated fouling resistance based on cleanliness factor.

In a shell-and-tube heat exchanger, the additional surface can be provided either by increasing the length of tubes or by increasing the number of tubes (hence the shell diameter). Such a change will affect the design conditions such as flow velocities, number of cross passes, or baffle spacing. Therefore, the new design with the increased surface should be related to achieve optimum conditions.

Example 5.1

A double-pipe heat exchanger is used to condense steam at a rate of 120 kg/h at 45°C. Cooling water (seawater) enters through the inner tube at a rate of 1.2 kg/s at 15°C. The tube with 25.4-mm O.D. and 22.1-mm I.D. is made of mild steel, $k = 45$ W/m·K. The heat transfer coefficient on the steam side, h_o, is 7000 W/m²·K. Calculate the overall heat transfer coefficient under clean and fouled conditions.

SOLUTION

From Appendix B (Table B.2), the properties of water at 15 and 45°C are $c_p = 4.18$ kJ/kg·K and $h_{fg} = 2.392$ kJ/kg. The exit temperature of the water can be obtained from the heat balance:

$$Q = \dot{m}_o h_{fg} = \dot{m}_i c_p (T_{out} - T_{in})$$

$$Q = \frac{120}{3600}(2392 \times 10^3) = 79.73 \text{ kW}$$

$$T_{out} = T_{in} + \frac{Q}{\dot{m}_i c_p} = 15 + \frac{79.73 \times 10^3}{1.2 \times 4.18 \times 10^3} = 30.89 \text{ °C}$$

At the mean temperature of 23°C, the properties of water are $\rho = 997.207$ kg/m^3, $k = 0.605$ W/m·K, $\mu = 9.09 \times 10^{-4}$ Pa·s, and $Pr = 6.29$:

$$Re = \frac{u_m d_i \rho}{\mu} = \frac{4\dot{m}_i}{\mu \pi d_i} = \frac{4 \times 1.2}{9.09 \times 10^{-4} \pi (0.0221)} = 76056$$

The flow is turbulent.

The heat transfer coefficient inside the tube can be calculated by the use of the correlation given for turbulent flow in Chapter 3. The Gnielinski correlation for constant properties, Equation (3.31), is used here:

$$f = (1.58 \ln Re - 3.28)^{-2}$$

$$f = (1.58 \ln 76056 - 3.28)^{-2} = 0.0047$$

$$Nu_b = \frac{(f/2)(Re_b - 1000)Pr_b}{1 + 2.7(f/2)^{1/2}(Pr_b^{2/3} - 1)}$$

$$Nu_b = \frac{0.0047/2(76056 - 1000)6.29}{1 + 12.7(0.0047/2)^{1/2}(6.29^{2/3} - 1)} = 447$$

$$h_i = \frac{Nu_b k}{d_i} = \frac{447 \times 0.605}{0.0221} = 12236.9 \text{ W/m}^2 \cdot \text{K}$$

The overall heat transfer coefficients for the clean and fouled surfaces based on the outside surface area of the tube are

$$U_c = \frac{1}{\dfrac{d_o}{d_i}\dfrac{1}{h_i} + \dfrac{d_o \ln(d_o/d_i)}{2k} + \dfrac{1}{h_o}}$$

and

$$U_f = \frac{1}{\dfrac{d_o}{d_i}\dfrac{1}{h_i} + \dfrac{d_o}{d_i}R_{fi} + \dfrac{d_o \ln(d_o/d_i)}{2k} + R_{fo} + \dfrac{1}{h_o}}$$

The inside and the outside fouling resistances of the inner tube are obtained from Table 5.9 as:

$$R_{fi} = 0.088 \text{ m}^2 \cdot \text{K/kW} \text{ for seawater}$$

$$R_{fo} = 0.088 \text{ m}^2 \cdot \text{K/kW} \text{ for condensate}$$

Water side velocity is calculated by:

$$U_m = \frac{m_i}{\rho_i A_i} = \frac{1.2}{(997.207)\frac{\pi}{4}(0.022)^2} = 3.17 \text{ m/s}$$

Determine mean temperature difference by:

$$\Delta T_m = \frac{\Delta T_1 - \Delta T_2}{\ln \frac{\Delta T_1}{\Delta T_2}} = \frac{(45-15)-(45-30.89)}{\ln \frac{(45-15)}{(45-30.89)}} = 21.06°\text{C}$$

The surface area is calculated from Equation (5.1). The distribution of resistances, the outside heat transfer coefficients, and surface area can be tabulated as:

Distribution of Resistances	Clean ($m^2 \cdot K/W$)	%	Fouled ($m^2 \cdot K/W$)	%
Water side, h_i = 12236.9 $W/m^2 \cdot K$	0.939×10^{-4}	34.01	0.939×10^{-4}	20.18
Inside fouling	0		1.011×10^{-4}	21.73
Tube wall	0.393×10^{-4}	14.23	0.393×10^{-4}	8.45
Outside fouling	0		0.88×10^{-4}	18.92
Steam side, h_i = 7000 $W/m^2 \cdot K$	1.429×10^{-4}	51.76	1.429×10^{-4}	30.72
Total resistance, $m^2 \cdot K/W$	2.761×10^{-4}		4.652×10^{-4}	
Overall heat transfer coefficient, $W/m^2 \cdot K$	3621.9		2149.6	
Mean temperature difference, K	21.06		21.06	
Surface area, m^2	1.045		1.761	

For assumed values of fouling resistances, the overall heat transfer coefficient, under fouled conditions, is about 60% of that under clean conditions. In the preceding example, the heat transfer surface area should be increased by 76% to obtain the desired heat transfer rate under fouled conditions. Increasing the surface area by such a magnitude may be expensive. An alternate solution would be to design with lower fouling resistances and arrange for cleaning or mitigation techniques to control fouling.

Cleanliness Factor

If a cleanliness factor of 0.85 (typical value) is used, U_f can be obtained from Equation (5.17):

$$U_f = 0.85 \times 3621.9 = 3079 \text{ W/m}^2 \cdot \text{K}$$

From Equation (5.2) this is equivalent to a total fouling resistance of 4.867×10^{-5} $m^2 \cdot K/W$, which is rather small compared with the value obtained from Table 5.9. The fouled area is obtained as resulting in excess area of only 23%.

Percent Over Surface

If 25% over surface (typical value) is used, from Equation (5.20) the total fouling resistance is obtained as 6.895×10^{-5} $m^2 \cdot K/W$, which is about 39% of the TEMA value.

Seemingly different methods for providing an allowance for fouling essentially do the same thing. They result in an increase in the heat transfer surface area. Total fouling resistance, cleanliness factor, and percentage over surface can be related to each other through Equations (5.3) and (5.17) to (5.20). Such a relationship for the preceding Example is presented in Table 5.12.

TABLE 5.12

Relationship between R_{ft}, CF, and OS

R_{ft} · m²K/W	Cf	OS (%)
0.00005	0.85	17.7
0.00010	0.74	35.3
0.00015	0.65	53.0
0.00020	0.59	70.7
0.00025	0.53	88.4
0.00030	0.49	106.0
0.00035	0.45	123.7
0.00040	0.41	141.4

Note: U_c = 3534 W/m²·K.

Example 5.2

A double-pipe heat exchanger is used to condense steam at a rate of 113.68 kg/h at 50°C. Cooling water enters through the tubes at a rate of 0.9 kg/s at 10°C. The tube with 25.4-mm O.D. and 22.1-mm I.D. is made of mild steel (k = 45 W/m·K). The heat transfer coefficient on the steam side, h_o is 10,000 W/m²·K, and on the inside h_i is 8000 W/m²·K.

To increase the heat transfer surface area, the plain tube is replaced by a low-finned tube L with fins either on the inside or the outside. Assume that the use of fins increases the available surface area by a factor of 2.9. Assume 100% fin efficiency, identical wall resistance, and inside fouling for both plain and finned tubes of 0.000176 m²·K/W. Fouling on the shell side is negligible.

Table 5.13 shows various conditions and results of the preceding example. The heat transfer rate is reflected by the product $U \times A$. It is seen that the finned tube results in a

TABLE 5.13

Fouling in Plain and Finned Tube

Quantity	Plain	Finned Inside	Finned Outside
d_o	25.4 mm	25.4 mm	25.4 mm
d_i	22.1 mm	22.1 mm	22.1 mm
h_i	8,000 W/m²·K	8,000 W/m²·K	8,000 W/m²·K
h_o	10,000 W/m²·K	10,000 W/m²·K	10,000 W/m²·K
k	45 W/mK	45 W/mK	45 W/mK
Area ratio			
A_o/A_{op}	1.0	1.0	2.9
A_i/A_{ip}	1.0	2.9	1.0
A_o/A_i	1.1493	0.3963	3.3333
U_c	3534 W/m²·K	5296 W/m²·K	1586 W/m²·K
U_cA_o/L	282 W/K/m	423 W/K/m	367 W/K/m
% Increase	0	50	30
R_{fi}	0.000176 m²·K/W	0.000176 m²·K/W	0.000176 m²·K/W
U_f	2061 W/m²·K	3867 W/m²·K	822 W/m²·K
U_fA_o/L	164 W/K/m	309 W/K/m	190 W/K/m
% Increase	0	88	16

higher heat transfer than the plain tube. For clean conditions, the heat transfer increases by 50% if fins are used on the inside and by 30% if fins are used on the outside. Fins are more effective on the inside because of the relatively smaller heat transfer coefficient (hence higher thermal resistance) on the inside. Under fouled conditions, the increase in heat transfer is 88 and 16% for fins on the inside and the outside, respectively. In the preceding example enhancement is favorable if done on the inside of the tube. External fins are not very effective because the heat transfer increases by only 16% even though the surface area is increased by 290%. As a general rule, enhancement is effective if done on the side with dominant thermal resistance. Identical fouling resistance was used in the preceding example. Such an assumption may not be realistic.

5.6 Operations of Heat Exchangers Subject to Fouling

The specifications of excess area in the heat exchanger, due to the anticipated fouling, lead to operational problems and may be the cause of fouling of the surface beyond that which was used in the design. Due to the time-dependent nature of fouling, the operating conditions will usually be different than the design conditions. The following example illustrates the effect of fouling on operation of the heat exchanger for two different situations.

Example 5.3

The condenser in Example 5.2 is designed with a total fouling resistance of 0.000176. The tube is made of copper. Consider its operation (1) if the water flow rate is maintained at the design value of 0.9 kg/s and (2) if the heat transfer rate is maintained at design value of 75240 W.

For simplicity, the wall resistance and thickness of the tube are ignored. Therefore, the clean surface overall heat transfer coefficient, U_c, is

$$\frac{1}{U_c} = \frac{1}{8000} + \frac{1}{10000}$$

or

$$U_c = 4444.4 \text{ W/m}^2\cdot\text{K}$$

The design surface area can be obtained from:

$$Q = U_f A_f \Delta T_{mf}$$

$$Q = 75240 \text{ W},\ U_f = 2493.8 \text{ W/m}^2\cdot\text{K},\ \Delta T_{mf} = 28.85°\text{C}$$

Therefore, A_f =1.0456 m², (A_c = 0.5868 m², oversurface = 78.2%). The heat exchanger area is fixed at 1.0456 m².

If the water flow rate is maintained at the design value, the heat transfer will be highest when the surface is clean. A higher heat transfer results in a higher water outlet temperature.

However, the overall heat transfer coefficient will decrease as the fouling builds up, and the outlet water temperature decreases toward the design outline.

Fouling resistance used in the design brings about a 78% increase in the surface area over that required if there was no fouling resistance. Therefore, to achieve the design heat transfer rate when the heat exchanger is placed in operation ($R_f = 0$), the water flow rate needs to be adjusted. A common practice (though not recommended) adapted by plant operators is to reduce the mass flow rate. The resulting change in velocity affects the tube-side heat transfer coefficient, h_i, and the water outlet temperature, T_{out}. The water flow rate and the exit temperature to achieve the design heat transfer rate can be computed from Equations (5.21) and (5.22). However, the tube-side heat transfer coefficient, h_i, should be related to the design heat transfer coefficient, h_d (= 8000 W/m²·K), as:

$$Q = U_f A_f \Delta T_m = \dot{m}_i c_p (T_o - T_i) \tag{5.21}$$

$$\frac{1}{U_f} = \frac{1}{U_c} + R_{ft} \tag{5.22}$$

Both higher water temperatures and lower water flow rates increase the tendency of the surface exposed to water to foul. This demonstrates that the adding of heat transfer surface area to allow for anticipated fouling tends to accelerate the rate of fouling.

It would be desirable, if during the initial operating period, the water velocity could be maintained at the design value. This can be accomplished by recirculating some of the water or flooding the condenser. Additional capital costs of pumps and piping required for recirculation should be balanced against advantage of reducing fouling in the heat exchanger.

Performance of the heat exchanger should be monitored during operation. Based on these observations, a proper cleaning schedule can be established. If proper fouling data are available, the cleaning schedule can be based on a rational economic criterion. Somerscales[20] provides a discussion of some of these techniques.

5.7 Techniques to Control Fouling

There are a number of strategies to control fouling. Additives that act as fouling inhibitors can be used while the heat exchanger is in operation. If it is not possible to stop fouling, it becomes a practical matter to remove it. Surface cleaning can be done either on-line or off-line. Table 5.14[21] provides a summary of the various techniques used to control fouling. Following is a discussion of some of these techniques.

5.7.1 Surface Cleaning Techniques

If prior arrangement is made, cleaning can be done on-line. At other times, off-line cleaning must be used. Cleaning methods can be classified as continuous cleaning or periodic cleaning.

TABLE 5.14

Various Techniques to Control Fouling

On-Line Techniques	Off-Line Techniques
Use and control of appropriate additives	Disassembly and manual cleaning
Inhibitors	
Antiscalants	Lances
Dispursants	Liquid jet
Acids	Steam
	Air jet
On-Line Cleaning	Mechanical cleaning
Sponge balls	Drills
Brushes	Scrapers
Sonic horns	
Soot blowers	
Chains and scrapers	Chemical cleaning
Thermal shock	
Air bumping	

From Chenoweth, J. M. (1988) In *Fouling Science and Technology*, pp. 477–494. Kluwer, Dordrecht, The Netherlands. With permission.

Continuous Cleaning

Two of the most common techniques are the sponge–ball and brush systems. The sponge–ball system recirculates rubber balls through a separate loop feeding into the upstream end of the heat exchanger. The system requires extensive installation and therefore is limited to large facilities. The brush system has capture cages at the ends of each tube. It requires a flow-reversal valve that may be expensive.

Periodic Cleaning

Fouling deposits may be removed by mechanical or chemical means. The mechanical methods of cleaning include high-pressure water jets, steam, brushes, and water guns. High-pressure water works well for most deposits, but frequently a thin layer of the deposit is not removed, resulting in greater affinity for fouling when the bundle is returned to service.[5,22] High-temperature steam is useful for removal of hydrocarbon deposits. Brushes or lances are scraping devices attached to long rods and sometimes include a water or steam jet for flushing and removing the deposit.

Chemical cleaning is designed to dissolve deposits by a chemical reaction with the cleaning fluid. The advantage of chemical cleaning is that a hard-to-reach area can be cleaned (e.g., finned tubes). However, the solvent selected for chemical cleaning should not corrode the surface.

5.7.2 Additives

Chemical additives are commonly used to minimize fouling. The effect of additives is best understood for water. For various types of fouling Strauss and Puckorious[23] provide the following observations.

Crystallization Fouling

Minerals from the water are removed by softening. The solubility of the fouling compounds is increased by using chemicals such as acids and polyphosphates. Crystal modification by chemical additives is used to make deposits easier to remove.

Particulate Fouling

Particles are removed mechanically by filtration. Flocculants are used to aid filtration. Dispersants are used to maintain particles in suspension.

Biological Fouling

Chemical removal using continuous or periodic injection of chlorine and other biocides is most common.

Corrosion Fouling

Additives are used to produce protective films on the metal surface.

A number of additive options may be available depending on the type of fluid and application. Additional information on additives is given by Marner and Suitor.[22]

Nomenclature

A	heat transfer surface area, m^2
A_{cr}	cross-sectional area, m^2
C_b	concentration in the fluid, kg/m^3
c_p	specific heat at constant pressure, J/kg·K
C_s	concentration at the heat transfer surface, kg/m^3
CF	cleanliness factor defined by Equation (5.17)
D	diffusion coefficient, m^2/s
d	tube diameter, m
f	Fanning friction factor
h	heat transfer coefficient, $W/m^2 \cdot K$
h_D	mass transfer coefficient, m/s
h_{fg}	heat of vaporization, J/kg
k	thermal conductivity, W/m·K
L	length of the heat transfer surface, m
$\dot{m}$	fluid mass flow rate, kg/s
Nu	Nusselt number, hd/k
OS	oversurface defined in Equation (5.20), %

ΔP	pressure drop, Pa
Pr	Prandtl number, $\mu c_p/k$
Q	heat transfer rate, W
Re	Reynolds number, $\rho u_m d/\mu$
R_f	fouling resistance, $m^2 \cdot K/kW$
R_{ft}	total fouling resistance, $m^2 \cdot K/kW$
R_f^*	asymptotic fouling resistance, $m^2 \cdot K/kW$
Sh	Sherwood number, $h_D d/D$
T	temperature,°C, K
ΔT_m	effective mean temperature difference,°C, K
t	time, s
t_f	fouling thickness, m
t_D	delay time, s
U	overall heat transfer coefficient, $W/m^2 \cdot K$
u_m	average fluid velocity, m/s

Greek Symbols

ρ	fluid density, kg/m^3
μ	viscosity, kg/m·s
ϕ_d	fouling deposition function, $m^2 \cdot K/kJ$
ϕ_r	fouling removal function, $m^2 \cdot K/kJ$
θ	time constant, s

References

1. Kakaç, S. (Ed.) (1991) *Boilers, Evaporators and Condensers*, Wiley, New York.
2. Kakaç, S., Shah, R. K., and Aung, W. (Eds.) (1987) *Handbook of Single-Phase Convective Heat Transfer*, Wiley, New York.
3. Chenoweth, J. M. (1987) Fouling problems in heat exchangers. In *Heat Transfer in High Technology and Power Engineering*, W.K. Yang and Y. Mori (Eds.), pp. 406–419. Hemisphere, New York.
4. Garrett-Price, B. A., Smith, S. A., Watts, R. L., Knudsen, J. G., Marner, W. J., and Suitor, J. W. (1985) *Fouling of Heat Exchangers: Characteristics, Costs, Prevention, Control, and Removal*, Noyes, Park Ridge, NJ.
5. Taborek, J., Akoi, T., Ritter, R. B., and Palen, J. W. (1972) Fouling: The major unresolved problem in heat transfer. *Chem. Eng. Prog.*, Vol. 68, 59–67.
6. Epstein, N. (1978) Fouling in heat exchangers. In *Heat Transfer 1978*, Vol. 6, pp. 235–254. Hemisphere, New York.
7. Somerscales, E. F. C. and Knudsen, J. G. (Eds.) (1981) *Fouling of Heat Transfer Equipment*, Hemisphere, New York.
8. Melo, L. F., Bott, T. R., and Bernardo, C. A. (Eds.) (1988) *Fouling Science and Technology*, Kluwer, Dordrecht, The Netherlands.

9. Epstein, N. (1983) Thinking about heat transfer fouling: A 5 × 5 matrix. *Heat Transfer Eng.*, Vol. 4, 43–56.
10. Epstein, N. (1981) Fouling in heat exchangers. In *Low Reynolds Number Flow Heat Exchangers*, S. Kakaç, R. K. Shah, and A. E. Bergles (Eds.), Hemisphere, New York.
11. Friedlander, S. K. (1977) *Smoke, Dust and Haze*, Wiley, New York.
12. Whitmore, P. J. and Meisen, A. (1977) Estimation of thermo- and diffusio- phoretic particle deposition. *Can. J. Chem. Eng.*, Vol. 55, 279–285.
13. Nishio, G., Kitani, S., and Tadahashi, K. (1974) Thermophoretic deposition of aerosol particles in a heat-exchanger pipe. *Ind. Eng. Chem. Proc. Design Dev.*, Vol. 13, 408–415.
14. Taborek, J., Aoki, T., Ritter, R. B., and Palen, J. W. (1972) Predictive methods for fouling behavior. *Chem. Eng. Progr.*, Vol. 68, 69–78.
15. McCabe, W. L. and Robinson, C. S. (1924) Evaporator scale formation. *Ind. Eng. Chem.*, Vol. 16, 478–479.
16. Kern, D. Q. and Seaton, R. E. (1959) A theoretical analysis of thermal surface fouling. *Br. Chem. Eng.*, Vol. 4, 258–262.
17. Knudson, J. G. (1984) Fouling of heat exchangers: are we solving the problem? *Chem. Eng. Prog.*, Vol. 80, 63–69.
18. Garret-Price, B. A., Smith, S. A., Watts, R. L., and Knudsen, J. G. (1984) Industrial Fouling: Problem Characterization, Economic Assessment, and Review of Prevention, Mitigation, and Accomodation Techniques. Report PNL-4883, Pacific Northwest Laboratory, Richland, WA.
19. *Standards of the Turbulent Exchanger Manufacturers Association*, (1988) 7th ed., TEMA, Tarrytown, New York.
20. Somerscales, E. F. C. (1988) Fouling. In *Two-Phase Flow Heat Exchangers: Thermo–Hydraulic Fundamentals and Design*, S. Kakaç, A. E. Bergles, and E. O. Fernandes (Eds.), Chapter 21. Wiley, New York.
21. Chenoweth, J. M. (1988) General design of heat exchangers for fouling conditions. In *Fouling Science and Technology*, L. F. Melo, T. R. Bott, and C. A. Bernardo (Eds.), pp. 477–494. Kluwer, Dordrecht, The Netherlands.
22. Marner, W. J., and Suitor, J. W. (1987) Fouling with convective heat transfer. In *Handbook of Single-Phase Convective Heat Transfer*, S. Kakaç, R. K. Shah, and W. Aung (Eds.), Chapter 21. Wiley, New York.
23. Strauss, S. D. and Puckorious, P. R. (1984) Cooling-water treatment for control of scaling, fouling and corrosion. *Power*, Vol. 128, p S1–S24.

Problems

5.1. The heat transfer coefficient of a steel (k = 43 W/m·K) tube (1.9-cm I.D. and 2.3-cm O.D.) in a shell-and-tube heat exchanger is 500 W/m^2·K on the inside and 120 W/m^2·K on the shell side, and it has a deposit with a total fouling factor of 0.000176 m^2·K/W. Calculate:

a. The overall heat transfer coefficient

b. The cleanliness factor, and percentage oversurface

5.2. Consider a shell-and-tube heat exchanger having one-shell and four-tube passes. The fluid in the tubes enters at 200°C and leaves at 100°C. The temperature of the fluid is 20°C entering the shell and 90°C leaving the shell. The overall heat transfer coefficient based on the clean surface area of 12 m^2 is 300 W/m^2·K. During the operation of this heat exchanger for 6 months, assume that a deposit with a fouling factor of 0.000528 m^2·K/W has built up. Estimate the percentage decrease in the heat transfer rate between the fluids.

5.3. It is proposed to preheat the water for a boiler using flue gases from the boiler stack. The flue gases are available at the rate of 0.25 kg/s at 150°C, with a specific heat of 1000 J/kg·K. The water entering the exchanger at 15°C at the rate of 0.05 kg/s is to be heated to 90°C. The heat exchanger is to be of the type with one-shell pass and four-tube passes. The water flows inside the tubes, which are made of copper (2.5-cm I.D., 3.0-cm O.D.). The heat transfer coefficient on the gas side is 115 W/m²·K, while the heat transfer coefficient on the water side is 1150 W/m²·K. A scale on the water side and gas side offer an additional total thermal resistance of 0.000176 m²·K/W.

a. Determine the overall heat transfer coefficient based on the outer tube diameter.
b. Determine the appropriate mean temperature difference for the heat exchanger.
c. Estimate the required tube length.
d. Calculate the percentage oversurface design and the cleanliness factor.

5.4. A horizontal shell-and-tube heat exchanger is used to condense organic vapors. The organic vapors condense on the outside of the tubes, while water is used as the cooling medium on the inside of the tubes. The condenser tubes are 1.9-cm O.D. and 1.6-cm I.D. copper tubes 2.4 m in length. There are a total of 768 tubes. The water makes four passes through the exchanger. Test data obtained when the unit was first placed into service are as follows:

Water rate = 3700 l/min

Inlet water temperature = 29°C

Outlet water temperature = 49°C

Organic-vapor condensation temperature = 118°C.

After 3 months of operation, another test made under the same conditions as the first (i.e., same water rate and inlet temperature, and same condensation temperature) showed that the exit water temperature was 46°C.

a. What is the tube-side fluid (water) velocity?
b. By assuming no changes in the inside heat transfer coefficient or the condensing coefficient, negligible shell-side fouling, and no fouling at the time of the first test, estimate the tube-side fouling coefficient at the time of the second test.

5.5. In a double-pipe heat exchanger, deposits of calcium carbonate with a thickness of 1.12 mm and magnesium phosphate with a thickness of 0.88 mm on the inside and outside of the inner tube, respectively, have formed over time. Tubes (I.D. = 1.9 cm, O.D. = 2.3 cm) are made of carbon steel (k = 43 W/m·K). Calculate the total fouling resistance based on the outside surface area of the heat exchanger.

5.6. In a shell-and-tube heat exchanger, water at a flow rate of 3 kg/s is heated from 20 to 90°C. On the shell side, steam condenses to heat the water resulting in an overall heat transfer coefficient of 2100 W/m²·K. After a period of 6 months of continuous operation, the temperature of the tube inside water drops to 80°C. Calculate:

a. Overall heat transfer coefficient under 6 months operating conditions
b. Total fouling resistance under operating conditions

5.7. In problem 5.1, if shell-side fluid is refined lube oil and shell fluid is seawater with a velocity of 2 m/s, what value of the overall heat transfer coefficient should be used for design proposes if the shell-side heat transfer coefficient remains unchanged? Calculate the oversurface design. Is this acceptable?

5.8. A counterflow double-pipe heat exchanger is designed to cool lubricating oil for a large industrial gas turbine engine. The flow rate of cooling water through the inner tube is 0.2 kg/s, and enters the tubes at 30°C while the flow rate of oil through the outlet annulus is 0.4 kg/s. The oil enters and leaves the heat exchanger at temperatures of 60 and 30°C, respectively. The heat transfer coefficients in the annulus and in the inner tube have been estimated as 8 $W/m^2 \cdot K$ and 2445 $W/m^2 \cdot K$, respectively. The O.D. of the inner tube is 25 mm, and the I.D. of the outer tube is 45 mm. The total length of the double-pipe heat exchanger (total length per hairpin) is 15 m. In the analysis, the tube wall resistance and the curvature of the wall are neglected. What is the total value of the fouling resistance used in this design?

5.9. A shell-and-tube type condenser with one shell pass and four tube passes is used to condense organic vapor. The condensation occurs on the shell side, while the coolant water flows inside the tubes that are 1.9-cm O.D. and 1.6-cm I.D. copper tubes. The length of the heat exchanger is 3 m long. The total number of tubes is 840. The initial data of the condenser are recorded as:

Water rate, 70 kg/s

Water inlet temperature, 20°C

Water outlet temperature, 45°C

After 4 months of operation, under the same conditions, the exit temperature of water drops to 40°C. By assuming shell-side fouling is negligible, there is no fouling at the time of the first operation, and the inside and outside heat transfer coefficients are unchanged, estimate the tube-side fouling factor after the operation of 4 months.

5.10. In a single-phase double-pipe heat exchanger, water is to be heated from 35 to 95°C. The water flow rate is 4 kg/s. Condensing steam at 200°C in the annulus is used to heat the water. The overall heat transfer coefficient used in the design of this heat exchanger is 3500 $W/m^2 \cdot K$. After an operation of 6 months, the outlet temperature of the hot water drops to 90°C. The maintaining of the outlet temperature of the water is essential for the purpose of this heat exchanger; therefore fouling is not acceptable. Calculate the total fouling factor under these operations and comment if the cleaning cycle must be extended.

6

Double-Pipe Heat Exchangers

6.1 Introduction

A typical double-pipe heat transfer exchanger consists of one pipe placed concentrically inside another of a larger diameter pipe with appropriate fittings to direct the flow from one section to the next, as shown in Figures 1.8 and 6.1. One fluid flows through the inner pipe (tube side); the other flows through the annular space (annulus). The inner pipe is connected by U-shaped return bends enclosed in a return–bend housing. Double-pipe heat exchangers can be arranged in various series and parallel arrangements to meet pressure drop and mean temperature difference (MTD) requirements. The major use of double-pipe heat exchanger is for sensible heating or cooling of process fluids where small heat transfer areas (up to 50 m^2) are required. This configuration is also suitable for one or both of the fluids at high pressure because of the smaller diameter of the pipes. The major disadvantage is that they are bulky and expensive per unit of heat transfer surface area.

These double-pipe heat exchangers are also called hairpin heat exchangers, and they can also be used when one stream is a gas or viscous liquid, or small in volume. These heat exchangers can be used under highly fouling conditions because of the ease of cleaning and maintenance. The outer surface of the inner tube can be finned and then the tube can be placed concentrically inside a large pipe (Figures1.8 and 6.2). In another type, there are multitubes finned or bare inside a larger pipe (Figure 6.3). They are available with as many as finned tubes inside a large pipe shell. The fins increase the heat transfer surface per unit length and reduce the size and therefore the number of hairpins required for a given heat duty. Integrally resistance welded longitudinal fins (usually fabricated in U-shape) have proved to be the most efficient for double-pipe heat exchangers. Fins are most efficient when the film coefficient is low.

Double-pipe heat exchangers are based on modular principles, individual hairpin sections arranged in proper series (Figure 6.4), parallel, or series–parallel combinations (Figures 6.5 and 6.6) to achieve the required duty. When heat duties change, more hairpins can be easily added to satisfy the new requirements.

For fluids with a lower heat transfer coefficient such as oil or gas, outside finned inner tubes are justified; and this type of fluids stream will flow through the annulus.

Double-pipe heat exchangers are used as counterflow heat exchangers and can be considered as an alternative to shell-and-tube heat exchangers.

Double-pipe heat exchangers have an outer pipe I.D. of 50 to 400 mm at a nominal length of 1.5 to 12.0 m per hairpin. The O.D. of the inner tube may vary between 19 and 100 mm.

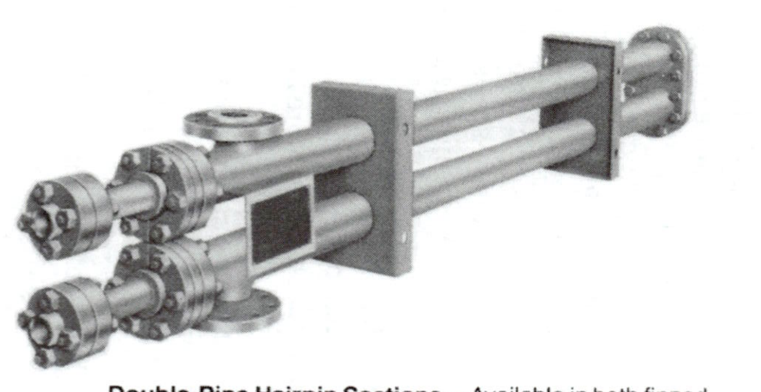

Double-Pipe Hairpin Sections—Available in both finned and bare tube designs. Shell sizes range from 2″ to 6″ IPS. Applications: Ideal for all severe operating conditions. A simple, low-cost section is also available for standard petroleum, petrochemical or chemical service.

Multi-Tube Hairpin Sections—Available in both finned and bare tube designs. Shell sizes range from 3″ to 16″ IPS. Applications: Designs are available to meet most heat exchanger requirements; from heating heavy oils and asphalts to light chemicals, gasoline, butanes, and other hydrocarbons. Also available are designs for high-pressure-gas heat transfer and for processing lethal or hard-to-contain fluids and gases such as hydrogen, Dowtherm and acids.

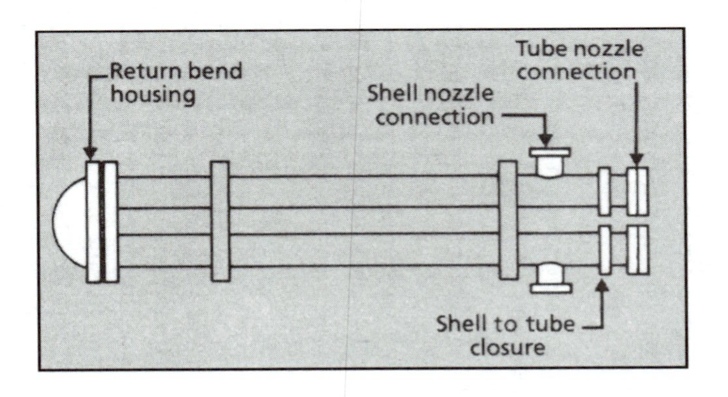

FIGURE 6.1
Double-pipe hairpin heat exchanger. (Courtesy of Brown Fintube, Inc.)

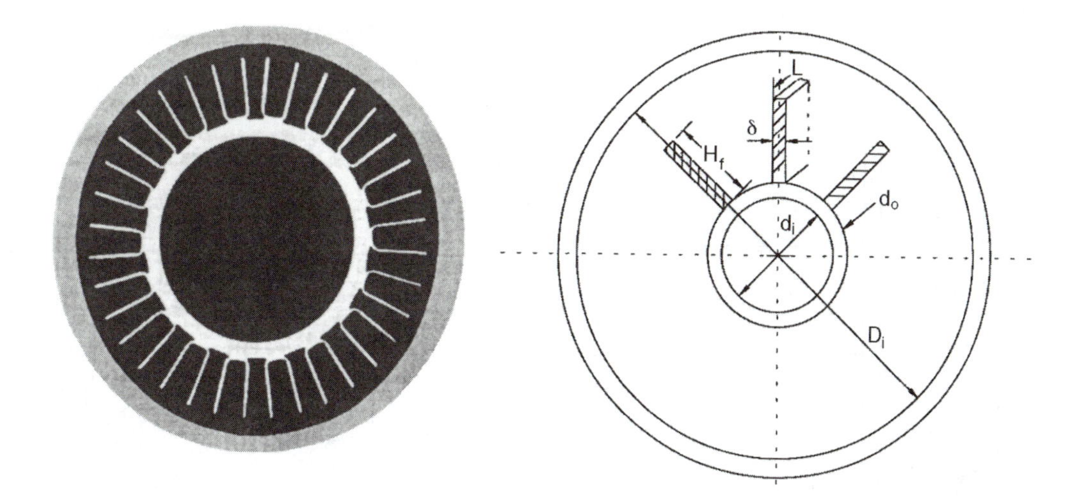

FIGURE 6.2
Flow cross section of a longitudinally finned inner-tube heat exchanger.

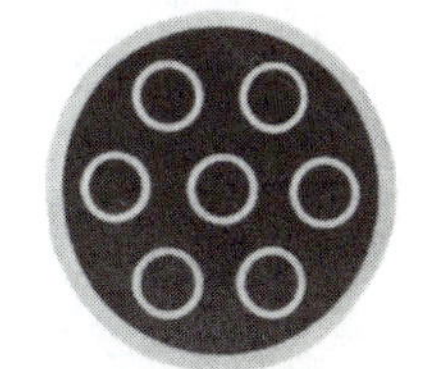

FIGURE 6.3
Multitube hairpin heat exchanger. (Courtesy of Brown Fintube, Inc.)

6.2 Thermal and Hydraulic Design of Inner Tube

Correlations available in the literature are used to calculate heat transfer coefficients inside the inner tube. Important correlations are given in Chapter 3. The frictional pressure drop for flow inside a circular tube is calculated by the use of the Fanning Equation (4.3):

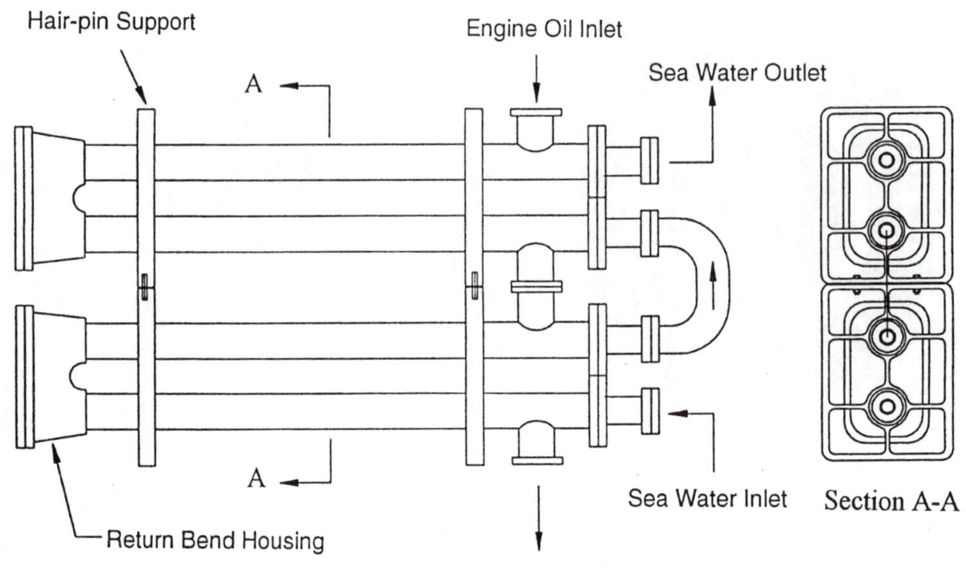

FIGURE 6.4
Two hairpin sections arranged in series.

$$\Delta p = 4f\frac{L}{d_i}\rho\frac{u_m^2}{2} \tag{6.1}$$

or

$$\Delta p = 4f\frac{L}{d_i}\frac{G^2}{2\rho} \tag{6.2}$$

Hairpin exchangers are used both for single- and two-phase flows; our analysis in this chapter is limited to single-phase convective heat transfer. We will also consider the case where the outside of the inner tubes is longitudinally finned to facilitate mechanical cleaning.

6.3 Thermal and Hydraulic Analysis of Annulus

Heat transfer and pressure calculations for flow inside the annulus shell are similar to those for the tube-side flow, provided that the hydraulic diameter of the annulus is used instead of the tube I.D. The hydraulic (equivalent) diameter is given by Equation (3.14):

$$D_h = 4\left\{\frac{\text{net free - flow area}}{\text{wetted perimeter}}\right\} \tag{6.3}$$

It should be noted that only a portion of the wetted perimeter is heated or cooled. Therefore, the hydraulic (equivalent) diameter for the heat transfer calculations is not the same as that used in pressure–drop calculations that is given by Equation (3.18).

$$D_h = \frac{4A_c}{P_w} \tag{6.4}$$

Equation (6.4) is used for the calculation of the Reynolds number, and the equivalent diameter based on heat transfer is given by Equation (3.19):

$$D_e = \frac{4A_c}{P_h} \tag{6.5}$$

where A_c is the net cross-sectional area for flow, and P_w and P_h are the wetted perimeters for the pressure–drop and heat transfer calculations, respectively. Equation (6.5) will be used to calculate the heat transfer coefficient from the Nusselt number and evaluate the Grashof number.

6.3.1 Hairpin Heat Exchanger with Bare Inner Tube

In the case of double pipe with a bare inner tube of O.D., d_o, and an outer tube of the I.D., D_i, we get:

$$D_h = \frac{4\left(\frac{\pi D_i^2}{4} - \frac{\pi d_o^2}{4}\right)}{\pi D_i + \pi d_o} = D_i - d_o \tag{6.6}$$

$$D_e = \frac{4\left(\frac{\pi D_i^2}{4} - \frac{\pi d_o^2}{4}\right)}{\pi d_o} = \frac{D_i^2 - d_o^2}{d_o} \tag{6.7}$$

Example 6.1

Water at a flow rate of 5000 kg/h will be heated from 20 to 35°C by hot water at 140°C. A 15°C hot water temperature drop is allowed. A number of 3.5-m hairpins of 3 in. (I.D. = 0.0779 m) by 2 in. (I.D. = 0.0525 m, O.D. = 0.0603 m) double-pipe heat exchangers with annuli and pipes each connected in series will be used. Hot water flows through the inner tube. Fouling factors are $R_i = 0.000176$ m²·K/W and $R_o = 0.000352$ m²·K/W. Assume that the pipe is made of carbon steel ($k = 54$ W/m·K). The heat exchanger is insulated against heat losses.

a. Calculate the number of hairpins.
b. Calculate the pressure drops.

SOLUTION

Inner tube-side heat transfer coefficient — We first calculate the Reynolds number to determine if the flow is laminar or turbulent.

From Appendix B (Table B.2), the properties of hot water at $T_b = 132.5°C$ are

$$\rho = 932.53 \text{ kg/m}^3, \quad c_p = 4268 \text{ kJ/kg}\cdot\text{K}$$

$$k = 0.687 \text{ W/m}\cdot\text{K}, \quad \mu = 0.207\times10^{-3} \text{Pa}\cdot\text{s}$$

$$Pr = 1.28$$

Calculate hot water mass flow rate by :

$$Q = (\dot{m}c_p)_c \Delta T_c = (\dot{m}c_p)_h \Delta T_h$$

$$\dot{m}_h = \frac{(\dot{m}c_p)_c \Delta T_c}{c_p \Delta T_h} = \frac{(5000/3600)\times4.179\times(35-20)}{4.268\times(15)} = 1.36 \text{ kg/s}$$

where $c_p = 4.179$ kJ/kg·K for cold water at $T_b = 27.5°C$.

Determine the velocity and Reynolds number as:

$$u_m = \frac{\dot{m}_h}{\rho_h A_c} = \frac{1.36}{(932.53)\left(\frac{\pi}{4}\right)(0.0525)^2} = 0.673 \text{ m/s}$$

$$Re = \frac{\rho u_m d_i}{\mu} = \frac{4\dot{m}_h}{\pi\mu d_i} = \frac{4\times1.36}{\pi\times0.207\times10^{-3}\times0.0525} = 159,343$$

Hence, the flow is turbulent; we can select a correlation from Chapter 3. Prandtl's correlation is used here with constant properties:

$$Nu_b = \frac{(f/2)(Re_b)Pr_b}{1+8.7(f/2)^{1/2}(Pr_b-1)}$$

where

$$f = (1.58 \ln Re - 3.28)^{-2}$$

$$= [1.58 \ln (159343) - 3.28]^{-2}$$

$$= 4.085\times10^{-3}$$

$$Nu_b = \frac{(0.004085/2)(159343)(1.28)}{1+8.7\left(\frac{0.004085}{2}\right)^{1/2}(1.28-1)} = 3753$$

$$h_i = \frac{Nu_b k}{d_i} = \frac{375.3\times0.687}{0.0525} = 4911 \text{ W/m}^2\cdot\text{K}$$

Heat transfer coefficient in annulus — Properties at T_b = 27.5°C from Appendix B (Table B.2) are

$$\rho = 996.4\ \text{kg/m}^3, \qquad c_p = 4.179\ \text{kJ/kg}\cdot\text{K}$$

$$k = 0.609\ \text{W/m}\cdot\text{K}, \qquad \mu = 0.841\times10^{-6}\text{Pa}\cdot\text{s}$$

$$Pr = 5.77$$

The velocity of cold water, u_m, through the annulus:

$$u_m = \frac{\dot{m}}{A_c\rho} = \frac{5000/3600}{(0.0779^2 - 0.0603^2)(996.4)} = 0.729\ \text{m/s}$$

$$D_h = \frac{4A_c}{P_w} = D_i - d_o = 0.0779 - 0.0603 = 0.0176\ \text{m}$$

$$Re_b = \frac{\rho u_m D_h}{\mu} = \frac{996.4\times0.729\times0.0176}{0.841\times10^{-3}} = 15201$$

Therefore the flow is turbulent. Prandtl's correlation is also used for the annulus:

$$f = (3.64\ \log_{10}Re_b - 3.28)^{-2} = [3.64\ \log_{10}(15201) - 3.28]^{-2} = 7.021\times10^{-3}$$

$$f/2 = 3.51\times10^{-3}$$

$$Nu_b = \frac{(f/2)(Re_b)Pr_b}{1+8.7(f/2)^{1/2}(Pr_b - 1)}$$

$$= \frac{(3.51\times10^{-3})(15201)(5.77)}{1+8.7(3.51\times10^{-3})^{1/2}(5.77-1)} = 89.0$$

The hydraulic diameter (equivalent diameter) for heat transfer from Equation (6.7) is

$$D_e = \frac{D_i^2 - d_o^2}{d_o} = \frac{0.0779^2 - 0.0603^2}{0.0603} = 0.0403\ \text{m}$$

then

$$Nu_b = \frac{h_o D_e}{k}$$

and

$$h_o = \frac{Nu_b k}{D_e} = \frac{89.0\times0.609}{0.0403} = 1345\ \text{W/m}^2\cdot\text{K}$$

The overall heat transfer coefficient based on the outside area of the inner tube:

$$\frac{1}{U_f} = \frac{d_o}{d_i h_i} + \frac{d_o R_{fi}}{d_i} + \frac{d_o \ln(d_o/d_i)}{2k} + R_{fo} + \frac{1}{h_o}$$

$$= \frac{0.0603}{0.0525 \times 4911} + \frac{0.0603 \times 1.76 \times 10^{-4}}{0.0525} + \frac{0.0603 \ln(603/525)}{2 \times 54} + 3.52 \times 10^{-4} + \frac{1}{1345}$$

$$U_f = 622 \text{ W/m}^2 \cdot \text{K}$$

The heat transfer surface is

$$A_o = \frac{Q}{U_o \Delta T_m}$$

since

$$\Delta T_m = \Delta T_1 = \Delta T_2 = 105°\text{C}$$

and the heat duty of the heat exchanger is

$$Q = (\dot{m}c_p)_c \Delta T_c = 1.389 \times 4.179 \times 105 = 609.49 \text{ kW}$$

so

$$A_o = \frac{609.49 \times 1000}{622 \times 105} = 9.332 \text{ m}^2$$

The heat transfer area per hairpin is

$$A_{hp} = 2\pi d_o L = 2\pi \times 0.0603 \times 3.5 = 1.325 \text{ m}^2$$

$$\frac{A_o}{A_{hp}} = \frac{9.332}{1.325} = 7.64$$

Therefore, the number of the hairpins, $N_{hp} = 8$.

The clean heat transfer coefficient based on the outside heat transfer area is

$$\frac{1}{U_c} = \frac{d_o}{d_i h_i} + \frac{d_o \ln(d_o/d_i)}{2k} + \frac{1}{h_o}$$

$$= \frac{0.0603}{0.0525 \times 4911} + \frac{0.0603 \ln(603/525)}{2 \times 54} + \frac{1}{1345}$$

$$U_c = 948 \text{ W/m}^2 \cdot \text{K}$$

Cleanliness factor is obtained by:

$$C_i = \frac{U_f}{U_c} = \frac{622}{948} = 0.66$$

Calculate percentage oversurface by:

$$OS = 100 U_c R_{ft}$$

$$R_{ft} = \frac{1-CF}{U_c CF} = \frac{1-0.66}{948 \times 0.66} = 0.543 \times 10^{-3}\ \mathrm{m^2 \cdot K/W}$$

$$OS = 100 \times 948 \times 0.543 \times 10^{-3} = 51.5\%$$

The frictional pressure drop in the tube side can be calculated from Equation (4.3):

$$\Delta p_t = 4f \frac{2L}{d_i} N_{hp} \frac{\rho u_m^2}{2}$$

$$= 4 \times 4.085 \times 10^{-3} \frac{2 \times 3}{0.0525} \times 8 \times 932.53 \frac{(0.673)^2}{2} = 3681.04\ \mathrm{Pa}$$

Pumping power for the tube stream is given by Equation (4.25):

$$P_t = \frac{\Delta p_t \cdot \dot{m}_h}{\eta_p \rho_h} = \frac{3681.04 \times 1.36}{0.80 \times 932.33} = 6.72\ \mathrm{W}$$

Pressure drop in the annulus side is calculated by:

$$\Delta p_a = 4f \frac{2L}{D_h} \rho \frac{u_m^2}{2} N_{hp}$$

$$= 4 \times 7.02 \times 10^{-3} \frac{2 \times 3.5}{0.0176} \times 996.41 \frac{(0.719)^2}{2} \times 8 = 23011.15\ \mathrm{Pa}$$

Determine pumping power for the annulus stream by:

$$P_a = \frac{\Delta p_a \dot{m}_c}{\eta_p \rho_c} = \frac{23011.15 \times 1.389}{0.80 \times 996.41} = 40.10\ \mathrm{W}$$

6.3.2 Hairpin Heat Exchanger with Multitube-Finned Inner Tubes

The total wetted perimeter of the annulus with longitudinally finned inner tubes can be written as (see Figure 6.2):

$$P_w = \pi (D_i + d_o N_t) + 2(H_f N_f N_t) \tag{6.8}$$

The heat transfer perimeter of the annulus can be calculated from:

$$P_h = \pi d_o N_t + 2(H_f N_f N_t) \tag{6.9}$$

It is assumed that the outside of the annulus is insulated against heat losses and the heat transfer occurs through the wall and the fins of the inner tube.

The net cross-sectional free-flow area in the annulus with longitudinal finned inner tubes is given by:

$$A_c = \frac{\pi}{4}(D_i^2 - d_o^2 N_t) - (\delta H_f N_t) N_f \tag{6.10}$$

From the relations in Equations (6.4), (6.5), (6.8), (6.9), and (6.10), the hydraulic diameter for the Reynolds number and the pressure drop:

$$D_h = \frac{4A_c}{P_w}$$

and the equivalent diameter for heat transfer (for Nusselt and Grashof numbers):

$$D_e = \frac{4A_c}{P_h}$$

can be calculated.

The validity of the equivalent diameter and hydraulic diameter approach has been substantiated by the results of experiments with finned annuli.[1]

For a double-pipe heat exchanger of length L, the unfinned (bare) and the fin areas are, respectively:

$$A_u = 2N_t(\pi d_o L - N_f L\delta) \tag{6.11}$$

$$A_f = 2N_t N_f L(2H_f + \delta) \tag{6.12}$$

The total outside heat transfer surface area of a hairpin is the sum of the unfinned and fin surfaces given by Equations (6.11) and (6.12) (see Figure 6.2):

$$A_t = A_u + A_f = 2N_t L(\pi d_o + 2N_f H_f) \tag{6.13}$$

The overall heat transfer coefficient based on the outside area of the inner tube is given by Equation (2.17). It should be noticed that A_o is replaced by A_t:

$$U_o = \frac{1}{\frac{A_t}{A_i}\frac{1}{h_i} + \frac{A_t}{A_i}R_{fi} + A_t R_w + \frac{R_{fo}}{\eta_o} + \frac{1}{\eta_o h_o}} \tag{6.14}$$

where $\eta_o = \left[1 - (1 - \eta_f)\frac{A_f}{A_t}\right]$, which is defined by Equation (2.15). As can be seen from

Equation (6.14), area ratios A_t/A_i and A_f/A_t are to be calculated to find the overall heat transfer coefficient based on the outside surface area of the inner tube. It should be noted that R_w is calculated for bare tubes using Equation (2.12b).

The fin efficiency is given by Equation (2.18):

$$\eta_f = \frac{\tanh(mH_f)}{mH_f} \tag{6.15}$$

where

$$m = \sqrt{\frac{2h}{\delta k_f}}$$

Equation (6.15) gives the efficiency of a rectangular continuous longitudinal fin[2] with a negligible heat loss from the tip of the fin. In practice, fins generally have no uniform cross sections, and therefore the analysis of fin thermal behavior is more complex. For complex geometry, fin efficiencies are available in graphic forms.[3–5]

As can be seen from Equation (6.15), the finned area efficiency compared with the bare surface is determined by the fin thickness, height, thermal conductivity, and (most importantly) fluid heat transfer coefficient. Therefore, to have the fluid with poorest heat transfer properties on the finned side is the best way to utilize the maximum area of the fins.

If the hairpin heat exchanger units operate as purely countercurrent or parallel flow, then the correction factor F is unity. However, for series–parallel systems, a correction must be included in the calculations given by Equation (6.18).

In a design problem, the heat duty, Q, of the heat exchanger will be known and the overall heat transfer coefficient and ΔT_m are determined as outlined earlier. The surface area in this case will be the external total surface area of the inner tubes and is calculated from Equation (2.32):

$$Q = UN_{hp}A_t\Delta T_m \tag{6.16}$$

where N_{hp} is the number of hairpins, and $A_t = 2\pi d_o L N_t$.

If the length of the hairpin is fixed, then the number of hairpins, N_{hp}, can be calculated for a specific heat duty.

The overall heat transfer coefficients can also be based on the internal area of the inner tube, A_i. In this case the heat transfer surface area will be the inner surface area of the inner tube, which is

$$A_i = 2(\pi d_i L N_t)$$

Then the heat duty is

$$Q = U_i N_{hp} A_i \Delta T_m \tag{6.17}$$

In design calculation, the length of hairpin, L, can be first assumed; then the number of hairpins is calculated.

Example 6.2

The objective of this example is to design an oil cooler with sea water using finned tube double-pipe heat exchanger.

Engine oil at a rate of 3 kg/s will be cooled from 65 to 55°C by sea water at 20°C. The sea water outlet temperature is 30°C and it flows through the inner tube. The properties are given as:

Fluid	Annulus Fluid Oil	Tube-Side Fluid Sea Water
Density, kg/m^2	885.27	1013.4
Specific heat, kJ/kg·K	1.902	4.004
Viscosity, kg/m·s	0.075	9.64×10^{-4}
Thermal conductivity, W/m·K	0.1442	0.639
Prandtl number, *Pr*	1050	6.29

The following design data are selected:

Length of the hairpin = 4.5 m

Annulus nominal diameter = 2 in. (schedule 40)

Nominal diameter of the inner tube = 3/4 in. (schedule 40)

Fin height, $H_f = 0.0127$ m

Fin thickness, $\delta = 0.9$ mm

Number of fins = 30

Material throughout = carbon steel ($k = 52$ W/m·K)

Number of tubes inside the annulus, $N_t = 1$

Proper fouling factors are to be selected; and the surface area of the heat exchanger, the number of hairpins including pressure drops, and the pumping powers for both streams are to be calculated.

SOLUTION

From Table 8.2 for the nominal diameter 3/4 in., $d_o = 0.02667$ m, and $d_i = 0.02093$ m. Annulus nominal diameter is 2 in., and $D_i = 0.0525$ m.

The sea water mass flow rate can be calculated from heat balance as:

$$\dot{m}_c = \frac{(\dot{m}c_p)_h \Delta T_h}{c_{pc}\Delta T_c}$$

$$= \frac{3 \times 1.902 \times (10)}{4.004 \times (10)} = 1.425 \text{ kg/s}$$

Annulus — The net cross-sectional area in the annulus with longitudinal finned tubes is given by Equation (6.10):

$$A_c = \frac{\pi}{4}(D_i^2 - d_o^2 N_t) - (\delta H_f N_f N_t)$$

$$= \frac{\pi}{4}[(0.0525)^2 - (0.0266)^2] - (0.9 \times 10^{-3})(0.0127)(30)(1)$$

$$= 1.263 \times 10^{-3} \text{ m}^2$$

Determine hydraulic diameter by:

$$D_h = \frac{4A_c}{P_w}$$

where P_w is given by Equation (6.8):

$$P_w = \pi(D_i + d_o N_t) + 2H_f N_f N_t$$

$$= \pi(0.0525 + 0.0266) + 2(0.0127)(30)$$

$$= 1.011 \text{ m}$$

$$D_h = \frac{4A_c}{P_w} = \frac{4(1.263 \times 10^{-3})}{1.011} = 5.0 \times 10^{-3} \text{ m}$$

The wetted perimeter for heat transfer can be calculated using Equation (6.9):

$$P_h = \pi d_o N_t + 2N_f H_f N_t$$

$$= \pi(0.0266)(1) + 2(30)(0.0127)(1) = 0.845 \text{ m}$$

The equivalent diameter for heat transfer is

$$D_e = \frac{4A_c}{P_h} = \frac{4(1.263 \times 10^{-3})}{0.845} = 5.98 \times 10^{-3} \text{ m}$$

Inner tube of sea water — The velocity is

$$u_m = \frac{\dot{m}_c}{\rho \pi \frac{d_i^2}{4}} = \frac{4 \times 1.425}{1013.4 \times \pi (0.0209)^2} = 4.1 \text{ m/s}$$

It is a rather high velocity in practical applications.

Reynolds number:

$$Re = \frac{\rho u_m d_i}{\mu} = \frac{1013.4 \times 4.1 \times 0.02093}{9.64 \times 10^{-4}} = 90{,}082$$

Hence the flow is turbulent. One of the turbulent forced convection corrections given in Chapter 3 can be selected.

Petukhov–Kirillov correlation is used here:

$$f = (1.58 \ln Re - 3.28)^{-2} = (1.58 \ln 90082 - 3.28)^{-2} = 0.0046$$

$$Nu_b = \frac{(f/2)(Re_b)Pr_b}{1.07 + 12.7(f/2)^{1/2}(Pr_b^{2/3} - 1)}$$

$$= \frac{(0.0023)(90082)(6.29)}{1.07 + 12.7(0.0023)^{1/2}(6.29^{2/3} - 1)} = 129.43$$

$$h_i = \frac{Nu \cdot k}{d_i} = \frac{(129.43 \times 0.639)}{0.02093} = 3952 \text{ W/m}^2 \cdot \text{K}$$

Annulus of oil — The velocity is

$$u_m = \frac{\dot{m}_h}{\rho A_c} = \frac{3}{885.27 \times 1.263 \times 10^{-3}} = 2.68 \text{ m/s}$$

$$Re = \frac{\rho u_m D_h}{\mu} = \frac{885.27 \times 2.68 \times 5 \times 10^{-3}}{0.075} = 158.17$$

It is a laminar flow.

The Sieder–Tate correlation (Equation 3.24) can be used to calculate the heat transfer coefficient in the annulus:

$$Nu_b = 1.86\left(Re_b Pr_b \frac{D_h}{L}\right)^{1/3}\left(\frac{\mu_b}{\mu_w}\right)^{0.14}$$

$$Re_b Pr_b \frac{D_h}{L} = (158.17)(1050)\frac{5 \times 10^{-3}}{4.5} = 184.5$$

$$T_w \approx \frac{1}{2}\left(\frac{65 + 55}{2} + \frac{20 + 30}{2}\right) = 42.5°\text{C}$$

$$\mu_w = 0.197 \text{ Pa} \cdot \text{s}$$

$$\left(Re_b Pr_b \frac{D_h}{L}\right)^{1/3}\left(\frac{\mu_b}{\mu_w}\right)^{0.14} = (184.5)^{1/3}\left(\frac{0.075}{0.197}\right)^{0.14} = 3.9 > 2$$

Therefore, the Sieder–Tate correlation is applicable.

$$Nu_b = 186(184.5)^{1/3}\left(\frac{0.075}{0.197}\right)^{0.14} = 9.25$$

$$h_o = \frac{Nu \cdot k}{D_e} = \frac{9.15 \times 0.1442}{5.98 \times 10^{-3}} = 223 \text{ W/m}^2 \cdot \text{K}$$

Note that Equation (3.25) is satisfied to use Equation (3.24). Finned and unfinned heat transfer areas are given by Equations (6.11) and (6.12):

$$A_f = 2N_t N_f L(2H_f + \delta) = (2)(1)(30)(4.5)(2 \times 0.0127 + 0.9 \times 10^{-3})$$

$$= 7.101 \text{ m}^2$$

$$A_u = 2N_t(\pi d_o L - N_f L\delta) = (2)(1)[\pi(0.0266)(4.5) - (30)(4.5)(0.9 \times 10^{-3})]$$

$$= 0.509 \text{ m}^2$$

The total area of a hairpin is

$$A_t = A_u + A_f = 7.101 + 0.509 = 7.610 \text{ m}^2$$

Fin efficiency from Equation (6.15) is

$$\eta_f = \frac{\tanh(mH_f)}{mH_f}$$

$$m = \sqrt{\frac{2h_o}{\delta k_f}} = \sqrt{\frac{2(223.05)}{(0.9 \times 10^{-3})(52)}} = 97.63$$

$$\eta_f = \frac{\tanh(97.63 \times 0.0127)}{97.63 \times 0.0127} = 0.682$$

Overall surface efficiency from Equation (2.14) is

$$\eta_o = \left[1 - (1 - \eta_f)\frac{A_f}{A_t}\right] = \left[1 - (1 - 0.682)\frac{7.101}{7.610}\right] = 0.703$$

The overall heat transfer coefficient (fouled) is

$$U_{o_f} = \frac{1}{\dfrac{A_t}{A_i}\dfrac{1}{h_i} + \dfrac{A_o}{A_i} + A_o R_w + \dfrac{R_{fo}}{\eta_o} + \dfrac{1}{\eta_o h_o}}$$

$$= \frac{1}{\dfrac{A_t}{A_i}\dfrac{1}{h_i} + \dfrac{d_o}{d_i}R_{fi} + \dfrac{A_t \ln(d_o/d_i)}{2\pi k \cdot 2L} + \dfrac{R_{fo}}{\eta_o} + \dfrac{1}{\eta_o h_o}}$$

By using the fouling resistances given in Tables 5.5 and 5.11:

$$R_{fo} = 0.176 \times 10^{-3} \text{m}^2 \cdot \text{K/W} \quad \text{(engine oil)}$$

$$R_{fi} = 0.088 \times 10^{-3} \text{m}^2 \cdot \text{K/W} \quad \text{(seawater)}$$

Thus

$$U_{o_f} = \left[\frac{7.610}{(0.592)(3952)} + \frac{7.610}{0.592}(0.088 \times 10^{-3}) + \frac{7.610 \ln\left(\frac{0.0266}{0.0209}\right)}{2(52) \times \pi \times (2 \times 4.5)} + \frac{0.176 \times 10^{-3}}{0.703} + \frac{1}{(0.703)(223)} \right]^{-1}$$

$$= 85.94 \text{ W/m}^2 \cdot \text{K}$$

Under clean conditions:

$$U_{oc} = \frac{1}{\frac{A_t}{A_i h_i} + \frac{A_t \ln(d_o/d_i)}{2\pi k \cdot 2L} + \frac{1}{\eta_o h_o}}$$

$$= \left[\frac{7.610}{(0.592)(3952)} + \frac{7.610 \ln\left(\frac{0.0266}{0.0209}\right)}{2(52) \times \pi \times (2 \times 4.5)} + \frac{1}{(0.703)(223)} \right]^{-1}$$

$$= 92.92 \text{ W/m}^2 \cdot \text{K}$$

Obtain the cleanliness factor by:

$$CF = \frac{U_{o_f}}{U_{oc}} = \frac{85.94}{92.92} = 0.925$$

Determine total heat transfer surface area by:

$$A_o = \frac{Q}{U_o \Delta T_m}$$

$$Q = (\dot{m}c_p)_h (T_{h_1} - T_{h_2}) = (3)(1.902 \times 10^3)(65 - 55) = 57060 \text{ W}$$

$$\Delta T_m = T_1 = \Delta T_2 = 35°\text{C}$$

Without fouling:

$$A_{oc} = \frac{Q}{U_{oc} \Delta T_m} = \frac{57060}{(92.92)(35)} = 17.54 \text{ m}^2$$

With fouling:

$$A_{o_f} = \frac{Q}{U_{o_f} \Delta T_m} = \frac{57060}{(85.94)(35)} = 18.97 \text{ m}^2$$

Calculate the hairpin heat transfer surface area by:

$$A_{hp} = \mathrm{A_t} = 7.610\ \mathrm{m}^2$$

The number of hairpins can be determined by:

$$N_{hp} = \frac{A_o}{A_{hp}}$$

$$N_{hp} = \frac{A_{of}}{A_{hp}} = \frac{18.97}{7.610} = 2.493$$

One may select three hairpins with a shorter length or two hairpins with a longer length. However, in this case, the analysis must be repeated.

Pressure drops and pumping powers:

For the inner tube:

$$\Delta p_t = 4f\frac{2L}{d_i}\rho\frac{u_m^2}{2}N_{hp}N_t$$

$$= 4(0.0046)\frac{4.5\times 2}{0.0209}(1013.4)\frac{(4.1)^2}{2}(3)(1) = 0.202\ \mathrm{MPa}$$

$$P_t = \frac{\Delta p_t \dot{m}}{\eta_p \rho_t} = \frac{(0.202\times 10^6)(1.425)}{(0.80)(1013.4)} = 355.88\ \mathrm{W}$$

For the annulus:

$$f_{cp} = \frac{16}{Re} = \frac{16}{158.17} = 0.1011$$

From Equation (3.21b) with the use of Table 3.2, we have:

$$f = f_{cp}\left(\frac{\mu_b}{\mu_w}\right)^{-0.58} = 0.1011\left(\frac{0.075}{0.197}\right)^{-0.58} = 0.1771$$

$$\Delta p_a = 4f\frac{2L}{\Delta h}\rho\frac{u_m^2}{2}N_{hp}$$

$$= 4(0.1771)\frac{4.5\times 2}{(5\times 10^{-3})}(885.27)\frac{(2.68)^2}{2}(3) = 12.162\ \mathrm{MPa}$$

$$P_a = \frac{\Delta p_a \dot{m}_a}{\eta_p \rho_a} = \frac{(12.162\times 10^6)(3)}{(0.80)(885.27)} = 51.5\ \mathrm{kW}$$

Figure 6.5 shows two hairpin sections arranged in series with main components.

6.4 Parallel–Series Arrangements of Hairpins

When the design dictates a large number of hairpins in a double-pipe heat exchanger design for a given service, it may not always be possible to connect both the annulus and the tubes in series for pure counterflow arrangement. If there is a large quantity of fluid through the tube or annulus, it may result in high pressure drops as a result of high velocities that may exceed available pressure drop. In such circumstances, the mass flow may be separated into a number of parallel streams and the smaller mass flow rate side can be connected in series. Then the system results in a series–parallel arrangement (Figure 6.5). Two double-pipe units in series on the annulus (shell) side and parallel on the tube side are shown in Figure 6.6. The inner-pipe fluid is divided into two streams that face with hotter and colder fluids in units (2) and (1). For the inner pipe, a new heat transfer coefficient must be calculated. Therefore, in each hairpin section, different amounts of heat is transferred; the true mean temperature difference, ΔT_m, will be significantly different from the log mean temperature difference (LMTD). In this case, the true mean temperature difference in Equation (6.16) is given by:[3,6]

$$\Delta T_m = S(T_{h_1} - T_{c_1}) \tag{6.18}$$

where the dimensionless quantity S is defined as:

$$S = \frac{(\dot{m}c_p)_h (T_{h_1} - T_{c_2})}{UA(T_{h_1} - T_{c_1})} \tag{6.19}$$

For n-hairpins, the value of S depends on the number of hot and cold streams and their series–parallel arrangement. In the simplest form, either the cold fluid is divided between the n-hairpins in parallel or the mass flow rate of the hot fluid is equally divided between the n-hairpins.

For one-series hot fluid and n_1-parallel cold streams:

$$S = \frac{1 - P_1}{\left(\dfrac{n_1 R_1}{R_1 - 1}\right) \ln\left[\left(\dfrac{R_1 - 1}{R_1}\right)\left(\dfrac{1}{P_1}\right)^{1/n_1} + \dfrac{1}{R_1}\right]} \tag{6.20}$$

where

$$P_1 = \frac{T_{h_2} - T_{c_1}}{T_{h_1} - T_{c_1}}, \quad R_1 = \frac{T_{h_1} - T_{h_2}}{n_1 (T_{c_2} - T_{c_1})} \tag{6.21}$$

For one-series cold stream and n_2-parallel hot streams:

$$S = \frac{1 - P_2}{\left(\dfrac{n_2}{1 - R_2}\right) \ln\left[(1 - R_2)\left(\dfrac{1}{P_2}\right)^{1/n_2} + R_2\right]} \tag{6.22}$$

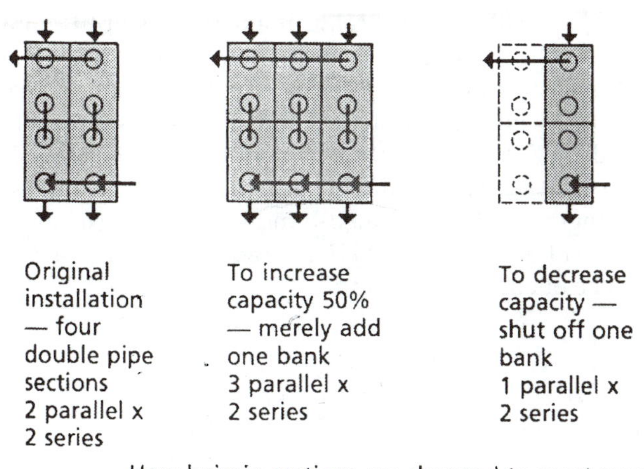
Original installation — four double pipe sections
2 parallel x 2 series
To increase capacity 50% — merely add one bank
3 parallel x 2 series
To decrease capacity — shut off one bank
1 parallel x 2 series
How hairpin sections are changed to meet new requirements

FIGURE 6.5
Parallel–series combinations. (Courtesy of Brown Fintube, Inc.)

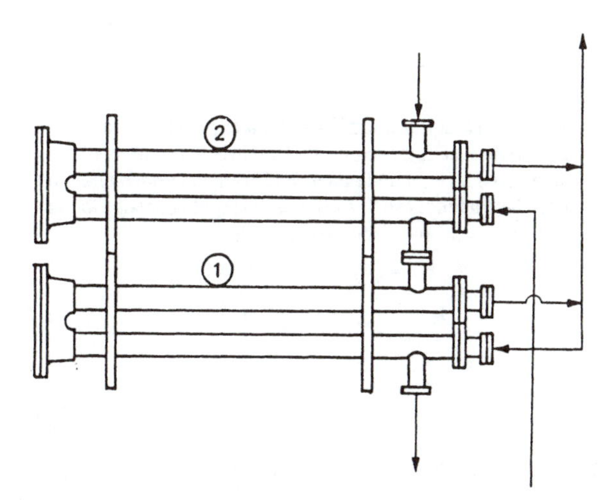
2
1

FIGURE 6.6
Two double-pipe units in series on the annulus (shell) side and parallel on the tube side.

where

$$P_2 = \frac{T_{h_1} - T_{c_2}}{T_{h_1} - T_{c_1}}, \; R_2 = \frac{n_2(T_{h_1} - T_{h_2})}{T_{c_2} - T_{c_1}} \tag{6.23}$$

The preceding relations are derived in Kern.[3]

The total rate of heat transfer can be expressed as:

$$Q = UAS(T_{h_1} - T_{c_1}) \tag{6.24}$$

It is assumed that the parallel fluid stream mass flow is split equally between the n-hairpins, the overall heat transfer coefficient is U, the specific heat of two fluids are constant, and the heat transfer areas of the units are equal. One can calculate the LMTD and apply a correction factor, F, depending on the number of series–parallel combinations; for these, graphs are available.[3,5,10] However when the number of the tube-side parallel path is equal to the number of the shell-side parallel path, then normal LMTD should be used.[10]

In the design of a longitudinal-flow, finned-bundle-type heat exchanger (Figure 6.7), the correction factor, F, must be applied to LMTD. Because the flow is not purely countercurrent, the F factor will be obtained from Figures 2.7 to 2.11.[7,8]

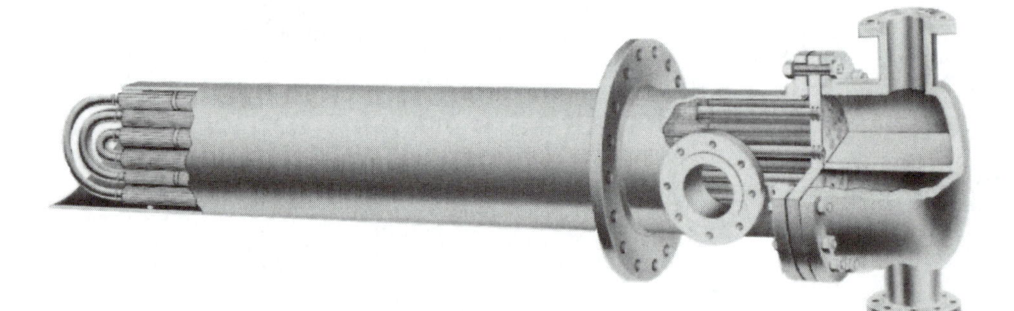

Process Heat Exchangers—Available in shell sizes from 8⅝″ to 36″ O.D. Applications: Used in a wide variety of services on both light and heavy fluids ranging from higher heat transfer materials such as gases and light gasolines to heavy lube oil stocks and viscous chemicals.

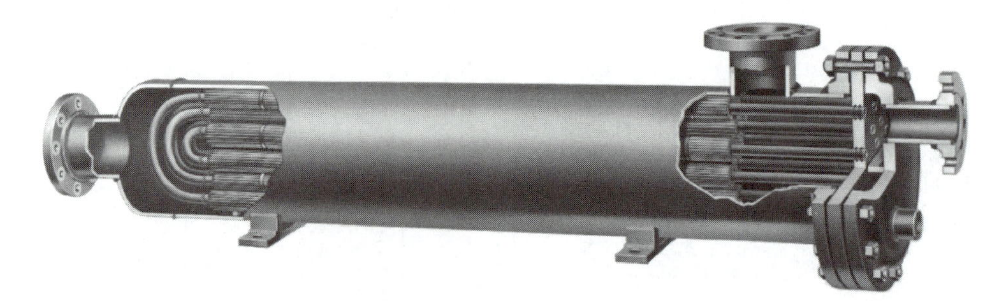

Tank Suction and Line Heaters—Available in shell sizes from 8⅝″ to 48″ O.D. Applications: For heating viscous fluids as they are removed from storage or pumped over long distances. The suction heater is installed with its open end extending into the tank, and only the fluid removed is raised to the desired temperature. Line heaters can be used on either the suction or the pressure side of pumps depending upon design criteria.

FIGURE 6.7
The longitudinally finned-bundle exchanger. (Courtesy of Brown Fintube, Inc.)

6.5 Total Pressure Drop

The total pressure drop in a heat exchanger includes fluid frictional pressure drop, entrance and exit pressure drops, static head, and momentum change pressure drop.

The frictional pressure drop is calculated from Equation (6.1) or Equation (6.2):

$$\Delta p_f = 4f\left(\frac{2L}{D_h}\right)\rho\left(\frac{u_m^2}{2}\right)N_p \tag{6.25}$$

or

$$\Delta p_f = 4f\left(\frac{2L}{D_h}\right)\left(\frac{G^2}{2\rho}\right)N_p \tag{6.26}$$

If the equivalent straight length of the U-bend is known, this should be added to L in tube-side pressure drop ($D_h = d_i$) calculations. The correlations for friction factor for smooth tubes are given in Table 4.1 and for commercial tubes in Figure 4.1. It should be noted that if there are large temperature differences between the wall and the fluid, the effect of property variations on friction coefficient should be included using Equations (3.21b) or (3.22b).[7,9] Friction factors for helical, spirals, and bends are discussed in Chapter 4.

There will be pressure losses through the inlet and outlet nozzles that can be estimated from:

$$\Delta p_n = K_c \frac{\rho u_m^2}{2} \tag{6.27}$$

where $\rho u_m^2/2$ is the dynamic head, $K_c = 1.0$ at the inlet, and $K_c = 0.5$ at the outlet nozzle.[4]

The static head is calculated as:

$$\Delta p_f = \rho \Delta H \tag{6.28}$$

For horizontal hairpins, ΔH is the total difference in elevation between the inlet and outlet nozzles.

For fully developed conditions, the momentum–change pressure drop is calculated from:

$$\Delta p_m = G^2\left(\frac{1}{\rho_{out}} - \frac{1}{\rho_{in}}\right) \tag{6.29}$$

It is important to note that available (or permissible) pressure drop is usually a dictating factor in heat exchanger design.

In longitudinally finned-tube heat exchangers, the heat transfer coefficient can be enhanced by a "cut-and-twist" at regular intervals. The fins are radially cut from the edge of the root, and the leading edge is twisted. This has a mixing effect and is normally accounted for by using the cut-and-twist pitch as the developed length in the film coefficient calculation.[10] The optimum cut-and-twist pitch lies between 300 and 1000 mm. In this case, the pressure drop will increase. The cut-and-twist loss is accounted for by multiplying the frictional pressure drop by a factor, F_{ct}, defined by the following empirical relationship:

$$F_{ct} = 1.58 - 0.525 L_{ct} \tag{6.30}$$

where L_{ct} is the cut-and-twist length in meters. The optimum value of L_{ct} is about 0.3 to 1 m. If cut-and-twist is provided, then in calculating heat transfer coefficient in laminar flow $L = L_{ct}$, which resulted in enhancement of heat transfer.

6.6 Design and Operational Features

As explained in the previous sections, in hairpin exchangers, two double pipes are joined at one end by a U-bend welded to the inner pipes and a return bend housing on the annulus (shell) side, which has a removable cover to allow withdrawal of the inner pipes (Figures 6.3 and 6.8).

Double-pipe heat exchangers have four key design components:

1. Shell nozzles
2. Tube nozzles
3. Shell-to-tube closure
4. Return bend housing

The opposite site of the inner tubes has a closure arrangements. Return bend housing, cover plate, and shell-to-tube closure arrangements are shown in Figure 6.8.

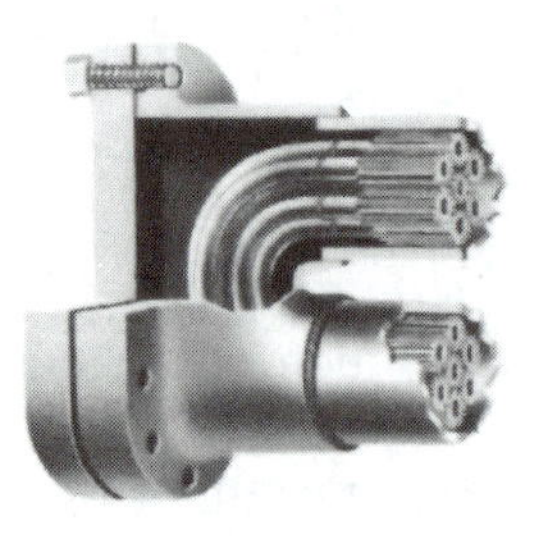

Return bend housing and cover plate

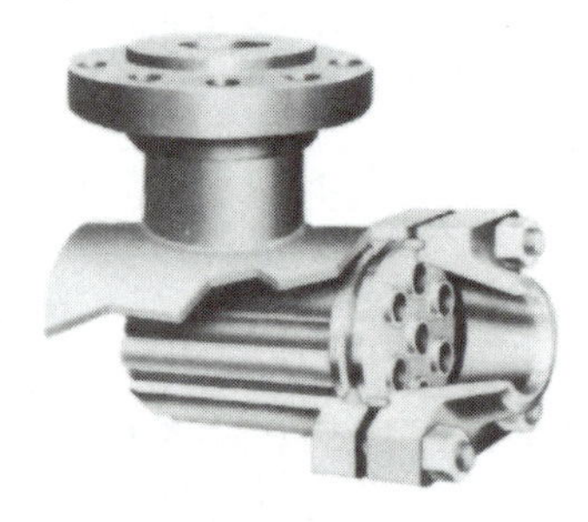

4-bolt standard shell to tube closure and tubeside joint with ASA Lap Joint shell flange (Pressures to 500 psig)

FIGURE 6.8
(a) Return bend housing and the cover plate; (b) shell to tube closure and tube-side joint. (Courtesy of Brown Fintube, Inc.)

The inner tube can be removed from the tail end of the unit for cleaning and maintenance.

The longitudinally finned-bundle exchanger shown in Figure 6.7 is another version of the double-pipe heat exchanger. It is composed of a bundle of finned tubes connected by U-bends and mounted in a shell. This configuration provides a single shell pass and two tube passes. It is used to heat viscous liquids, such as asphalt and fuel oils to reduce viscosity for easier pumping.

The fins are welded onto the inner pipe using an electric resistance Slam–Welding technique. This technique is suitable for carbon steel and stainless steel. Combination of materials can also be used such as a stainless steel inner pipe with carbon steel fins. Fins made from brass or similar materials can be soldered to copper, nickel, admiralty brass, or aluminum brass tubes. The fins are normally 0.5 mm thick for fin height up to 12.7 mm, and 0.8 mm thick for fin height over 12.7 mm.

Double-pipe heat exchangers are light and easily handled with a minimum of lifting equipment. Multiple units can be easily bolted together. Annulus sides connected in series only need a connecting gasket, and the tube side uses simple return bend.

The typical geometries of double-pipe heat exchangers are given in Table 6.1.[10]

Double-pipe heat exchangers are easier to clean in case of fouling buildup. They are flexible and have a low cost of installation. Simple constructions, easily reached bolting, lightweight fin–tube elements, and a minimum number of parts all contribute to minimizing costs.

Nomenclature

A	heat transfer surface area, m^2
A_c	net free flow area, m^2
c_p	specific heat at constant pressure, J/kg·K
D_e	equivalent diameter for heat transfer, m
D_h	hydraulic diameter for pressure drop, m
d	diameter of the inner tube, m
D	diameter of the annulus, m
F	correction factor for LMTD
f	Fanning friction factor
G	mean mass velocity, $kg/m^2 \cdot s$
g	acceleration due to gravity, 9.81 m/s^2
H	difference in elevation between inlet and outlet nozzles, m
h	heat transfer coefficient, $W/m^2 \cdot K$
H_f	height of the fin, m
K_c	velocity-head coefficient
k	thermal conductivity, W/m·K
L	nominal length of exchanger section, also length of longitudinal fin, m
L_{ct}	cut-and-twist length, m
$\dot{m}$	mass flow rate, kg/s
N_f	number of fins per tube

N_{hp} number of exchanger section hairpins

N_t number of tubes in one leg of hairpin exchanger; number of tube holes in tube sheet of bundle exchanger (twice the number of U-bends), or number of U-tubes in bundle exchanger

n_1, n_2 number of parallel cold and hot streams; Equations (6.20) and (6.22)

Nu Nusselt number, hd_i/k or hD_e/k

P perimeter, m

P_1, P_2 temperature ratios defined by Equations (6.21) and (6.23)

Δp pressure drop, kPa

Pr Prandtl number, $\mu c_p/k$

R thermal resistance, $m^2 \cdot K/W$

R_1, R_2 temperature ratios defined by Equations (6.21) and (6.23)

Re Reynolds number, GD_h/μ or Gd_i/μ

Q heat load or duty, W

S dimensionless parameter for true temperature difference of series–parallel arrangement defined by Equations (6.18) and (6.19)

T absolute temperature,°C, K

ΔT temperature difference,°C, K

tanh hyperbolic tangent

U_o overall heat-transfer coefficient based on total external surface area, $W/m^2 \cdot K$

u_m mean stream velocity, m/s

Greek Symbols

η_f fin efficiency

η_o overall surface efficiency

η_p pump efficiency

μ dynamic viscosity, Pa

ρ mass density, kg/m^3

δ fin thickness, m

Subscripts

b bulk

c cold, cross section

e equivalent

f fins, fin side, or frictional

h hot, heat transfer

h hydraulic

hp hairpins

i inside of tubes

TABLE 6.1

Typical Double-Pipe Geometries with One-Finned Tube

Shell IPS (mm O.D.)	Tube O.D. (mm)	Tube Wall (mm)	Fin Height (mm)	No. of Fins (max.)	NFA (mm^2)	D_h (mm)	$\frac{A_f}{A_o}$	$\frac{A_o}{A_i}$	Surface Area (m^2/m)
Sch. 40 Shell (Standard Pressure Units)									
2 in. (60)	25.4	2.77	11.1	20	1445	8.41	0.85	8.40	1.05
3 in. (89)	25.4	2.77	23.8	20	3826	12.0	0.92	16.5	2.07
3 in. (89)	48.3	3.68	12.7	36	2510	7.65	0.86	8.3	2.13
3.5 in. (102)	48.3	3.68	19.05	36	3916	8.66	0.90	11.86	3.05
3.5 in. (102)	60.3	3.91	12.7	40	3039	8.18	0.84	7.31	2.41
4 in. (114)	48.3	3.68	25.4	36	5548	9.63	0.92	15.42	3.96
4 in. (114)	60.3	3.91	19.05	40	4652	9.14	0.89	10.39	3.43
4 in. (114)	73.0	5.16	12.7	48	3452	7.80	0.84	7.35	2.90
Sch. 80 Shell (High-Pressure Units)									
2 in. (60)	25.4	2.77	9.53	20	1228	8.47	0.83	7.39	0.92
3 in. (89)	48.3	5.08	11.1	36	2050	6.93	0.84	7.95	1.90
3.5 in. (102)	48.3	5.08	15.9	36	3368	8.64	0.88	10.82	2.59
3.5 in. (102)	60.3	5.54	11.1	40	2447	7.29	0.82	6.97	2.16
4 in. (114)	48.3	5.08	22.2	36	4851	9.42	0.91	14.64	3.50
4 in. (114)	60.3	5.54	15.9	40	3947	8.99	0.87	9.43	2.92
4 in. (114)	73.0	7.01	9.53	48	2787	7.70	0.80	6.17	2.29

Note: Fin thickness is 0.889 mm (i.e., weldable steels and alloys); surface area per meter is total for both legs.
From Guy, A. R. (1983) In *Heat Exchanger Design Handbook*, Chapter 3.2. Hemisphere, New York. With permission.

in	inlet
m	momentum
n	nozzles
o	outside of bare tubes
out	outlet
s	shell side or fin side
s-p	series–parallel arrangement
t	total
u	unfinned
w	wall, wetted

References

1. Delorenzo, N. and Anderson, E. D. (1945) Heat transfer and pressure drop of liquids in double pipe fintube exchangers. *Trans. ASME*, Vol. 67, 697.
2. Kakaç, S. and Yener, Y. (1993) *Heat Conduction*, 3rd ed., Taylor & Francis, New York.
3. Kern, D. Q. (1950) *Process Heat Transfer*, McGraw-Hill, New York.
4. Fraas, A. P. (1989) *Heat Exchanger Design*, 2nd ed., Wiley, New York.
5. Hewitt, G. F., Shires, G. L., and Bott, T. R. (1994) *Process Heat Transfer*, CRC Press, Boca Raton, FL.
6. Cherrmisinoff, N. P. (Ed.) (1986) *Handbook of Heat and Mass Transfer*, Vol. 1, Chapter 28, pp. 889–903. McGraw-Hill, New York.
7. Kakaç, S. (Ed.) (1991) *Boilers, Evaporators and Condensers*, Chapter 2. Wiley, New York.
8. *Standards of Tubular Exchanger Manufacturers Association* (1988), TEMA, Tarrytown, NewYork.
9. Kakaç, S., Shah, R. K., and Aung, W. (1987) *Handbook of Single-Phase Convective Heat Transfer*, Wiley, New York.
10. Guy, A. R. (1983) Double-pipe heat exchangers. In *Heat Exchanger Design Handbook*, E. U. Schlünder (Ed.), Chapter 3.2. Hemisphere, New York.

Problems

6.1. A counterflow double-pipe heat exchanger is used to cool the lubricating oil for a large industrial gas turbine engine. The flow rate of cooling water through the inner tube is $m_c = 0.2$ kg/s, while the flow rate of oil through the outer annulus is $m_h = 0.4$ kg/s. The oil and water enter at temperatures of 60 and 30°C, respectively. The heat transfer coefficient in the annulus is calculated to be 15 W/m²·K. The I.D. of the tube is 25 mm and the I.D. of the outer annulus is 45 mm. The outlet temperature of the oil is 40°C. Take $c_p = 4178$ J/kg·K for water and $c_p = 2006$ J/kg·K for oil. The tube wall resistance and the curvature of the wall are neglected. Calculate the length of the heat exchanger if fouling is neglected.

6.2. In Problem 6.1, the heat transfer coefficient in the annulus is given as 15 W/m²·K. Is this value acceptable? Assume that the water used is city water with a fouling resistance of 0.000176 m²·K/W inside the tube. The oil-side deposit is neglected.

If the length of the hairpin used is 4 m, calculate the heat transfer area and the number of hairpins. If the space to accommodate this heat exchanger is limited, what are the alternatives?

6.3. Engine oil (raffinate) with a flow rate of 5 kg/s will be cooled from 60 to 40°C by seawater at 20°C in a double-pipe heat exchanger. The water flows through the inner tube whose outlet is heated to 30°C. The inner tube O.D. and I.D. are d_o = 1.315 in. (= 0.3340 m) and d_i = 1.049 in. (= 0.02664 m), respectively. For the annulus, D_o= 4.5 in. (= 0.1143 m) and D_i = 4.206 in. (= 0.10226 m). The length of the hairpin is fixed at 3 m. The wall temperature is 35°C. The number of tubes in the annulus is 3. The thermal conductivity of the tube wall is 43 W/m·K.

a. Calculate the heat transfer coefficient in the annulus.

b. Calculate the overall heat transfer coefficient.

c. Calculate the pressure drop in the annulus and in the inner tube (only straightsections will be considered).

d. What is your decision as an engineer?

6.4. The objective of this problem is to design an oil cooler with seawater. The decision was made to use a hairpin heat exchanger.

Fluid	Annulus Fluid — Engine Oil	Tube-Side Fluid — Seawater
Flow rate, kg/s	4	—
Inlet temperature, °C	65	20
Outlet temperature, °C	55	30
Density, kg/m³	885.27	1013.4
Specific heat, kJ/kg· °C	1.902	4.004
Viscosity, kg/m·s	0.075	9.64×10^{-4}
Prandtl number (*Pr*)	1050	6.29
Thermal conductivity, W/m·K	0.1442	0.6389

Length of the hairpin	= 3 m
Annulus nominal diameter	= 2 in.
Nominal diameter of the inner tube	= 3/4 in.
Fin height, H	= 0.00127 m
Fin thickness, δ	= 0.9 mm
Number of fins and tubes	= 18
Material throughout	= carbon steel
Thermal conductivity, k	= 52 W/m·K
Number of tubes inside the annulus	= 3

Select the proper fouling factors. Calculate:

a. Velocities in the tube and in the annulus

b. Overall heat transfer coefficient for clean and fouled heat exchanger

c. Total heat transfer area of the heat exchanger with and without fouling; *OS* design

d. Surface area of a hairpin and number of hairpins

e. Pressure drop inside the tube and in the annulus

f. Pumping powers for both streams

6.5. Repeat Problem 6.4 with unfinned inner tubes. Write your comments.

6.6. A double-pipe heat exchanger is designed as an engine oil cooler. The flow rate of oil is 5 kg/s and it will be cooled from 60 to 40°C through annulus (I.D. = 0.10226 m, O.D. = 0.1143 m). Seawater flows through the tubes (I.D. = 0.02664 m, O.D. = 0.03340 m) and is heated from 10 to 30°C. The number of bare tubes in the annulus is 3 and the length of the hairpin is 3 m. Assume that the tube wall temperature is 35°C. Design calculations give the number of hairpins as 85. Allowable pressure drop in the heat exchanger for both stream is 20 $lb_f/in.^2$. Is this design acceptable? Outline your comments.

6.7. Assume that the mass flow rate of oil in Problem 6 is doubled (10 kg/s) that may result in an unacceptable pressure drop. It is decided that two hairpins will be used in which the hot fluid flows in parallel between two units and the cold fluid flows in series. Calculate:

a. Pressure drop in the annulus for series arrangements for both streams
b. Pressure drop in the annulus if the hot fluid is split equally and flows in parallel between these two units
c. Heat transfer area and length of the heat exchanger

6.8. Seawater at 30°C flows on the inside of a 25-mm I.D. steel tube with a 0.8-mm wall thickness at a flow rate of 0.4 kg/s. The tube forms the inside of a double-pipe heat exchanger. Engine oil (refined lubricating oil) at 100°C flows in the annular space with a flow rate of 0.1 kg/s. The outlet temperature of the oil is 60°C. The material is carbon steel. The I.D. of the outer tube is 45 m and the I.D. and O.D. of the inner tube are 25 and 27 mm, respectively.

Calculate:

a. Heat transfer coefficient in the annulus
b. Heat transfer coefficient inside the tube
c. Overall heat transfer coefficient with fouling
d. Area of the heat exchanger; and by assuming the length of a hairpin to be 4 m, the number of hairpins
e. Pressure drops and pumping powers for both streams

6.9. A counterflow double-pipe heat exchanger is used to cool the lubricating oil for a large industrial gas turbine engine. The flow rate of oil through the annulus is m_h = 0.4 kg/s. The oil and water enter at temperatures of 60 and 30°C, respectively. The heat transfer coefficient in the inner tube is calculated to be 3000 $W/m^2 \cdot K$. The I.D. of the tube is 25 mm and the I.D. of the outer tube is 45 mm. The outlet temperature of oil and water are 40 and 50°C, respectively. Take cp = 4178 J/kg·K for water and c_p = 20000 J/kg·K for oil. The tube wall resistance and the curvature of the wall are neglected.

Assume the length of the double-pipe heat exchanger as 4 m.

Calculate:

a. Heat transfer coefficient in the annulus.
b. Heat transfer area and the number of hairpins.
c. Because of the space limitation, assume that maximum number of hairpins should not be more than 10. Is this condition satisfied? If not, suggest a new design for a solution for the use of double-pipe heat exchanger.

6.10. In Example 6.1, assume that the mass flow rate of cold fluid is increased to 20,000 kg/h. The remaining geometric parameter and the process specifications remain the same. In this case pressure drop will increase. Assume that the allowable pressure drop in the annulus is 10 kPa. The velocity of cold water through the annulus will be around 3 m/s, which would require a large pressure drop to drive the cold fluid through the inner tube. The pressure drop could be reduced by using several units with cold water flowing in four parallel units and hot water flowing through the inner tube in series.

Calculate:

a. Number of hairpins

b. Pressure drop in the annulus

6.11. A sugar solution (ρ = 1080 kg/m^3, C_p = 3601 J/kg·K, k_f = 0.5764 W/m·K, μ = 1.3×10^{-3} N·s/m^2) flows at rate of 2 kg/s and is to be heated from 25 to 50°C. Water at 95°C is available at a flow rate of 1.5 kg/s (C_p = 40004 J/kg·K). Sugar solution flows through the annulus. Design a double-pipe heat exchanger for this process.

The following geometric parameters are proposed:

Length of the hairpin = 3 m
D_o (annulus) = 6 cm
D_i (annulus) = 5.25 cm
d_o (Inner tube) = 2.6 cm
d_i (Inner tube) = 2.09 cm
Number of tubes inside the annulus = 1
Number of the fins = 20
Fin height = 0.0127 m
Fin thickness = 0.0009 m
Thermal conductivity of the material, k = 52 W/m·K

a. Calculate the hydraulic diameter of the annulus for pressure drop analysis.

b. Calculate the equivalent diameter for heat transfer analysis.

c. Assume that D_e = 0.0009 m and D_h = 0.0075 m, overall surface efficiency = 0.90, and no fouling. The tube-side heat transfer coefficient for water is 4000 W/m^2·K. Calculate the overall heat transfer coefficient.

d. Calculate the heat transfer surface area of this heat exchanger and number of hairpins.

Design Project 6.1

Oil Cooler for Finned-Tube Double-Pipe Heat Exchanger

The objective of this example is to design an oil cooler with sea water. The decision was made to use a hairpin heat exchanger.

Fluid	Annulus Fluid—Engine Oil	Tube-Side Fluid—Seawater
Flow rate, kg/s	5	—
Inlet temperature, °C	65	20
Outlet temperature, °C	55	30
Density, kg/m^3	885.27	1013.4
Specific heat, kJ/kg· °C	1.902	4.004
Viscosity, kg/m·s	0.075	9.64×10^{-4}
Thermal conductivity, W/m·K	0.1442	0.639
Prandtl number (*Pr*)	1050	6.29

The length of the hairpin = 3 m
Annulus nominal diameter = 2 in.
Nominal diameter of inner tube = 3/4 in.
Fin height H_f = 0.0127 m
Fin thickness δ = 0.9 mm
Number of fins = 30
Material throughout = carbon steel
Thermal conductivity, k = 52 W/m·K
Number of tubes inside annulus (varied, starting with $N_t = 1$)

Select proper fouling factors. The geometric information is provided to initiate the analysis. These will be taken as variable parameters to come up with a suitable design; one can start with one inner tube and complete the hand calculations then by checking geometric parameters to find the effects of these changes on the design. Cost analysis of the selected design will be made The final report will include material selection, mechanical design, and technical drawings of the components and the assembly. Calculate:

a. Velocities in the tube and in the annulus
b. Overall heat transfer coefficient for clean and fouled heat exchanger
c. Total heat transfer area of the heat exchanger with and without fouling; *OS* design
d. Surface area of a hairpin and number of hairpins
e. Pressure drop inside the tube and in the annulus
f. Pumping powers for both streams
g. Technical drawings

Design Project 6.2

Oil Cooler with Bare Inner Tube

The following parameters are fixed.

Fluid	SAE-30 Oil	Sea Water
Inlet temperature,°C	65	20
Outlet temperature,°C	55	30
Pressure drop limitations, kPa	140	5
Total mass flow rate, kg/s	2.5	—

Design assumptions for hand calculations follow:

Assume length of heat exchanger is 3-m long. Assume nominal pipe sizes of $3\,{}^1/_2$ and 2 in. as outer and inner tubes, respectively. In the design the geometric parameters must be taken as variable to come up with an acceptable design. In the hand calculation for the given data, the steps indicated in Design Project 6.1 may be followed.

7

Design Correlations for Condensers and Evaporators

7.1 Introduction

Condensation and evaporation heat transfer occurs in many engineering applications. Power condensers, boilers, and steam generators are important components in conventional and nuclear power stations. Condensers are used in process industries. Evaporators and condensers are also an essential part of vapor–compression refrigeration cycles. In the design of these condensers and evaporators, proper correlations must be selected to calculate the condensing and boiling-side heat transfer coefficients.

When a saturated vapor comes in contact with a surface at a lower temperature, condensation occurs.

When evaporation occurs at a solid–liquid interface, it is named boiling. The boiling process occurs when the temperature of the surface, T_w, exceeds the saturation temperature, T_{sat}, corresponding to the liquid pressure.

7.2 Condensation

The most common type of condensation involved in heat exchangers is surface condensation where a cooled wall, at a temperature less than the local saturation temperature of the vapor, is placed in contact with the vapor. In this situation, the vapor molecules that strike the cold surface may stick to it and condense into liquid. The resulting liquid (i.e., condensate) will accumulate in one of two ways. If the liquid "wets" the cold surface, the condensate will form a continuous film, and this mode of condensation is referred to as *filmwise condensation.* If the liquid does not wet the cold surface, it will form into numerous microscopic droplets. This mode of condensation is referred to as *dropwise condensation* and results in much larger heat transfer coefficients than during filmwise conditions. Because long-term dropwise condensation conditions are very difficult to sustain, all surface condensers today are designed to operate in the filmwise mode. More information on the fundamentals of condensation is given in References 1 and 2.

7.3 Film Condensation on a Single Horizontal Tube

7.3.1 Laminar Film Condensation

Nusselt[3] treated the case of laminar film condensation of a quiescent vapor on an isothermal horizontal tube as depicted in Figure 7.1. In this situation, the motion of the condensate is determined by *a balance of gravitational and viscous forces*. In the Nusselt analysis, convection terms in the energy equation are neglected; thus the local heat transfer coefficient around the tube can be written simply as:

$$h(\phi) = \frac{k_l}{\delta(\phi)} \tag{7.1}$$

Here δ is the film thickness that is a function of the circumferential angle ϕ. Clearly at the top of the tube, where the film thickness is a minimum, the heat transfer coefficient is maximum but falls to zero as the film thickness increases to infinity. The Nusselt theory yields the following *average heat transfer coefficient*:[3]

$$\frac{h_m d}{k_l} = 0.728\left[\frac{\rho_l(\rho_l - \rho_g)g i_{lg} d^3}{\mu_l(T_{sat} - T_w)k_l}\right]^{1/4} \tag{7.2}$$

Example 7.1

Quiescent refrigerant 134-A vapor at a saturation temperature of 47°C condenses on a horizontal smooth copper tube whose outside wall temperature is maintained constant at 40°C. The outside tube diameter is 19 mm. Calculate the average condensation heat transfer coefficient on the tube.

SOLUTION

The *average heat transfer coefficient* can be calculated using the Nusselt expression, Equation (7.2).

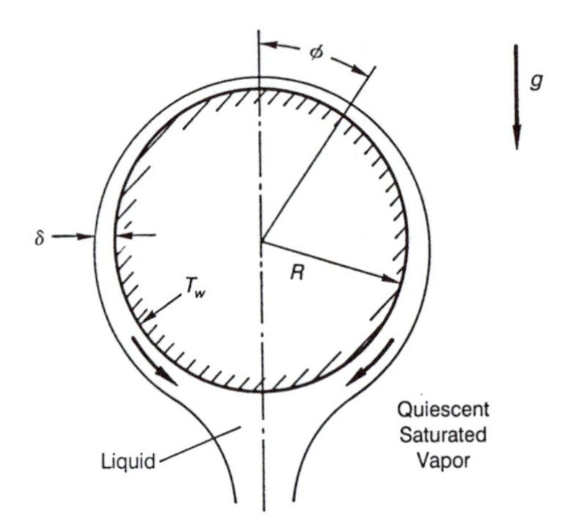

FIGURE 7.1
Film condensation profile on a horizontal tube. (Adapted from Marto, P. J. [1991] In *Boilers, Evaporators and Condensers*, Wiley, New York.)

From Appendix B (Table B.10) the thermophysical properties of R-134a at 47°C are

$$\rho_l = 1117.3 \text{ kg/m}^3$$

$$\rho_g = 62.5 \text{ kg/m}^3$$

$$k_l = 0.068 \text{ W/m}\cdot\text{K}$$

$$\mu_l = 1.72\times10^{-4}\text{Pa}\cdot\text{s}$$

$$i_g = 154.6 \text{ kJ/kg}$$

Substitution of these in Equation (7.2) yields:

$$h_m = 0.728\,\frac{(0.068)}{(0.019)}\left[\frac{(1117.3)(1117.3-62.5)(9.81)(154.6\times10^3)(0.019)^3}{(1.72\times10^{-4})(47-40)(0.068)}\right]^{1/4}$$

$$h_m = 1620.8 \text{ W/m}^2\cdot\text{K}$$

7.3.2 Forced Convection

When the vapor surrounding a horizontal tube is moving at a high velocity, the analysis for film condensation is affected in two important ways: (1) the *surface shear stress* between the vapor and the condensate and (2) the effect of *vapor separation.*

The early analytic investigations of this problem were extensions of Nusselt's analysis to include the interfacial shear boundary condition at the edge of the condensate film. Shekriladze and Gomelauri[4] assumed that the primary contribution to the surface shear was due to the change in momentum across the interface. Their simplified solution for an *isothermal* cylinder without separation and with *no body forces* is

$$Nu_m = \frac{h_m d}{k_l} = 0.9\,\tilde{R}e^{1/2} \tag{7.3}$$

where $\tilde{R}e$ is defined as a two-phase Reynolds number involving the vapor velocity and condensate properties, $u_g d/\nu_l$. When both gravity and velocity are included, they recommend the relationship:

$$\frac{Nu_m}{\tilde{R}e^{1/2}} = 0.64(1+(1+1.69\,F)^{1/2})^{1/2} \tag{7.4}$$

where

$$F = \frac{g d \mu_l i_{lg}}{u_g^2 k_l \Delta T} \tag{7.5}$$

Equation (7.4) neglects vapor separation, which occurs somewhere between 82° and 180° from the stagnation point of the cylinder. After the separation point, the condensate film rapidly thickens and, as a result, heat transfer is deteriorated. A conservative approach suggested by Shekriladze and Gomelauri[4] is to assume that there is no heat transferred beyond the separation point. If the minimum separation angle of 82° is then chosen, a most conservative equation results and the heat transfer decreases by approximately 35%. Therefore, Equation (7.3) reduces to:

$$Nu_m = 0.59\,\tilde{R}e^{1/2} \tag{7.6}$$

An interpolation formula based on this conservative approach, which satisfies the extremes of gravity-controlled and shear-controlled condensation, was proposed by Butterworth[5]:

$$\frac{Nu_m}{\tilde{R}e^{1/2}} = 0.416[1 + (1 + 9.47\,F)^{1/2}]^{1/2} \tag{7.7}$$

Vapor boundary layer effects, especially separation, and the pressure gradient effect around the lower part of the tube provide significant difficulties in arriving at an accurate analytic solution. As a result, approximate, conservative expressions are used. Figure 7.2 compares the predictions of Equations (7.6) and (7.7) along with the Nusselt equation, Equation (7.2), with experimental data from Rose.[6] In general, Equation (7.7) is conservative and can be used with reasonable confidence.

Example 7.2

Suppose that the refrigerant-134A vapor in Example 7.1 were moving downward over the tube at a velocity of 10 m/s instead of being quiescent as originally stated. Calculate the average heat transfer coefficient in this situation.

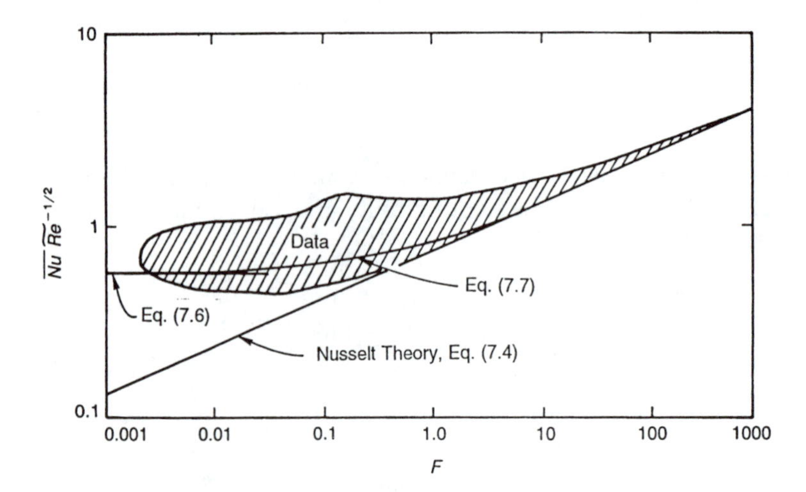

FIGURE 7.2
Condensation in downflow over horizontal tubes. (From Marto, P. J. [1991] In *Boilers, Evaporators and Condensers*, Wiley, New York. With permission.)

SOLUTION

For downward moving vapor over a horizontal tube, the average heat transfer coefficient can be calculated using Equations (7.3), (7.4), and (7.7). First, using Equation (7.7):

$$\frac{h_m d}{k_l} = 0.416(1+(1+9.47\,F)^{1/2})^{1/2}\,\tilde{R}e^{1/2}$$

where F is given by Equation (7.5):

$$F = \frac{g d \mu_l i_{lg}}{u_g^2 k_l \Delta T}$$

On substitution of R-134A properties into Equation (7.5), as listed in Example 7.1, we get:

$$F = \frac{(9.81)(0.019)(1.72\times10^{-4})(1.546\times10^{5})}{(10^2)(0.068)(47-40)} = 0.104$$

Then the two-phase Reynolds number is

$$\tilde{R}e = \frac{\rho_l u_g d}{\mu_l} = \frac{(1117.3)(10)(0.019)}{(1.72\times10^{-4})} = 1.23\times10^{6}$$

When these values for F and $\tilde{R}e$ are substituted into Equation (7.7), this yields:

$$h_m = 0.416\frac{(0.068)}{(0.019)}[1+(1+9.47\times0.104)^{1/2}]^{1/2}(1.23\times10^{6})^{1/2}$$
$$= 2563\ \mathrm{W/m^2\cdot K}$$

which represents a 58% increase over the quiescent vapor case of Example 7.1.

Now recalculate using Equation (7.3):

$$h_m = 0.9\frac{k_l}{d}\tilde{R}e^{1/2}$$
$$= 0.9\frac{(0.068)}{(0.019)}(1.23\times10^{6})^{1/2}$$
$$= 3572\ \mathrm{W/m^2\cdot K}$$

which represents a 120% increase over the quiescent vapor case of Example 7.1 and a 39% higher value than the one from Equation (7.7).

Once again, recalculate by using Equation (7.4):

$$h_m = 0.64\frac{k_l}{d}[1+(1+1.69F^{1/2})]^{1/2}\tilde{Re}^{1/2}$$

$$= 0.64\frac{(0.068)}{(0.019)}[1+(1+1.69\times 0.104)^{1/2}]^{1/2}(1.23\times 10^6)^{1/2}$$

$$= 3629\ \text{W/m}^2\cdot\text{K}$$

which represents a 124% increase over the quiescent vapor case of Example 7.1 and a 42% higher value than the one from Equation (7.7) and a 1.4% higher value than the one from Equation (7.3).

7.4 Film Condensation in Tube Bundles

During film condensation in tube bundles, the conditions are much different than for a single tube. The presence of neighboring tubes creates several added complexities as depicted schematically in Figure 7.3.[1,2] In the idealized case (Figure 7.3a) the condensate from a given tube is assumed to drain by gravity to the lower tubes in a continuous, laminar sheet. In reality, depending on the spacing-to-diameter ratio of the tubes and depending on whether they are arranged in a staggered in-line configuration, the condensate from one tube may not fall on the tube directly below it but instead may flow sideways (Figure 7.3b). Also, it is well known experimentally that condensate does not drain from a horizontal tube in a continuous sheet but in discrete droplets along the tube axis. When these droplets strike the lower tube, considerable splashing can occur (Figure 7.3c), causing ripples and turbulence in the condensate film. Perhaps most important of all, large vapor velocities can create significant shear forces on the condensate, stripping it away, independent of gravity (Figure 7.3d).

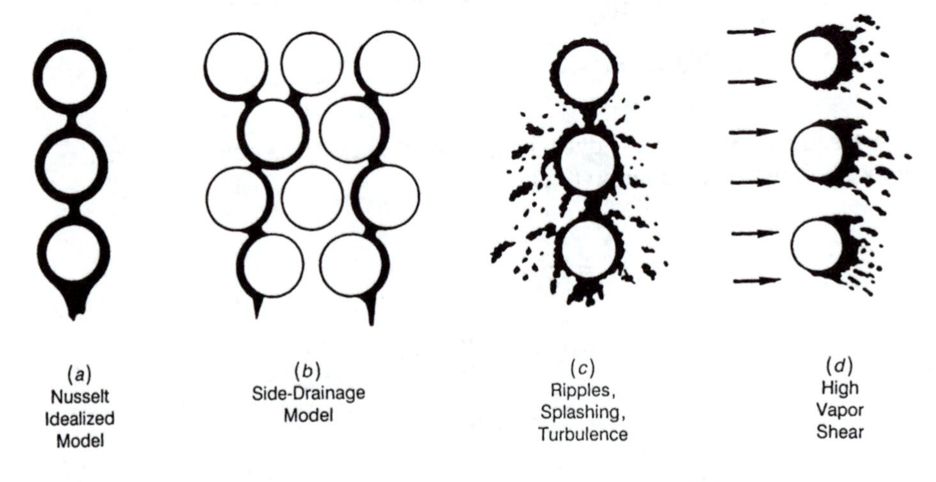

FIGURE 7.3
Schematic representation of condensate flow. (Adapted from Marto, P. J. [1988] In *Two-Phase Flow Heat Exchangers: Thermal–Hydraulic Fundamental and Designs*, pp. 221–291. Kluwer, Dordrecht, The Netherlands.)

7.4.1 Effect of Condensate Inundation

In the absence of vapor velocity, as condensate flows by gravity onto lower tubes in a bundle, the condensate thickness around the lower tubes should increase and the condensation heat transfer coefficient should therefore decrease.

Nusselt[3] extended his analysis to include film condensation on a vertical in-line column of horizontal tubes. He assumed that all the condensate from a given tube drains as a continuous laminar sheet directly onto the top of the tube below it. With this assumption, together with the $(T_{sat} - T_w)$ remaining the same for all tubes, he showed that the average coefficient for a vertical column of N tubes, compared with the coefficient for the first tube (i.e., the top tube in the row), is

$$\frac{h_{m,N}}{h_1} = N^{-1/4} \tag{7.8}$$

In Equation (7.8), h_1 is calculated using Equation (7.2). In terms of the local coefficient for the Nth tube, the Nusselt theory gives:

$$\frac{h_N}{h_1} = N^{3/4} - (N-1)^{3/4} \tag{7.9}$$

Kern[7] proposed a less conservative relationship:

$$\frac{h_{m,N}}{h_1} = N^{-1/6} \tag{7.10}$$

or in terms of the local value:

$$\frac{h_N}{h_1} = N^{5/6} - (N-1)^{5/6} \tag{7.11}$$

Eissenberg[8] experimentally investigated the effects of condensate inundation by using a staggered tube bundle. He postulated a side-drainage model that predicts a less severe effect of inundation:

$$\frac{h_N}{h_1} = 0.60 + 0.42N^{-1/4} \tag{7.12}$$

Numerous experimental measurements have been made in studying the effect of condensate inundation. The data, however, are very scattered. As a result, it is not too surprising that there is no successful theoretical model today that can predict accurately the effect of condensate inundation on the condensation performance for a variety of operating conditions. For design purposes, the Kern expressions either Equation (7.10) or (7.11) are conservative, and have been recommended by Butterworth.[8]

Example 7.3

Suppose that refrigerant-134A in Example 7.1 condenses under quiescent conditions on the shell side of a bundle of 41 tubes. The bundle can be configured in a square, in-line arrangement or in a triangular, staggered arrangement as shown in Figure 7.4. Find the average shell-side coefficient for each of the configurations.

SOLUTION

To find the average heat transfer coefficient for the bundle, correct the Nusselt expression for a single tube; use Equation (7.2) and the Kern relationship in Equation (7.10); then use Equation (7.8) and finally Equation (7.12).

Square, in-line arrangement — With this configuration, there are five columns of seven tubes each and two columns of three tubes each. This arrangement would be approximately equivalent to seven columns of six tubes each. Therefore, $N = 6$.

By using $h_1 = 1620.8$ W/m²·K from Equation (7.2) in Example 7.1, and Equation (7.10) we get:

$$h_{m,N} = h_1 (N)^{-1/6}$$

$$h_{m,6} = h_1 (6)^{-1/6} = 1620.8\,(0.7418)$$

$$= 1202 \text{ W/m}^2 \cdot \text{K}$$

Now by using Equation (7.8) we have:

$$h_{m,N} = h_1 (N)^{-1/4}$$

$$h_{m,6} = h_1 (6)^{-1/4} = 1620.8\,(0.6389)$$

$$= 1036 \text{ W/m}^2 \cdot \text{K}$$

Recalculation by using Equation (7.12) yields:

$$h_{m,N} = h_1 (0.60 + 0.42\,N^{-1/4})$$

$$h_{m,6} = 1620.8\,(0.60 + 0.42(6)^{-1/4})$$

$$= 1407 \text{ W/m}^2 \cdot \text{K}$$

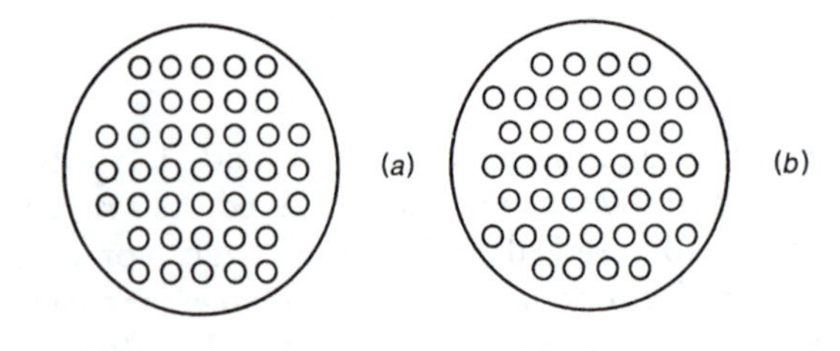

FIGURE 7.4
Tube bundle layout. (a) Square, in-line arrangement; (b) triangular, staggered arrangement. (Adapted from Marto, P. J. [1991] In *Boilers, Evaporators and Condensers*, Wiley, New York.)

Therefore, Equation (7.8) and (7.12) provide the most and least conservative values of $h_{m,6}$, respectively. The value from Equation (7.12) is 36% higher than that of Equation (7.8). The Kern relation gives an intermediate value of $h_{m,6}$.

Triangular, staggered arrangement — With this configuration, assume that the condensate falls straight down and not sideways; and that there are seven columns of three tubes, four columns of four tubes, and two columns of two tubes. This arrangement would be equivalent to approximately 13 columns of three tubes each. Therefore, $N = 3$.

Once again, by using $h_1 = 1620.8$ W/m²·K from Equation (7.2) in Example 7.1 and Equation (7.10), we get:

$$h_{m,N} = h_1(N)^{-1/6}$$

$$h_{m,3} = h_1(3)^{-1/6} = 1620.8\,(0.8327)$$

$$= 1350\ \mathrm{W/m^2\cdot K}$$

By using Equation (7.8) we have:

$$h_{m,N} = h_1(N)^{-1/4}$$

$$h_{m,3} = h_1(3)^{-1/4} = 1620.8\,(0.7598)$$

$$= 1232\ \mathrm{W/m^2\cdot K}$$

Equation (7.12) gives:

$$h_{m,N} = h_1(0.60 + 0.42\,N^{-1/4})$$

$$h_{m,3} = 1620.8\,[0.60 + 0.42(3)^{-1/4}]$$

$$= 1490\ \mathrm{W/m^2\cdot K}$$

Here Equation (7.12) gives a value of $h_{m,3}$ that is 21% greater than Equation (7.8), and once again the Kern relation Equation (7.10) yields an intermediate value of $h_{m,3}$. Therefore, the results of this example show that the staggered arrangement of tubes yields an average heat transfer coefficient that is generally larger than that for in-line arrangement.

One way of preventing inundation of condensate on lower tubes is to incline the tube bundle with respect to horizontal. Shklover and Buevich[9] conducted an experimental investigation of steam condensation in an inclined bundle of tubes. They found that inclination of the bundle increases the average heat transfer coefficient over the horizontal bundle result by as much as 25%. These favorable results led to the design of an inclined bundle condenser with an inclination angle of 5°.

7.4.2 Effect of Vapor Shear

In tube bundles, the influence of vapor shear has been measured by Kutateladze et al.,[10] Fujii et al.,[11] and Cavallini et al.[12] Fujii et al.[11] found that there was little difference between the downward-flow and horizontal-flow data obtained, but the upward-flow data were as much as 50% lower in the range $0.1 < F < 0.5$. What was arrived at is the following empirical expression that correlated the downward-flow and horizontal-flow data reasonably well:

$$\frac{Nu_m}{\tilde{R}e^{1/2}} = 0.96\,F^{1/5} \tag{7.13}$$

for $0.03 < F < 600$. Cavallini et al.[12] compared their data with the prediction of Shekriladze and Gomelauri,[4] Equation (7.4), and found the prediction to be conservative.

In a tube bundle, it is not clear which local velocity should be used to calculate vapor shear effects. Butterworth[5] points out that the use of the maximum cross-sectional area gives a conservative prediction. Shklover and Buevich[9] have used the mean local velocity through the bundle. They calculated this velocity based on a mean flow width that is given by:

$$w = \frac{p_L p_t - \pi d^2/4}{p_L} \tag{7.14}$$

where p_L and p_t are the tube pitches (i.e., centerline-to-centerline distance) in the longitudinal and transverse directions, respectively.

7.4.3 Combined Effects of Inundation and Vapor Shear

Initially, the effects of inundation and vapor shear were treated separately. The combined average heat transfer coefficient for condensation in a tube bundle was simply written as:

$$h_{m,N} = h_1 C_N C_{ug} \tag{7.15}$$

where h_1 represents the average coefficient for a single tube from Nusselt theory, Equation (7.2), and C_N and C_{ug} are correction factors to account for inundation and vapor shear, respectively.

However, in a tube bundle, a strong interaction exists between vapor shear and condensate inundation; and local heat transfer coefficients are very difficult to predict. Butterworth[5] proposed a relationship for the local heat transfer coefficient in the Nth tube row that separates out the effects of vapor shear and condensate inundation. A slightly modified form of his equation is

$$h_N = \left[\frac{1}{2}h_{sh}^2 + \left(\frac{1}{4}h_{sh}^4 + h_1^4\right)^{1/2}\right]^{1/2} \times \left[N^{5/6} - (N-1)^{5/6}\right] \tag{7.16}$$

where h_{sh} is from Equation (7.6):

$$h_{sh} = 0.59\frac{k_l}{d}\tilde{R}e^{1/2}$$

and h_1 is obtained using Equation (7.2).

McNaught[13] has suggested that shell-side condensation may be treated as two-phase forced convection. He therefore proposed the following relationship for the local coefficient for the Nth tube row:

$$h_N = (h_{sh}^2 + h_G^2)^{1/2} \tag{7.17}$$

where h_G is given by Equation (7.11):

$$h_G = h_1[N^{5/6} - (N-1)^{5/6}]$$

and h_{sh} is given as:

$$h_{sh} = 1.26\left[\frac{1}{X_{tt}}\right]^{0.78} h_l \tag{7.18}$$

In Equation (7.18), X_{tt} is the Lockhart-Martinelli parameter, defined as:

$$X_{tt} = \left(\frac{1-x}{x}\right)^{0.9}\left(\frac{\rho_g}{\rho_l}\right)^{0.5}\left(\frac{\mu_l}{\mu_g}\right)^{0.1} \tag{7.19}$$

and h_l is the liquid-phase forced convection heat transfer coefficient across a bank of tubes. This is generally expressed as:

$$h_l = C\frac{k_l}{d} Re_l^m Pr_l^n \tag{7.20}$$

where C, m, and n depend on the flow conditions through the tube bank. One of the correlations given in Chapter 3 can be selected. The numerical values in Equation (7.18) were obtained for steam condensing in a bank of in-line or staggered tubes under the following conditions:[9]

$$\frac{p}{d} = 1.25$$

$$10 \le G \le 70 \text{ kg/(m}^2 \cdot \text{s)}$$

$$0.025 \le x \le 0.8$$

$$0.008 \le X_{tt} \le 0.8$$

The correlation includes the effect of condensate inundation. McNaught[13] found that Equations (7.17) and (7.18) correlated 90% of the steam data within ±25%. Care must be taken to avoid using this correlation when the operating conditions fall outside of the ranges indicated earlier.

Example 7.4

Steam at saturation temperature of 100°C is condensing in a bundle of 320 tubes within a 0.56-m wide duct. The tubes are arranged in a square, in-line pitch ($p = 35.0$ mm) as shown in Figure 7.5. The bundle is made of up to 20 rows of tubes with 3-cm O.D. and with 16 tubes in each row. The tube wall temperature in each row is kept constant at 93°C. The steam flows downward in the bundle, and at the sixth row of tubes the local mass flow rate of vapor is $\dot{m}_g = 14.0$ kg/s.

Find the local heat transfer coefficient for this sixth row of tubes using the method of Butterworth.[5]

SOLUTION

From Appendix B (Table B.2) the thermophysical properties of steam at 100°C:

$$\rho_l = 957.9 \text{ kg/m}^3$$

$$\rho_g = 0.598 \text{ kg/m}^3$$

$$k_l = 0.681 \text{ W/m}\cdot\text{K}$$

$$\mu_l = 2.79 \times 10^{-4} \text{ kg/m}\cdot\text{s}$$

$$\mu_g = 1.2 \times 10^{-5} \text{ kg/m}\cdot\text{s}$$

$$i_{lg} = 2.257 \times 10^3 \text{ kJ/kg}$$

$$c_{pl} = 4.219 \text{ kJ/kg}\cdot\text{K}$$

$$Pr_l = 1.73$$

Calculation of local steam velocity

$$u_g = \frac{\dot{m}_g}{\rho_g A_m}$$

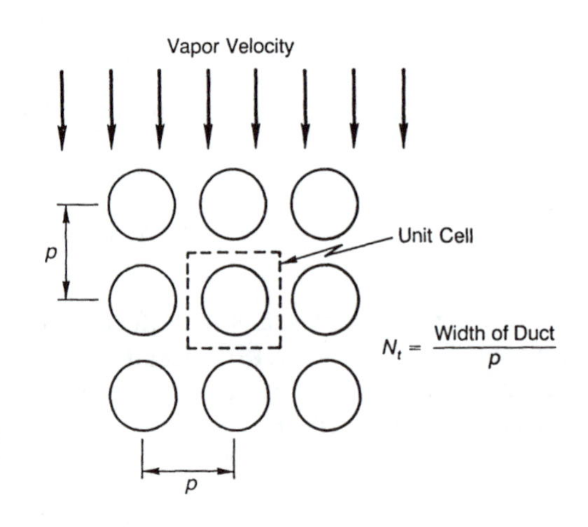

FIGURE 7.5
Schematic of square, in-line tube arrangement. (Drawn after Marto, P. J. [1991] In *Boilers, Evaporators and Condensers*, Wiley, New York.)

where A_m is the mean flow area. This mean flow area can be written in terms of the number of unit cells N_t and the mean width per cell w:

$$A_m = wN_tL$$

where w is given by Equation (7.14):

$$w = \frac{p_L p_t - \pi d^2/4}{p_L}$$

In this example, $p_L = p_t = p = 0.035$ m. Therefore

$$w = \frac{(0.035)^2 - \pi(0.03)^2/4}{0.035} = 0.0148 \text{ m}$$

The mean flow area is then:

$$A_m = (0.0148)(16)(4) = 0.947 \text{ m}^2$$

The local steam velocity is

$$u_g = \frac{(14.0)}{(0.598)(0.947)} = 24.7 \text{ m/s}$$

Method of Butterworth — The local heat transfer coefficient is given by Equation (7.16):

$$h_N = \left[\frac{1}{2}h_{sh}^2 + \left(\frac{1}{4}h_{sh}^4 + h_1^4\right)^{1/2}\right]^{1/2} \times [N^{5/6} - (N-1)^{5/6}]$$

where

$$h_{sh} = 0.59\frac{k_l}{d}\tilde{R}e^{1/2}$$

and

$$\frac{h_m d}{k_l} = 0.728\left[\frac{\rho_l(\rho_l - \rho_g)g i_{lg} d^3}{\mu_l(T_{sat} - T_w)k_l}\right]^{1/4}$$

The two-phase Reynolds number is

$$\tilde{R}e = \frac{\rho_l u_g d}{\mu_l}$$

$$\tilde{Re} = \frac{(957.9)(24.7)(0.03)}{(2.79\times10^{-4})} = 2.54\times10^{6}$$

Therefore

$$h_{sh} = 0.59\frac{(0.681)}{(0.03)}(2.54\times10^{6})^{1/2} = 21,345\ \text{W/m}^2\cdot\text{K}$$

From Equation (7.2):

$$h_1 = 0.728\ \frac{(0.681)}{(0.03)}\left[\frac{(957.9)(957.9-0.598)(9.81)(2.257\times10^{6})(0.03)^3}{(2.79\times10^{-4})(100-93)(0.681)}\right]^{1/4} = 13,241\ \text{W/m}^2\cdot\text{K}$$

Therefore, on substitution into Equation (7.16) with $N = 6$, this yields:

$$h_N = \left\{\frac{1}{2}(21345)^2 + \left[\frac{1}{4}(21345)^4 + (13241)^4\right]^{1/2}\right\}^{1/2} \times\left[6^{5/6} - (6-1)^{5/6}\right] = 14,242\ \text{W/m}^2\cdot\text{K}$$

Numerical modeling must be used to accurately predict the average shell-side heat transfer coefficient in a tube bundle. Marto[14] has reviewed numerical methods to predict shell-side condensation. The vapor flow must be followed throughout the bundle and within the vapor flow lanes. By calculating the local vapor velocity, the local vapor pressure and temperature, and the distribution of condensate from other tubes (and the local concentration of any noncondensable gases), then it is possible to predict a local heat transfer coefficient in the bundle that can be integrated to arrive at the overall bundle performance. Early efforts to model the thermal performance of condensers were limited essentially to one-dimensional routines in the plane perpendicular to the tubes.[15–18] Now, more sophisticated two-dimensional models exist that have been used successfully to study the thermal-hydraulic performance of complex tube bundle geometries.[19–24]

7.5 Condensation Inside Tubes

During film condensation inside a tube, a variety of flow patterns can exist as the flow passes from the tube inlet to the exit. These different flow patterns can alter the heat transfer considerably, so that local heat transfer coefficients must be calculated along the length of the tube. During horizontal in-tube condensation, the transition from annular to stratified flow is most important. Details of in-tube condensation can be found in Reference 25.

Different heat transfer models are used to calculate the heat transfer coefficient depending on whether vapor shear or gravitational forces are more important.

The previously mentioned flow patterns have been studied by numerous investigators in recent years.[25–30] The transition from one flow pattern to another must be predicted to make the necessary heat transfer calculations. Breber et al.[29] have proposed a simple method of predicting flow pattern transitions that depend on the dimensionless mass velocity, j_g^*, defined as:

$$j_g^* = \frac{xG}{[gd\rho_g(\rho_l - \rho_g)]^{1/2}} \tag{7.21}$$

and the Lockhart–Martinelli parameter, X_{tt}, given by Equation (7.19). Their flow pattern criteria are

$$j_g^* > 1.5 \qquad X_{tt} < 1.0 : \text{mist and annular} \tag{7.22a}$$

$$j_g^* < 0.5 \qquad X_{tt} < 1.0 : \text{wavy and stratified} \tag{7.22b}$$

$$j_g^* < 0.5 \qquad X_{tt} > 1.5 : \text{slug} \tag{7.22c}$$

$$j_g^* > 1.5 \qquad X_{tt} < 1.5 : \text{bubble} \tag{7.22d}$$

During condensation inside horizontal tubes, different heat transfer models are used, depending on whether vapor shear or gravitational forces are more important. When the vapor velocity is low (i.e., j_g^* is less than 0.5), flow will be dominated by gravitational forces and stratification of the condensate will occur. At high vapor velocities (i.e., $j_g^* > 1.5$) where interfacial shear forces are large, gravitational forces may be neglected and condensate flow will be annular. When the flow is stratified, the condensate forms a thin film on the top portion of the tube walls. This condensate drains toward the bottom of the tube where a stratified layer exists as shown schematically in Figure 7.6. The stratified layer flows axially due to vapor shear forces. In this circumstance, the Nusselt theory for laminar flow is generally valid over the top, thin-film region of the tube. However, Butterworth[31] points out that if the axial vapor velocity is high, turbulence may occur in this thin film and the Nusselt analysis is no longer valid. In the stratified layer, heat transfer is generally negligible. For laminar flow, the average heat transfer coefficient over the entire perimeter may be expressed by a modified Nusselt result:

$$h_m = \Omega \left\{ \frac{k_l^3 \rho_l (\rho_l - \rho_g) g i_{fg}}{\mu_l d \Delta T} \right\}^{1/4}$$

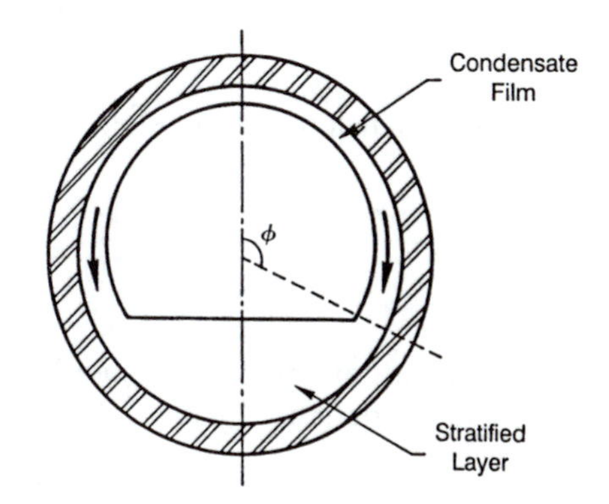

FIGURE 7.6
Idealized condensate profile inside a horizontal tube. (Drawn after Marto, P. J. [1991] In *Boilers, Evaporators and Condensers*, Wiley, New York.)

where the coefficient Ω depends on the fraction of the tube that is stratified. Jaster and Kosky[32] have shown that Ω is related to the void fraction of the vapor α_g :

$$\Omega = 0.728\,\alpha_g^{3/4} \tag{7.23}$$

where

$$\alpha_g = \frac{1}{1+[(1-x)/x](\rho_g/\rho_l)^{2/3}} \tag{7.24}$$

Generally, laminar flow models (based on a Nusselt analysis) predict heat transfer coefficients that are too low, and turbulent flow models must be used. These turbulent flow models either are based on empirical dimensionless correlations or are based on heat transfer-momentum analogy. Some of the models are rather cumbersome to use and result in expressions that are inconvenient for design purposes.

Several correlations available in the literature have been verified for use with refrigerants. Some of these correlations are described in the following discussion. They are the correlation by Traviss et al.,[33] the correlation by Cavallini and Zecchin,[34] and the correlation by Shah.[35] For heat transfer in the superheated and subcooled regions of the condenser, the single-phase correlations discussed earlier during in-tube evaporation are applicable.

The correlation by Traviss et al.[33] was originally developed as part of an extensive study of condensation of R-12 and R-22. The correlation was also verified successfully by using these same R-12 and R-22 data. This in-tube condensation equation was derived by applying the momentum and heat transfer analogy to an annular flow model. The velocity distribution in the annular film was described by the von Karman universal velocity distribution. Radial temperature gradients in the vapor core were neglected, and a saturation temperature was assumed at the liquid–vapor interface. The resulting two-phase heat transfer coefficient is

$$Nu = \frac{h_{TP} d}{k_l} = Pr_l Re_l^{0.9} \frac{F_1(X_{tt})}{F_2(Re_l, Pr_l)} \tag{7.25}$$

where the liquid Reynolds number is

$$Re_l = \frac{G(1-x)d}{\mu_l} \tag{7.26}$$

and the nondimensional parameter, F_1, is

$$F_l = 0.15\left[\frac{1}{X_{tt}} + \frac{2.85}{X_{tt}^{0.476}}\right] \tag{7.27}$$

Three functions are given for F_2, with the choice of which function to use in the correlation being dependent on the Reynolds number range. The three functions are

$$F_2 = 0.707\,Pr_l\,Re_l^{0.5} \quad \text{for } Re_l \le 50 \tag{7.28}$$

$$F_2 = 5\,Pr_l + 5\ \ln\,[1 + Pr_l(0.09636\,Re_l^{0.585} - 1)] \quad \text{for } 50 < Re_l \le 1125 \tag{7.29}$$

$$F_2 = 5\,Pr_l + 5\ \ln\,(1 + 5Pr_l) + 2.5\ \ln\,(0.00313\,Re_l^{0.812}) \quad \text{for } Re_l > 1125 \tag{7.30}$$

where Reynolds number is defined in Equation (7.26).

The preceding correlation has been used extensively in the past. However, two more recent correlations, described next, are simpler to implement; in addition, they have been shown to correlate experimental data just as well.[36]

Cavallini and Zecchin[34] developed a semiempirical equation that is simple in form and correlates refrigerant data quite well. Data for several refrigerants — including R-11, R-12, R-21, R-22, R-113, and R-114 — were used to derive and verify the correlation. The basic form of the correlation was developed from a theoretical analysis similar to that used by Traviss et al.[33]

The working equation suggested by Cavallini and Zecchin is

$$h_{TP} = 0.05\ Re_{eq}^{0.8}\ Pr_l^{0.33}\,\frac{k_l}{d} \tag{7.31}$$

where the equivalent Reynolds number, Re_{eq}, is defined by:

$$Re_{eq} = Re_v \left(\frac{\mu_v}{\mu_l}\right)\left(\frac{\rho_l}{\rho_v}\right)^{0.5} + Re_l \tag{7.32}$$

The equation for Re_l was presented earlier in Equation (7.26), and Re_v is defined similarly as:

$$Re_v = \frac{G\,x\,d}{\mu_v} \tag{7.33}$$

The Cavallini–Zecchin correlation is very similar in form to any one of several single-phase turbulent correlations (e.g., the well-known Dittus–Boelter equation). Cavallini and Zecchin also suggest that their equation can be used to calculate the average heat transfer coefficients between the condenser inlet and outlet, provided that the thermophysical properties and the temperature difference between the wall and fluid do not vary considerably along the tube.

The Shah[35] correlation was developed from a larger group of fluids, including water, than the previous correlations. It was developed by establishing a connection between condensing and Shah's earlier correlations for boiling heat transfer without nucleate boiling. The resulting correlation in terms of h_l is

$$h_{TP} = h_l\left(1 + \frac{3.8}{Z^{0.95}}\right) \tag{7.34}$$

where

$$Z = \left(\frac{1-x}{x}\right)^{0.8} p_r^{0.4} \tag{7.35}$$

and h_l is the liquid only heat transfer coefficient and can be found using the Dittus–Boelter equation:

$$h_l = 0.023\left[\frac{G(1-x)d}{\mu_l}\right]^{0.8}\frac{Pr_l^{0.4}k_l}{d} \tag{7.36}$$

Shah also suggested integrating these equations over a length of tubing to obtain the mean heat transfer coefficient in the condensing region:

$$h_{TP_m} = \frac{1}{L}\int_0^L h_{TP}dL \tag{7.37}$$

For the case of a linear quality variation over a 100 to 0% range, the result is

$$h_{TP_m} = h_l\left(0.55 + \frac{2.09}{p_r^{0.38}}\right) \tag{7.38}$$

The results from this equation differ by only 5% from the value obtained when a mean quality 50% is used in the local heat transfer correlation, Equation (7.34).

The local heat transfer coefficients for the previous three correlations are compared in Figure 7.7.[37] As with the comparisons of in-tube evaporation correlations presented earlier, an ozone-safe refrigerant, namely, R-134a, is used. Flow rate, temperature, tube diameter, and tube length conditions similar to a typical condenser were selected for this comparison. Over most of the quality range, the correlations agree with each other to within 20%. The local heat transfer coefficients decrease as the quality decreases, which is the result of the annular film thickness increasing as condensation proceeds from the high-quality inlet of the condenser to the low-quality exit.

Average heat transfer coefficients, obtained by using Equation (7.37) to integrate the local value over tube length, are plotted as a function of mass flux in Figure 7.8.[37] All three correlations show good agreement; however, two of the correlations agree to within 10% of each other. Average coefficient data from several past experimental studies have been predicted to within ± 20% by these three correlations.[35–37]

7.5.1 Condensation in Vertical Tubes

Condensation heat transfer in vertical tubes depends on the flow direction and its magnitude. For downward-flowing vapor, at low velocities, the condensate flow is controlled by gravity and the heat transfer coefficient may be calculated by Nusselt theory on a vertical surface:

$$Nu_m = \frac{h_m L}{k_l} = 0.943\left\{\frac{\rho_l(\rho_l-\rho_g)g\,i_{lg}L^3}{\mu_l \Delta T\, k_l}\right\} \tag{7.39}$$

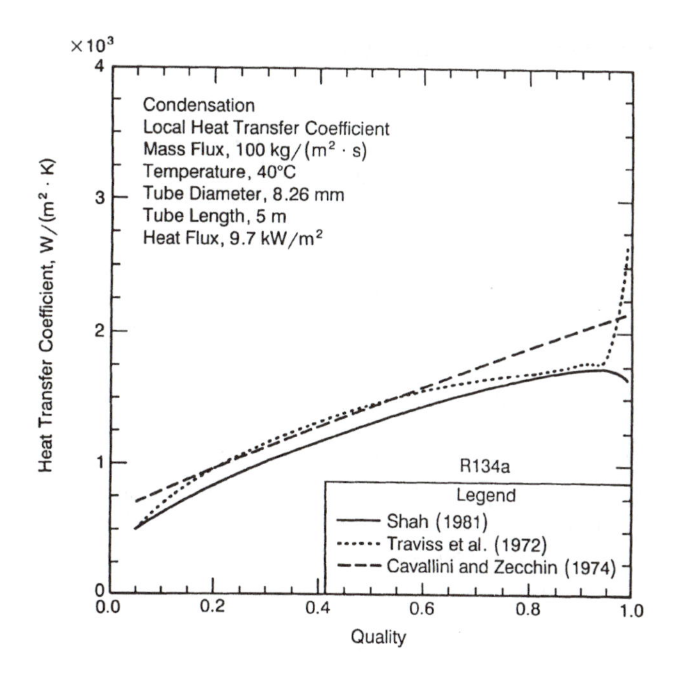

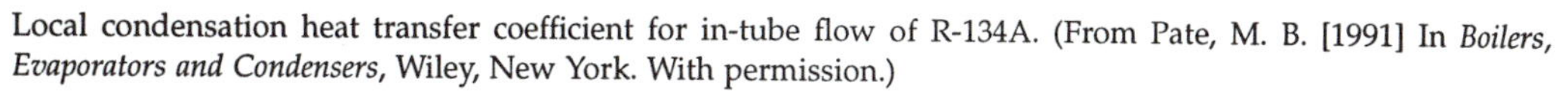

FIGURE 7.7
Local condensation heat transfer coefficient for in-tube flow of R-134A. (From Pate, M. B. [1991] In *Boilers, Evaporators and Condensers*, Wiley, New York. With permission.)

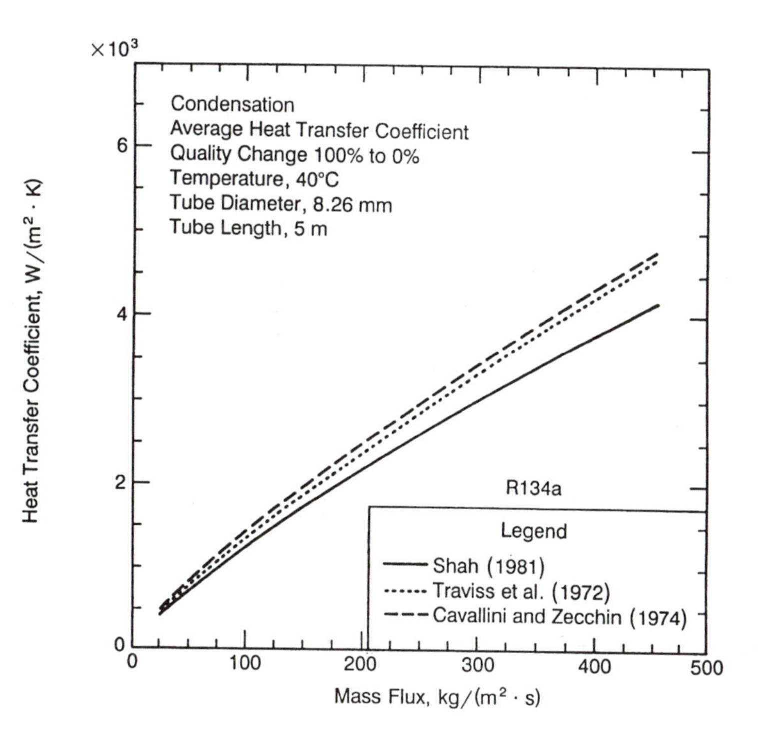

FIGURE 7.8
Local condensation heat transfer coefficient for in-tube flow of R-134A. (From Pate, M. B. [1991] In *Boilers, Evaporators and Condensers*, Wiley, New York. With permission.)

If the condensate film proceeds from laminar, wave-free to wavy conditions, a correction to Equation (7.39) can be applied:

$$\frac{h_{m,c}}{h_m} = 0.8\left(\frac{Re_F}{4}\right)^{0.11} \tag{7.40}$$

where the film Reynolds number $Re_F > 30$. If turbulent conditions exist, then the average heat transfer coefficient can be calculated by one of the methods described by Marto.[1] If the vapor velocity is very high, then the flow is controlled by shear forces and the annular flow models may be used.[1,2]

For upward-flowing vapor, interfacial shear will retard the drainage of condensate. As a result, the condensate film will thicken and the heat transfer coefficient will decrease. In this case, Equation (7.39) may be used with a correction factor of 0.7 to 1.0 applied, depending on the magnitude of the vapor velocity. Care must be exercised to avoid vapor velocities that are high enough to cause "flooding," which occurs when the vapor shear forces prevent the downflow of condensate. One criterion to predict the onset of flooding, due to Wallis,[38] is based on air–water systems:

$$(v_g^*)^{1/2} + (v_l^*)^{1/2} = C \tag{7.41}$$

where

$$v_g^* = \frac{v_g \rho_g^{1/2}}{[g d_i (\rho_l - \rho_g)]^{1/2}} \tag{7.42}$$

$$v_l^* = \frac{v_l \rho_l^{1/2}}{[g d_i (\rho_l - \rho_g)]^{1/2}} \tag{7.43}$$

The velocities v_g and v_l should be calculated at the bottom of the tube (where they are at their maximum values). Wallis[38] determined the parameter C to be 0.725 based on his measurements of air and water. Butterworth[31] suggests that C should be corrected for surface tension and for tube end effects using the relationship:

$$C_2 = 0.53\, F_\sigma F_g \tag{7.44}$$

where F_σ is a correction factor for surface tension and F_g depends on the geometry of the end of the tube.

7.6 Flow Boiling

The study of flow boiling is of fundamental importance in the nuclear, chemical, and petrochemical industries. In the first case the main parameters of concern are the onset of nucleate boiling (ONB), the void fraction in the subcooled and saturated boiling regions that is required for pressure drop, the transient reactor response and instability prediction, and the heat transfer crises and postdryout.

7.6.1 Subcooled Boiling

Consider a heated channel with flowing coolant as shown in Figure 7.9; the subcooled boiling region commences at the point where bubbles begin to grow at the wall.

The bubbles will form on the heated surface, become detached, and pass into the bulk on the liquid where they may condense if the bulk temperature of the liquid is below its boiling temperature. This process of heat transfer is called *subcooled nucleate boiling*. Otherwise it is called *saturated nucleate boiling* or *two-phase convective boiling*. As the amount of superheat, $(T_w - T_s)$, increases, the number of nucleation sites where bubbles grow increases; and consequently there is a rapid increase in the rate of heat transfer. The ONB is indicated on Figure 7.9.[39]

If nucleate boiling has already begun in flow over any form of heated surface, two-phase boiling begins as soon as the bulk of the liquid reaches its boiling temperature. The formation of vapor increases the velocity of flow past the heated surface, thus facilitating the transfer of heat. Therefore, calculation of heat transfer in two-phase convection, in nucleate boiling, and in combination of the two is important; and will be explained by solved examples in the coming sections. It is important to note that at the ONB bubbles merely grow and collapse at, or very close to, the wall. There is, however, an increase in heat transfer coefficient either through simultaneous evaporation and condensation (i.e., the bubbles act as heat pipes) or through the increased bubble agitation from growth and collapse. Later bubbles slide along the wall but do not survive in the bulk liquid. Not until some farther distance downstream (see Figure 7.9) is there any significant net vapor generation (SNVG) resulting in an increase in void fraction as bubbles depart and survive in the bulk liquid.

The superheat for ONB and the corresponding radius are given by Davis and Anderson:[40]

$$\Delta T_{\text{ONB}} = \Delta T_{\text{WONB}} - T_s = \left(\frac{8\sigma T_s q''}{k\rho_v \Delta h_v} \right)^{1/2} \tag{7.45}$$

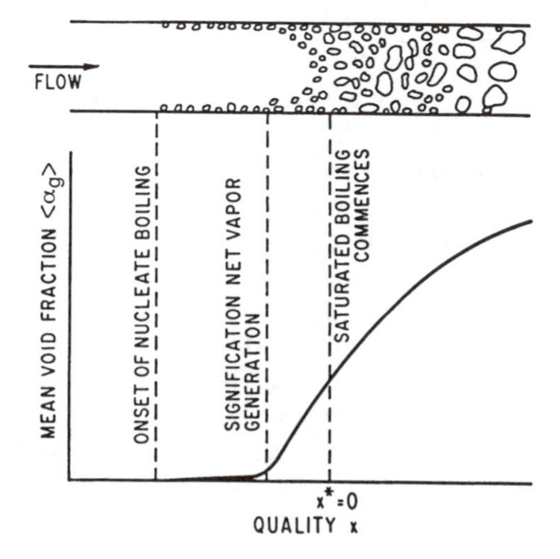

FIGURE 7.9
Variation of void fraction along the heated channel (x^* = thermodynamic equilibrium quality). (From Butterworth, D. and Shock, R. A. [1982] Flow boiling. *Heat Transfer 1982, Proc. 7th Int. Heat Transfer Conf.*, Hemisphere, New York. With permission.)

and the corresponding radius is:

$$r_c = \left(\frac{2k\sigma T_s}{\Delta h_v \rho_v q''} \right)^{1/2} \tag{7.46}$$

Most developments of such equations use the Clausius–Clapeyron equation and differ only in the assumptions used to obtain the solution. Many data are available in the literature and many of these have been compared with Equation (7.45); a detailed literature survey is given in Reference 39. It is found that for water and other nonwetting fluids Equations (7.45) and (7.46) are reliable, whereas ΔT_{ONB} is underpredicted and the corresponding r is overpredicted for well-wetting fluids, which include many organic substances.[40]

Bergles and Rohsenow[41] formulated Equation (7.45) with the properties of water at pressures of 1 bar and obtained the following equation:

$$q'' = 1120 p^{1.156} (1.8\ \Delta T_{ONB})^{2.16/p^{0.0234}} \tag{7.47}$$

which can be used in design. Equation (7.47) has been extensively tested in the range 1 to 138 bar.

7.6.2 Flow Pattern

The local hydrodynamic and heat transfer behavior is related to the distribution of liquid and vapor, referred to as the flow patterns or flow regimes. It is helpful to briefly discuss flow patterns even though experience has been that reasonably accurate correlations for pressure drop and heat transfer coefficient can be obtained without consideration of the flow pattern.

In a traditional once-through boiler the flow enters as subcooled liquid and exits as superheated vapor. Subcooled boiling is observed before the fluid reaches a bulk saturated condition; the flow pattern is *wall bubbly*. The "bubble boundary layer" thickens because of the accumulation of uncondensed vapor, which is promoted by the decreasing condensation potential. The fluid is in a nonequilibrium state, with superheated liquid near the wall and subcooled liquid in the core; and the vapor would condense if the flow were brought to rest and mixed. At some point the bulk enthalpy is at the saturated liquid condition ($x = 0$).

As the vapor volumetric fraction increases, the bubbles agglomerate and *slug flow* is observed (slug flow may also be observed in the subcooled region). Agglomeration of the slug flow bubbles leads to a transition regime termed *churn flow* where the nominally liquid film and the nominally vapor core are in a highly agitated state. The subsequent flow pattern, *annular flow*, has the phases more clearly separated spacewise. However, the film may contain some vapor bubbles and the core may contain liquid in the form of drops or streamers. The film thickness usually varies with time, with a distinct wave motion; and there is an interchange of liquid between the film and core.

There is a gradual depletion of the liquid due to evaporation. At some point before complete evaporation, the wall becomes nominally dry due to net entrainment of the liquid or abrupt disruption of the film. Beyond this dryout condition, *drop flow* prevails. A nonequilibrium condition again occurs, but in this case the vapor becomes superheated to provide the temperature difference required to evaporate the vapor. Eventually, beyond the point where the bulk enthalpy is at the saturated vapor condition ($x = 1$), the liquid evaporates and normal superheated vapor is obtained.

The fluid and wall temperature profiles shown in Figure 7.10[42] pertain to uniform heat flux, as might be approximated in a fired boiler with complete vaporization. The temperature difference between the wall and the fluid is inversely proportional to the heat transfer coefficient. The normal single-phase coefficient is observed near the entrance of the tube, perhaps with an increase right at the inlet due to flow or thermal development. The coefficient increases rapidly as subcooled boiling is initiated because of the intense agitation of the bubbles, but levels off in established boiling over the subcooled and low-quality regions. The coefficient is usually expected to increase in annular flow, because of the thinning of the liquid film. At the dryout point, the coefficient decreases rapidly as a result of the transition from a basically solid–liquid heat transfer to a solid–gas heat transfer. Droplets striking the surface elevate the heat transfer coefficient above what it would be for pure vapor.

Clearly, there are many similarities between flow boiling in vertical and in horizontal tubes. Not surprisingly, therefore, these similarities are reflected in the literature. Indeed, some authors have correlated data for both horizontal and vertical tubes without remarking on the possible differences between the two.

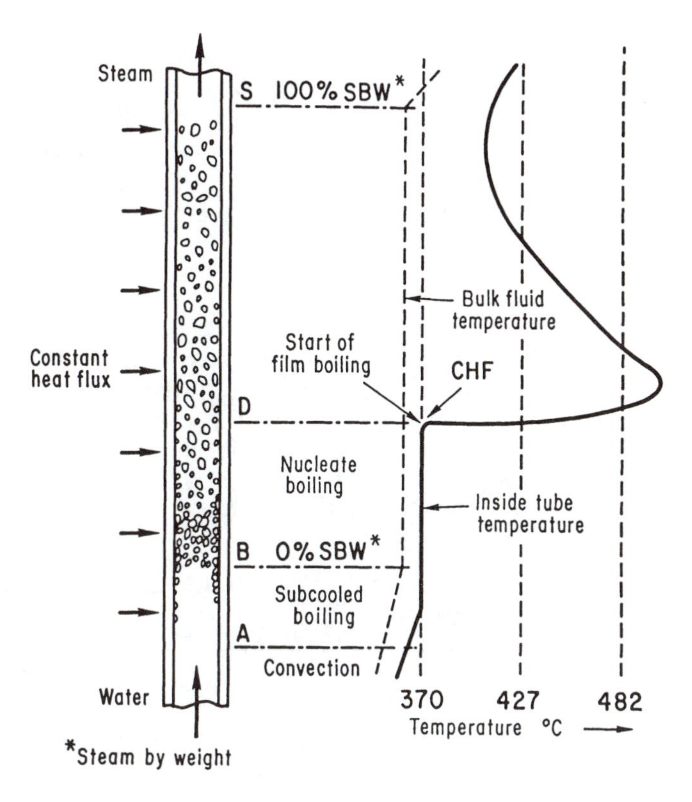

FIGURE 7.10
Temperature profiles in a vertical evaporator tube. (From Kitto, J. B. and Albrecht, M. J. [1988] In *Two-Phase Flow Heat Exchangers: Thermal–Hydraulic Fundamentals and Design*, pp. 221–291. Kluwer, Dordrecht, The Netherlands. With permission.)

However, there is a very important difference between vertical and horizontal tubes, this is, of course, the stratifying effect of gravity. Many experimenters have observed the two-phase flow patterns, and some have tried to interpret the heat transfer results in terms of these observations. Also, a few of the correlators of data have tried to include additional terms in their correlation to account for stratification.

Much of the literature on horizontal flow would, however, suggest to the casual reader that there are very great differences between boiling in vertical and horizontal tubes. This arises because the two topics have tended to be researched by different people with the aim of serving a different industry. The research on vertical tubes has mainly been for the power industry, particularly the nuclear industry, while that on horizontal tubes has mainly been for the refrigeration and air-conditioning industry. Collier[43] suggested that the flow development in a horizontal evaporator channel is as given in Figure 7.11.

During horizontal in-tube boiling, stratification of the flow could occur leaving a region at the top of the tube with a very low local heat transfer coefficient. The coefficient averaged around the circumference would be correspondingly low. The flow pattern inside the horizontal tube will depend on the velocity of flow. The figure shown by Collier would seem to be for a moderately high flow in the tube. With very high flows, one would expect less stratification while with very low flows stratified flow would be the predominant flow pattern, occurring almost directly beyond the point of initial vapor formation. The stratification will reduce the heat transfer coefficient. Therefore, the designer needs to know under what conditions heat transfer deterioration due to stratification may be expected. This topic will be discussed in the following sections with worked examples.

In recent years several correlations have been developed to deal with the different stages of boiling and the effects of two-phase flow. Correlations for in-tube boiling developed by Chen,[44] Shah,[45,46] Güngör and Winterton,[47] and Kandlikar[48,49] will be discussed herein. The full range of boiling conditions for which the correlations apply includes single-phase vapor entering the evaporator and two-phase flow occurring in sections within the evaporator coils. The flow is characterized by nucleate boiling at the inlet.

7.6.3 Flow Boiling Correlations

Established forced convection subcooled boiling is subject to the same surface and fluid variables as pool boiling. However, it was shown by Brown[50] that the surface effects

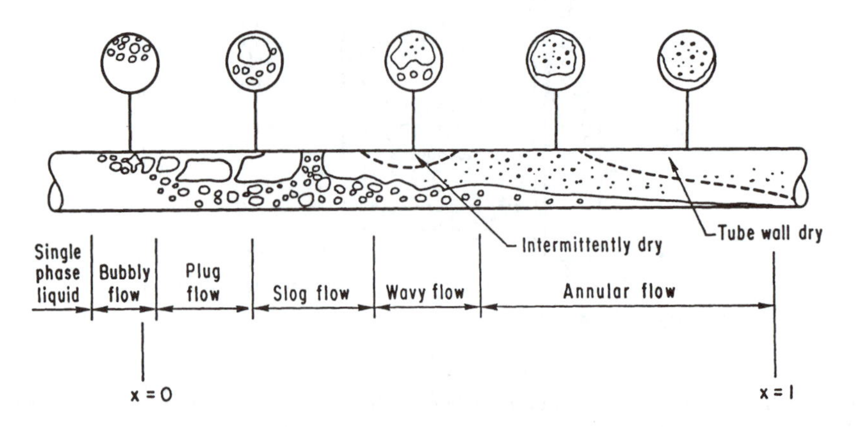

FIGURE 7.11
Flow patterns in a horizontal tube evaporator. (From Collier, J. G. [1981] *Convective Boiling and Condensation*, 2nd ed., McGraw-Hill, London. With permission.)

become less pronounced as the levels of velocity and subcooling increase. Jens and Lottes[51] suggested the following equation for water boiling from stainless steel or nickel surfaces:

$$q'' = 3.91 \times 10^5\, e^{0.065p}\, (T_w - T_s)^4 \tag{7.48}$$

where p is the absolute pressure in bar, $(T_w - T_s)$ is in K, and q'' is in MW/m². No correlations based on a variety of surfaces are available for other fluids. The pressure effect has not been established for other fluids.

An extensive review of correlations for subcooled boiling of water over a wide range of pressures was conducted by Guglielmini et al.[52] For the partial boiling regions they examined several interpolation techniques and without giving supporting evidence recommended that of Pokhvalov et al.,[53] which is proposed by its authors for water and other fluids.

For the fully developed region Guglielmini et al. compared 17 correlations, dating between 1955 and 1971. These include the often-used method of Thom et al.:[54]

$$\Delta T_s = 22.52\, e^{-0.115p}\, (q''/106)^{1/2} \tag{7.49}$$

(where p is the pressure in bar) resulting from data in flow boiling over a large range of pressures. Over a pressure range 1 to 100 bar there is a large scatter in predicted ΔT_s from the various correlations, which may have been due to the (1) differences in surface finish between tubes used by various workers or (2) differing degrees of accuracy in measuring ΔT_s, often in the range 5 to 10°C. Over the whole pressure range Guglielmini et al.[52] recommend, after comparison with a database covering nine experimental programs with round, rectangular, and annulus test sections, the methods of Rassokhin et al.:[55]

$$\begin{aligned} &\Delta T_s = m p^n q''^{1/3} \\ &1 \le p \le 80 \text{ bar} \qquad m = 0.304 \qquad n = -0.25 \\ &80 < p \le 200 \text{ bar} \qquad m = 34.7 \qquad n = -1.25 \end{aligned} \tag{7.50}$$

and Forster and Zuber:[56]

$$\Delta T_s = \frac{1 - 0.0045p}{3.4p^{0.18}}\, q''^{1/3} \tag{7.51}$$

As shown in Figure 7.10, nucleate boiling prevails over the low-quality region and the heat transfer coefficient is essentially at the established subcooled boiling level. After annular flow is developed, however, it is frequently observed that the heat transfer coefficient increases with increasing quality. Nucleation is suppressed because of the decreasing film thickness and the resultant increase in the heat transfer coefficient. Numerous correlations have been provided for this convective vaporization region as noted by Collier.[43]

The most popular correlations incorporate both low-quality and higher quality behavior in additive fashion. The widely used Chen[44] correlation, in which convective and nucleate boiling contributions are calculated separately and then added, was originally devised for saturated boiling but can easily be extended to subcooled boiling by the superposition method.

$$h_{TP} = h_{cb} + h_{nb} = h_{LO}\, F + h_p\, S \tag{7.52}$$

where h_{LO} is the heat transfer coefficient if the liquid were flowing on its own at that point in the channel (single-phase convective correlation) and h_p is the pool boiling coefficient for the same wall superheat. F (boiling enhancement factor) allows for the increase in velocity of the liquid vapor and S (suppression factor) allows for the decrease in bubble activity due to the steepening of the temperature gradient.

The contributions of convective boiling and nucleate boiling are estimated separately. Chen's procedure for estimating the convective and nucleate boiling coefficients is described here. His method for obtaining h_{nb} is based on the analysis of Forster and Zuber[56] where he uses the pool boiling correlation with a suppression factor, S.

By using experimental data from different sources, Chen determined experimental values of the convective boiling enhancement factor, F, and based it on the Martinelli parameter, X_{tt}.

$$\frac{1}{X_{tt}} = \frac{x}{(1-x)^{0.9}}\left(\frac{\rho_l}{\rho_v}\right)^{0.5}\left(\frac{\mu_v}{\mu_l}\right)^{0.1} \tag{7.53}$$

The enhancement factor therefore can be determined by the following equation to fit the curve proposed by Chen:[44]

$$\text{For } 1/X_{tt} \le 0.1,\ F = 1$$

$$\text{For } 1/X_{tt} > 0.1,\ F = 2.35\ (0.213 + 1/X_{tt})^{0.736} \tag{7.54}$$

As vaporization takes place, the amount of liquid in the tube decreases; then h_{LO} will also decrease. If one assumes that the heat transfer coefficient for the liquid flowing alone, h_{LO}, is proportional to the velocity raised to the power of 0.8 (in Dittus–Boelter correlation), the value of F_o is related to F and quality x by:[49]

$$F_o = F\ (1 - x) \tag{7.55}$$

The convection boiling heat transfer coefficient, h_{cb}, can then be found by multiplying the liquid heat transfer coefficient, h_{LO}, by the enhancement factor, F_o. To simplify the calculations of h_{LO}, a liquid heat transfer coefficient for zero vapor quality, h_{LO}, can be found to be constant throughout the length of the tube. By using the Dittus–Boelter correlation (or one of the new correlations given in Chapter 3 may be used):

$$h_{LO} = 0.023\ Re^{0.8}\ Pr^{0.4}\ k_l / d \tag{7.56}$$

The convection boiling coefficient can then be found by:

$$h_{cb} = F_o\ h_{LO} \tag{7.57}$$

where F_o, is the enhancement factor to account for the varying qualities given by Equation (7.55).

At low-vapor qualities, the enhancement due to nucleate boiling must be considered. Chen's correlation for the nucleate component is as follows:

$$h_{vo} = h_{nb} = 0.00122\frac{k_l^{0.079}\ C_{PL}^{0.45}\ \rho_L^{0.49}\ Sq_B^{0.24}\ Dp_v^{0.75}}{s^{0.5}\ m_L^{0.29}\ (i_{lg}\ \rho_g)^{0.24}} \tag{7.58}$$

where S is a suppression factor dependent on the two-phase flow Reynold's number:

$$S = \frac{1}{1 + 2.53 \times 10^{-6} (Re_{TP})^{1.17}} \quad (7.59)$$

and

$$\theta_B = \Delta T_{sat} = T_w - T_s \quad (7.60)$$

where θ_B is the wall superheat, T_s is the saturation temperature, and T_w is the wall temperature. The two-phase Reynolds number is found by:

$$Re_{TP} = (\dot{m}\, d / \mu_l)\,(1 - x)\, F^{1.25} \quad (7.61)$$

and Δp_v can be written from the Clapeyron equation as:

$$\Delta p_v = \Delta T_S\, i_{lg}\, \rho_v / \, T_s \quad (7.62)$$

The wall temperature is assumed at first to calculate an initial h_{NUCL}, and then later checked as shown in Example 7.6. A reiterative process is used until the assumed wall superheat agrees with the final value.

The other class of correlation for mixed nucleation and convection boiling is that in which the multiplier h_{TP}/h_{LO} or h_{TP}/h_{LV} is a function of the Martinelli parameter, $1/X_{tt}$, to characterize convection and of the boiling number, Bo:

$$\frac{h_{TP}}{h_{LO}} = C_1 \left\{ Bo + C_2 \left(\frac{1}{x_{tt}} \right)^{c_3} \right\}^{c_4} \quad (7.63)$$

where, for example, Smith[58] gives $C_1 = 6700$, $C_2 = 3.5 \times 10^{-4}$, $C_3 = 0.67$, and $C_4 = 1.0$. A relationship of the type of Equation (7.63) based on data for R-12, including entrance effects, was published by Mayinger and Ahrens:[57]

$$\frac{h_{TP}}{h_{LO}} = 0.85 \times 10^4 \left\{ Bo + 4.5 \times 10^{-4} \left(\frac{1}{X_{tt}} \right)^{0.35} \right\} \left(1 + \frac{d}{z} \right)^{[1/X_{tt}]^{0.41}} \quad (7.64)$$

Equation (7.64) suggests h_{TP} proportional to q'' in the nucleation-dominated region although close examination of the data shows $h_{TP} \sim (q'')^{0.78}$. The Shah[45] method is valid for vertical and horizontal flow and is discussed in the following pages. Close analysis shows that in the nucleate boiling region it gives $h_{TP} \sim (q'')^{0.5}$.

Shah's[45] correlation to calculate the heat transfer coefficient of boiling flow through pipes is based on four dimensionless parameters — Fr_L, Co, Bo, and F_o — which are the *Froude number, convection number, boiling number,* and *enhancement factor,* respectively. These dimensionless parameters characterize the flow and are employed to estimate the two-phase convective contribution to heat transfer in boiling.

The Froude number defined as:

$$Fr_l = \frac{G^2}{\rho_l^2 g d_i} \tag{7.65}$$

determines whether stratification effects are important or not. A Froude number greater than 0.04 ($Fr_l > 0.04$) signifies that stratification effects are negligible and inertial forces are dominant compared with gravitational forces. For low Froude numbers, the Shah method is recommended because it allows for effects of stratification. A correction factor, K_{FR}, is used when $Fr < 0.04$:[58]

$$K_{FR} = (25\ Fr)^{-0.3} \tag{7.66}$$

When $Fr > 0.04$, $K_{FR} = 1$ for horizontal or inclined pipes whereas $K_{FR} = 1$ for vertical pipes at all rates as long as, there is no liquid deficiency.

The convection number, Co, is defined as:[58]

$$Co = [(1 - x)/x]^{0.8} (\rho_v / \rho_l)^{0.5}\ K_{FR} \tag{7.67}$$

with the additional multiplication factor to Shah's original correlation to account for stratification effects for horizontal tubes when necessary.

The boiling number, Bo, is defined:

$$Bo = q''/\dot{m} i_{lg} \tag{7.68}$$

determines the enhancement due to nucleate boiling. The heat flux, q'', is the initial estimated value and i_{lg} is the latent heat of vaporization. A boiling number less than 1.9×10^{-5} signifies that there is no enhancement due to nucleation.

The enhancement factor, F_o, is dependent on the characteristics of the boiling. It is the ratio of heat transfer for two-phase flow to liquid only flow:

$$F_o = h_{TP}/h_{LO} \tag{7.69}$$

For pure convection boiling that occurs at high vapor qualities and low boiling numbers, the convection boiling factor is defined by:[58]

$$F_{cb} = 1.8\ Co^{-0.8} \qquad Co < 1.0 \tag{7.70}$$

and for lower vapor, qualities where $Co > 1.0$:

$$F_{cb} = 1.0 + 0.8 \exp [1 - (Co)^{0.5}] \qquad Co > 1.0 \tag{7.71}$$

where $F = F_{cb}$, and substituting it in Equation (7.55), F_o, the enhancement factor at varying vapor qualities can be found.

The convection boiling heat transfer coefficient is given by Equation (7.57):

$$h_{cb} = F_o \, h_{LO} \tag{7.72}$$

where h_{LO} is the liquid only heat transfer coefficient and is found using the Dittus–Boelter correlation as recommended by Shah. One can use better correlations given in Chapter 3.

In the nucleate boiling regime, $Bo > 1.9 \times 10^{-5}$ for very low vapor qualities, and the nucleate boiling factor is[46]

$$F_{nb} = 231\ Bo^{0.5} \qquad Co > 1.0 \tag{7.73}$$

where the nucleate boiling effects are dominant and $F = F_{nb}$ in Equation (7.55).

At higher vapor qualities, and $0.02 < Co < 1.0$, combined nucleate and convective boiling effects are considered[58] and the enhancement factor is determined from:

$$F_{cnb} = F_{nb}\,(0.77 + 0.13\,F_{cb}) \tag{7.74}$$

which is developed empirically by Smith from Shah's experimental data.[58]

Whenever combined nucleate and convective boiling is applicable, the enhancement factor, F, is determined by choosing the greatest value of F_{nb} for $Co > 1.0$; or the greatest value of F_{cnb} and F_{cb} for $0.02 < Co < 1.0$ will be used for F in Equation (7.55).

The method of calculating the two-phase heat transfer coefficient, h_{TP}, for considering *pure convection boiling* is as follows:

1. Calculate the liquid only heat transfer coefficient, h_{LO}.
2. Determine the convection number, Co, at the desired vapor quality.
3. Find the convection boiling factor at the corresponding Co using Equation (7.70) or (7.71).
4. Determine the enhancement factor, F_o, by using proper F in Equation (7.55).
5. The *convection boiling heat transfer coefficient* for tube-side flow is calculated by Equation (7.72).

To check the validity of the assumption that nucleate boiling is not present, the wall temperature for the ONB, T_{WONB} can be determined by the correlation developed by Davis and Anderson:[40]

$$T_{WONB} = \left(\frac{8\sigma q'' T_s}{k_l\, \Delta h_v \rho_v} \right)^{1/2} + T_s \tag{7.75}$$

The heat flux, q'', is defined as:

$$q'' = U\,(T_H - T_s) \tag{7.76}$$

where T_H is the temperature of the heating fluid and T_s is the saturated fluid temperature. The overall heat transfer coefficient is U.

The wall temperature, T_w, can be calculated from the heat flux and two-phase heat transfer coefficient:

$$T_w = \frac{q''}{h_{cb}} + T_s \tag{7.77}$$

where h_{cb} in the case of *pure convection boiling* is equal to the two-phase coefficient, h_{TP}.

If $T_{WONB} > T_W$, then nucleate boiling is not a concern and the assumption made is correct; but if $T_W > T_{WONB}$, then nucleate boiling should be considered and the calculations should be repeated for *combined nucleate and convective boiling*.

The method of calculating the boiling flow coefficient for combined convection and nucleation boiling is as follows:

1. Calculate the heat flux, q', by assuming no nucleate boiling as delineated previously.
2. To determine if enhancement due to nucleate boiling is significant, the boiling number, Bo, must first be calculated. If $Bo < 1.9 \times 10^{-5}$, the nucleate boiling is negligible and these steps discontinued; otherwise nucleate boiling should be considered.
3. Find the enhancement factor for nucleate boiling, F_{nb} or F_{cnb}, by using Equations (7.73) or (7.74), respectively.
4. The boiling heat transfer coefficient at zero quality is then found by:

$$h_{vo} = F_{nb}\, h_{LO} \tag{7.78}$$

5. The enhancement factor at different vapor qualities, x, for $Co > 1.0$ can then be determined by making F in Equation (7.78) equal to the greater of F_{nb} or F_{cb} as given by Equations (7.73) and (7.71), respectively.

   For $F_{cb} > F_{nb}$ the initial value of q''' that neglected nucleate boiling is correctly assumed and enhancement due to nucleation is negligible. If F_{nb}, however, is greater, then a more accurate value of heat flux must be calculated.

   If $0.02 < Co < 1.0$, then the enhancement factor is the greater of F_{cnb} or F_{cb} where F_{cnb} is calculated by Equation (7.74) and F_{cb}, by Equation (7.70). For $F_{cnb} > F_{cb}$, again a more accurate value of heat flux, q'', must be calculated for nucleation enhancement effects with the new enhancement factor, $F = F_{cnb}$; otherwise nucleation boiling enhancement is negligible.
6. The new boiling heat transfer coefficient at zero quality is defined as:

$$h_{vo} = F\, h_{LO} \tag{7.79}$$

Shah's method of determining heat transfer during saturated boiling is applicable for saturated boiling inside pipes of all Newtonian fluids (except metallic fluids). Shah identified the Chart correlation[46] in which two regions are defined: a nucleate boiling region, $Co > 1.0$, where heat transfer is determined by the boiling number only; and a region of convection boiling, $Co < 1.0$, where nucleation bubbles are completely suppressed.

For the purpose of simplifying the use of the Chart correlation, Shah's method was introduced in equation form as outlined in Shah's second article[46] and in Smith's[58] summary of Shah's method.

A comparative study of Shah correlations to data points for 11 regrigerants revealed a mean deviation of ±14%.

The Güngör and Winterton[47] correlation was developed from a large database of saturated boiling points of halocarbon refrigerants.

The form of the correlation is as follows:

$$h_{TP} = E\,h_l + S\,h_p \tag{7.80}$$

where E is an enhancement factor defined as:

$$E = 1 + 2.4 \times 10^4\,Bo^{1.16} + 1.37\,(1/X_{tt})^{0.88} \tag{7.81}$$

and S, a suppression factor:

$$S = [1 + 1.15 \times 10^{-6}\,E^2\,Re_l^{1.17}] \tag{7.82}$$

The pool boiling term, proposed by Cooper,[59] is defined as:

$$h_p = 55\,Pr^{0.12}\,(-\log Pr)^{-0.55}\,m^{-0.5}\,q''^{0.67} \tag{7.83}$$

The values of h_L, the liquid only heat transfer coefficient, and $1/X_{tt}$, the Martinelli parameter, are determined by Equations (7.55) and (7.53), respectively. The liquid Reynolds number is defined as:

$$Re_l = G\,(1 - x)\,d/\mu_l \tag{7.84}$$

For $Fr < 0.05$, Güngör and Winterton[47] recommend that the enhancement factor be multiplied by the correction factors shown below:

$$E_2 = Fr^{(0.1 - 2Fr)} \tag{7.85}$$

$$S_2 = Fr \tag{7.86}$$

When the correlation was compared with actual experimental values, a mean deviation of ±21.3% was reported. The properties used to calculate the correlation should be determined at the saturated boiling temperature. Knowledge of q'', the heat flux, is also necessary to use the equations.

In 1990, Kandlikar[49] derived a correlation for flow boiling of pure liquids in smooth tubes. It was developed by using an extensive data bank of more than 10,000 data points with water, refrigerants, and cryogens. This correlation was able to represent the trends in h versus quality x, mass flux G, and heat flux q″. Subsequently in 1992, Kandlikar[48] suggested modifications by incorporating the Petukhov-Kirillov[60] and Gnielinski[61] correlations for calculating the single phase heat transfer coefficient with all flow liquid, h_{LO}. The flow boiling correlation for pure liquids is as follows

$$h_{TP} = \text{larger of} \begin{cases} h_{TP,\,NBD} \\ h_{TP,\,CBD} \end{cases} \tag{7.87}$$

Where the subscribe NBD and CBD refer to the nucleate boiling dominant and the convective boiling dominant regions for which the h_{TP} values are given by:

$$h_{TP,NBD} = 0.6683\,Co^{-0.2}\,(1-x)^{0.8}\,h_{LO} + 1058.0\,Bo^{0.7}\,(1-x)^{0.8}\,F_{fl}h_{LO} \tag{7.88}$$

and

$$h_{TP,CBD} = 1.136\,Co^{-0.9}\,(1-x)^{0.8}\,h_{LO} + 667.2\,Bo^{0.7}\,(1-x)^{0.8}\,F_{fl}h_{LO} \tag{7.89}$$

For horizontal tubes with Froude number, Fr_{LO}, defined as $Fr_{LO} = G^2/(\rho_l^2 gd)$, less than 0.04, a multiplier $(25Fr_{LO})^{0.324}$ is applied to the first terms in Equations (7.88) and (7.89). For $Fr_{LO} > 0.04$ and for vertical tubes, no correlation is needed. This correction is usually not needed for the range of mass fluxes employed in the refrigerant evaporators. Bo is the boiling number, $q''/(Gh_{fg})$, Co is the convection number given by $Co = (\rho_v/\rho_l)^{0.5}[(1-x)/x]^{0.8}$.

F_{fl} is fluid-surface parameter related to the nucleation characteristics. Table 7.1 lists its value for several fluids. The single-phase heat transfer coefficient, h_{LO}, is obtained from the Petukhov and Kirillov,[60] Equation (3.29), and Gnielinski,[61] Equation (3.31), correlations.

Petukhov and Kirillov[69] for $0.5 \le Pr_l \le 2000$ and $10^4 \le Re_{LO} \le 5 \times 10^6$

$$Nu_{LO} = h_{LO}D/k_l = Re_{LO}\,Pr_l(f/2)/[1.07 + 12.7(Pr_l^{2/3} - 1)(f/2)^{0.5}] \tag{7.90}$$

Gnielinski[70] for $0.5 \le Pr_l \le 2000$ and $2300 \le Re_{LO} \le 10^4$

$$Nu_{LO} = h_{LO}D/k_l = (Re_{LO} - 1000)Pr_l(f/2)/[1.0 + 12.7\,(Pr_l^{2/3} - 1)(f/2)^{0.5}] \tag{7.91}$$

The friction factor f in Equations (7.90) and (7.91) is given by

$$f = [1.58\ln(Re_{LO}) - 3.28]^{-2} \tag{7.92}$$

The first terms in Equations (7.88) and (7.89) for NBD and CBD regions respectively represent the convective components, while the second terms, which include the heat flux, represent the nucleate boiling component. The demarcation between NBD and CBD regions is made automatically by comparing h_{TP} predicted by Equations (7.88) and (7.89) respectively, and taking the larger of the two as indicated by Equation (7.87).

Kandlikar[62] compared various correlations for the parametric trends among quality, heat flux, mass flux, and heat transfer coefficient. He found that the Kandlikar[49] correlation was able to predict the correct trends for refrigerants and water. Based on these observations, Kandlikar[62] proposed a new flow boiling map to represent the entire saturated flow boiling region. The correlation was recently extended to compact evaporators, microfin tubes, and binary mixtures as well.

Three correlations given by Shah (1982), Kandlikar (1987), and Güngör and Winterton (1986) are compared in Figures 7.12 and 7.13.

Example 7.5

Refrigerant-22 (R-22) boiling at 250 K flows through a tubular heat exchanger. Tubes are arranged horizontally with an I.D. of 0.0172 m. The mass velocity of the R-22 is 200 kg/m^2·s. Heat is supplied by the condensing of the fluid outside at –12°C. Hot fluid-side

TABLE 7.1

Fluid Dependent Parameter F_{fl} for Refrigerants in the Kandlikar[49] Correlation

Fluid	F_{fl}
Water	1.00
R-11	1.30
R-12	1.50
R-13B1	1.31
R-22	2.20
R-113	1.30
R-114	1.24
R-124	1.9
R-134a	1.63
R-152a	1.10

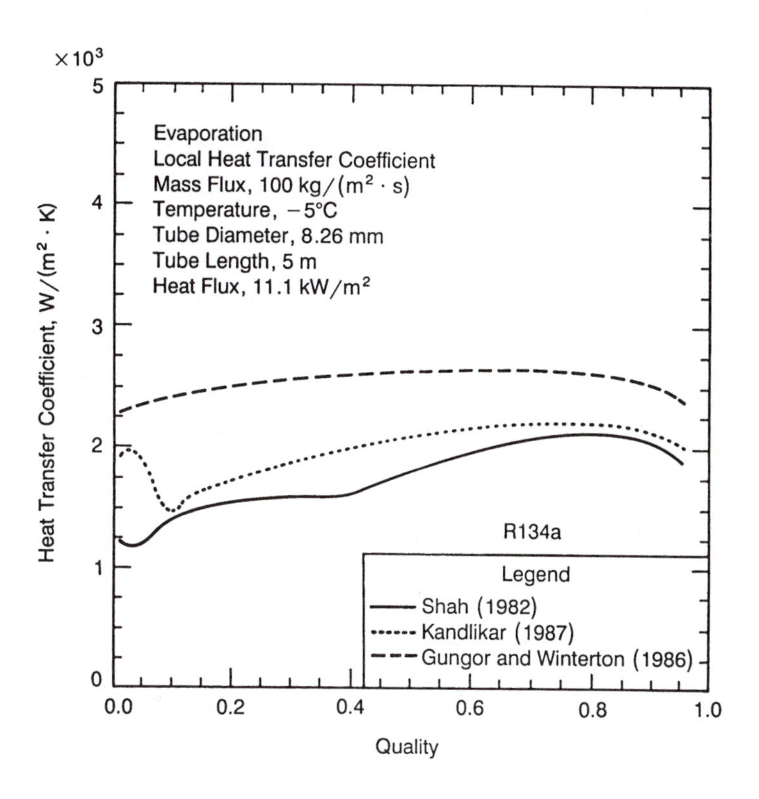

FIGURE 7.12
Local evaporation heat transfer coeffficients for in-tube flow of R-134A. (From Pate, M. B. [1991] In *Boilers, Evaporators and Condensers*, Wiley, New York. With permission.)

heat transfer coefficient, h_h, is 5400 W/m²·K. The tube is made of copper, and wall resistance and fouling are neglected. By the use of both Chen's method and Shah's method:

1. Calculate convection heat transfer coefficient for R-22 for various values of quality by assuming no nucleate boiling.
2. Calculate heat flux, q'' (W/m²).

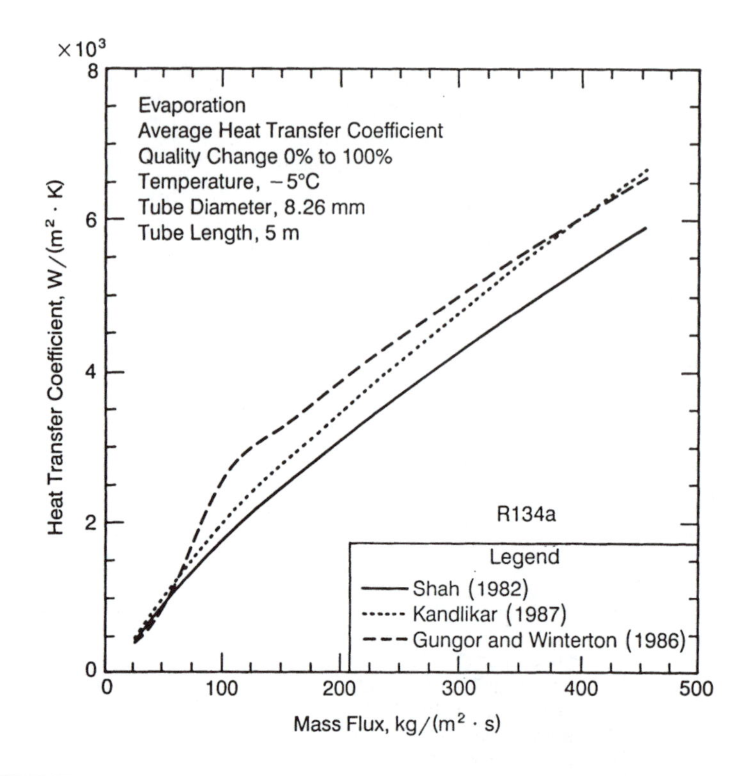

FIGURE 7.13
Average evaporation heat transfer coeffficients for in-tube flow of R-134A. (From Pate, M. B. [1991] In *Boilers, Evaporators and Condensers*, Wiley, New York. With permission.)

3. Determine tube wall temperature (T_{WONB}) for the onset of nucleate boiling. Is the assumption of no nucleate boiling valid? It may be so at certain values of x_v.

SOLUTION

Refrigerant-22 properties from Appendix B (Table B8) at 250 K (–23°C) are

$$p_s = 217.4 \text{ kPa (2.174 bars)}$$

$$\Delta h_v = 221900 \text{ J/kg}$$

$$\rho_v = 9.6432 \text{ kg/m}^3$$

$$\rho_l = 1360 \text{ kg/m}^3$$

$$c_{pl} = 1122 \text{ J/kg·K}$$

$$k_l = 0.112 \text{ W/m·K}$$

$$\sigma = 0.0155 \text{ N/m}$$

$$\mu_v = 0.110 \times 10^{-4} \text{ Pa·s}$$

$$\mu_l = 0.282 \times 10^{-3} \text{ Pa·s}$$

Prandtl number can be calculated by:

$$Pr = \frac{\mu_l c_{p_l}}{k_l}$$

$$= \frac{(0.282 \times 10^{-3})(1122)}{0.112} = 2.825$$

Reynolds number for the liquid phase can be calculated by:

$$Re_{LO} = \frac{Gd_i}{\mu_l}$$

$$= \frac{(200)(0.0172)}{0.000282} = 12.200$$

Nusselt number at $x = 0$ can be calculated by the Gnielinski correlation:

$$Nu_{LO} = \frac{(f/2)(Re_{LO} - 1000)\,Pr_l}{1 + 12.7(f/2)^{1/2}(Pr_l^{2/3} - 1)}$$

where

$$f = (1.58 \ln Re_l - 3.28)^{-2}$$

$$f = [1.58 \ln (12200) - 3.28]^{-2}$$

$$Nu_{LO} = \frac{(0.00372)(12200 - 1000)(2.825)}{1 + 12.7(0.00372)^{1/2}(2.825^{2/3} - 1)} = 66.43$$

Heat transfer coefficient is calculated by:

$$h_{LO} = \frac{Nu_{LO} k_l}{d_i}$$

$$= \frac{(66.43)(0.112)}{(0.0172)} = 432.55 \text{ W/m}^2 \cdot \text{K}$$

By using the Shah method[45,46] define first whether the effect of stratification is important or not. For this calculate Froude number:

$$Fr_{LO} = \frac{G^2}{\rho_l^2 g d_i} = \frac{(200)^2}{(1360)^2(9.81)(0.0172)} = 0.1282$$

$Fr > 0.04$; therefore, stratification effects are negligible and both Shah's and Chen's methods may be used.

For Chen's method at $x = 0.05$, calculate the Martinelli parameter:

$$\frac{1}{X_{tt}} = \left(\frac{x}{(1-x)}\right)^{0.9}\left(\frac{\rho_l}{\rho_v}\right)^{0.5}\left(\frac{\mu_v}{\mu_l}\right)^{0.1}$$

$$= \left(\frac{0.05}{0.95}\right)^{0.9}\left(\frac{1360}{9.64}\right)^{0.5}\left(\frac{110}{2820}\right)^{0.1} = 0.6069$$

The enhancement factor can be calculated from Equation (7.55):

$$F_o = F\,(1 - x_v)^{0.8}$$

where F from Equation (7.54) is

$$F = 2.35\left(0.213 + \frac{1}{X_{tt}}\right)^{0.736} = 2.031$$

Therefore

$$F_o = 2.031\,(1 - 0.05)^{0.8} = 1.949$$

and h_{cb} can be calculated from Equation (7.72):

$$h_{cb} = F_o\,(h_{LO})$$

$$= 1.949\,(432.55) = 842.7 \text{ W/m}^2\cdot\text{K}$$

Calculate the overall heat transfer coefficient, with negligible wall resistance and fouling:

$$U = \left(\frac{1}{h_{cb}} + \frac{1}{h_h}\right)^{-1}$$

$$= \left(\frac{1}{842.7} + \frac{1}{5400}\right)^{-1} = 729.0 \text{ W/m}^2\cdot\text{K}$$

Heat flux can be calculated by:

$$q'' = U\,(T_H - T_s)^{-1}$$

$$= 729.0\,[-12 - (-23)] = 8019 \text{ W/m}^2$$

To determine if nucleate boiling exists, the wall temperature for the onset of nucleate boiling can be determined by Equation (7.45):

$$q'' = U(T_H - T_s)^{-1} = 729.0\ (-12 - (-23)) = 8019\ \text{W/m}^2$$

On the other hand, the wall temperature, T_w, can be calculated from the heat flux and heat transfer coefficient:

$$T_w = \frac{q''}{h_{cb}} + T_s$$

$$= \frac{8019}{842.7} + 250 = 259.52\ \text{K}$$

For $x = 0.05$, $T_W > T_{WONB}$, nucleate boiling is present and should be considered. This is demonstrated in the following solved problem, Problem 7.6.

Shah's method accounts for stratification effects at $x = 0.05$.

Convection number can be determined from Equation (7.67):

$$Co = \left(\frac{1-x}{x}\right)^{0.8} \left(\frac{\rho_g}{\rho_l}\right)^{0.5} K_{FR}$$

where $K_{FR} = 1$ because at $Fr > 0.04$, inertial forces are dominant compared with gravity forces.

$$Co = \left(\frac{1-0.05}{0.05}\right)^{0.8} \left(\frac{9.6432}{1360}\right)^{0.5} (1) = 0.8878$$

$$F_{cb} = \text{convection boiling}$$

$$F_{cb} = 1.0 + 0.8\ \exp\left(1 - \sqrt{Co}\right) \quad Co > 1.0$$

$$F_{cb} = 1.8\ \exp\left(1 - \sqrt{Co}\right) \quad Co < 1.0$$

Therefore

$$F_{cb} = 1.8\ Co^{-0.8}$$

$$= 1.8\ (0.8878)^{-0.8} = 1.9798$$

Because of the enhancement factor from Equation (7.55):

$$F_o = F_{cb}(1-x)^{0.8}$$

$$= 1.9798(1 - 0.05)^{0.8} = 1.90$$

$$h_{cb} = F_o(h_{LO})$$

$$= 1.90(432.55) = 822.0\ \text{W/m}^2 \cdot \text{K}$$

Determine overall heat transfer coefficient, with negligible wall resistance and fouling:

$$U = \left(\frac{1}{h_{cb}} + \frac{1}{h_h}\right)^{-1}$$

$$= \left(\frac{1}{822} + \frac{1}{5400}\right)^{-1} = 713.0 \text{ W/m}^2 \cdot \text{K}$$

Calculate heat flux:

$$q'' = U\,(T_H - T_s)$$

$$= 713.0\,(-12 - (-23)) = 7847 \text{ W/m}^2$$

The wall temperature for the onset of nucleate boiling can be determined by Equation (7.45) :

$$T_{\text{WONB}} = \left(\frac{8\sigma q'' T_s}{k_l \Delta h_v \rho_v}\right)^{1/2} + T_s$$

$$= \left[\frac{8(0.0155)(7847)(250)}{(0.112)(221900)(9.64)}\right]^{1/2} + 250 = 251.01 \text{K}$$

On the other hand, the wall temperature, T_w , can be calculated from the heat flux and heat transfer coefficient:

$$T_w = q''/h_{cb} + T_s$$

$$= 7847/822 + 250 = 259.55 \text{ K}$$

For $x = 0.05$, $T_W > T_{\text{WONB}}$ and nucleate boiling is present, and will be considered in Example 7.6. The calculations for each value of quality x can be performed in similar way to calculate the various quantities as a function of quality.

Example 7.6

Repeat Example 7.5 with Refrigerant-22 at 250 K, now considering combined convection and nucleate boiling where necessary.

SOLUTION

From Example 7.5:

$$T_H = -12°\text{C}$$

$$T_s = -23°\text{C}$$

$$G = 200 \text{ kg/m}^2\text{·s}$$

$$h_H = 5400 \text{ W/m}^2\text{·K}$$

Refrigerant-22 properties at 250 K (–23°C) from Appendix B (Table B.8) are

$$p_{sat} = 217.4 \text{ kPa (2.174 bars)}$$

$$\Delta h_v = 221900 \text{ J/kg}$$

$$\rho_v = 9.6432 \text{ kg/m}^3$$

$$\rho_l = 1360 \text{ kg/m}^3$$

$$c_{pl} = 1122 \text{ J/kg·K}$$

$$k_l = 0.112 \text{ W/m·K}$$

$$\sigma = 0.0155 \; N/\text{m}$$

$$\mu_v = 0.110 \times 10^{-4} \text{ Pa·s}$$

$$\mu_l = 0.282 \times 10^{-3} \text{ Pa·s}$$

At $x = 0.05$ (from Example 7.5):

$$h_{cb} \text{ (Chen)} = 842.7 \text{ W/m}^2\text{·K}$$

$$h_{cb} \text{ (Shah)} = 822.0 \text{ W/m}^2\text{·K}$$

$$q'' \text{ (Shah)} = 7847 \text{ W/m}^2$$

By using Chen's method at $x = 0.05$ determine wall superheat:

$$\theta_B = \Delta T_{sat} = T_W - T_s$$

where T_s is the saturation temperature and $T_W = -18.5°C$ (assumed wall temperature):

$$\theta_B = -18.5 - (-23) = 4.5°C$$

Calculate Reynolds number for two-phase flow from Equation (7.61):

$$Re_{TP} = \frac{\dot{m} d_i}{\mu_l}(1 - x)F^{1.25}$$

where $F_{0.05} = 2.031$ is obtained in Example 7.5:

$$Re_{TP} = \frac{(200)(0.0172)}{(0.000282)}(0.95)(2.031)^{1.25} = 28.079$$

$$\Delta p_v = \Delta T_{sat} \Delta h_v \rho_v / T_2 = 4.5(221900)(9.643)/250 = 38.5 \text{ kPa}$$

Determine suppression factor dependent on Re_{TP} from Equation (7.59):

$$S = [1 + 2.53 \times 10^{-6} (Re_{TP})^{1.17}]^{-1}$$

$$= [1 + 2.53 \times 10^{-6} (28079)^{1.17}]^{-1} = 0.71$$

By using Chen's equation for the nucleate component, h_{vo}:

$$h_{nb} = h_{vo} = 0.00122 \frac{k_l^{0.79} c_{p_l}^{0.49} S\theta_B^{0.24} \Delta p_v^{0.75}}{\sigma^{0.5} \mu_l^{0.29} (\Delta h_v \rho_v)^{0.24}}$$

$$F_{cnb} = F_{nb}(0.77 + 0.13F_{cb}) = 3.06[0.77 + 0.13(1.98)] = 3.14$$

Thus $F_{cb} < F_{cnb}$, and enhancement due to nucleate boiling is present; therefore:

$$F = F_{cnb} = 3.14$$

Now go back and recalculate q'' with the new value of F for enhancement due to nucleate boiling:

$$h_{nb} = F\, h_{LO} = 3.14\ (432.55) = 1358\ \text{W/m}^2\cdot\text{K}$$

Overall heat transfer coefficient, neglecting wall resistance and fouling, is:

$$U_o = (1/h_{nb} + 1/h_H)^{-1}$$
$$= (1/1345 + 1/5400)^{-1} = 1085.0\ \text{W/m}^2\cdot\text{K}$$

Calculate heat flux:

$$q'' = U_o\,(T_H - T_s) = 1085\ [-12 - (-23)] = 11.938\ \text{W/m}^2\cdot\text{K}$$

By using Shah's method at $x = 0.05$ determine boiling number, Bo:

$$Bo = q''/G\Delta h_v = 7847/(200)\ (221900) = 1.77 \times 10^{-4}$$

This is the approximate value of the boiling number and $Bo > 1.9 \times 10^{-5}$, therefore nucleation enhancement should be considered.
Calculate the value of the nucleate boiling factor F for nucleation:

$$F_{nb} = 230\ Bo^{1/2} = 230\ (1.77 \times 10^{-4})^{1/2} = 3.06$$

Calculate heat transfer coefficient for boiling at zero quality:

$$h_{nb} = F_{nb}\, h_{LO} = 3.06\ (432.55) = 1345\ \text{W/m}^2\cdot\text{K}$$

where $h_{LO} = 432.55$ W/m²·K (see Example 7.7).
To calculate F:

$Co > 1.0$, $\quad F = F_{cb}$ or F_{nb} whichever is greater

$0.02 > Co > 1.0$, $\quad F = F_{cb}$ or F_{nb} whichever is greater

At $x = 0.05$, $Co = 0.8878$ and $F_{cb} = 1.9798$ (see Example 7.5):

$$h_{nb} = 0.00122 \frac{(0.112)^{0.79}(1122)^{0.45}(1360)^{0.49}(0.71)(4.5)^{0.24}(38500)^{0.75}}{(0.0155)^{0.5}(0.000282)^{0.29}(221900 \times 9.64)^{0.24}}$$

$$= 1297 \ \mathrm{W/m^2 \cdot K}$$

Another way to calculate h_{nb} is by considering:

$$h_{nb} = f_{nbi} \ S \ \theta_B$$

where f_{nbi} is calculated by the equation given in theory as 405.0 W/m²·K.

The combined convection and nucleate boiling coefficient:

$$h_{cnb} = h_{cb} + h_{nb}$$

$$q'' = h_{cb} \ \theta_B + f_{nbi} S_B^2$$

$$= (842.7)\ (4.5) + (405)\ (0.71)\ (4.5)^2 = 9629 \ \mathrm{W/m^2}$$

where $\theta_B = T_W - T_s$.

The temperature difference across the boiling fluid, $(T_H - T_s)$ must be determined by trial and error. The value q'' must satisfy:

$$T_H - T_s = q''/h_H + \theta_B$$

$$= 9629/5400 + 4.5 = 6.28°\mathrm{C}$$

where H stands for the hot fluid. However, from the statement of the problem, $T_H - T_s =$ 11°C; therefore $\theta_B = 4.5$ is too small.

The first assumption for θ_B was much too small; a new value of $\theta_B = T_W - T_s = 7.2$ can be assumed, and by repeating the calculations it can be shown that is acceptable.

The heat flux calculated for combined nucleate boiling and convective boiling for $x = 0.05$ is much higher than the heat flux for convection boiling only (2.5 times higher).

Nomenclature

- A area, m²
- Bo boiling number $= q''/\Delta h_v \dot{m}$
- C constant
- Co convection number $= [(1-x)/x]^{0.8} (\rho_v/\rho_l)^{0.5}$
- c_p constant pressure specific heat, J/kg·K
- d tube diameter, m
- E enhancement factor
- F correction factor
- Fr Froude number $= G^2/g\, D_i\, \rho_h$
- G mass velocity, kg/m²·s
- g gravitational acceleration, m/s²

h heat transfer coefficient, W/m²·K
i_{lg} latent heat of vaporization ($h_g - h_f$), J/kg
i enthalpy, J/kg
j_g^* dimensionless mass velocity
K correction factor
k thermal conductivity
L length, m
m constant
$\dot{m}$ mass flow rate, kg/s
N number of tubes in a vertical column
n constant
Nu Nusselt number = hd/k
p pressure, N/m²
Pr Prandtl number = $c_p \mu/k$
Q cumulative heat release rate, W
q' heat input per unit tube length, W/m
q'' heat flux, W/m²
r radius of curvature, bubble radius, m
Re Reynolds number = $\rho U/\mu$, $\rho ud/\mu$
Re_F film Reynolds number = $4F/\mu_L$
$\tilde{R}e$ two-phase Reynolds number = $u_g D/\nu_l$
S suppression factor
T temperature,°C, K
t time, s
ΔT temperature difference, K
ΔT_s wall superheat,°C, K
ΔT_{sub} subcooling,°C, K
U overall heat transfer coefficient, W/m²·K
u velocity, m/s
ν specific volume, m³/kg
w mean flow width, m
X_{tt} Lockhart–Martinelli parameter, defined by Equation (7.53)
x vapor quality
Z coefficient
z distance from heater inlet, m

Greek Symbols

α void fraction of vapor
Γ film flow rate per unit length, kg/(m·s)

δ film thickness, m
μ dynamic viscosity, Pa·s
ν kinematic viscosity = μ/ρ, m^2/s
ρ density, kg/m^3
σ surface tension, N/m
τ shear stress, N/m^2
θ_B wall superheat,°C, K

Subscripts

b bulk
c cold, critical point
cb convective boiling
cnb combined convection and nucleate boiling
f liquid
g gas phase
H heating fluid
LO liquid alone
l liquid phase
m mean
N corresponding to N tubes
nb, n nucleate boiling
ONB onset of nucleate boiling
p pool boiling
TP two-phase
S saturated fluid
sat saturation
sh vapor shear
v vapor phase, saturated vapor
w wall

References

1. Marto, P. J. (1991) Heat transfer in condensation. In *Boilers, Evaporators and Condensers*, S. Kakaç (Ed.), Wiley, New York.
2. Marto, P. J. (1988) Fundamentals of condensation. In *Two-Phase Flow Heat Exchangers: Thermal–Hydraulic Fundamentals and Design*, S. Kakaç, A. E. Bergles, and E. O. Fernandes (Eds.), pp. 221–291. Kluwer, Dordrecht, The Netherlands.
3. Nusselt, W. (1916) The condensation of steam on cooled surfaces. *Z. Ver. Dtsch. Ing.*, *60*, 541–546 and 569–575 (translated into English by D. Fullarton [1982] *Chem. Eng. Fund.*, Vol. 1, no. 2, 6–19.

4. Shekriladze, I. G. and Gomelauri, V. I. (1966) Theoretical study of laminar film condensation of flowing vapor. *Int. J. Heat Mass Transfer*, Vol. 9, 581–591.
5. Butterworth, D. (1977) Developments in the design of shell and tube condensers. *ASME Winter Annual Meeting*, Atlanta, ASME Preprint 77-WA/HT-24.
6. Rose, J. W. (1988) Fundamentals of condensation heat transfer: laminar film condensation. *JSME Int. J.*, Vol. 31, 357–375.
7. Kern, D. Q. (1958) Mathematical development of loading in horizontal condensers. *AIChE J.*, Vol. 4, 157–160.
8. Eissenberg, D. M. (1972) An Investigation of the Variables Affecting Steam Condensation on the Outside of a Horizontal Tube Bundle, Ph.D. Thesis, University of Tennessee, Knoxville.
9. Shklover, G. G. and Buevich, A. V. (1978) Investigation of steam condensation in an inclined bundle of tubes. *Thermal Eng.*, Vol. 25, no. 6, 49–52.
10. Kutateladze, S. S., Gogonin, N. I., Dorokhov, A. R., and Sosunov, V. I. (1979) Film condensation of flowing vapor on a bundle of plain horizontal tubes. *Thermal Eng.*, Vol. 26, 270–273.
11. Fujii, T., Uehara, H., Hirata, K., and Oda, K. (1972) Heat transfer and flow resistance in condensation of low pressure steam flowing through tube banks. *Int. J. Heat Mass Transfer*, Vol. 15, 247–260.
12. Cavallini, A., Frizzerin, S., and Rossetto, L. (1986) Condensation of R-11 vapor flowing downward outside a horizontal tube bundle. *Proc. 8th Int. Heat Transfer Conf.*, San Francisco, Vol. 4, 1707–1712.
13. McNaught, J. M. (1982) Two-phase forced convection heat transfer during condensation on horizontal tube bundles. *Proc. 7th Int. Heat Transfer Conf.*, Munich, Vol. 5, 125–131.
14. Marto, P. J. (1984) Heat transfer and two-phase flow during shell-side condensation. *Heat Transfer Eng.*, Vol. 5, no. 1–2, 31–61.
15. Barness, E. J. (1963) Calculation of the performance of surface condensers by digital computer. ASME Paper No. 63-PWR-2, *National Power Conf.*, Cincinnati, OH, September.
16. Emerson, W. H. (1969) The application of a digital computer to the design of surface condenser. *Chem. Eng.*, Vol. 228, no. 5, 178–184.
17. Wilson, J. L. (1972) The design of condensers by digital computers. I. *Chem. Eng. Symp. Ser.*, no. 35, 21–27.
18. Hafford, J. A. (1973) ORCONI: A Fortran Code for the Calculation of a Steam Condenser of Circular Cross Section. ORNL-TM-4248, Oak Ridge National Laboratory, Oak Ridge, TN.
19. Hopkins, H. L., Loughhead, J., and Monks, C. J. (1983) A computerized analysis of power condenser performance based upon an investigation of condensation. In *Condensers: Theory and Practice*, *I. Chem. Eng. Symp. Series*, no. 75, 152–170, Pergamon Press, London.
20. Shida, H., Kuragaska, M., and Adachi, T. (1982) On the numerical analysis method of flow and heat transfer in condensers. *Proc. 7th Int. Heat Transfer Conf.*, Munich, Vol. 6, 347–352.
21. Al-Sanea, S., Rhodes, N., Tatchell, D. G., and Wilkinson, T. S. (1983) A computer model for detailed calculation of the flow in power station condensers. In *Condensers: Theory and Practice*, *I. Chem. E. Symp. Series*, no. 75, 70–88, Pergamon Press, London.
22. Caremoli, C. (1983) Numerical computation of steam flow in power plant condensers. In *Condensers: Theory and Practice*, *I. Chem. E. Symp. Ser.*, no. 75, 89–96, Pergamon Press, London.
23. Beckett, G., Davidson, B. J., and Ferrison, J. A. (1983) The use of computer programs to improve condenser performance. In *Condensers: Theory and Practice*, *I. Chem. Eng. Symp. Ser.*, no. 75, 97–110, Pergamon Press, London.
24. Zinemanas, D., Hasson, D., and Kehat, E. (1984) Simulation of heat exchangers with change of phase. *Comput. Chem. Eng.*, Vol. 8, 367–375.
25. Kakac, S. (Ed.) (1991) *Boilers, Evaporators and Condensers*, Wiley, New York.
26. Soliman, H. M. and Azer, N. Z. (1974) Visual studies of flow patterns during condensation inside horizontal tubes. *Proc. 5th Int. Heat Transfer Conf.*, Tokyo, Vol. 3, 241–245.
27. Rahman, M. M., Fathi, A. M., and Soliman, H. M. (1985) Flow pattern boundaries during condensation: new experimental data. *Can. J. Chem. Eng.*, Vol. 63, 547–552.
28. Soliman, H. M. (1986) Flow pattern transitions during horizontal in-tube condensation. In *Encyclopedia of Fluid Mechanics*, Chapter 12. Gulf Publishing, Houston.

29. Breber, G., Palen, J. W., and Taborek, J. (1980) Prediction of horizontal tubeside condensation of pure components using flow regime criteria. *J. Heat Transfer*, Vol. 102, 471–476.
30. Tandon, T. N., Varma, H. K., and Gupta, C. P. (1982) A new flow regime map for condensation inside horizontal tubes. *J. Heat Transfer*, Vol. 104, 763–768.
31. Butterworth, D. (1983) Film condensation of pure vapor. In *Heat Exchanger Design Handbook*, E. U. Schlünder (Ed.), Vol. 2, Section 2.6.2. Hemisphere, New York.
32. Jaster, H. and Kosky, P. G. (1976) Condensation heat transfer in a mixed flow regime. *Int. J. Heat Mass Transfer*, Vol. 19, 95–99.
33. Traviss, D. P., Rohsenow, W. M., and Baron, A. B. (1972) Forced convection condensation inside tubes: a heat transfer equation for condenser design. *ASHRAE Trans.*, Vol. 79, 157–165.
34. Cavallini, A. and Zecchin, R. (1974) A dimensionless correlation for heat transfer in forced convection condensation. *Proc. 5th Int. Heat Transfer Conf.*, Tokyo, Japan, September 3–7, 309–313.
35. Shah, M. M. (1979) A general correlation for heat transfer during film condensation inside pipes. *Int. J. Heat Mass Transfer*, Vol. 22, 547–556.
36. Schlager, L. M., Pate, M. B., and Bergles, A. E. (1990) Performance predictions of refrigerant-oil mixtures in smooth and internally finned tubes. II. design equations. *ASHRAE Trans.*, Vol. 96, no. 1.
37. Pate, M. B. (1991) Evaporators and condensers, for refrigeration and air conditioning. In *Boilers, Evaporators and Condensers*, S. Kakaç (Ed.), Wiley, New York.
38. Wallis, G. B. (1961) Flooding Velocities for Air and Water in Vertical Tubes, UKAEA Report AEEW-R123.
39. Butterworth, D. and Shock, R. A. (1982) Flow boiling. *Heat Transfer 1982, Proc. 7th Int. Heat Transfer Conf.*, Hemisphere, New York.
40. Davis, E. J. and Anderson, G. H. (1966) The incipience of nucleate boiling in convective flow, *AIChE J.*, Vol. 12, 774–780.
41. Bergles, A. E. and Rohsenow, W. M. (1964) The determination of forced-convection surface-boiling heat transfer, *J. Heat Transfer*, Vol. 86, 356–372.
42. Kitto, J. B. and Albrecht, M. J. (1988) Elements of two-phase flow in fossil boilers. In *Two-Phase Flow Heat Exchangers: Thermal–Hydraulic Fundamentals and Design*, S. Kakaç, A. E. Bergles, and E. O. Fernandes (Eds.), pp. 221–291. Kluwer, Dordrecht, The Netherlands.
43. Collier, J. G. (1981) *Convective Boiling and Condensation*, 2nd ed., McGraw-Hill, London.
44. Chen, J. C. (1966) A correlation for boiling heat transfer to saturated fluids in convective flow. *Ind. Eng. Chem. Process Des. Dev.*, Vol. 5, 322–329.
45. Shah, M. M. (1976) A new correlation for heat transfer during boiling flow through pipes. *ASHRAE Trans.*, Vol. 82, 66–86.
46. Shah, M. M. (1982) Chart correlation for saturated boiling heat transfer: equations and further study. *ASHRAE Trans.*, Vol. 88, 185–196.
47. Güngör, K. E. and Winteron, R. H. S. (1986) A general correlation for flow boiling in tubes and annuli. *Int. J. Heat Mass Transfer*, Vol. 19, no. 3, 351–358.
48. Kandlikar, S. G. (1991) A model for correlating flow boiling heat transfer in augmented tubes and compact evaporators. *J. Heat Transfer*, Vol. 113, 966–972.
49. Kandlikar, S. G. (1990) A general correlation for saturated two-phase flow boiling heat transfer inside horizontal and vertical tubes. *J. Heat Transfer*, Vol. 112, 219–228.
50. Brown, W. T., Jr. (1967) A Study of Flow Surface Boiling, Ph.D. Thesis in Mechanical Engineering, Massachusetts Institute of Technology, Cambridge, MA.
51. Jens, W. H. and Lottes, P. A. (1951) An Analysis of Heat Transfer, Burnout, Pressure Drop, and Density Data for High Pressure Water, Argonne National Laboratory Report ANL-4627.
52. Guglielmini, G., Nannei, E., and Pisoni, C.(1980) *Warme Stoffübertragung*, Vol. 13, 177–185.
53. Pokhvalov, Y. E., Kronin, G. H., and Kurganova, I. V. (1966) *Teploenergetika*, Vol. 13, 63–68.
54. Thom, J. R. S., Walker, V. M., Fallon, J. S., and Reising, G. F. S. (1965/1966) *Proc. Inst. Mech. Eng.*, Vol. 3C/80, 226–246.
55. Rassokhin, N. G., Shvetsvov, N. K., and Kuzmin, A. V. (1970) *Thermal Eng.*, Vol. 9, 86–90.
56. Forster, H. K. and Zuber, N. (1955) Dynamics of vapor bubbles and boiling heat transfer. *AIChE J.*, Vol. 1, no. 4, 531–535.

57. Mayinger, F. and Ahrens, K. H. (1978) Boiling heat transfer in the transition region from bubble flow to annular flow. *Int. Semin. Momentum Heat Mass Transfer in Two-Phase Energy and Chemical Systems*, Dubrovnik, Yugoslavia.
58. Smith, R. A. (1976) *Vaporisers: Selection Design & Operation*, Longman Scientific & Technical and Wiley, New York.
59. Cooper, M. G. (1984) Saturation Nucleate Pool Boiling: A simple correlation. *1st U.K. National Conference on Heat Transfer. I. Chem. Engineering Symposium Ser.*, no. 86, Vol. 2, 785–793.
60. Petukhov, B. S. and Popov, V. N. (1963) Theoretical Calculation of Heat Exchange and Frictional Resistance in Turbulent Flow in Tubes of an Incompressible Fluid with Variable Physical Properties. *Teplofiz Vysok. Temperature (High Temperature Heat Physics)*, Vol. 1, no. 1.
61. Gnielinski, V. (1976) New equations for heat and mass transfer in turbulent pipe and channel flow. *Int. Chem. Eng.*, Vol. 16, pp. 359–368.
62. Kandlikar, S. G. (1991) Development of a Flow Boiling Map for Subcooled and Saturated Flow Boiling of Different Fluids in Circular Tubes. In *Heat Transfer with Phase Change*, Habib, I.S. et al., Eds., ASME HTD-Vol. 114, pp. 51–62, 1990. Also published in *Transactions of ASME, J. Heat Transfer*, Vol. 113, pp. 190–200.

Problems

7.1. Repeat Example 7.1 for the refrigerant-22 and compare with R-134A.

7.2. Repeat Example 7.2 for the refrigerant-22 and compare with R-134A.

7.3. Under the conditions given in Example 7.4, calculate the local heat transfer coefficient for the 20th row of tubes using the method given by Butterworth.

7.4. Calculate the average heat transfer coefficient for film-type condensation of water at pressure of 10 kPa for:

a. An outside surface of 19-mm O.D. horizontal tubes-2 m long

b. A 12-tube vertical bank of 19-mm horizontal tubes 2-m long

It is assumed that the vapor velocity is negligible and the surface temperatures are constant at 10°C below saturation temperature.

7.5. A horizontal 2-cm O.D. tube is maintained at a temperature of 27°C on its outer surface. Calculate the average heat transfer coefficient if saturated steam at 6.22 kPa is condensing on this tube.

7.6. In a shell-and-tube type of steam condenser, assume that there are 81 tubes arranged in a square pitch with 9 tubes per column. The tubes are made of copper with an O.D. of 1 in. The length of the condenser is 1.5 m. The shell-side condensation occurs at temperature pressure. Water flows inside tubes with a mass flow rate of 4 kg/s. The tube outside wall temperature I.D. 90°C.

7.7. A water-cooled, shell-and-tube freon condenser with in-tube condensation will be designed to satisfy the following specifications:

Cooling load of the condenser	125 kW
Refrigerant	R-22
Condensing temperature	37°C
Coolant water	City water
	Inlet temperature, 18°C
	Outlet temperature, 26°C
	Mean pressure, 0.4 mPa

Heat transfer matrix	3/4 in. O.D., 20 BWG
	Brass tubes

If it is proposed that the following heat exchanger parameters are fixed: one-tube pass with shell diameter of 15 1/4 in., pitch size is 1 in. with baffle spacing of 35 cm. The number of tubes is 137.

a. Calculate the shell-and-tube size heat transfer coefficients.

b. By assuming proper fouling factors, calculate the length of the condenser.

c. The space available is 6 m. Is this design acceptable?

7.8. R-134a flows in a horizontal 8 mm diameter circular tube. The mass flux is 400 kg/m^2s, entering quality is 0.0 and the exiting quality is 0.8. The tube length is 3 m. Assuming a constant saturation temperature of 20°C, plot the variation of h_{TP} as a function of x.

8

Shell-and-Tube Heat Exchangers

8.1 Introduction

Shell-and-tube heat exchangers are the most versatile type of heat exchangers. They are used in process industries, in conventional and nuclear power stations as condensers, in steam generators in pressurized water reactor power plants, in feed water heaters, and in some air-conditioning and refrigeration systems. They are also proposed for many alternative energy applications including ocean, thermal, and geothermal.

Shell-and-tube heat exchangers provide relatively large ratios of heat transfer area to volume and weight and they can be easily cleaned.

Shell-and-tube heat exchangers offer great flexibility to meet almost any service requirement. The reliable design methods and shop facilities are available for their successful design and construction. Shell-and-tube heat exchangers can be designed for high-pressures relative to the environment and high-pressure differences between the fluid streams.

8.2 Basic Components

Shell-and-tube heat exchangers are built of round tubes mounted in a cylindrical shell with the tubes parallel to the shell. One fluid flows inside the tubes, while the other fluid flows across and along the axis of the exchanger. The major components of this exchanger are tubes (tube bundle), shell, front-end head, rear-end head, baffles, and tube sheets. Typical parts and connections, for illustrative purposes only, are shown in Figure 8.1.[1]

8.2.1 Shell Types

Various front and rear head types and shell types have been standardized by Tubular Exchanger Manufacturers Association (TEMA). They are identified by an alphabetic character as shown in Figure 8.2.[1]

Figure 8.3 shows the most common shell types as condensers (v shows the location of the vent).[2] The E-shell is the most common due to its cheapness and simplicity. In this shell, the shell fluid enters at one end of the shell and leaves at the other end; that is, there is one pass on the shell side. The tubes may have a single or multiple passes and are supported by transverse baffles. This shell is the most common for single-phase shell fluid applications. With a single-tube pass, a nominal counterflow can be obtained.

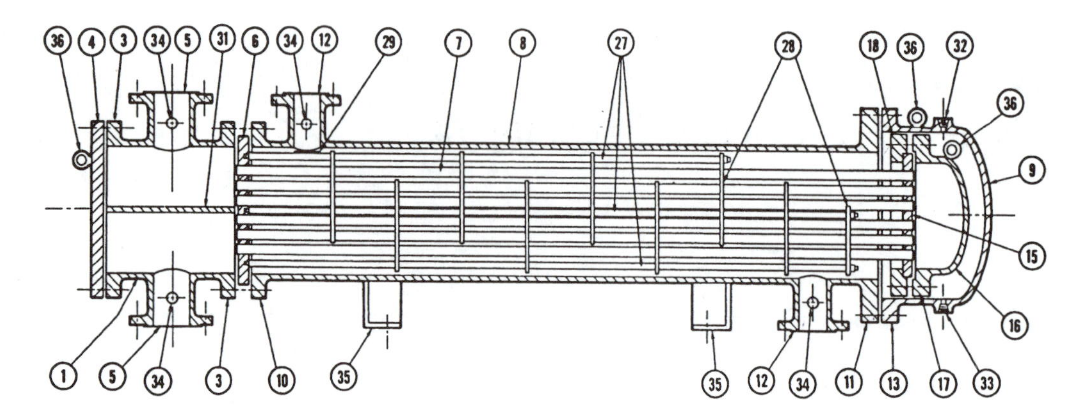

FIGURE 8.1
Constructional parts and connections. (1) Stationary head — channel; (2) stationary head — bonnet; (3) stationary head flange — channel or bonnet; (4) channel cover; (5) stationary head nozzle; (6) stationary tube sheet; (7) tubes; (8) shell; (9) shell cover; (10) shell flange — front head end; (11) shell flange — rear head end; (12) shell nozzle; (13) shell cover flange; (14) expansion joint; (15) floating tube sheet; (16) floating head cover; (17) floating head backing device; (18) floating head backing device; (19) split shear ring; (20) slip-on backing flange; (21) floating head cover — external; (22) floating tube sheet skirt; (23) packing box; (24) packing; (25) packing gland; (26) lantern ring; (27) tie rods and spacers; (28) transverse baffle or support plates; (29) impingement plate; (30) longitudinal baffle; (31) pass partition; (32) vent connection; (33) drain connection; (34) instrument connection; (35) support saddle; (36) lifting lug; (37) support bracket; (38) weir; (39) liquid level connection. (From Standards of the Tubular Exchanger Manufacturers Association (1988), New York. With permission.)

To increase the effective temperature differences and hence exchanger effectiveness, a pure counterflow arrangement is desirable for a two-tube-pass exchanger. This is achieved by the use of F-shell having a longitudinal baffle and resulting in two-shell passes. It is used when units in series are required, with each shell pass representing one unit. The pressure drop is much higher than the pressure drop of a comparable E-shell.

Other important ones are the J-shell and X-shell. In the divided flow J-shell, fluid entry is centrally located and split into two parts. The single nozzle is at the midpoint of tubes and two nozzles are near the tube ends. This shell is used for low-pressure drop design applications such as a condenser in vacuum, because J-shell has approximately 1/8 the pressure drop of a comparable E-shell. When it is used for a condensing shell fluid, it will have two inlets for the vapor phase, and one central outlet for the condensate.

The X-shell has a centrally located fluid entry and outlet, usually with a distributor dome. The two fluids are over the entire length of the tubes and are in crossflow arrangement. No baffles are used in this shell tube. Consequently, the pressure drop is extremely low. It is used for vacuum condensers and low-pressure gases.

The split flow shells such as the G-shell and H-shell are used for specific applications.

The split flow G-shell has horizontal baffles with ends removed; the shell nozzles are 180° apart at the midpoint of tubes. The G-shell has the same pressure drop as that for the E-shell, but the log mean temperature difference (LMTD) factor F and hence the exchanger effectiveness is higher for the same surface area and flow rates. The G-shell can be used for single-phase flows, but is often used as a horizontal thermosiphon reboiler. In this case, the longitudinal baffle serves to prevent flashing out of the lighter components of the shell fluids and provides increased mixing.

The double split flow H-shell is similar to the G-shell, but with two outlet nozzles and two horizontal baffles.

The K-shell is a kettle reboiler with the tube bundle in the bottom of the shell covering about 60% of the shell diameter. The liquid covers the tube bundle and the vapor occupies

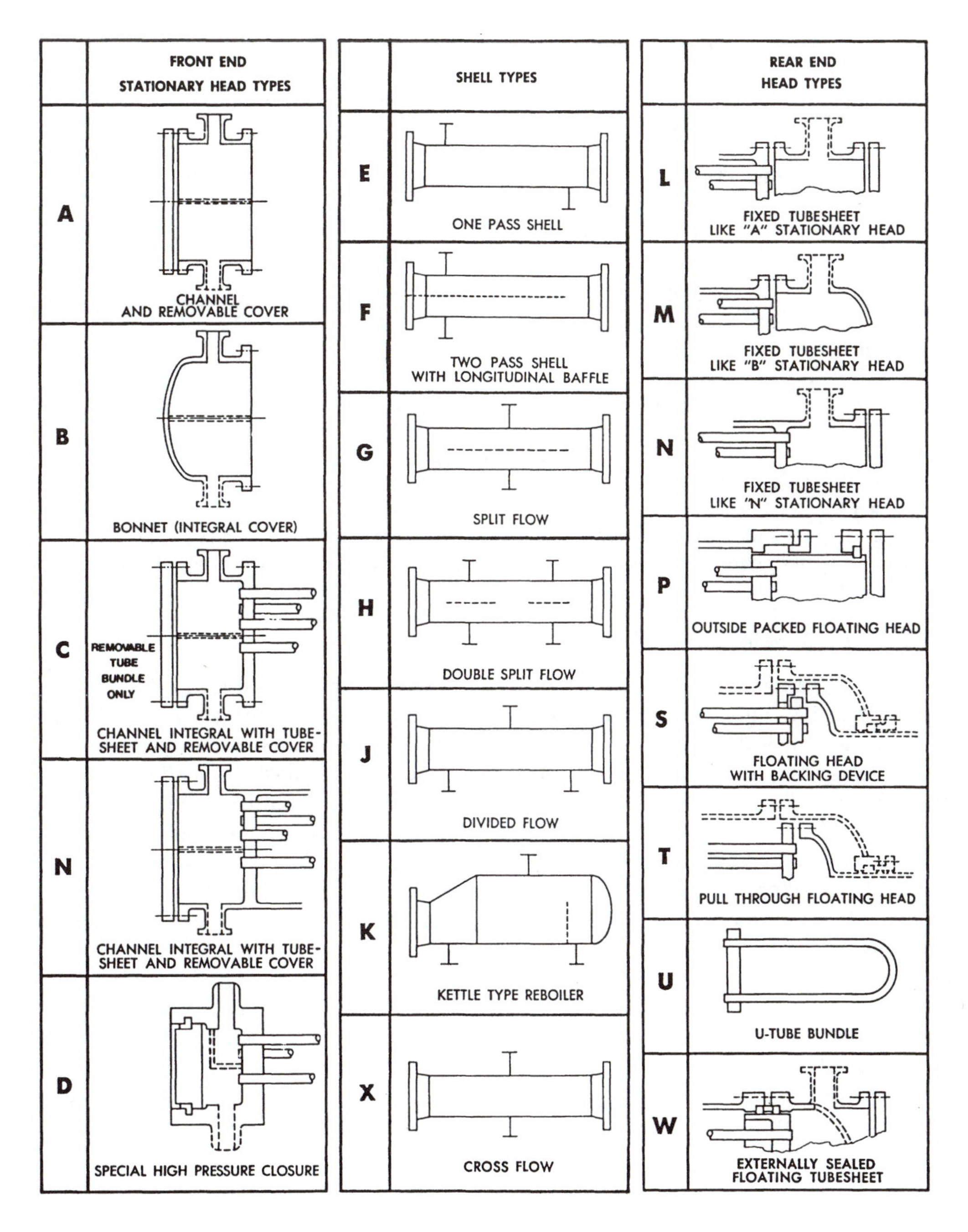

FIGURE 8.2
Standard shell types, and front-end and rear-end head types. (TEMA Standards, (1988): Tubular Exchanger Manufacturers Association, New York. With permission.)

the upper space without tubes. This shell is used when a portion of a stream needs to be vaporized, typically to a distillation column. The feed liquid enters the shell at the nozzle near the tube sheet, the nearly dry vapor exits out the top nozzle, and nonvaporized liquid overflows the end weir and exits through the right-hand nozzle. The tube bundle is commonly a U-tube configuration.

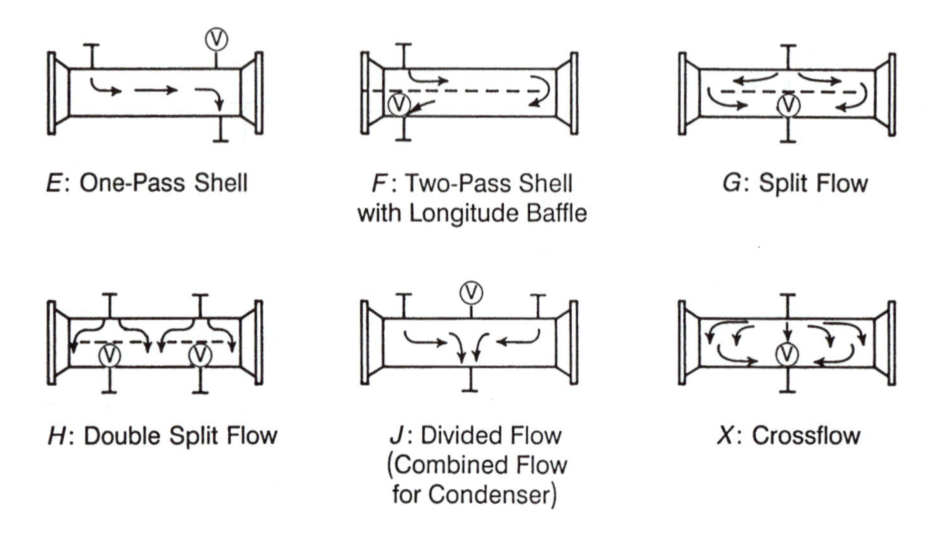

FIGURE 8.3
Schematic sketches of most common TEMA shell types. (From Butterworth, D. [1988] In *Two-Phase Flow Heat Exchangers — Thermal Hydraulic–Fundamentals and Design*, Kluwer, Dordrecht, The Netherlands. With permission.)

8.2.2 Tube Bundle Types

The most representative tube bundle types are shown in Figures 8.4 through 8.6. The main design objectives here are to accommodate thermal expansion, to furnish ease of cleaning, or to provide the least expensive construction if other features are of no importance.

One design variation that allows independent expansion of tubes and shell is the U-tube configuration (Figure 8.4). Therefore, thermal expansion is unlimited. The U-tube is the least expensive construction because only one tube sheet is needed. The tube side cannot be cleaned by mechanical means because of the *U-bend* shape. Only an even number of tube passes can be accommodated. Individual tubes cannot be replaced except in the outer row.

A *fixed tube sheet* configuration is shown in Figure 8.5. The shell is welded to the tube sheets and there is no access to the outside of the tube bundle for cleaning. This low-cost option has only limited thermal expansion, which can be somewhat increased by expansion bellows. Individual tubes are replaceable. Cleaning of tube is mechanically easy.

Several designs have been developed that permit the tube sheet to float. That is to move with thermal expansion. A classic type of pull-through *floating head design* is shown in Figure 8.6. The bundle can be removed with minimum disassembly that is important for heavy fouling units.

8.2.3 Tubes and Tube Passes

Only E-shell with one tube pass and F-shell with two tube passes result in nominal counterflow. All other multiple tube passes require a temperature profile correction (factor *F*); or in some cases, simply cannot deliver the desired temperatures because of temperature cross. The next resort is to use multiple units in series.

Generally, a large number of tube passes are used to increase tube-side fluid velocity and the heat transfer coefficient (within the available pressure drop), and to minimize fouling. If, for architectural reasons, the tube-side fluid must enter and exit on the same side, an even number of tube passes is mandatory.

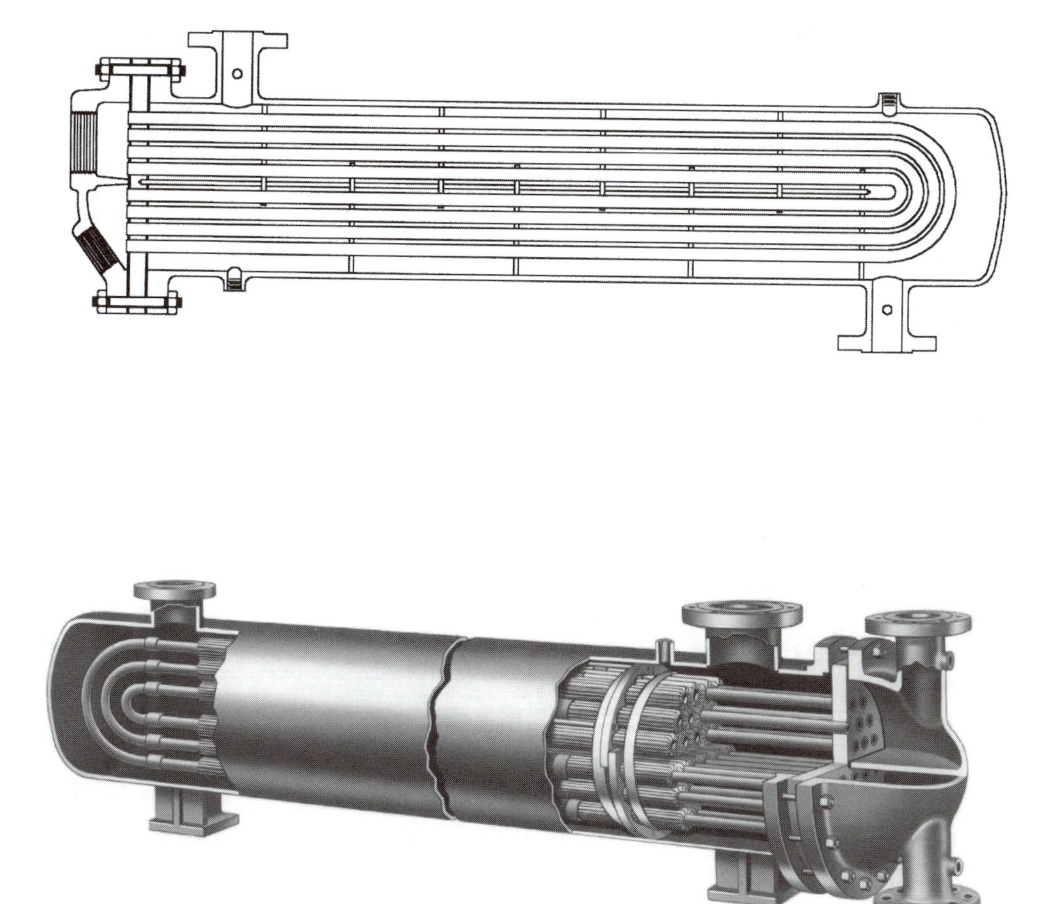

FIGURE 8.4
(a) A bare U-tube, baffled single-pass shell, shell-and-tube heat exchanger (courtesy of Patterson-Kelley Co.); and (b) finned U-tube shell-and-tube heat exchanger (courtesy of Brown-Fintube).

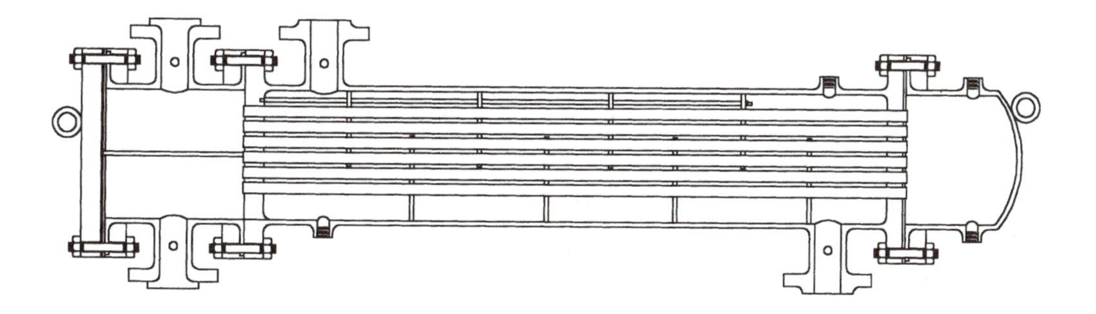

FIGURE 8.5
A two-pass tube, baffled single-pass shell, shell-and-tube heat exchanger designed for mechanical cleaning of the inside of the tubes. (Courtesy of Patterson-Kelley Co.)

Tube metal is usually low carbon steel, low alloy steel, stainless steel, copper, Admiralty, cupronickel, inconel, aluminum (in the form of alloys), or titanium. Other materials can also be selected for specific applications. The wall thickness of heat exchanger tubes is standardized in terms of the Birmingham wire gauge (BWG) of the tube. Tables. 8.1 and 8.2 give data on heat exchanger tubes.

Small tube diameters (8 to 15 mm) are preferred for greater area and volume density, but are limited for purposes of in-tube cleaning to 20 mm (3/4 in.). Larger tube diameters are often required for condensers and boilers.

The tubes either may be bare or have low fins on the outside. Low fin tubes are used when the fluid on the outside of the tubes has a substantially lower heat transfer coefficient than the fluid on the inside of the tubes.

Tube length affects the cost and operation of heat exchangers. Basically, the longer the tube (for any given total surface), the fewer tubes are needed; fewer holes are drilled; and shell diameter decreases, resulting in lower cost. There are, of course, several limits to this general rule, best expressed that the shell-diameter-to-tube-length ratio should be within limits of about 1/5 to 1/15. Maximum tube length is sometimes dictated by architectural layouts and ultimately by transportation to about 30 m.

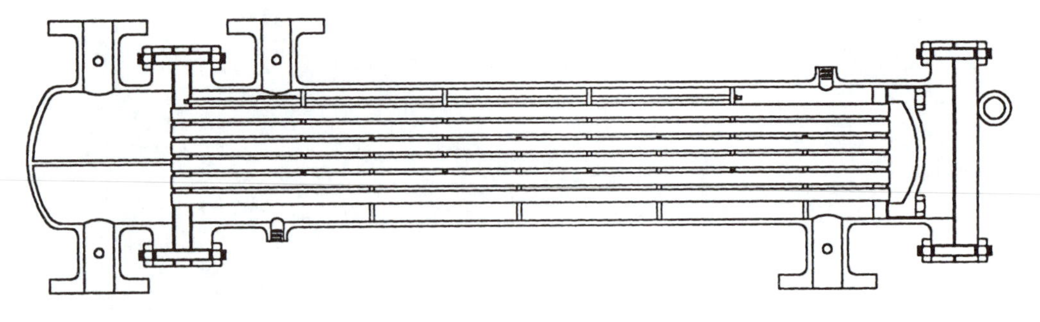

FIGURE 8.6
A heat exchanger similar to that of Figure 8.5 except of that this one is designed with a floating head to accommodate differential thermal expansion between the tubes and the shell. (Courtesy of Patterson-Kelley Co.)

TABLE 8.1
Dimensional Data for Commercial Tubing

O.D. of Tubing (in.)	BWG Gauge	Thickness (in.)	Internal Flow Area (in. ²)	External Surface per Foot Length (ft²)	Internal Surface per Foot Length (ft²)	Weight per Ft Length, Steel (lb)	I.D. Tubing (in.)	O.D./I.D. (in)
1/4	22	0.028	0.0295	0.0655	0.0508	0.066	0.194	1.289
1/4	24	0.022	0.0333	0.0655	0.0539	0.054	0.206	1.214
1/4	26	0.018	0.0360	0.0655	0.0560	0.045	0.214	1.168
3/8	18	0.049	0.0603	0.0982	0.0725	0.171	0.277	1.354
3/8	20	0.035	0.0731	0.0982	0.0798	0.127	0.305	1.233
3/8	22	0.028	0.0799	0.0982	0.0835	0.104	0.319	1.176
3/8	24	0.022	0.0860	0.0982	0.0867	0.083	0.331	1.133
1/2	16	0.065	0.1075	0.1309	0.0969	0.302	0.370	1.351
1/2	18	0.049	0.1269	0.1309	0.1052	0.236	0.402	1.244
1/2	20	0.035	0.1452	0.1309	0.1126	0.174	0.430	1.163
1/2	22	0.028	0.1548	0.1309	0.1162	0.141	0.444	1.126
5/8	12	0.109	0.1301	0.1636	0.1066	0.602	0.407	1.536
5/8	13	0.095	0.1486	0.1636	0.1139	0.537	0.435	1.437
5/8	14	0.083	0.1655	0.1636	0.1202	0.479	0.459	1.362

TABLE 8.1 (CONTINUED)

Dimensional Data for Commercial Tubing

O.D. of Tubing (in.)	BWG Gauge	Thickness (in.)	Internal Flow Area (in. ²)	External Surface per Foot Length (ft²)	Internal Surface per Foot Length (ft²)	Weight per Ft Length, Steel (lb)	I.D. Tubing (in.)	O.D./I.D. (in)
5/8	15	0.072	0.1817	0.1636	0.1259	0.425	0.481	1.299
5/8	16	0.065	0.1924	0.1636	0.1296	0.388	0.49s	1.263
5/8	17	0.058	0.2035	0.1636	0.1333	0.350	0.509	1.228
5/8	18	0.049	0.2181	0.1636	0.1380	0.303	0.527	1.186
5/8	19	0.042	0.2298	0.1636	0.1416	0.262	0.541	1.155
5/8	20	0.035	0.2419	0.1636	0.1453	0.221	0.555	1.136
3/4	10	0.134	0.1825	0.1963	0.1262	0.884	0.482	1.556
3/4	11	0.120	0.2043	0.1963	0.1335	0.809	0.510	1.471
3/4	12	0.109	0.2223	0.1963	0.1393	0.748	0.532	1.410
3/4	13	0.095	0.2463	0.1963	0.1466	0.666	0.560	1.339
3/4	14	0.083	0.2679	0.1963	0.1529	0.592	0.584	1.284
3/4	15	0.072	0.2884	0.1963	0.1587	0.520	0.606	1.238
3/4	16	0.065	0.3019	0.1963	0.1623	0.476	0.620	1.210
3/4	17	0.058	0.3157	0.1963	0.1660	0.428	0.634	1.183
3/4	18	0.049	0.3339	0.1963	0.1707	0.367	0.652	1.150
3/4	20	0.035	0.3632	0.1963	0.1780	0.269	0.680	1.103
7/8	10	0.134	0.2892	0.2291	0.1589	1.061	0.607	1.441
7/8	11	0.120	0.3166	0.2291	0.1662	0.969	0.635	1.378
7/8	12	0.109	0.3390	0.2291	0.1720	0.891	0.657	1.332
7/8	13	0.095	0.3685	0.2291	0.1793	0.792	0.685	1.277
7/8	14	0.083	0.3948	0.2291	0.1856	0.704	0.709	1.234
7/8	16	0.065	0.4359	0.2291	0.1950	0.561	0.745	1.174
7/8	18	0.049	0.4742	0.2291	0.2034	0.432	0.777	1.126
7/8	20	0.035	0.5090	0.2291	0.2107	0.313	0.805	1.087
1	8	0.165	0.3526	0.2618	0.1754	1.462	0.670	1.493
1	10	0.134	0.4208	0.2618	0.1916	1.237	0.732	1.366
1	11	0.120	0.4536	0.2618	0.1990	1.129	0.760	1.316
1	12	0.109	0.4803	0.2618	0.2047	1.037	0.782	1.279
1	13	0.095	0.5153	0.2618	0.2121	0.918	0.810	1.235
1	14	0.083	0.5463	0.2618	0.2183	0.813	0.834	1.199
1	15	0.072	0.5755	0.2618	0.2241	0.714	0.856	1.167
1	16	0.065	0.5945	0.2618	0.2278	0.649	0.870	1.119
1	18	0.049	0.6390	0.2618	0.2361	0.496	0.902	1.109
1	20	0.035	0.6793	0.2618	0.2435	0.360	0.930	1.075
1 1/4	7	0.180	0.6221	0.3272	0.2330	2.057	0.890	1.404
1 1/4	8	0.165	0.6648	0.3272	0.2409	1.921	0.920	1.359
1 1/4	10	0.134	0.7574	0.3272	0.2571	1.598	0.982	1.273
1 1/4	11	0.120	0.8012	0.3272	0.2644	1.448	1.010	1.238
1 1/4	12	0.109	0.8365	0.3272	0.2702	1.329	1.032	1.211
1 1/4	12	0.095	0.8825	0.3272	0.2773	1.173	1.060	1.179
1 1/4	14	0.083	0.9229	0.3272	0.2838	1.033	1.084	1.153
1 1/4	16	0.065	0.9852	0.3272	0.2932	0.823	1.120	1.116
1 1/4	18	0.049	1.042	0.3272	0.3016	0.629	1.152	1.085
1 1/4	20	0.035	1.094	0.3272	0.3089	0.456	1.180	1.059
1 1/2	10	0.134	1.192	0.3927	0.3225	1.955	1.232	1.218
1 1/2	12	0.109	1.291	0.3927	0.3356	1.618	1.282	1.170
1 1/2	14	0.083	1.398	0.3927	0.3492	1.258	1.334	1.124
1 1/2	16	0.065	1.474	0.3927	0.3587	0.996	1.370	1.095
2	11	0.120	2.433	0.5236	0.4608	2.410	1.760	1.136
2	13	0.095	2.573	0.5236	0.4739	1.934	1.810	1.105
2 1/2	9	0.148	3.815	0.6540	0.5770	3.719	2.204	1.134

Note: Courtesy of Tubular Exchanger Manufacturers Association.

TABLE 8.2

Heat Exchanger and Condenser Tube Data

Nominal Pipe Size (in)	O.D. (in.)	Schedule Number or Weight	Wall Thickness (in.)	I.D. (in.)	Surface Area		Cross-Sectional Area	
					Outside (ft²/ft)	Inside (ft²/ft)	Metal Area (in.²)	Flow Area (in.²)
		40	0.113	0.824	0.275	0.216	0.333	0.533
3/4	1.05	80	0.154	0.742	0.275	0.194	0.434	0.432
		40	0.133	1.049	0.344	0.275	0.494	0.864
1	1.315	80	0.179	0.957	0.344	0.250	0.639	0.719
		40	0.140	1.38	0.434	0.361	0.668	1.496
1 1/4	1.660	80	0.191	1.278	0.434	0.334	0.881	1.283
		40	0.145	1.61	0.497	0.421	0.799	2.036
1 1 /2	1.900	80	0.200	1.50	0.497	0.393	1.068	1.767
		40	0.154	2.067	0.622	0.541	1.074	3.356
2	2.375	80	0.218	1.939	0.622	0.508	1.477	2.953
		40	0.203	2.469	0.753	0.646	1.704	4.79
2 1 /2	2.875	80	0.276	2.323	0.753	0.608	2.254	4.24
		40	0.216	3.068	0.916	0.803	2.228	7.30
3	3.5	80	0.300	2.900	0.916	0.759	3.106	6.60
		40	0.226	3.548	1.047	0.929	2.680	9.89
3 1/2	4.0	80	0.318	3.364	1.047	0.881	3.678	8.89
		40	0.237	4.026	1.178	1.054	3.17	12.73
4	4.5	80	0.337	3.826	1.178	1.002	4.41	11.50
		10 S	0.134	5.295	1.456	1.386	2.29	22.02
5	5.563	40	0.258	5.047	1.456	1.321	4.30	20.01
		80	0.375	4.813	1.456	1.260	6.11	18.19
		10 S	0.134	6.357	1.734	1.664	2.73	31.7
6	6.625	40	0.280	6.065	1.734	1.588	5.58	28.9
		80	0.432	5.761	1.734	1.508	8.40	26.1
		10 S	0.148	8.329	2.258	2.180	3.94	54.5
8	8.625	30	0.277	8.071	2.258	2.113	7.26	51.2
		80	0.500	7.625	2.258	1.996	12.76	45.7
		10 S	0.165	10.420	2.81	2.73	5.49	85.3
10	10.75	30	0.279	10.192	2.81	2.67	9.18	81.6
		Extra heavy	0.500	9.750	2.81	2.55	16.10	74.7
		10 S	0.180	12.390	3.34	3.24	7.11	120.6
	12.75	30	0.330	12.09	3.34	3.17	12.88	114.8
		Extra heavy	0.500	11.75	3.34	3.08	19.24	108.4
		10	0.250	13.5	3.67	3.53	10.80	143.1
14	14.0	Standard	0.375	13.25	3.67	3.47	16.05	137.9
		Extra heavy	0.500	13.00	3.67	3.40	21.21	132.7
		10	0.250	15.50	4.19	4.06	12.37	188.7
16	16.0	Standard	0.375	15.25	4.19	3.99	18.41	182.7
		Extra heavy	0.500	15.00	4.19	3.93	24.35	176.7
		10 S	0.188	17.624	4.71	4.61	10.52	243.9
18	18.0	Standard	0.375	17.25	4.71	4.52	20.76	233.7
		Extra heavy	0.500	17.00	4.71	4.45	27.49	227.0

Note: Courtesy of Tubular Exchanger Manufacturers Association.

8.2.4 Tube Layout

Tube layout is characterized by the included angle between tubes, as shown in Figure 8.7. Layout of 30° results in the greatest tube density and is therefore used, unless other requirements dictate otherwise. For example, clear lanes (1/4 in. or 7 mm) are required because of external cleaning using a square 90° or 45° layout. Tube pitch, P_T, is usually chosen so that the pitch ratio P_T/d_o, is between 1.25 and 1.5. When the tubes are too close, the tube sheet

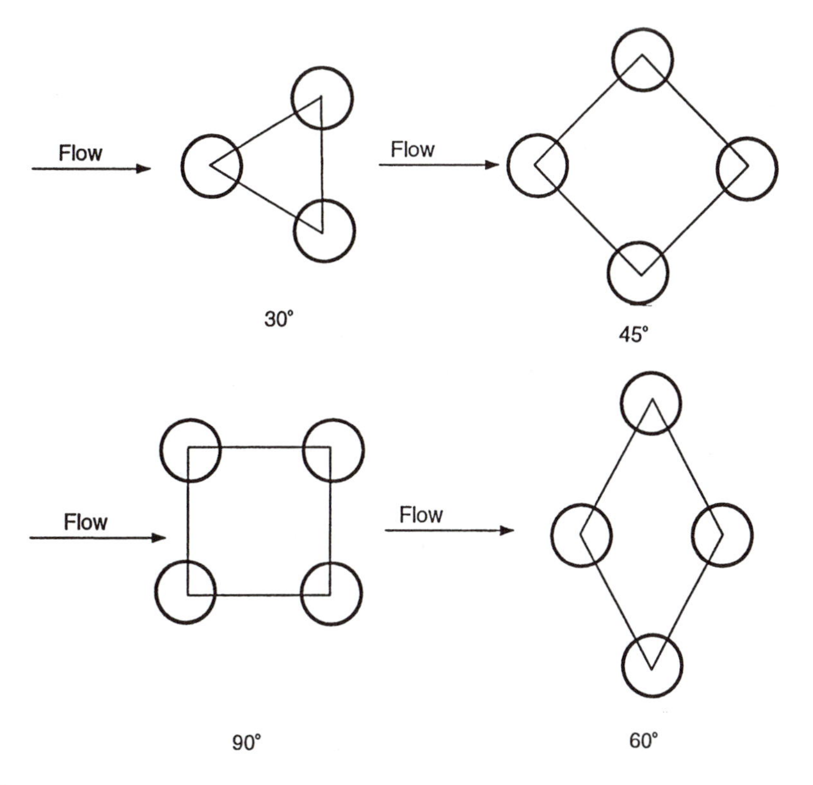

FIGURE 8.7
Tube layout angles.

becomes structurally weak. The tube layout and the tube locations have been standardized. The number of tubes (tube counts) that can be placed within a shell depend on tube layout, tube O.D., pitch size, number of passes, and shell diameter. Tube counts are given in Table 8.3,[3] which actually gives the maximum number of tubes that can be accommodated under the conditions specified.

For example, consider that we have 1 in.-O.D. tubes laid out on a 1 1/4-in. square pitch with a shell diameter of 31 in. If the tube arrangement is that for one pass (1-P), then Table 8.3 gives the maximum tubes as 406. If the heat exchanger is that for two passes (2-P), then the maximum number of tubes is 398.

8.2.5 Baffle Type and Geometry

Baffles serve two functions: first and most importantly to support the tubes for structural rigidity, preventing tube vibration and sagging; and second to divert the flow across the bundle to obtain a higher heat transfer coefficient.

Baffles may be classified as transverse and longitudinal types (e.g., F-shell has longitudinal baffle). The transverse baffles may be classified as *plate* baffles and *rod* baffles.

The most commonly used plate baffle types are shown in Figure 8.8, and a brief description follows.[4]

The *single* and *double segmental* baffles are most frequently used. They divert the flow most effectively across the tubes. The baffle spacing, however, must be chosen carefully. Optimum baffle spacing is somewhere between 0.4 and 0.6 of the shell diameter, and a baffle cut of 25 to 35% is usually recommended. The *triple* and *no-tubes-in-window* segmental

TABLE 8.3

Tube-Shell Layouts (Tube Counts)

Shell I.D. (in.)	1-P	2-P	4-P	6-P	8-P
3/4-in. O.D. Tubes on 1-in. Triangular Pitch					
8	37	30	24	24	
10	61	52	40	36	
12	92	82	76	74	70
13 1/4	109	106	86	82	74
15 1/4	151	138	122	118	110
17 1/4	203	196	178	172	166
19 1/4	262	250	226	216	210
21 1/4	316	302	278	272	260
23 1/4	384	376	352	342	328
25	470	452	422	394	382
27	559	534	488	474	464
29	630	604	556	538	508
31	745	728	678	666	640
33	856	830	774	760	732
35	970	938	882	864	848
37	1074	1044	1012	986	870
39	1206	1176	1128	1100	1078
1-in. O.D. Tubes on 1 1/4-in. Triangular Pitch					
8	21	16	16	14	
10	32	32	26	24	
12	55	52	48	46	44
13 1/4	68	66	58	54	50
15 1/4	91	86	80	74	72
17 1/4	131	118	106	104	94
19 1/4	163	152	140	136	128
21 1/4	199	188	170	164	160
23 1/4	241	232	212	212	202
25	294	282	256	252	242
27	349	334	302	296	286
29	397	376	338	334	316
31	472	454	430	424	400
33	538	522	486	470	454
35	608	592	562	546	532
37	674	664	632	614	598
39	766	736	700	688	672
3/4-in. O.D. Tubes on 1-in. Square Pitch					
8	32	26	20	20	
10	52	52	40	36	
12	81	76	68	68	60
13 1/4	97	90	82	76	70
15 1/4	137	124	116	108	108
17 1/4	177	166	158	150	142
19 1/4	224	220	204	192	188
21 1/4	277	270	246	240	234
23 1/4	341	324	308	302	292
25	413	394	370	356	346
27	481	460	432	420	408
29	553	526	480	468	456

TABLE 8.3 (CONTINUED)

Tube-Shell Layouts (Tube Counts)

Shell I.D. (in.)	1-P	2-P	4-P	6-P	8-P
31	657	640	600	580	560
33	749	718	688	676	648
35	845	824	780	766	748
37	934	914	886	866	838
39	1049	1024	982	968	948
1-in. O.D. Tubes on 1 1/4-in. Square Pitch					
8	21	16	14		
10	32	32	26	24	
12	48	45	40	38	36
13 1/4	61	56	52	48	44
15 1/4	81	76	68	68	64
17 1/4	112	112	96	90	82
19 1/4	138	132	128	122	116
21 1/4	177	166	158	152	148
23 1/4	213	208	192	184	184
25	260	252	238	226	222
27	300	288	278	268	260
29	341	326	300	294	286
31	406	398	380	368	358
33	465	460	432	420	414
35	522	518	488	484	472
37	596	574	562	544	532
39	665	644	624	612	600
3/4-in. O.D. Tubes on 15/16-in. Triangular Pitch					
8	36	32	26	24	18
10	62	56	47	42	36
12	109	98	86	82	78
13 1/4	127	114	96	90	86
15 1/4	170	160	140	136	128
17 1/4	239	224	194	188	178
19 1/4	301	282	252	244	234
21 1/4	361	342	314	306	290
23 1/4	442	420	386	378	364
25	532	506	468	446	434
27	637	602	550	536	524
29	721	692	640	620	594
31	847	822	766	722	720
33	974	938	878	852	826
35	1102	1068	1004	988	958
37	1240	1200	1144	1104	1072
39	1377	1330	1258	1248	1212
1 1/4-in. O.D. Tubes on 1 9/16-in. Square Pitch					
10	16	12	10		
12	30	24	22	16	16
13 1/4	32	30	30	22	22
15 1/4	44	40	37	35	31
17 1/4	56	53	51	48	44
19 1/4	78	73	71	64	56
21 1/4	96	90	86	82	78

TABLE 8.3 (CONTINUED)

Tube-Shell Layouts (Tube Counts)

Shell I.D. (in.)	1-P	2-P	4-P	6-P	8-P
23 1/4	127	112	106	102	96
25	140	135	127	123	115
27	166	160	151	146	140
29	193	188	178	174	166
31	226	220	209	202	193
33	258	252	244	238	226
35	293	287	275	268	258
37	334	322	311	304	293
39	370	362	348	342	336
1 1/2-in. O.D. Tubes on 1 7/8-in. Square Pitch					
12	16	16	12	12	
13 1/4	22	22	16	16	
15 1/4	29	29	24	24	22
17 1/4	29	39	34	32	29
19 1/4	50	48	45	43	39
21 1/4	62	60	57	54	50
23 1/4	78	74	70	66	62
25	94	90	86	84	78
27	112	108	102	98	94
29	131	127	120	116	112
31	151	146	141	138	131
33	176	170	164	160	151
35	202	196	188	182	176
37	224	220	217	210	202
39	252	246	237	230	224
1 1/2-in. O.D. Tubes on 1 7/8-in. Triangular Pitch					
12	18	14	14	12	12
13 1/4	27	22	18	16	14
15 1/4	26	34	32	30	27
17 1/4	48	44	42	38	36
19 1/4	61	58	55	51	48
21 1/4	76	78	70	66	61
23 1/4	95	91	86	80	76
25	115	110	105	98	95
27	136	131	125	118	115
29	160	154	147	141	136
31	184	177	172	165	160
33	215	206	200	190	184
35	246	238	230	220	215
37	275	268	260	252	246
39	307	299	290	284	275
1 1/4-in. O.D. Tubes on 1 9/16-in. Triangular Pitch					
10					
10	20	18	14		
12 1/4	32	30	26	22	20
13 1/4	38	36	32	28	26
15 1/4	54	51	45	42	38
17 1/4	69	66	62	58	54
19 1/4	95	91	86	78	69
21 1/4	117	112	105	101	95

TABLE 8.3 (CONTINUED)

Tube-Shell Layouts (Tube Counts)

Shell I.D. (in.)	1-P	2-P	4-P	6-P	8-P
23 1/4	140	136	130	123	117
25	170	164	155	150	140
27	202	196	185	179	170
29	235	228	217	212	202
31	275	270	255	245	235
33	315	305	297	288	275
35	357	348	335	327	315
37	407	390	380	374	357
39	449	436	425	419	407

From Kern, D. Q. (1950) *Process Heat Transfer*, McGraw Hill, New York. With permission.

baffles are used for low-pressure drop applications, which are approximately 0.5 and 0.3 of the segmental value.

No-tubes-in-window construction eliminates the tubes that are otherwise supported only by every second baffle, thus minimizing tube vibration.

Disk-and-ring (doughnut) baffles are composed of alternating outer rings and inner disks, which direct the flow radially across the tube field. The potential bundle-to-shell bypass stream is thus eliminated; there are some indications that the baffle type is very effective in pressure drop to heat transfer conversion. At present these baffles are rarely used in the United States, but they are popular in Europe.

Another type of plate baffle is the orifice baffle in which shell-side fluid flows through the clearance between tube O.D. and baffle hole diameter.

Rod or grid baffles are formed by a grid of rod or strip supports (Figure 8.9). The flow is essentially longitudinal, resulting in very low pressure drops. Because of the close baffle spacing, the tube vibration danger is virtually eliminated. This construction can be used effectively for vertical condensers and reboilers.

8.2.6 Allocation of Streams

A decision must be made as to which fluid will flow through the tubes and which one, through the shell. In general, the following considerations apply:

- The more seriously fouling fluid flows through the tube, because the tube side is easier to clean, especially if mechanical cleaning is required.
- The high-pressure fluid flows through the tubes. Because of their small diameter, normal thickness tubes are available to withstand with higher pressures and only the tube-side channels and other connections need to be designed to withstand high pressure.
- The corrosive fluid must flow through the tubes; otherwise both the shell and tubes will be corroded. Special alloys are used to resist corrosion, and it is much less expensive to provide special alloy tubes than to provide both special tubes and special alloy shell.
- The stream with the lower heat transfer coefficient flows on the shell side, because it is easy to design outside finned tubes. In general, it is better to put the stream with lower mass flow rate on the shell side. Turbulent flow is obtained at lower Reynolds numbers on the shell side.

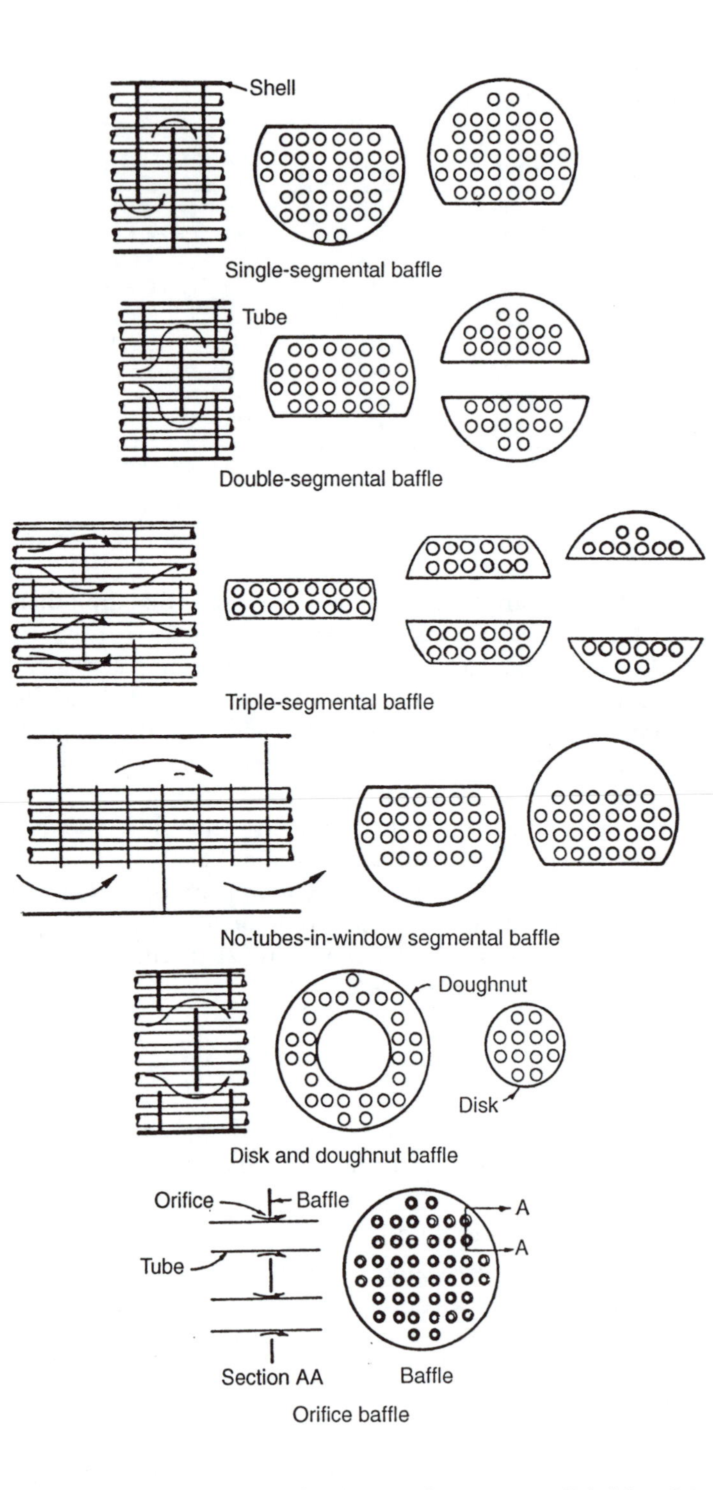

FIGURE 8.8
Plate baffle types. (Adapted from Kakaç, S., Bergles, A. E., and Mayinger, F. [Eds.] [1981] *Heat Exchangers — Thermal–Hydraulic Fundamentals and Design*, Taylor & Francis, Washington, D.C.)

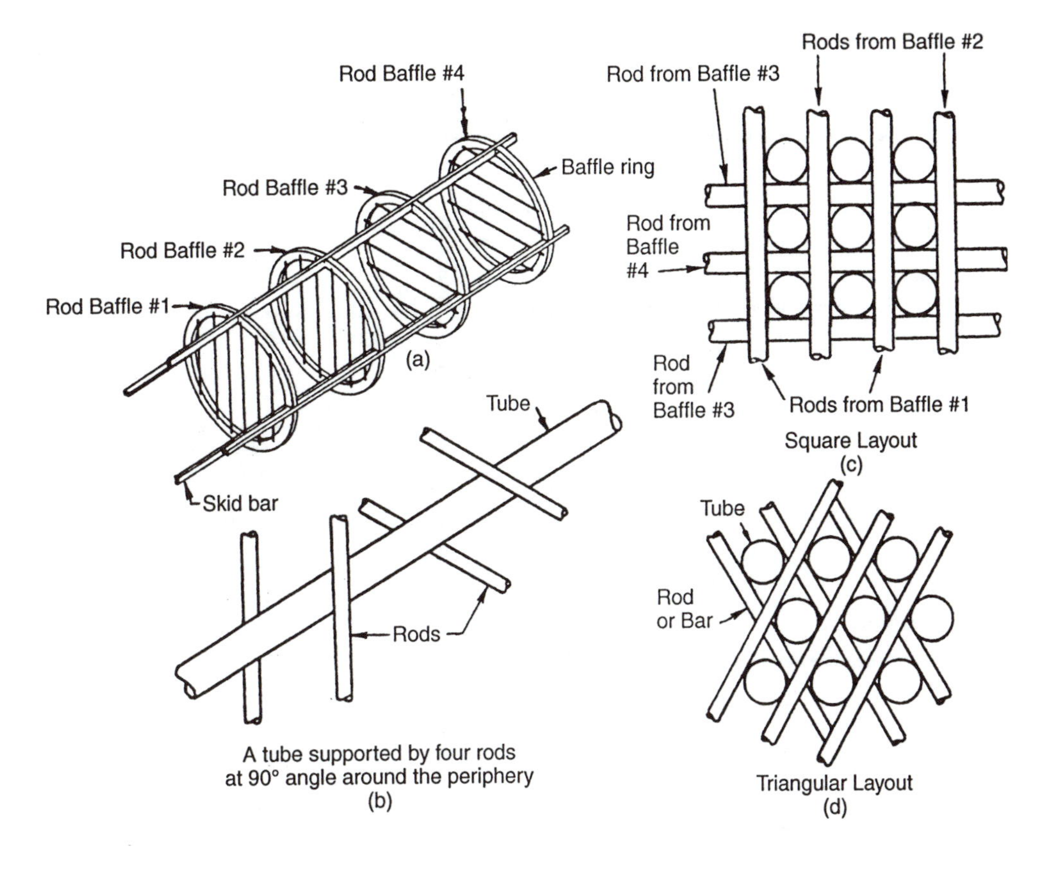

FIGURE 8.9
(a) Four rod baffles held by skid bars (no tube shown), (b) a tube supported by four rods, (c) a square layout of tubes with rods, and (d) a triangle layout of tubes with rods. (Adapted from Kakaç, S., Bergles, A. E., and Mayinger, F. [Eds.] [1981] *Heat Exchangers — Thermal–Hydraulic Fundamentals and Design*, Taylor & Francis, Washington, D.C.)

Problems arise when the preceding requirements are in conflict. Then the designer must estimate trade-offs, and find the most economical choices.

8.3 Basic Design Procedure of a Heat Exchanger

A selected shell-and-tube heat exchanger must satisfy the process requirements with the allowable pressure drops until the next scheduled cleaning of the plant. The basic logical structure of the process heat exchanger design procedure is shown in Figure 8.10.[5]

First, the problem must be identified as completely as possible. Matters not only like flow rates and compositions (condensation or boiling), inlet and outlet temperatures, pressures of both streams, but also like the exact requirements of the process engineer and the additional information needed for the design engineer, must be discussed in detail. The main duty of the process engineer is to supply all the information to the heat exchanger designer.

At this point in the design process, the basic configuration of the heat exchanger must be tentatively selected (if one is not unconditionally desired) whether it is to be U-tube, baffled single-pass shell, tube pass, baffled single-pass shell with fixed tubes, or shell-and-tube

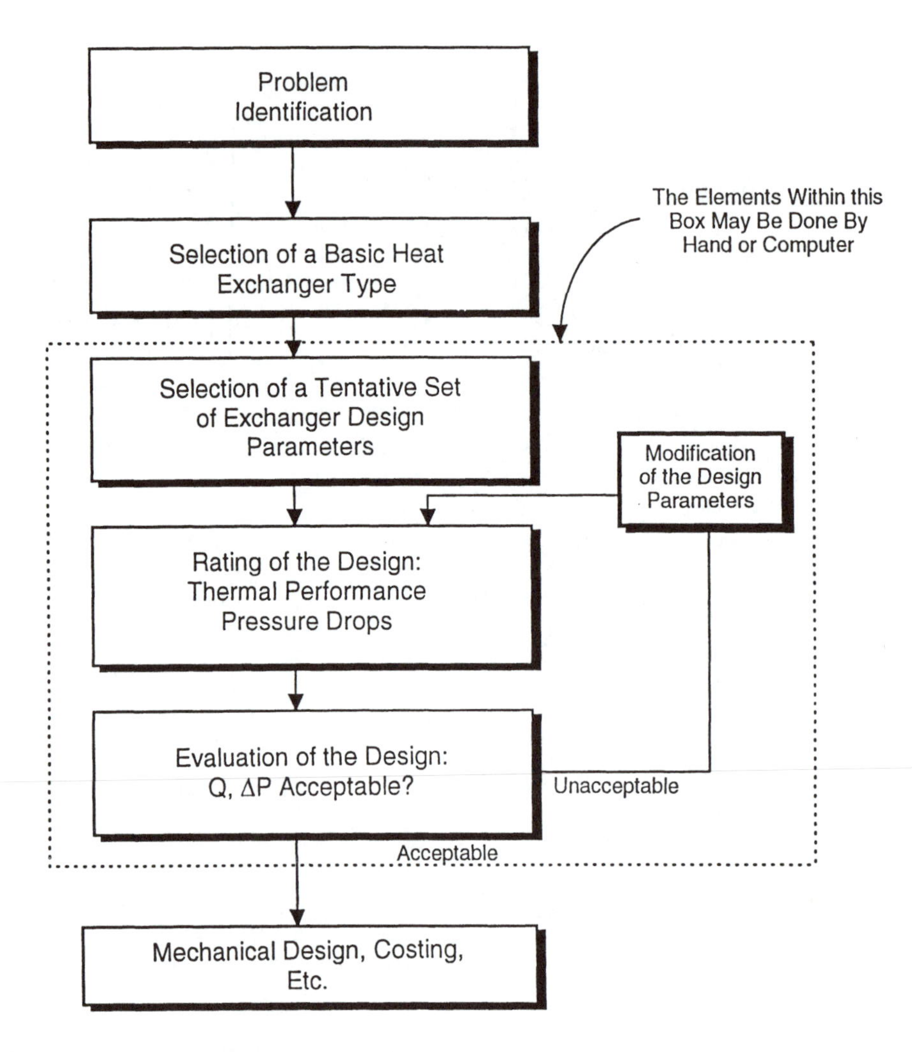

FIGURE 8.10
Basic logic structure for process heat exchanger design. (Drawn after Bell, K. J. [1981] *Heat Exchangers — Thermal–Hydraulic Fundamentals and Design*, Taylor & Francis, Washington, D.C.)

heat exchanger with a floating head to accommodate differential thermal expansion between the tube and the shell. The next step is to select a tentative set of exchanger design parameters. A preliminary estimate of the heat exchanger size can be made as outlined in Section 8.3.1. Then the initial design will be rated; that is, the thermal performance and the pressure drops for both streams will be calculated for this design.

8.3.1 Preliminary Estimation of Unit Size

The size of the heat transfer exchanger can be obtained from Equation (2.36):

$$A_o = \frac{Q}{U_o \Delta T_m} = \frac{Q}{U_o F \Delta T_{lm,cf}} \tag{8.1}$$

where A_o is the outside heat transfer surface area based on the O.D. of the tube, and Q is the heat duty of the exchanger.

First we estimate the individual heat transfer coefficients with fouling factors. Tables for the estimation of individual heat transfer coefficients or overall heat transfer coefficients are available in various handbooks, such as Tables 8.4 and 8.5. The estimation of heat transfer coefficients is preferable for estimating the overall heat transfer coefficient; then the designer can get a feel for the relative magnitude of the resistances.[6,7]

The overall heat transfer coefficient, U_o, based on the O.D. of tubes can be estimated from the estimated values of individual heat transfer coefficients, the wall and fouling resistance, and the overall surface efficiency using Equation (2.17).

TABLE 8.4

Typical Film Heat Transfer Coefficients for Shell-and-Tube Heat Exchangers

	Fluid Condition	W/(m²·K)
Sensible Heat Transfer		
Water	Liquid	5,000–7,500
Ammonia	Liquid	6,000–8,000
Light organics	Liquid	1,500–2,000
Medium organics	Liquid	750–1,500
Heavy organics	Liquid	
	Heating	250–750
	Cooling	150–400
Very heavy organics	Liquid	
	Heating	100–300
	Cooling	60–150
Gas	1–2 bar abs	80–125
Gas	10 bar abs	250–400
Gas	100 bar abs	500–800
Condensing Heat Transfer		
Steam, ammonia	No noncondensable	8,000–12,000
Light organics	Pure component, 0.1 bar abs, no noncondensable	2,000–5,000
Light organics	0.1 bar, 4% noncondensable	750–1,000
Medium organics	Pure or narrow condensing range, 1 bar abs	1,500–4,000
Heavy organics	Narrow condensing range, 1 bar abs	600–2,000
Light multicomponent mixture, all condensable	Medium condensing range, 1 bar abs	1,000–2,500
Medium multicomponent mixutre, all condensable	Medium condensing range, 1 bar abs	600–1,500
Heavy multicomponent mixture, all condensable	Medium condensing range, 1 bar abs	300–600
Vaporizing Heat Transfer		
Water	Pressure < 5 bar abs, ΔT = 25 K	5,000–10,000
Water	Pressure 5–100 bar abs, ΔT = 20 K	4,000–15,000
Ammonia	Pressure < 30 bar abs, ΔT = 20 K	3,000–5,000
Light organics	Pure component, pressure < 30 bar abs, ΔT = 20 K	2,000–4,000
Light organics	Narrow boiling range, pressure 20–150 bar abs, ΔT = 15–20 K	750–3,000
Medium organics	Narrow boiling range, pressure < 20 bar abs, ΔT_{max} = 15 K	600–2,500
Heavy organics	Narrow boiling range, pressure < 20 bar abs, ΔT_{max} = 15 K	400–1,500

TABLE 8.5

Approximate Overall Heat Transfer Coefficients for Preliminary Analysis

Fluids	U (W/m²·K)
Water to water	1300–2500
Ammonia to water	1000–2500
Gases to water	10–250
Water to compressed air	50–170
Water to lubricating oil	110–340
Light organics ($\mu < 5 \times 10^{-4}$ Ns/m²) to water	370–750
Medium organics ($5 \times 10^{-4} < \mu < 10 \times 10^{-4}$ Ns/m²) to water	240–650
Heavy organics ($\mu > 10 \times 10^{-4}$ Ns/m²) to lubricating oil	25–400
Steam to water	2200–3500
Steam to ammonia	1000–3400
Water to condensing ammonia	850–1500
Water to boiling Freon-12	280–1000
Steam to gases	25–240
Steam to light organics	490–1000
Steam to medium organics	250–500
Steam to heavy organics	30–300
Light organics to light organics	200–350
Medium organics to medium organics	100–300
Heavy organics to heavy organics	50–200
Light organics to heavy organics	50–200
Heavy organics to light organics	150–300
Crude oil to gas oil	130–320
Plate heat exchangers: water to water	3000–4000
Evaporators: steam/water	1500–6000
Evaporators: steam/other fluids	300–2000
Evaporators of refrigeration	300–1000
Condensers: steam/water	1000–4000
Condensers: steam/other fluids	300–1000
Gas boiler	10–50
Oil bath for heating	30–550

$$\frac{1}{U_o} = \frac{A_o}{A_i}\left(\frac{1}{\eta_i h_i} + \frac{R_{fi}}{\eta_i}\right) + A_o R_W + \frac{R_{fo}}{\eta_o} + \frac{1}{\eta_o h_o} \tag{8.2}$$

At this stage, it is useful to determine the distribution of the thermal resistances under clean and fouled conditions.

For the single-tube pass, purely countercurrent heat exchanger, $F = 1.00$. For preliminary design shell with any even number of tube-side passes, F may be estimated as 0.9.

Heat load can be estimated from the heat balance as:

$$Q = (\dot{m}c_p)_c(T_{c_2} - T_{c_1}) = (\dot{m}c_p)_h(T_{h_1} - T_{h_2}) \tag{8.3}$$

If one stream changes phase:

$$Q = mh_{fg} \tag{8.4}$$

where m is the mass of the stream changing phase per unit time and h_{fg} is the latent heat of the phase change.

We need to calculate the LMTD for countercurrent flow from the four given inlet and outlet temperatures. If three temperatures are known, the fourth one can be found from the heat balance:

$$\Delta T_{lm,cf} = \frac{(T_{h_1} - T_{c_2}) - (T_{h_2} - T_{c_1})}{\ln \dfrac{T_{h_1} - T_{c_2}}{T_{h_2} - T_{c_1}}} \tag{8.5}$$

The problem now is to convert the area calculated from Equation (8.1) into reasonable dimensions of the first trial. The objective is to find the right number of tubes of diameter, d_o, and the shell diameter, D_s, to accommodate the number of tubes, N_t, with given tube length, L.

$$A_o = \pi d_o N_t L \tag{8.6}$$

One can find the shell diameter, D_s, which would contain the right number of tubes, N_t, of diameter, d_o.

The total number of the tubes, N_t, can be predicted in fair approximation as function of the shell diameter by taking the shell circle and dividing it by the projected area of the tube layout (Figure 8.7) pertaining to a single tube A_1:

$$N_t = (CTP)\frac{\pi D_s^2}{4A_1} \tag{8.7}$$

where CTP is the tube count calculation constant that accounts for the incomplete coverage of the shell diameter by the tubes, due to necessary clearances between the shell and the outer tube circle and tube omissions due to tube pass lanes for multitube pass design.

Based on fixed tube sheet the following values are suggested :

one-tube pass:	$CTP = 0.93$
two-tube pass:	$CTP = 0.9$
three-tube pass:	$CTP = 0.85$

$$A_1 = (CL)\,P_T^2 \tag{8.8}$$

where CL is the tube layout constant:

$CL = 1.0$ for 90° and 45°

$CL = 0.87$ for 30° and 60°

Equation (8.7) can be written as:

$$N_t = 0.875\left(\frac{CTP}{CL}\right)\frac{D_s^2}{(PR)^2 d_o^2} \tag{8.9}$$

where PR is tube pitch ratio ($= P_T/d_o$).

By substituting N_t from Equation (8.6) into Equation (8.9), an expression for the shell diameter in terms of main constructional diameters can be obtained as:[7]

$$D_s = 0.637\sqrt{\frac{CL}{CTP}}\left[\frac{A_o (PR)^2 d_o}{L}\right]^{1/2} \tag{8.10}$$

Example 8.1

A heat exchanger is to be designed to heat raw water by the use of condensed water at 67°C and 0.2 bar, which will flow in the shell side with a mass flow rate of 50,000 kg/h. The heat will be transferred to 30,000 kg/h of city water coming from a supply at 17°C (c_p = 4184 J/kg·K). A single-shell and a single-tube pass is preferable. A fouling resistance of 0.000176 m²·K/W is suggested with 35% surface over design, whichever is smaller. A maximum coolant velocity of 1.5 m/s is suggested to prevent erosion. A maximum tube length of 5 m is required because of space limitations. The tube material is carbon steel (k = 60 W/m·K). Raw water will flow inside of 3/4-in. straight tubes (19-mm O.D. with 16-mm I.D.). Tubes are laid out on a square pitch with a pitch ratio of 1.25. The baffle spacing is approximated by 0.6 of shell diameter and the baffle cut is set to 25%. The permissible maximum pressure drop on the shell side is 1.5 lb_f/in.². Water outlet temperature should not be less than 40°C. Perform the preliminary analysis.

SOLUTION

Preliminary analysis *— The cold water outlet temperature of 40°C minimum determines the exchanger configuration to be considered. Heat duty can be calculated from the fully specified cold stream:*

$$Q = (\dot{m}c_p)_c (T_{c_2} - T_{c_1})$$

$$= \frac{30{,}000}{3600} \times 4179(40-17) = 801 \text{ kW}$$

The hot water outlet temperature becomes:

$$T_{h_2} = T_{h_1} - \frac{Q}{(\dot{m}c_p)_h} = 67 - \frac{801\times10^3}{\frac{50{,}000}{3600}\times 4184} = 53.2°\text{C}$$

First, we have to estimate the individual heat transfer coefficients from Table 8.4. We can assume the shell-side heat transfer coefficient and the tube-side heat transfer coefficient as 5000 W/m²·K and 4000 W/m²·K, respectively. By assuming bare tubes, one can estimate the overall heat transfer coefficient from Equation (8.2) as:

$$\frac{1}{U_f} = \frac{1}{h_o} + \frac{r_o}{r_i}\frac{1}{h_i} + R_{ft} + r_o\frac{\ln(r_o/r_i)}{k}$$

$$U_f = \left[\frac{1}{5{,}000} + \frac{19}{16}\frac{1}{4{,}000} + 0.000176 + \frac{0.019}{2}\frac{\ln(19/16)}{60}\right]^{-1} = 1428.40 \text{ W/m}^2\cdot\text{K}$$

and

$$\frac{1}{U_c} = \frac{1}{h_o} + \frac{r_o}{r_i}\frac{1}{h_i} + r_o\frac{\ln(r_o/r_i)}{k}$$

$$U_c = \frac{1}{5,000} + \frac{19}{16}\frac{1}{6,000} + \frac{0.019}{2}\frac{\ln(19/16)}{60} = 1908.09 \text{ W/m}^2\cdot\text{K}$$

We need to calculate ΔT_m from the four inlet and outlet temperatures:

$$\Delta T_{lm,cf} = \frac{\Delta T_1 - \Delta T_2}{\ln(\Delta T_1/\Delta T_2)} = \frac{27 - 36.2}{\ln\left(\frac{27}{36.2}\right)} = 31.4°\text{C}$$

By assuming $F = 0.90$, then:

$$\Delta T_m = 0.90\Delta T_{lm,cf} = 0.90 \times 3.14 \approx 28°\text{C}$$

Next we can estimate the required areas A_f and A_c:

$$A_f = \frac{Q}{U_f\Delta T_m} = \frac{801.93 \times 10^3}{1428.40 \times 28} = 19.87 \text{ m}^2$$

$$A_c = \frac{Q}{U_c\Delta T_m} = \frac{801.93 \times 10^3}{1908.09 \times 28} = 14.87 \text{ m}^2$$

The overall surface design is $A_f/A_c = 1.34$ (34%), which is acceptable.
Shell diameter can be calculated from Equation (8.10), where $d_o = 0.019$ m, $PR = 1.25$, $CTP = 0.93$, $CL = 1.0$, and let us assume $L = 3$ m.

$$D_s = 0.637\sqrt{\frac{CL}{CTP}}\left[\frac{A_o(PR)^2 d_o}{L}\right]^{1/2}$$

$$= 0.637\sqrt{\frac{1.0}{0.93}}\left[\frac{19.87 \times (1.25)^2 \times 0.019}{3}\right]^{1/2}$$

$$= 0.293 \text{ m, round off to } 0.30 \text{ m}$$

The number of tubes can be calculated from Equation (8.9) as:

$$N_t = 0.785\left(\frac{CTP}{CL}\right)\frac{D_s^2}{(PR)^2 d_o^2}$$

$$= \frac{0.785 \times 0.93 \times (0.35)^2}{1.0 \times (1.25)^2 \times (0.019)^2} = 116.48 \approx 117$$

Therefore, the preliminary estimation of the unit size is

Shell diameter	$D_s = 0.3$ m
Tube length	$L = 3$ m
Tube diameter	O.D. = 19 mm, I.D. 16 mm
Baffle spacing	$B = 0.20$ m, baffle cut 25%
Pitch ratio	$P_T/d_o = 1.25$, square pitch

Then a rating analysis must be performed that is presented in the following sections.

8.3.2 Rating of Preliminary Design

After determining the tentative selected and calculated constructional design parameters (i.e., after a heat exchanger is available with process specifications), then these data can be used as inputs into a computer rating program or for manual calculations. The rating program is shown schematically in Figure 8.11.[5]

In some cases, a heat exchanger may be available and the performance analysis for this available heat exchanger needs to be done. In this case, a preliminary design analysis is not needed. If the calculation shows that the required amount of heat cannot be transferred to satisfy specific outlet temperatures or if one or both allowable pressure drops are exceeded, it is necessary to select a different heat exchanger and re-rate it (see Example 8.2).

For the rating process, all the preliminary geometric calculations must be conducted as the input into the heat transfer and the pressure drop correlations. When the heat exchanger is available, then all geometric parameters are also known. In the rating process, the other two basic calculations are the calculations of heat transfer coefficients and the pressure drops for each stream specified. If the length of the heat exchanger is fixed, then the rating program calculates the outlet temperatures of both streams. If the heat duty (heat load) is fixed, then the result from the rating program is the length of the heat exchanger required to satisfy the fixed heat duty of the exchanger. In both cases, the pressure drops for both streams in the heat exchanger are calculated.

The correlations for heat transfer and pressure drop are needed in quantitative forms that may be available from theoretical analyses or from experimental studies. The correlations for tube-side heat transfer coefficient and pressure drop calculations for single-phase flow are given in Chapters 3 and 4, respectively. The correlations for two-phase flow heat transfer are discussed in Chapter 7. The most involved ones are the correlations for the heat transfer and the pressure drop on the shell-side stream, which will be discussed in the following sections.

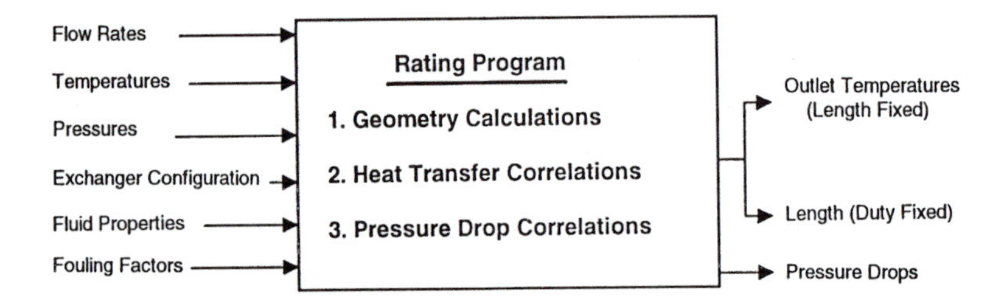

FIGURE 8.11
The rating program. (Drawn after Bell, K. J. [1981] *Heat Exchangers — Thermal–Hydraulic Fundamentals and Design*, Taylor & Francis, Washington, D.C.)

If the output of the rating analysis is not acceptable, a new geometric modification must be made. If, for example, the heat exchanger cannot deliver the amount of heat that should be transferred, then one should find means to increase the heat transfer coefficient or to increase the area of the exchanger. To increase the tube-side heat transfer coefficient, one can increase the tube-side velocity; thus one should increase the number of tube passes. One can decrease baffle spacing or decrease the baffle cut to increase shell-side heat transfer coefficient. To increase the area, one can increase the length of the heat exchanger, or increase the shell diameter; one can also go to multiple shells in series.

If the pressure drop on the tube side is greater than the allowable pressure drop, then the number of tube passes can be decreased or the tube diameter can be increased, which can decrease the tube length and increase the shell diameter and the number of tubes.

If the shell-side pressure drop is greater than the allowable pressure drop, then baffle spacing, tube pitch, and baffle cut can be increased; or one can change the type of the baffles.

8.4 Shell-Side Heat Transfer and Pressure Drop

To predict the overall heat transfer coefficient, we should calculate the tube-side and shell-side heat transfer coefficients from available correlations. For tubes in a shell-and-tube exchanger, the correlations given in Chapters 3 and 7 or from the available literature can be applied depending on the flow conditions, as was done for the double-pipe exchanger. The shell-side analysis described next is called the Kern method.[3]

8.4.1 Shell-Side Heat Transfer Coefficient

The heat transfer coefficient outside tube bundles is referred to as shell-side heat transfer coefficient. When the tube bundle employs baffles, the heat transfer coefficient is higher than the one with undisturbed flow conditions along the axis of tubes without baffles. If there are no baffles, the flow will be along the heat exchanger inside the shell. Then heat transfer coefficient can be based on the equivalent diameter, D_e, as it is done in a double-pipe heat exchanger; and the correlations of Chapter 3 are applicable. For baffled heat exchangers, the higher heat transfer coefficients result from the increased turbulence. In a baffled shell-and-tube heat exchanger, the velocity of fluid fluctuates because of the constricted area between adjacent tubes across the bundle. The correlations obtained for flow in tubes are not applicable for flow over tube bundles with segmental baffles.

Kern[3] suggested the following correlations for the shell-side heat transfer coefficient:

$$\frac{h_o D_e}{k} = 0.36\left(\frac{D_e G_s}{\mu}\right)^{0.55}\left(\frac{c_p \mu}{k}\right)^{1/3}\left(\frac{\mu_b}{\mu_w}\right)^{0.14}$$

$$\text{for } 2\times 10^3 < Re_s = \frac{G_s D_e}{\mu} < 1\times 10^6 \tag{8.11}$$

where h_o is the shell-side heat transfer coefficient, D_e is the equivalent diameter on the shell side, and G_s is the shell-side mass velocity.

The properties are evaluated at the average fluid temperature in the shell. In the preceding correlation, the equivalent diameter, D_e, is calculated along (instead of across) the long axes of the shell. The equivalent diameter of the shell is taken as four times net flow area as layout on the tube sheet (for any pitch layout) divided by the wetted perimeter:

$$D_e = \frac{4 \times \text{free - flow area}}{\text{wetted perimeter}} \tag{8.12}$$

As an example, Figure 8.12 shows a square- and a triangular-pitch layout. For each pitch layout, Equation (8.11) applies. For the square pitch, the perimeter is the circumference of a circle and the area is a square of pitch size (P_T^2) minus the area of a circle (the hatched section).

Thus for the square and triangular pitches, one can write the following for the square pitch:

$$D_e = \frac{4(P_T^2 - \pi d_o^2 / 4)}{\pi d_o} \tag{8.13}$$

and for the triangular pitch:

$$D_e = \frac{4\left(\dfrac{P_T^2\sqrt{3}}{4} - \dfrac{\pi d_o^2}{8}\right)}{\pi d_o / 2} \tag{8.14}$$

where d_o is the tube O.D.

There is no free-flow area on the shell side by which the shell-side mass velocity, G_s, can be calculated. For this reason, fictitious values of G_s can be defined based on the bundle crossflow area at the hypothetical tube row possessing the maximum flow area corresponding to the center of the shell.

The variables that affect the velocity are the shell diameter, D_s; the clearance, C, between adjacent tubes; the pitch size, P_T; and the baffle spacing, B. The width of the flow area at the tubes located at center of the shell is $(D_s/P_T)C$ and the length of the flow area is taken as the baffle spacing, B. Therefore, the bundle crossflow area, A_s, at the center of the shell is

$$A_s = \frac{D_s CB}{P_T} \tag{8.15}$$

where D_s is the I.D. of the shell. Then the shell-side mass flow rate is found with:

$$G_s = \frac{\dot{m}}{A_s} \tag{8.16}$$

8.4.2 Shell-Side Pressure Drop

The shell-side pressure drop depends on the number of tubes, the number of times the fluid passes the tube bundle between the baffles, and the length of each crossing. Suppose the length of a bundle is divided by four baffles; then all the fluid travels across the bundle five times.

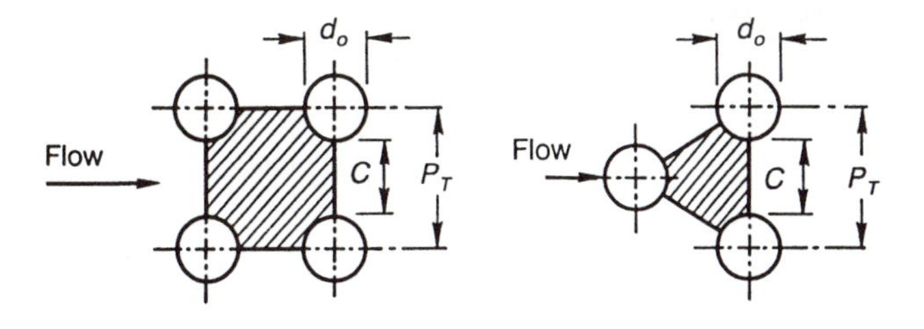

FIGURE 8.12
Square and triangular pitch–tube layouts.

A correlation has been obtained using the product of distance across the bundle, taken as the I.D. of the shell, D_s, and the number of times the bundle is crossed. The equivalent diameter used for calculating the pressure drop is the same as for heat transfer.

The pressure drop on the shell side is calculated by the following expression:[3,8]

$$\Delta p_s = \frac{fG_s^2(N_b + 1) \cdot D_s}{2\rho D_e \phi_s} \tag{8.17}$$

where $\phi_s = (\mu_b/\mu_w)^{0.14}$, N_b is the number of baffles, and $(N_b + 1)$ is the number of times the shell fluid passes the tube bundle. The friction factor, f, for the shell is calculated from:

$$f = \exp(0.576 - 0.19 \ln Re_s) \tag{8.18}$$

where

$$400 < Re_s = \frac{G_s D_e}{\mu} \leq 1 \times 10^6$$

The correlation has been tested based on data obtained on actual exchangers. The friction coefficient also takes entrance and exit losses into account.

8.4.3 Tube-Side Pressure Drop

The tube-side pressure drop can be calculated by knowing the number of tube passes, N_p, and the length, L, of the heat exchanger.

The pressure drop for the tube-side fluid is given by Equation (4.17):

$$\Delta p_t = 4f \frac{LN_p}{d_i} \rho \frac{U_m^2}{2} \tag{8.19}$$

or

$$\Delta p_t = 4f \frac{LN_p}{d_i} \frac{G_t^2}{2\rho} \tag{8.20}$$

The change of direction in the passes introduces an additional pressure drop, Δp_r, due to sudden expansions and contractions that the tube fluid experiences during a return that is accounted for allowing four velocity heads per pass:[8]

$$\Delta p_r = 4N_p \frac{\rho u_m^2}{2} \tag{8.21}$$

The total pressure drop of the tube side becomes:

$$\Delta p_{total} = \left(4f \frac{LN_p}{d_i} + 4N_p\right)\frac{\rho u_m^2}{2} \tag{8.22}$$

Example 8.2

In Problem 8.1, we estimated the size of the unit. By selecting a shell diameter of 15 1/4 in. according to TEMA standards from Table 8.3 with 124 tubes for 2-P shell-and-tube heat exchanger, re-rate this heat exchanger for the given process specifications by use of the Kern method. Note that heat duty is fixed; then the heat exchanger length and pressure drops for both streams can be calculated.

SOLUTION

The selected shell-and-tube heat exchanger for this purpose has the following geometric parameters:

Shell internal diameter	D_s = 15 1/4 in. (= 0.39 m)
Number of tubes	N_t = 124
Tube diameter	O.D. = 19 mm, I.D. = 16 mm
Tube material	k = 60 W/m^2·K
Baffle spacing	B = 0.25 m, baffle cut 25%
Pitch size	P_T = 0.024 m
Number of tube passes	N_p = 2

Heat duty is fixed with the assumed outlet temperature of 40°C

The properties of the shell-side fluid can be taken at $T_b = \frac{67 + 53.2}{2} = 60°\text{C}\ (= 333\ \text{K})$ from Appendix B (Table B.2):

ρ = 983.2 kg/m^3
c_p = 4184 J/kg·K
μ = 4.67 × 10^{-4} N·s/m^2
k = 0.652 W/m·K
Pr = 3.00

Properties of the tube-side water at 28.5°C (≈ 300 K) from Appendix B (Table B.2) are:

ρ = 996.8 kg/m^3
c_p = 4179 J/kg·K

$\mu = 8.2 \times 10^{-4}$ N·s/m^2

$k = 0.610$ W/m·K

$Pr = 5.65$

By use of specifications of maximum tube length $L_{max} = 5$ m, and maximum pressure drop on shell-side $\Delta p_s = 1.5$ lb$_f$/in.^{2}:

$$\frac{h_o D_e}{k} = 0.36\left(\frac{D_e G_s}{\mu}\right)^{0.55}\left(\frac{c_p \mu}{k}\right)^{1/3}\left(\frac{\mu_b}{\mu_w}\right)^{0.14}$$

for $2 \times 10^3 < Re_s < 10^6$.

For square pitch tube layout:

$$D_e = \frac{4(P_T^2 - \pi d_o^2/4)}{\pi d_o} = \frac{4[(0.024)^2 - \pi(0.019^2/4)]}{\pi(0.019)} = 0.0196 \text{ m}$$

$$C = P_T - d_o = 0.024 - 0.019 = 0.005 \text{ m}$$

$$A_s = \frac{D_s CB}{P_T} = \frac{(0.39 \text{ m})(0.005 \text{ m})(0.25 \text{ m})}{0.024 \text{ m}} = 0.0203 \text{ m}^2$$

$$G_s = \frac{\dot{m}}{A_s} = \frac{50,000 \text{ kg/h}}{0.0203 \text{ m}^2}\left(\frac{1 \text{ h}}{3600 \text{ s}}\right) = 684.18 \text{ kg/s}\cdot\text{m}^2$$

$$Re_s = \frac{G_s D_e}{\mu} = \frac{(684.2 \text{ kg/s}\cdot\text{m}^2)(0.0196 \text{ m})}{4.67 \times 10^{-4} N\cdot\text{s/m}^2} = 28,715$$

$$T_w = \frac{1}{2}\left(\frac{T_{c_1} + T_{c_2}}{2} + \frac{T_{h_1} + T_{h_2}}{2}\right) = \frac{1}{2}\left(\frac{17+40}{2} + \frac{67+53}{2}\right) = 44.25°\text{C}$$

$$T_{c_1} = 17°\text{C} \qquad T_{h_1} = 67°\text{C}$$

$$T_{c_2} = 40°\text{C} \qquad T_{h_2} = 53°\text{C}$$

From Table B.2 in Appendix B at the approximated well temperature of 331 K:

$$\mu_w = 6.04 \times 10^{-4} N\cdot\text{s/m}^2$$

$$\frac{h_o D_e}{k} = 0.36\left(\frac{(0.0196 \text{ m})(684.2 \text{ kg/s}\cdot\text{m}^2)}{4.67 \times 10^{-4} N\cdot\text{s/m}^2}\right)^{0.55}\left(\frac{(4184 \text{ J/kg}\cdot\text{K})(4.67 \times 10^{-4})}{0.652 \text{ W/m}\cdot\text{K}}\right)^{1/3}\left(\frac{4.67 \times 10^{-4}}{6.04 \times 10^{-4}}\right)^{0.14}$$

$$= 141.75$$

$$h_o = \frac{(141.75)(0.652)}{0.0196} = 4715.4 \text{ W/m}^2\cdot\text{K}$$

Calculate the tube-side heat transfer coefficient:

$$A_{tp} = \frac{\pi d_i^2}{4} \cdot \frac{N_T}{2} = \frac{\pi (0.016\text{ m})^2}{4} \times \frac{124}{2} = 1.246 \times 10^{-2}\text{ m}^2$$

$$u_m = \frac{\dot{m}_t}{\rho_t A_{tp}} = \frac{30{,}000\text{ kg/h}}{(996.8\text{ kg/m}^3)(1.246 \times 10^{-2}\text{ m}^2)} \left(\frac{1\text{ h}}{3600\text{ s}}\right) = 0.67\text{ m/s}$$

$$\text{Re} = \frac{\rho u_m d_i}{\mu} \frac{(996.8\text{ kg/m}^3)(0.67\text{ m/s})(0.016\text{ m})}{8.2 \times 10^{-4}\text{ N} \cdot \text{s/m}^2} = 13{,}049.9$$

Because $Re > 10^4$, the flow is turbulent.

By using Gnielinski's correlation:

$$Nu_b = \frac{(f/2)(Re_b - 1000)\,Pr}{1 + 12.7(f/2)^{1/2}(Pr^{2/3} - 1)}$$

$$f = (1.58 \ln Re - 3.28)^{-2} = [1.58 \ln (13{,}049.9) - 3.28]^{-2}$$

$$= 0.00731$$

$$Nu_b = \frac{(0.0037)(13049.9 - 1000)(5.65)}{1 + 12.7(0.0037)^{1/2}(5.65^{2/3} - 1)} = 94.06$$

$$h_i = \frac{Nu_b k}{d_i} = \frac{(94.06)(0.61\text{ W/m} \cdot \text{K})}{0.016\text{ m}} = 3586.1\text{ W/m}^2 \cdot \text{K}$$

Determine the overall heat transfer coefficient:

$$U_f = \frac{1}{\dfrac{d_o}{d_i h_i} + \dfrac{d_o R_{fi}}{d_i} + \dfrac{d_o \ln (d_o/d_i)}{2k} + R_{fo} + \dfrac{1}{h_o}}$$

$$= \frac{1}{\dfrac{0.019}{(0.016)(3586.1)} + \dfrac{(0.019)(0.000176)}{0.016} + \dfrac{(0.019) \ln (0.019/0.016)}{2(60)} + 0.000176 + \dfrac{1}{4715.4}}$$

$$= 1046.7\text{ W/m}^2 \cdot \text{K}$$

$$U_c = \frac{1}{\dfrac{d_o}{d_i h_i} + \dfrac{d_o \ln (d_o/d_i)}{2k} + \dfrac{1}{h_o}}$$

$$= \frac{1}{\dfrac{0.019}{(0.016)(3586.1)} + \dfrac{(0.019) \ln (0.019/0.016)}{2(60)} + \dfrac{1}{4715.4}}$$

$$= 1753.1\text{ W/m}^2 \cdot \text{K}$$

Calculate shell-side pressure drop by:

$$\Delta p_s = \frac{fG_s^2(N_b + 1)Ds}{2\rho D_e \phi_s}$$

$$f = \exp(0.576 - 0.19 \ln Re_s)$$

$$= \exp[0.576 - 0.19 \ln (28,715)] = 0.253$$

$$\phi_s = \left(\frac{\mu_b}{\mu_w}\right)^{0.14} = \left(\frac{4.67 \times 10^{-4}}{6.04 \times 10^{-4}}\right)^{0.14} = 0.9646$$

$$\Delta p_s = \frac{(0.253)(684.2^2)(2+1)(0.39)}{2(983.2)(0.0196)(0.9646)} = 3727.2 \text{ Pa} = 0.54 \text{ lb}_f/\text{in.}^2$$

Because $0.54 < 1.5$, shell-side pressure drop is acceptable.

Calculate the tube length:

$$Q = (\dot{m}c_p)_c(T_{c_2} - T_{c_1}) = (8.33 \text{ kg/s})(4184 \text{ J/kg K})(40 - 17)\text{K} = 801.6 \text{ kW}$$

$$A_{of} = \frac{Q}{U_{of}\Delta T_m}$$

$$\Delta T_{lm,cf} = \frac{\Delta T_1 - \Delta T_2}{\ln(\Delta T_1/\Delta T_2)} = \frac{(67-40)-(53-17)}{\ln(67-40)(53-17)} = 31.3 \text{ K}$$

$$\Delta T_m = F\Delta T_{lm,cf} = (0.9)(31.3) = 28.2 \text{ K}$$

$$A_{of} = \frac{801,600 \text{ W}}{(1046.7 \text{ W/m}^2\text{K})(28.2 \text{ K})} = 27.16 \text{ m}^2$$

$$A_o = \pi d_o L N_t$$

$$L = \frac{A_o}{\pi d_o N_t} = \frac{27.16 \text{ m}^2}{\pi(0.019 \text{ m})(124)} = 3.67 \text{ m}$$

It is rounded off to 4 m. Because $4 < 5$ m, the length of the heat exchanger is acceptable.

Calculate tube-side pressure drop by:

$$\Delta p_t = \left(4f\frac{LN_p}{d_i} + 4N_p\right)\frac{\rho u_m^2}{2}$$

$$= \left(4 \times 0.000731 \times \frac{4 \times 2}{0.016} + 4 \times 2\right) \times \frac{996.8 \times (0.67)^2}{2}$$

$$= 2116.95 \text{ Pa} = 0.307 \text{ lb}_f/\text{in.}^2$$

8.4.4 Bell–Delaware Method

The calculation given in Section 8.4 for shell-side heat transfer and pressure drop analysis (Kern method) is a simplified method. The shell-side analysis is not as straightforward as the tube-side analysis; this is because the shell flow is complex, combining crossflow and baffle window flow, as well as baffle-shell and bundle-shell bypass streams and complex flow pattern as shown in Figures 8.13 and 8.14.[5,6,9–11]

As indicated on Figure 8.13, five different streams are identified. The A-stream is leaking through the clearance between the tubes and the baffle. The B-stream is the main stream across the bundle. This is the stream desired on the shell-side of the exchanger. The C-stream is the bundle bypass stream flowing around the tube bundle between the outermost tubes in the bundle and the inside of the shell. The E-stream is the baffle-to-shell leakage stream flowing through the clearance between the baffles and the shell I.D.

Then there is the F-stream that flows through any channel within the tube bundle caused by the provision of pass dividers in the exchanger header for multiple tube passes. Figure 8.13 is an idealized representation of the streams. The streams shown can mix and interact with one another, and a more complete mathematical analysis of the shell-side flow would take this into account.[9]

The Bell–Delaware method takes into account the effects of various leakage and bypass streams on the shell-side heat transfer coefficient and pressure drop. The Bell–Delaware

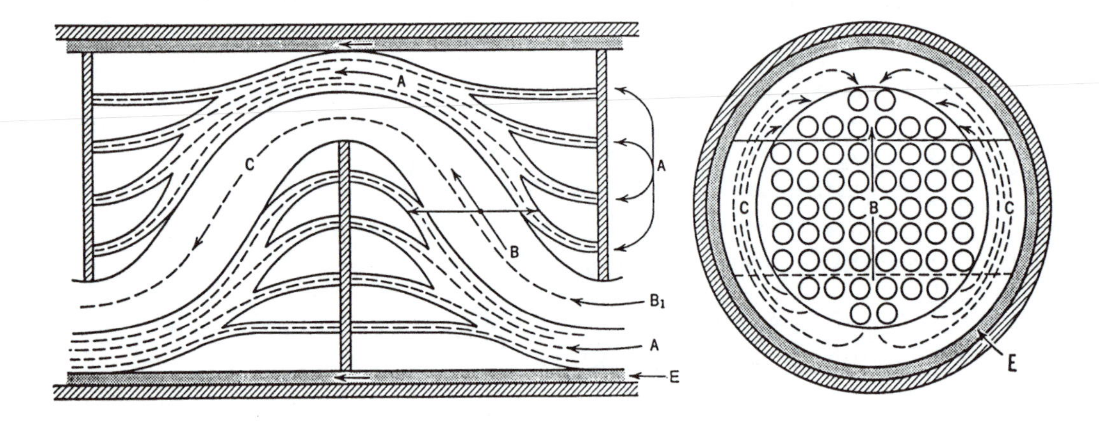

(a)

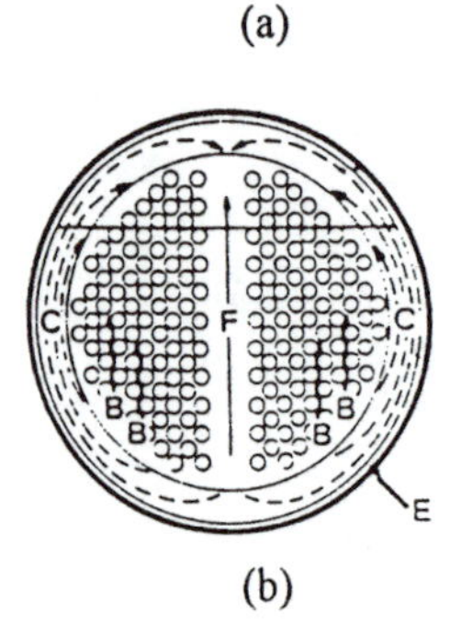

(b)

FIGURE 8.13
(a) Diagram indicating leaking paths for flow bypassing the tube matrix, both through the baffle clearances between the tube matrix and shell. (b) F-stream for two-tube pass exchanger. (Adapted from Taborek, J. [1983] In *Heat Exchanger Design Handbook*, Section 3.3. Hemisphere, New York. With permission from Begell House, NY.)

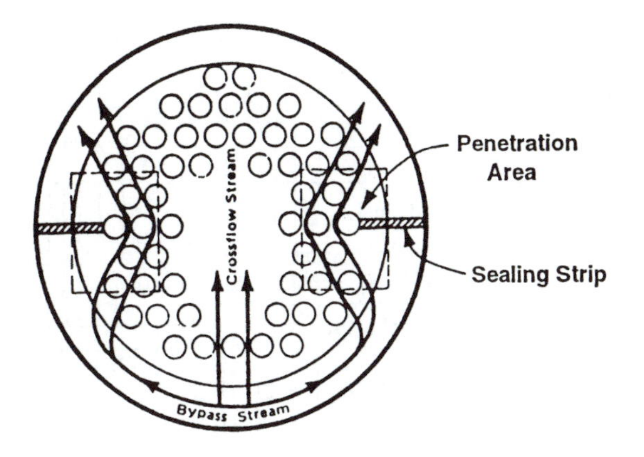

FIGURE 8.14
Radial baffles designed to reduce the amount of bypass flow through the gap between the side of the tube matrix and the shell. (Adapted from Taborek, J. [1983] In *Heat Exchanger Design Handbook*, Section 3.3. Hemisphere, New York. With permission from Begell House, NY.)

method is the most reliable method at present for the shell-side analysis. In the Bell–Delaware method the B-stream is the main essential stream. The other streams reduce the B-stream and alter the shell-side temperature profile resulting in a decrease in heat transfer coefficient.

A brief discussion of the Bell–Delaware method for shell-side heat transfer coefficient and the pressure drop analysis is given in this section.

Shell-Side Heat Transfer Coefficient

The basic equation for calculating the average shell-side heat transfer coefficient is given by:

$$h_o = h_{id} J_c J_l J_b J_s J_r \tag{8.23}$$

where h_{id} is the ideal heat transfer coefficient for pure crossflow in an ideal tube bank and is calculated from:

$$h_{id} = j_i c_{ps} \left(\frac{\dot{m}_s}{A_s} \right) \left(\frac{k_s}{c_{ps} \mu_s} \right)^{2/3} \left(\frac{\mu_s}{\mu_{s,w}} \right)^{0.14} \tag{8.24}$$

where j_i is the Colburn j-factor for an ideal tube bank, s stands for shell, and A_s is the crossflow area at the centerline of the shell for one crossflow between two baffles.

Graphs are available for j_i as a function of shell-side Reynolds number, $Re_s = d_o \dot{m}_s / \mu_o A_s$; tube layout; and pitch size. Such graphs are shown in Figures 8.15 through 8.17.[6] A_s is given by Equation (8.15), that is, the Reynolds number is based on the outside tube diameter and on the minimum cross-section flow area at the shell diameter. On the same graphs friction coefficients for ideal tube banks are also given for the pressure drop calculations. Therefore, depending on the shell-side constructional parameters, the correction factors must be calculated.[6,10]

Although the ideal values of j_i and f_i are available in graphic forms, for computer analysis a set of curve fit correlations are obtained in the following forms:[6,11]

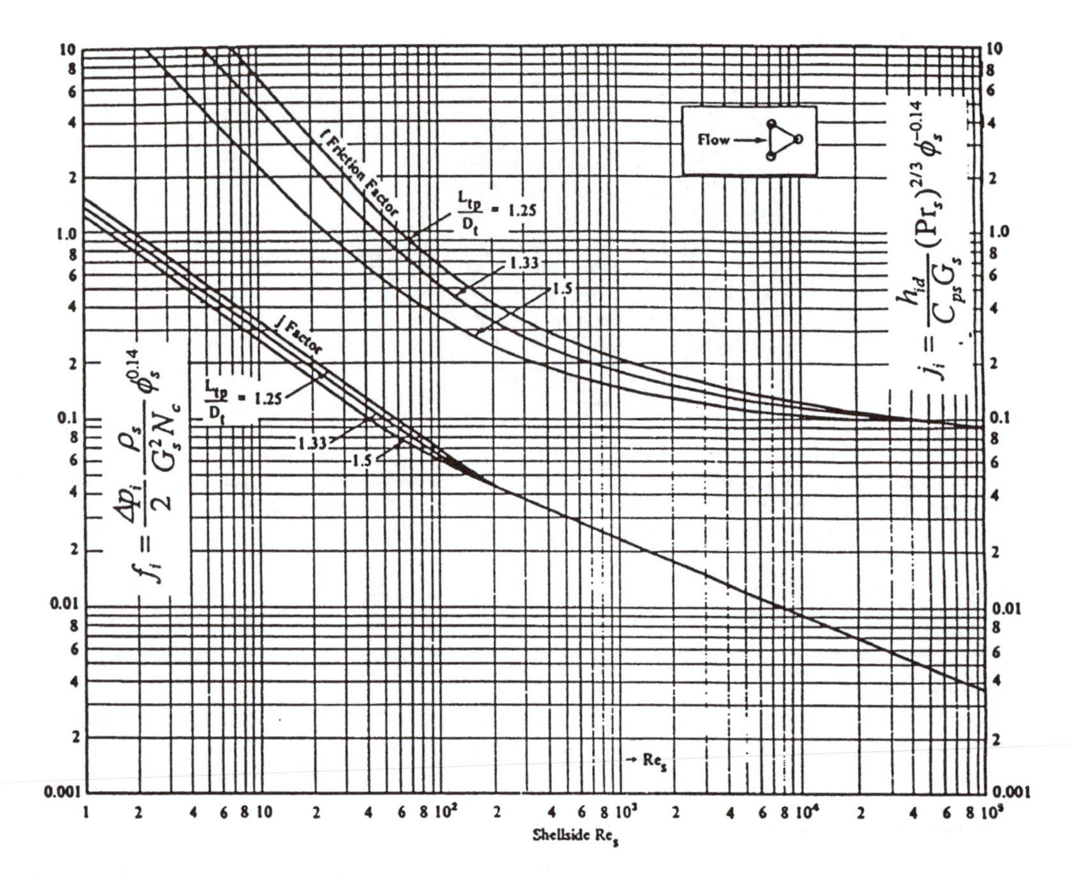

FIGURE 8.15
Ideal tube bank j_i and f_i factors for 30°C staggered layout. (From Taborek, J. [1983] In *Heat Exchanger Design Handbook*, Section 3.3. Hemisphere, New York. With permission from Begell House, NY.)

$$j_i = a_1 \left(\frac{1.33}{P_T / d_o} \right)^a (Re_s)^{a_2} \tag{8.25}$$

where

$$a = \frac{a_3}{1 + 0.14 (Re_s)^{a_4}}$$

and

$$f_i = b_1 \left(\frac{1.33}{P_T / d_o} \right)^b (Re_s)^{b_2} \tag{8.26}$$

where

$$b = \frac{b_3}{1 + 0.14 (Re_s)^{b_4}}$$

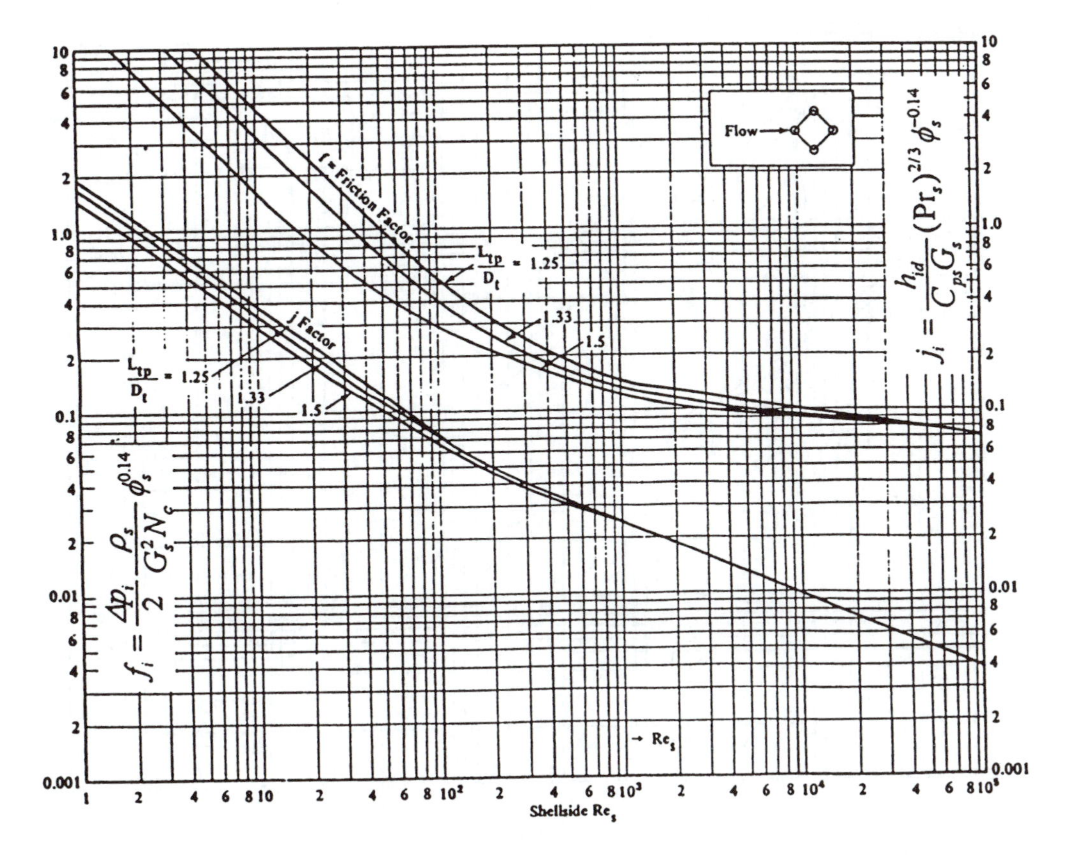

FIGURE 8.16
Ideal tube bank j_i and f_i factors for 45°C staggered layout. (From Taborek, J. [1983] In *Heat Exchanger Design Handbook*, Section 3.3. Hemisphere, New York. With permission from Begell House, NY.)

Table 8.6 gives the coefficients of Equations (8.25) and (8.26).[6]

J_c is the correction factor for baffle cut and spacing. This factor takes into account the heat transfer in the window and calculates the overall average heat transfer coefficient for the entire heat exchanger. It depends on the shell diameter and the baffle cut distance from the baffle tip to the shell I.D. For a large baffle cut, this value may decrease to a value of 0.53, and it is equal to 1.0 for a heat exchanger with no tubes in the window. It may increase to a value as high as 1.15 for small windows with a high window velocity.

J_l is the correlation factor for baffle leakage effects including tube-to-baffle and shell-to-baffle leakage (A and E streams). If the baffles are put too close together, then the fraction of the flow in the leakage streams increases compared with the crossflow. J_l is a function of the ratio of total leakage area per baffle to the crossflow area between adjacent baffles and also of ratio of the shell-to-baffle leakage area to the tube-to-baffle leakage area. A typical value of J_l is in the range of 0.7 and 0.8.

J_b is the correction factor for bundle bypassing effects due to the clearance between the outermost tubes and the shell and pass dividers (*C* and *F* streams). For relatively small clearance between the outermost tubes and the shell for fixed tube sheet construction $J_b \approx$ 0.90. For a pull-through floating head, larger clearance is required and $J_b \approx 0.7$. The sealing strips (see Figure 8.14) can increase the value of J_b.

J_s is the correction factor for variable baffle spacing at the inlet and outlet. Because of the nozzle spacing at the inlet and outlet and the changes in local velocities, the average heat

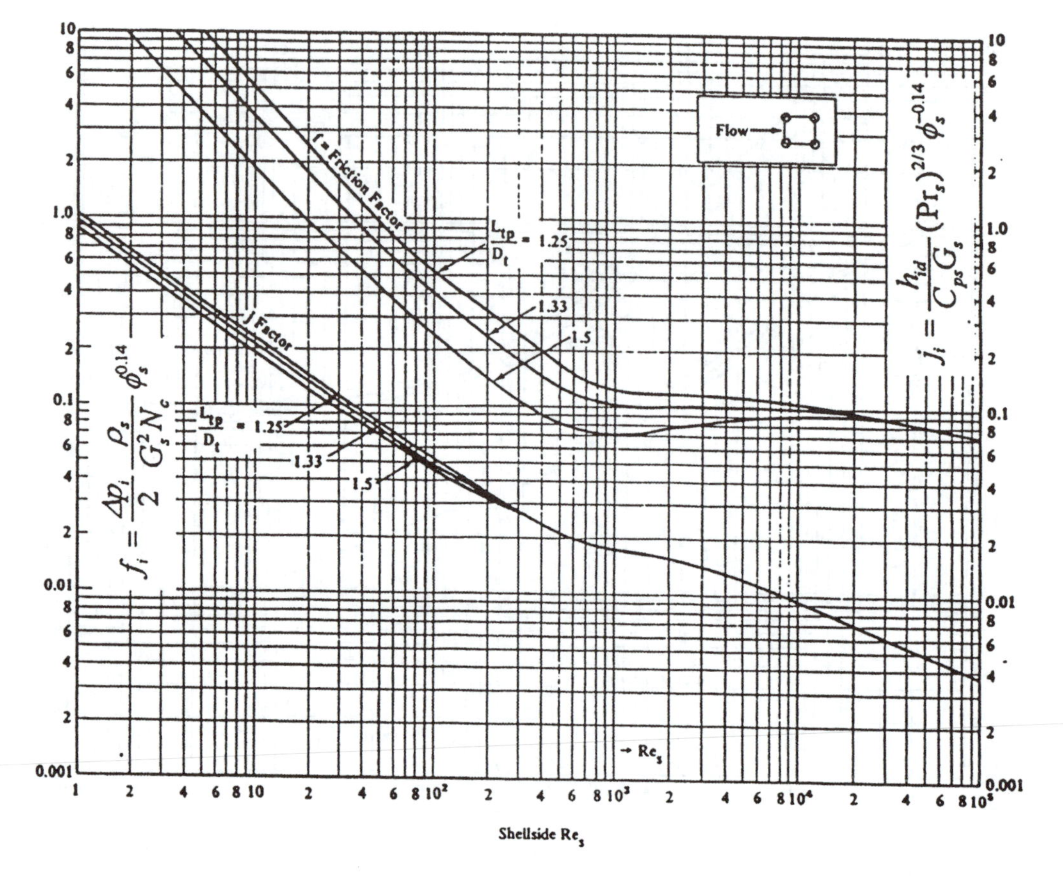

FIGURE 8.17
Ideal tube bank j_i and f_i factors for 90°C in-line layout. (From Taborek, J. [1983] In *Heat Exchanger Design Handbook*, Section 3.3. Hemisphere, New York. With permission from Begell House, NY.)

TABLE 8.6

Correlation Coefficients for j_i and f_i Equations (8.25) and (8.26)

Layout Angle	Reynolds Number	a_1	a_2	a_3	a_4	b_1	b_2	b_3	b_4
30°	10^5–10^4	0.321	–0.388	1.450	0.519	0.372	–0.123	7.00	0.500
	10^4–10^3	0.321	–0.388			0.486	–0.152		
	10^3–10^2	0.593	–0.477			4.570	–0.476		
	10^2–10	1.360	–0.657			45.100	–0.973		
	<10	1.400	–0.667			48.000	–1.000		
45°	10^5–10^4	0.370	–0.396	1.930	0.500	0.303	–0.126	6.59	0.520
	10^4–10^3	0.370	–0.396			0.333	–0.136		
	10^3–10^2	0.730	–0.500			3.500	–0.476		
	10^2–10	0.498	–0.656			26.200	–0.913		
	<10	1.550	–0.667			32.00	–1.000		
90°	10^5–10^4	0.370	–0.395	1.187	0.370	0.391	–0.148	6.30	0.378
	10^4–10^3	0.107	–0.266			0.0815	+0.022		
	10^3–10^2	0.408	–0.460			6.0900	–0.602		
	10^2–10	0.900	–0.631			32.1000	–0.963		
	10	0.970	–0.667			35.0000	–1.000		

transfer coefficient on the shell side will change. J_s value will usually be between 0.85 and 1.00.

J_r applies if the shell-side Reynolds number, Re_s, is less than 100. If $Re_s < 20$, it is fully effective. This factor is equal to 1.00 if $Re_s > 100$.

The combined effects of all these correction factors for a reasonable well-designed shell and tube heat exchanger is of the order of 0.60.[5,12]

Example 8.3

Distilled water with a flow rate of 50 kg/s enters a baffled shell-and-tube heat exchanger at 32°C and leaves at 25°C. Heat will be transferred to 150 kg/s of raw water coming from a supply at 20°C. You are requested to design the heat exchanger for this purpose. A single-shell and single-tube pass is preferable. The tube diameter is 3/4 in. (19-mm O.D. with 16-mm I.D.) and tubes are laid out on 1-in. square pitch. Maximum length of the heat exchanger 8 m is required because of space limitations. The tube material is 0.5 Cr alloy (k = 42.3 W/m·K). Assume a total fouling resistance of 0.000176 m²·K/W. Note that surface over design should not exceed 30%. The maximum flow velocity through the tube is also suggested to be 2 m/s to prevent erosion. Perform thermal and hydraulic analysis of the heat exchanger.

SOLUTION

Properties of the tube-side fluid, at 20°C, from Appendix B (Table B.2) are

$$c_p = 4182 \text{ J/kg·K}$$
$$k = 0.598 \text{ W/m}^2\text{·K}$$
$$Pr = 7.01$$
$$\rho = 998.2 \text{ kg/m}^3$$
$$\mu = 10.02 \times 10^{-4} \text{ N·s/m}^2$$

Properties of the shell-side fluid at average temperature, from Appendix B (Table B.2) are

$$c_p = 4179 \text{ J/kg·K}$$
$$k = 0.612 \text{ W/m}^2\text{·K}$$
$$Pr = 5.75$$
$$\rho = 995.9 \text{ kg/m}^3$$
$$\mu = 8.15 \times 10^{-4} \text{ N·s/m}^2$$

The steps of the solution follow.
Estimate the number of tubes:

$$\dot{m}_t = \rho u_m A_c N_t$$

$$N_T = \frac{\dot{m}}{\rho u_m A_c} = \frac{4\dot{m}_t}{\rho \pi d_i^2 u_m} = \frac{4 \times 150}{998.2 \times \pi \times (0.016)^2 \times 2} = 373.88$$

$$N_T \approx 374$$

Flow area through the tubes is

$$A_t = \frac{\pi d_i^2}{4} N_T = \frac{\pi (0.016)^2}{4} \times 374 = 0.075 \text{ m}^2$$

Now, we can estimate the shell diameter from Equation (8.9)

$$D_s^2 = \frac{N_T (CL)(PR)^2 d_o^2}{0.785(CTP)}$$

$$CTP = 0.93$$

$$CL = 1.0$$

$$D_s = \left[\frac{374 \times 1 \times \left(\frac{2.54 \times 10^{-2}}{1.9 \times 10^{-2}} \right)^2 (0.019)^2}{0.785 \times 0.93} \right]^{1/2} = 575 \text{ mm}$$

which is rounded off as 580 mm.

The selected and estimated constructional parameters can be summarized as:

Shell I.D.	$D_s = 0.58$ m
Number of tubes	$N_t = 374$
Tube O.D.	$d_o = 19$ mm
Tube I.D.	$d_i = 16$ mm
Square tube pitch	$P_t = 0.0254$ m
Baffle spacing	$B = 0.5$ m (25% cut)

Kern Method — Estimate the crossflow area at the shell diameter:

$$A_s = (D_s - N_{TC} d_o) B$$

where

$$N_{TC} = \frac{D_s}{P_T} = \frac{580}{25.4} = 22.83$$

$$A_s = (0.58 - 22.83 \times 0.019) \times 0.5 = 0.073 \text{ m}^2$$

Equivalent diameter can be calculated from Equation (8.13):

$$D_e = \frac{4\left(P_T^2 - \frac{\pi d_o^2}{4} \right)}{\pi d_o} = \frac{4\left[(2.54 \times 10^{-2})^2 - \frac{\pi (0.019)^2}{4} \right]}{\pi \times 0.019} = 0.024 \text{ m}$$

Calculate the Reynolds number:

$$Re = \left(\frac{\dot{m}_s}{A_s}\right)\frac{D_e}{\mu} = \frac{50}{0.073} \times \frac{0.024}{8.15 \times 10^{-4}} = 20169.76$$

By assuming constant properties, then heat transfer can be estimated from Equation (8.11):

$$h_o = \frac{0.36k}{D_e} Re^{0.55} Pr^{1/3} = \frac{0.36 \times 0.612}{0.024} (20169.76)^{0.55} (5.57)^{1/3} = 3793.6 \text{ W/m}^2 \cdot \text{K}$$

Taborek method — Taborek gives the following correlation for the shell-side heat transfer coefficient for turbulent flow:[7]

$$Nu = 0.2 Re_s^{0.6} Pr_s^{0.4}$$

where Reynolds number is based on the tube diameter and the velocity on the crossflow area at the diameter of the shell:

$$Re_s = \frac{\rho u_s d_o}{\mu} = \frac{\dot{m}_s}{A_s}\frac{d_o}{\mu} = \frac{50}{0.073}\,\frac{0.019}{8.15 \times 10^{-4}} = 15967.8$$

$$Nu = 0.2(15967.8)^{0.6}(5.57)^{0.4} = 132.2$$

$$h_s = \frac{Nu \cdot k}{d_o} = \frac{132.2 \times 0.612}{0.019} = 4259.09 \text{ W/m}^2 \cdot \text{K}$$

Bell–Delaware method — Shell-side heat transfer coefficient is given by Equation (8.24):

$$h_{id} = j_i c_{ps} \left(\frac{\dot{m}_s}{A_s}\right)\left(\frac{k_s}{c_{ps}\mu_s}\right)^{2/3}\left(\frac{\mu_s}{\mu_{s,w}}\right)^{0.14}$$

j_i can be obtained from Figure 8.17 or from the correlation (8.25) with Table 8.6:

$$j_i = 0.185\, Re_s^{-0.324}$$

where

$$Re_s = \frac{d_o \dot{m}}{\mu_s A_s} = 15967.8$$

$$j_i = 0.185(15967.8)^{-0.324} = 0.0080$$

$$h_{id} = 0.0080 \times 4179\left(\frac{50}{0.073}\right)\left(\frac{0.612}{4179 \times 8.15 \times 10^{-4}}\right)^{2/3}$$

$$= 7291.62 \text{ W/m}^2 \cdot \text{K}$$

It is assumed that the properties are constant. In the Bell–Delaware method, the correction factors due to bypass and leakage stream are provided in graphic forms depending on the constructional features of the heat exchanger. Let us assume that the combined effects of all these correction factors is 60%.

$$h_o = 0.60 \times 7291.62 = 4374.97 \text{ W/m}^2 \cdot \text{K}$$

Therefore, three methods have given comparable results to estimate the shell-side heat transfer coefficient.

To determine the tube-side heat transfer coefficient:

$$Re_t = \frac{\rho u_m d_i}{\mu} = \frac{998.2 \times 2 \times 0.016}{10.02 \times 10^{-4}} = 31878.6$$

Tube-side heat transfer coefficient, h_i, can be calculated from Petukhov-Kirillov correlation as given by Equation (3.29):

$$Nu_b = \frac{(f/2)Re_b Pr_b}{1.07 + 12.7(f/2)^{1/2}(Pr_b^{1/2} - 1)}$$

$$f = (158 \ln Re_b - 3.28)^{-2} = [1.58 \ln (31878.6 - 3.28]^{-2} = 0.0058$$

$$f/2 = 0.0029$$

$$Nu_b = \frac{(0.0029)(31878.6)(7.01)}{1.07 + 12.7(0.0029)^{1/2}(7.01^{2/3} - 1)} = 224.16$$

$$h_i = Nu_b \cdot \frac{k}{d_i} = (224.16)\frac{0.598}{0.016} = 8377.98 \text{ W/m}^2 \cdot \text{K}$$

Calculate the overall heat transfer coefficient for the clean surface:

$$\frac{1}{U_c} = \frac{1}{h_o} + \frac{1}{h_i}\frac{d_o}{d_i} + \frac{r_o \ln(r_o/r_i)}{k}$$

$$= \frac{1}{4375} + \frac{1}{8378}\frac{0.019}{0.016} + \frac{9.5 \times 10^{-3} \ln(19/16)}{42.3}$$

$$U_c = 2445.5 \text{ W/m}^2 \cdot \text{K}$$

and for the fouled surface:

$$\frac{1}{U_f} = \frac{1}{U_c} + R_{ft} = \frac{1}{2445.5} + 0.000176$$

$$U_f = 1709.7 \text{ W/m}^2 \cdot \text{K}$$

$$Q = \frac{(\dot{m}c_p)_h(T_{h_1} - T_{h_2})}{(\dot{m}c_p)_c} + T_{c_1}$$

$$T_{c_2} = \frac{50 \times 4179 \times (32 - 25)}{150 \times 4182} + 20 = 22.33°\text{C}$$

$$\text{LMTD} = \frac{(32 - 22.3) - (25 - 20)}{\ln\left(\dfrac{32 - 22.3}{25 - 20}\right)} = 7.09°\text{C}$$

$$Q = (\dot{m}c_p)_h(T_{h_1} - T_{h_2}) = 50 \times 4179 \times (32 - 25)$$

$$= 1462650 \text{ W}$$

$$= U_f A_f \Delta T_m$$

$$A_f = \frac{1462650}{1709.7 \times 7.09} = 120.66 \text{ m}^2$$

$$A_c = \frac{1462650}{2445.5 \times 7.09} = 84.36 \text{ m}^2$$

The over surface design can be calculated as:

$$OS = \frac{A_f}{A_c} = \frac{U_c}{U_f} = \frac{2445.5}{1709.7} = 1.4304\,(43\%)$$

which is the clean vs. fouling safety factor.

The over design should not be more than about 35%. Let us assume 20% surface over design; then cleaning scheduling must be arranged accordingly:

$$\frac{U_c}{U_f} = 1.20$$

$$U_f = \frac{U_c}{1.20} = \frac{2445.5}{1.20} \cong 2037.9 \text{ W/m}^2 \cdot \text{K}$$

The corresponding total resistance can be calculated from Equation (5.2):

$$\frac{1}{U_f} = \frac{1}{U_c} + R_{ft}$$

$$R_{ft} = 0.0000817 \text{ m}^2 \cdot \text{K/W}$$

For 20% oversurface design, the surface area of the exchanger becomes:

$$A_f = 1.20\,A_c$$

$$A_f = 1.20 \times 84.36 = 101.2 \text{ m}^2$$

The length of the heat exchanger is calculated by:

$$L = \frac{A_f}{N_T \pi d_o} = \frac{101.2}{374 \times \pi \times 0.019} = 4.54 \text{ m}$$

which is rounded off to $L = 5$ m.

Shell diameter can be recalculated from Equation (8.10):

$$D_s = 0.637 \sqrt{\frac{CL}{CTP}} \left[\frac{A_f (PR)^2 d_o}{L} \right]^{1/2}$$

$$= 0.637 \sqrt{\frac{1}{0.93}} \left[\frac{101.2 \times (2.54/1.9)^2 \times 19 \times 10^{-3}}{5} \right]^{1/2} = 0.548 \text{ m}$$

which is rounded of to $D_s = 0.60$ m.

Further analysis is suggested.

Now we have a new heat exchanger to be re-rated. After calculating D_s, we can compare with tube counts and shell diameter in Table 8.3. Tabulate new constructional parameters:

$$D_s = 0.60 \text{ m}$$
$$N_T = 374$$
$$L = 5 \text{ m}$$
$$d_o = 19 \text{ mm}$$
$$d_i = 16 \text{ mm}$$
$$P_T = 0.0254 \text{ m}$$
$$B = 0.50 \text{ m}$$

Calculate the pressure drops on the shell side and tube side by the use of Equations (8.17) and (8.22).

Shell-Side Pressure Drop

For a shell-and-tube type heat exchanger with bypass and leakage streams, the total nozzle-to-nozzle pressure drop is calculated as the sum of the following three components (Figure 8.18a, b, c):

1. By considering the pressure drop in the interior crossflow section (baffle tip to baffle tip), the combined pressure drop of all the interior crossflow section is[6,12]

$$\Delta p_c = \Delta p_{bi} (N_b - 1) R_l R_b \tag{8.27}$$

   where Δp_{bi} is the pressure drop in an equivalent ideal tube bank in one baffle compartment of central baffle spacing. R_l is the correction factor for baffle leakage effects (A and E streams). Typically, $R_l = 0.4$ to 0.5. R_b is the correction factor for bypass flow (C and F streams). Typically, $R_b = 0.5$ to 0.8, depending on the construction type and number of sealing strips. N_b is the number of baffles.

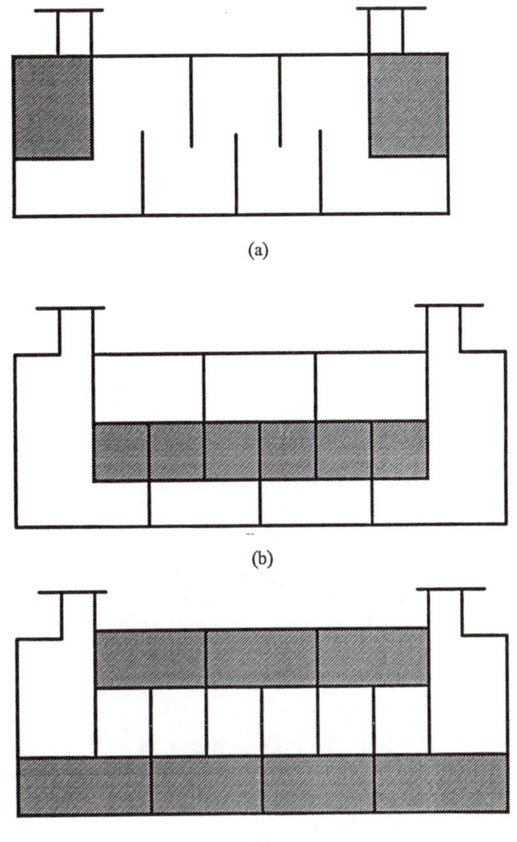

FIGURE 8.18
(a) Entrance; (b) internal; (c) window.

2. The pressure drop in the window is affected by leakage but not bypass. The combined pressure drop in all the windows is calculated from:

$$\Delta p_w = \Delta p_{wi} N_b R_l \tag{8.28}$$

   where Δp_{wi} is the pressure drop in an equivalent ideal tube bank in the window section.

3. The pressure drop in the entrance and exit sections is affected by bypass but not by leakage. Additionally, there is an effect due to variable baffle spacing. The combined pressure drop for the entrance and exit section is given by:

$$\Delta p_e = 2\Delta p_{bi} \frac{N_c + N_{cw}}{N_c} R_b R_s \tag{8.29}$$

   where N_c is the number of tube rows crossed during flow through one crossflow in the exchanger, and N_{cw} is the number of tube rows crossed in each baffle window. R_s is the correction factor for the entrance and exit section having a different baffle spacing than the internal sections due to the existence of the inlet and outlet nozzles. The correction factors are available in graphic forms in References 7, 10, 12, and 13.

The total pressure drop over the heat exchanger is:

$$\Delta p_T = \Delta p_c + \Delta p_w + \Delta p_e \tag{8.30}$$

$$\Delta p_T = [(N_b - 1)\,\Delta p_{bi} R_b + N_b \Delta p_{wi}] R_l + 2\Delta p_{bi}\left(1 + \frac{N_{cw}}{N_c}\right) R_b R_s \tag{8.31}$$

The pressure drops in the nozzles must be calculated separately and added to the total pressure drop.

In Equation (8.31), Δp_{bi} is calculated from:

$$\Delta p_{bi} = 4 f_i \frac{G_s^2}{2\rho_s}\left(\frac{\mu_{s,w}}{\mu_s}\right)^{0.14} N_c \tag{8.32}$$

Friction coefficients are given in Figures 8.14 through 8.16 and by Equation 8.25.

For an ideal baffle window section, Δp_{wi} is calculated from:

$$\Delta p_{wi} = \frac{\dot{m}_s^2 (2 + 0.6 N_{cw})}{2\rho_s A_s A_w} \tag{8.33}$$

if $Re_s \geq 100$, and

$$\Delta p_{wi} = 26 \frac{\mu_s \dot{m}_s}{\sqrt{A_s A_w \rho}}\left(\frac{N_{cw}}{p - d_o} + \frac{B}{D_w^2}\right) + \frac{\dot{m}_s}{A_s A_w \rho_s} \tag{8.34}$$

if $Re_s \leq 100$.

Calculation of equivalent diameter of window, D_w; area for flow through window, A_w; and the correction factors are given in References 7, 11, and 12.

The number of tube rows crossed in one crossflow section, N_c, can be estimated from:

$$N_c = \frac{d_i\left(1 - 2\frac{L_c}{d_i}\right)}{p_p} \tag{8.35}$$

p_p is defined in Figure 8.19 and can be obtained from Table 8.7, and L_e is the baffle cut distance from baffle tip to shell inside distance.

The number of effective crossflow rows in each window, N_{cw}, can be estimated from:

$$N_{cw} = \frac{0.8 L_c}{p_p} \tag{8.36}$$

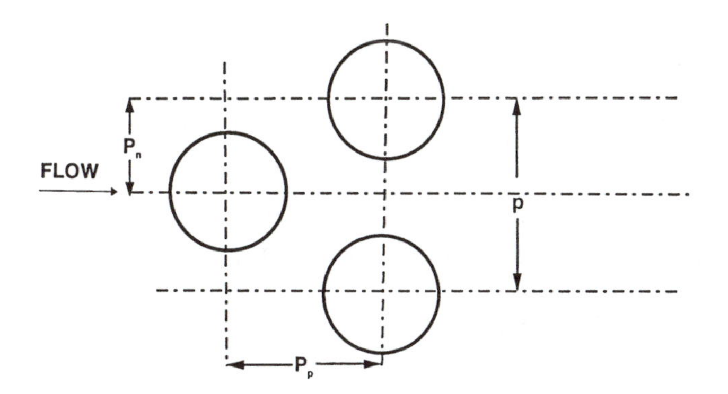

FIGURE 8.19
Tube pitches parallel and normal to flow (equilateral triangular arrangement shown).

TABLE 8.7
Tube Pitches Parallel and Normal to Flow

Tube O.D. (d_o, in.)	Tube Pitch (p, in.)	Layout	p_p (in.)	p_n (in.)
5/8 = 0.625	13/16 = 0.812	→ ◁	0.704	0.406
3/4 = 0.750	15/16 = 0.938	→ ◁	0.814	0.469
3/4 = 0.750	1.000	→ □	1.000	1.000
3/4 = 0.750	1.000	→ ◇	0.707	0.707
3/4 = 0.750	1.000	→ ◁	0.866	0.500
1	1 1/4 = 1.250	→ □	1.250	1.250
1	1 1/4 = 1.250	→ ◇	0.884	0.884
1	1 1/4 = 1.250	→ ◁	1.082	0.625

From Bell, K. J. (1981) In *Heat Exchangers — Thermal–Hydraulic Fundamentals and Design*, pp. 559–579. Taylor & Francis, Washington, D.C. With permission.

The number of baffles, N_b, can be calculated from:

$$N_b = \frac{L - B_i - B_o}{L} + 1 \tag{8.37}$$

The total shell-side pressure drop of a typical shell and-tube exchanger is on the order of 20 to 30 % of the pressure drop that would be calculated without taking into account baffle leakages and tube bundle bypass effects.[11]

Example 8.4

Distilled water with a mass flow rate of 80,000 kg/h enters an exchanger at 35°C and leaves at 25°C. The heat will be transferred to 140,000 kg/h of raw water coming from a supply at 20°C. The baffles will be spaced 12 in. apart. A single-shell and single-tube pass is preferable. The tubes are 18 BWG tubes with 1-in. O.D. (O.D. = 0.0254 m, I.D. = 0.0229 m), and they are laid out on 1-in. square pitch. Shell diameter is 15 1/4 in. A pitch size of 1.25 in. and a clearance of 0.25 in. are selected. Calculate the length of the heat exchanger and the pressure drop for each stream. If the shell-side allowable maximum pressure drop is 200 kPa , will this heat exchanger be suitable?

SOLUTION
Tube-side specifications are

O.D.	$d_o = 1$ in. = 0.0254 m
I.D.	$d_i = 0.902$ in. = 0.0229108 m
Flow area	$A_c = 0.639$ in.^{2} = 0.00041226 m^2
Wall thickness	$t_w = 0.049$ in. = 0.0012446 m

Calculate mass flow rate (from problem statement) by:

$$\dot{m}_t = \frac{14,000\ \text{kg/h}}{3,600\ \text{s/h}} = 38.89\ \text{kg/s}$$

Shell-side specifications are

Pitch size	$P_T = 1.25$ in. = 0.03175 m
Clearance	$C = 0.25 = 0.00635$ m
Baffle spacing	$B = 12$ in. = 0.3048 m
Shell diameter	$D_s = 15.25 = 0.38735$ m

Calculate mass ass flow rate (from problem statement) by:

$$\dot{m}_s = \frac{80,000\ \text{kg/h}}{3,600\ \text{s/h}} = 22.22\ \text{kg/s}$$

For a single-pass shell-and-tube heat exchanger with a diameter of 15.25 in. (from Table 8.3) the number of tubes, N_{ts}, in a 1.25-in. square pitch with an outer tube diameter of 1 in. is 81 tubes.

Shell-side fluid properties at 30°C average temperature from Appendix B (Table B.2) are

$$c_p = 4.1785\ \text{kJ/kg·K}$$
$$\mu_b = 0.000797\ \text{kg/m·s}$$
$$k = 0.614\ \text{W/m·K}$$
$$\rho = 995.7\ \text{kg/m}^3$$
$$Pr = 5.43$$

Water properties at 22.5°C are

$$c_p = 4.179\ \text{kJ/kg·K}$$
$$\mu_b = 0.00095\text{kg/m·s}$$
$$k = 0.6065\ \text{W/m·K}$$
$$\rho = 997\ \text{kg/m}^3$$
$$Pr = 6.55$$

Determine mean temperature difference by:

$$\Delta T_{lm,cf} = \frac{\Delta T_1 - \Delta T_2}{\ln \frac{\Delta T_1}{\Delta T_2}} = \frac{(35-25)-25-20)}{\ln\left(\frac{35-25}{25-20}\right)} = 7.21°\text{C}$$

$$\text{R} = \frac{T_{h_1} - T_{h_2}}{T_{c_2} - T_{c_1}} = \frac{(35-25)}{(25-20)} = 2$$

$$P = \frac{T_{c_1} - T_{c_2}}{T_{h_1} - T_{c_1}} = \frac{25-20}{35-20} = 0.333$$

The correction factor is assumed as $F \approx 1$.

Calculate shell-side heat transfer coefficient by:

$$A_s = \frac{(D_s CB)}{P_t} = \frac{(0.38735 \times 0.00635 \times 0.3048)}{0.03175} = 0.02361\ \text{m}^2$$

$$G_s = \frac{\dot{m}_s}{A_s} = \frac{22.22\ \text{kg/s}}{0.02361\ \text{m}^2} = 941.107\ \text{kg/m}^2 \cdot \text{s}$$

$$D_e = \frac{4(P_t^2 - \pi D_o^2/4)}{\pi D_o} = \frac{4(0.03175^2 - \pi \cdot 0.0254^2/4)}{\pi \cdot 0.0254} = 0.02513\ \text{m}$$

$$Re_s = \frac{D_e G_s}{\mu} = \frac{0.02513 \cdot 941.107}{0.000797} = 29673.8$$

Therefore, the flow of the fluid on shell side is turbulent. By using McAdam's correlation, Equation (8.11), we get the Nusselt number:

$$Nu = 0.36\left(\frac{D_e G_s}{\mu_b}\right)^{0.55}\left(\frac{c_p \mu_b}{k}\right)^{0.33}\left(\frac{\mu_b}{\mu_w}\right)^{0.14}$$

$$= 0.36\left(\frac{0.02513 \times 941.107}{0.000797}\right)^{0.55}\left(\frac{4178.5 \times 0.000797}{0.614}\right)^{0.33}\left(\frac{0.000797}{0.00086}\right)^{0.14}$$

$$= 179.39$$

It is assumed that the tube wall temperature is 26°C and $\mu_w = 0.00086$ kg/m·s. The shell-side heat transfer coefficient, h_o, is then calculated as:

$$h_o = \frac{Nu \cdot k}{D_e} = \frac{179.30 \times 0.614}{0.02513} = 4,383.09\ \text{W/m}^2 \cdot \text{K}$$

Calculate tube-side heat transfer coefficient by:

$$A_t = \frac{\pi d_i^2}{4} = \frac{\pi \times 0.02291^2}{4} = 0.0004122 \text{ m}^2$$

$$A_{tp} = \frac{N_t A_t}{\text{no. of passes}} = \frac{81 \times 0.0004122}{1} = 0.0339 \text{ m}^2$$

$$G_t = \frac{\dot{m}_t}{A_{tp}} = \frac{38.889}{0.03339} = 1{,}164.6 \text{ kg/m}^2 \cdot \text{s}$$

$$u_t = \frac{G_t}{\rho} = \frac{1164.6}{997} = 1.1682 \text{ m/s}$$

$$Re_t = \frac{u_t \rho d_i}{\mu} = \frac{1.1682 \times 997 \times 0.02291}{0.00095} = 28{,}087.5$$

Therefore, the flow of the fluid on tube side is turbulent. By using the Petukhov and Kirillov correlation:

$$Nu = \frac{(f/2) Re Pr}{1.07 + 12.7(f/2)^{1/2}(Pr^{2/3} - 1)}$$

where $f = (1.58 \ln Re - 3.28)^{-2} = [1.58 \times \ln (28{,}087.5) - 3.28]^{-2} = 0.0060$

$$Nu = \frac{(0.006/2) \times 28087.5 \times 6.55}{1.07 + 12.7 \times (0.006/2)^{1/2} \times (6.55^{2/3} - 1)} = 196.45$$

The tube-side heat transfer coefficient, h_i, is then found as:

$$h_i = \frac{Nu \cdot k}{d_i} = \frac{196.45 \times 0.6065}{0.02291} = 5{,}200.5 \text{ W/m}^2 \cdot \text{K}$$

Now, the overall heat transfer coefficient, U_o, is determined by the following equation:

$$U_o = \frac{1}{\dfrac{d_o}{d_i h_i} + \dfrac{d_o \ln(d_o/d_i)}{2k} + \dfrac{1}{h_o}}$$

$$= \frac{1}{\dfrac{0.0254}{0.02291 \times 5200.5} + \dfrac{0.0254 \times \ln(0.0254/0.02291)}{2 \times 54} + \dfrac{1}{4383.09}}$$

$$= 2{,}147.48 \text{ W/m}^2 \cdot \text{k}$$

To find the area and consequently the length of heat exchanger, the required heat transfer rate must first be determined by:

$$Q = \dot{m}_s c_p (T_{h_1} - T_{h_2}) = 22.22 \times 4178.5 \times (35 - 25) = 928.5 \text{ kW}$$

Now, the heat transfer rate is also defined as:

$$Q = U_o A F T_{lm,cf}$$

Therefore, the area can be determined:

$$A = \frac{Q}{U_o F T_{lm,cf}} = \frac{928.5 \times 1000}{2147.48 \times 1 \times 7.2135} = 59.93 \text{ m}^2$$

and the length

$$L = \frac{A}{N_t \pi D_o} = \frac{59.77}{81 \times \pi \times 0.0254} = 9.28 \text{ m}$$

The shell-side pressure drop can be calculated from Equation (8.17):

$$\Delta p_s = \frac{f G_s^2 (N_b + 1) D_s}{2 \rho D_e \phi_s}$$

$$N_b = L / B \approx 30 \text{ baffles}$$

$$f = \exp(0.576 - 0.19 \ln R_s)$$

$$f = \exp(0.576 - 0.19 \ln 29672.6) = 0.2514$$

$$\Delta p_s = \frac{0.2514 \times (941.107)^2 \times (30 + 1) \times 0.38735}{2 \times 995.7 \times 0.02513 \times \left(\frac{7.97}{8.6} \right)^{0.14}} = 53942 \text{ Pa} = 53.9 \text{ kPa}$$

$$\Delta p_s = 53.9 \text{ kPa} < 200 \text{ kPa}$$

Therefore, this exchanger is suitable.
The tube-side pressure drop can be calculated from Equation (8.22):

$$\Delta p_t = (4f \frac{L N_p}{d_i} + 4 N_b) \frac{\rho u_m^2}{2}$$

$$\Delta p_t = \left(\frac{4 \times 0.0060 \times 9.28 \times 1}{0.02291} + 4 \times 30 \right) \times 997 \times \frac{1.1682^2}{2}$$

$$= 88249.4 \text{ Pa} = 88.2 \text{ kPa}$$

The preceding analysis can be repeated by the Bell–Delaware method. The Bell–Delaware method assumes that in addition to flow specifications on the shell side, the shell-side geometric data must be known or specified. With this geometric information, all remaining geometric parameters needed in the shell-side calculations can be calculated or estimated by methods given in Reference 10 and the Example 8.4 can be repeated with the Bell–Delaware method (a thermal design project).

Nomenclature

A_o heat transfer area based on the outside surface area of tubes, m^2

A_i heat transfer area based on the inside surface area of tubes, m^2

A_s crossflow area at or near shell centerline, m^2

A_w area for flow through baffle window, m^2

B baffle spacing, m

B_i baffle spacing at the inlet, m

B_o baffle spacing at the outlet, m

C clearance between the tubes, m

D_s shell I.D., m

D_w equivalent diameter of baffle window, m

d_o tube O.D., m

d_i tube I.D., m

F correction factor to LMTD for non-counter-flow systems

f_i friction factor for flow across an ideal tube bank

G mass velocity, kg/m^2·s

h_i tube-side heat transfer coefficient, W/m^2·K

h_{id} shell-side heat transfer coefficient for ideal tube bank, W/m^2·K

h_o shell-side heat transfer coefficient for the exchanger, W/m^2·K

J_b bundle bypass correction factor for heat transfer

J_c segmental baffle window correction factor for heat transfer

J_l baffle leakage correction factor for heat transfer

J_r laminar flow heat transfer correction factor

J_s heat transfer correction factor for unequal end baffle spacings

J_i Colburn j-factor for an ideal tube bank

k_s thermal conductivity of shell-side fluid, W/m·K

k_w thermal conductivity of tube wall, W/m·K

L effective tube length of heat exchanger between tube sheets, m

$\dot{m}_s$ shell-side mass flow rate, kg/s

m_t tube-side mass flow rate, kg/s

N_b number of baffles in the exchanger

N_c number of tube rows crossed between baffle tips of one baffle compartment

N_{cw} number of tube rows crossed in one baffle window

N_t total number of tubes or total number of holes in tube sheet for U-tube bundle

P_T	pitch size, m
Q	heat duty of heat exchanger, W
R_b	bundle bypass correction factor for pressure drop
$R_{f,i}$	tube-side fouling resistance referred to inside tube surface, $m^2 \cdot K/W$
$R_{f,o}$	shell-side fouling resistance referred to outside tube surface, $m^2 \cdot K/W$
R_s	baffle end zones correction factor for pressure drop
Re_s	shell-side Reynolds number
T	temperature,°C, K
T_c	cold fluid temperature,°C, K
T_h	hot fluid temperature,°C, K
T_w	wall temperature,°C, K
U_c	overall heat transfer coefficient for clean surface based on the outside tube area, $W/m^2 \cdot K$
U_f	overall heat transfer coefficient for fouled surface based on the outside tube area, $W/m^2 \cdot K$
Δp_{bi}	pressure drop for one baffle compartment in crossflow, based on ideal tube bank, Pa
Δp_{wi}	pressure drop in one ideal window section of a segmentally baffles exchanger, Pa
$\Delta T_c, \Delta T_h$	cold and hot end terminal temperature differences,°C, K
ΔT_{lm}	log mean temperature differences,°C, K
ΔT_m	effective or true mean temperature difference,°C, K
μ_s	shell fluid dynamic viscosity at average temperature, mPa/s
μ_t	tube fluid dynamic viscosity at average temperature, mPa/s
ρ_s, ρ_t	shell- or tube-side fluid density, respectively, at average temperature of each fluid, kg/m^3
ϕ_s	viscosity correction factor for shell-side fluids, $(\mu_w/\mu_b)^{0.14}$

References

1. *Standards of the Tubular Exchanger Manufacturers Association* (1988), 7th ed., TEMA, Tarrytown, NY.
2. Butterworth, D. (1988) Condensers and their design. In *Two-Phase Flow Heat Exchangers — Thermal–Hydraulic Fundamentals and Design*, S. Kakaç, A. E. Bergles, and E. O. Fernandes (Eds.), Kluwer, Dordrecht, The Netherlands.
3. Kern, D. Q. (1950) *Process Heat Transfer*, McGraw-Hill, New York.
4. Kakaç, S., Bergles, A. E., and Mayinger, F. (Eds.) (1981) *Heat Exchangers — Thermal–Hydraulic Fundamentals and Design*, Taylor & Francis, Washington D.C.
5. Bell, K. J. (1981) Preliminary design of shell and tube heat exchangers. In *Heat Exchangers — Thermal–Hydraulic Fundamentals and Design*, S. Kakaç, A. E. Bergles, and F. Mayinger (Eds.), Taylor & Francis, Washington D.C., pp. 559–579.
6. Taborek, J. (1983) Shell-and-tube heat exchanger. In *Heat Exchanger Design Handbook*, E. U. Schlünder (Ed.), Section 3.3. Hemisphere, New York.

7. Taborek, J. (1991) Industrial heat exchanger design practices. In *Boiler, Evaporators, and Condensers*, S. Kakaç (Ed.), pp. 143–177. Wiley, New York.
8. Fraas, A. P. (1989) *Heat Exchanger Design*, Wiley, New York.
9. Tinker, T. (1951) Shell-side characteristics of shell-and-tube heat exchangers. *General Discussion on Heat Transfer*, pp. 97–116. Institute Mechanical Engineering and ASME, New York, London.
10. Palen, J. W. and Taborek, J. (1969) Solution of shell-side flow pressure drop and heat transfer by stream analysis method. *CEP Symp. Ser.* no. 92, 66, Heat Transfer, Philadelphia, pp. 53–63.
11. Bell, K. J. (1988) Delaware method of shell-side design. In *Heat Transfer Equipment Design*, R. K. Shah, E. C. Sunnarao, and R. A. Mashelkar (Eds.), Taylor & Francis, New York.
12. Bell, K. J. (1981) Delaware method for shell side design. In *Heat Exchangers — Thermal–Hydraulic Fundamentals and Design*, S. Kakac, A. E. Bergles, and F. Mayinger (Eds.), pp. 581–618. Taylor & Francis, Washington D.C.
13. Hewitt, G. F., Shires, G. L., and Bott, T. R. (1994) *Process Heat Transfer*, CRC Press, Boca Raton, FL.

Problems

8.1. Crude oil at a flow rate of 63.77 kg/s enters the exchanger at 102°C and leaves at 65°C. The heat will be transferred to 45 kg/s of tube water (city water) coming from a supply at 21°C. The exchanger data are: 3/4-in. O.D. and 18 BWG tubes on 1-in. square pitch, two-tube passes and four-tube passes will be considered. Tube material is carbon steel. The heat exchanger has one shell. Two different shell diameters of I.D. 35 and 37 in. should be studied. Baffle spacing is 275 mm. Calculate the length of the heat exchanger for clean and fouled surfaces. Calculate:

a. Tube-side velocity for 1 to 2 and 1 to 4 arrangements
b. Overall heat transfer coefficients for clean and fouled surfaces
c. Pressure drops
d. Pumping powers

The allowable shell-side and tube-side pressure drops are 60 and 45 kPa, respectively. The following properties are given:

	Shell side	Tube side
Specific heat, J/kg·K	2177	4186.8
Dynamic viscosity, N·s/m²	0.00189	0.00072
Thermal conductivity, W/m·K	0.122	0.605
Density, kg/m³	786.4	995
Prandtl number	33.73	6.29
Maximum pressure loss, Pa	60,000	45,000

8.2. Water at a flow rate of 60 kg/s enters a baffled shell-and-tube heat exchanger at 35°C and leaves at 25°C. The heat will be transferred to 150 kg/s of raw water coming from a supply at 15°C. You are requested to design the heat exchanger for this purpose. A single-shell and single-tube pass is preferable. The tube diameter is 3/4 in. (19-mm O.D. with 16-mm I.D.) and tubes are laid out on 1-in. square pitch. A maximum length of the heat exchanger of 8 m is required because of space limitations. The tube material is 0.5 Cr alloy. Assume a total fouling resistance of 0.000176 $m^2 \cdot K/W$. Note that surface over design should not exceed

30%. Reasonable design assumptions can be made along with the calculation if it is needed. Calculate shell diameter, number of tubes, fluid velocities, shell-side heat transfer coefficient, overall heat transfer coefficient, mean temperature difference, total area, and pressure drops. Is the final design acceptable? Discuss your findings.

8.3. Distilled water at a flow rate of 80,000 kg/h of enters an exchanger at 35 °C and leaves at 25 °C. The heat will be transferred to 140,000 kg/h of raw water coming from a supply at 20 °C. The baffles will be spaced 12 in. apart. Write a computer program to determine the effects of varying tube size and configuration, shell diameter on thermal characteristics, size and fluid flow characteristics of the shell, and tube type of heat exchanger. 1-P, 2-P, and 4-P configurations will be considered. The exchanger data are 3/4 in. O.D., 18 BWG tubes on 1-in. square pitch; and 3/4 in. O.D., 18 BWG tubes on 1-in. triangular pitch. Study three different shell diameters of 15 1/4, 17 1/4, and 19 1/4 in.

   a. Calculate the length of this heat exchanger for clean and fouled surfaces.
   b. A 120,000 N/m^2 pressure drop may be expended on both streams. Will this heat exchanger be suitable? For the pumping power, assume a pump efficiency of 80%.

8.4. A heat exchanger is available to heat raw water by the use of condensed water at 67 °C that flows in the shell side with a mass flow rate of 50,000 kg/h. Shell-side dimensions are I.D.$_s$ = D_s = 19 1/4 in., P_T = 1.25 in. (square), and baffle spacing = 0.3 m. The raw water enters the tubes at 17°C with mass flow rate of 30,000 kg/h. Tubes dimensions are d_o = 1 in. = 0.0254 m (18 BWG tubes) and d_i = 0.902 in. The length of the heat exchanger is 6 m with two passes. The permissible maximum pressure drop on the shell side is 1.5 lb$_f$ /in.^{2}. Water outlet temperature should not be less than 40°.

   a. Calculate the outlet temperatures.
   b. Calculate the heat load of the heat exchanger.
   c. Is the heat exchanger appropriate to be used for this purpose?

8.5. Distilled water at a flow rate of 20 kg/s enters an exchanger at 35 °C and leaves at 25°C. The heat will be transferred to 40 kg/s of raw water coming from supply at 20 °C. Available for this service is a 17 1/4-in. I.D. exchanger having 166, 3/4-in. O.D. tubes (18 BWG) and laid out on 1-in. square pitch. The bundle is arranged for two passes and baffles are spaced 12 in. apart.

   a. Calculate the length of this heat exchanger if all the surfaces are clean.
   b. Will this heat exchanger be suitable? Assume fouling factors. A 12 × 10^4 N/m^2 pressure drop may be expended on both streams.

8.6. A 1 to 2 baffled shell-and-tube type heat exchanger is used as an engine oil cooler. Cooling water flows through tubes at 25°C at a rate of 8.16 kg/s and exits at 35°C. The inlet and outlet temperatures of the engine oil are 65 and 55°C, respectively. The heat exchanger has 12.25-in. I.D. shell, and 18 BWG and 0.75-in. O.D. tubes. A total of 160 tubes are laid out on a 15/16-in. triangular pitch. By assuming $R_{fo} = 1.76 \times 10^{-4}$ m^2·W, $A_oR_w = 1.084 \times 10^{-5}$ m^2·K/W, h_o = 686 W/m^2·K, A_o/A_i =1.1476, and $R_{fi} = 8.8 \times 10^{-5}$ m^2·K/W, find the:

   1. Heat transfer coefficient inside the tubes
   2. Total surface area of the heat exchanger

8.7. A water-to-water system is used to test the effects of changing tube length, baffle spacing, tube pitch, pitch layout, and tube diameter. Cold water at 25°C and

100,000 kg/h is heated by hot water at 100°C, and also at 100,000 kg/h. The exchanger has a 31-in. I.D. shell. Perform calculations on this 1 to 2 shell-and-tube heat exchanger for the following conditions, as outlined in the examples; and put together an overall comparison chart.

a. Tubes are 3/4-in. O.D. and 12 BWG laid out on a 1-in. triangular pitch; and have three baffles per meter of tube length. Analyze the exchanger for tube lengths of 2, 3, 4, and 5 m.

b. Tubes are 3/4-in. O.D., 12 BWG laid out on a 1-in. triangular pitch, and 2 m long. Analyze for baffle placement of one baffle per meter of tube length, two baffles per meter of tube length, three baffles per meter of tube length, four baffles per meter of tube length, and five baffles per meter of tube length.

c. Tubes are 3/4-in. O.D., 12 BWG, and 2 m long; and have four baffles per meter of tube length. Analyze the exchanger for tube layouts of 15/16-in. triangular pitch, 1-in. triangular pitch, and 1-in. square pitch.

d. Tubes are 1-m O.D., 12 BWG, 4 m long; and have nine baffles. Analyze the exchanger for tube layouts of 1 1/4-in. triangular pitch, and 1 1/4-in. square pitch.

e. Tubes are 3/4-in. O.D., 12 BWG laid out on a 1-in. triangular pitch, and 2 m long; and have four baffles per meter of tube length. Analyze the exchanger for two, four, six, and eight tube passes; and compare to the case of true counterflow (i.e., one tube pass).

8.8. A sugar solution (ρ = 1080 kg/m^3, C_p = 3601 J/kg·K, k_f = 0.5764 W/m·K, μ = 1.3×10^{-3} N·s/m^2) flows at rate of 60,000 kg/h and is to be heated from 25 to 50°C. Water at 95°C is available at a flow rate of 75,000 kg/h (C_p = 40004 J/kg·K). It is proposed to use a one-shell-pass and two-tube-pass shell-and-tube heat exchanger containing 3/4-in. O.D., 16 BWG tubes. Velocity of the sugar solution through the tube is 1.5 m/s, and the length of the heat exchanger should not be more than 3 m because of the space limitations. Assume that the shell-side heat transfer coefficient is 700 W/m^2·K and thermal conductivity of the tube material is 52 W/m·K. Calculate:

a. Number of tubes and tube-side (sugar) heat transfer coefficient

b. Overall heat transfer coefficient

c. Length of this heat exchanger

Is this heat exchanger acceptable?

Design Project 8.1

Oil Cooler for Marine Applications

The following specifications are given:

Fluid	SAE-30 Oil	Sea Water
Inlet temperature, °C	65	20
Outlet temperature, °C	56	32
Pressure drop limit, kPa	140	40
Total mass flow rate, kg/s	20	—

A shell-and-tube heat exchanger type with geometric parameters can be selected. The heat exchanger must be designed and rated. Different configurations of shell-and-tube types can be tested. Parametric study is expected to come up with a suitable final design; mechanical design will be performed and the cost will be estimated. Follow the chart given in Figure 8.10.

Design Project 8.2

The shell-and-tube type of oil cooler for marine applications given in Project 8.1 can be repeated for a lubricating oil mass flow rate of 10 and 5 kg/s while keeping the other specifications constant.

Design Project 8.3

A shell-and-tube heat exchanger to be designed as a crude oil cooler: 120 kg/s of crude oil enters the shell side of the heat exchanger at 102°C and leaves at 65°C. The coolant to be used is city water entering the tube side at 21°C with a flow rate of 65 kg/s. Shell-side pressure drop is limited to 150 kPa. Design and rating expectations are the same as outlined in Design Project 8.1

Design Project 8.4

Problems 8.1, 8.2, 8.3, and 8.7 can be assigned as students' thermal design projects. Materials selection, parametric study, mechanical design, technical drawings of various components and the assembly, and cost analysis must be included.

9

Compact Heat Exchangers

9.1 Introduction

Compact heat exchangers of plate–fin and tube–fin types, tube bundles with small diameters, and regenerative type are used generally for applications where gas flows. A heat exchanger having a surface area density greater than about 700 m^2/m^3 is quite arbitrarily referred to as a *compact heat exchanger*. The heat transfer surface area is increased by fins to increase the surface area per unit volume and there are many variations.[1-3] The device is referred to as a microheat exchanger if the surface area density is about 10,000 m^2/m^3.

Compact heat exchangers are widely used in industry especially as gas-to-gas or liquid-to-gas heat exchangers; some examples are vehicular heat exchangers, condensers and evaporaters in air-conditioning and refrigeration industry, aircraft oil coolers, automative radiators, oil coolers, unit air heaters, intercoolers of compressors, and aircraft and space applications. Compact heat exchangers are also used in cryogenics process, electronics, energy recovery, conservation and conversion, and other industries. Some examples of compact heat exchanger surfaces are shown in Figure 9.1.[1,2]

Compact heat exchangers are plate-fin or tube-fin heat exchangers.

9.1.1 Plate–Fin Heat Exchangers

In this type, each channel is defined by two parallel plates separated by fins or spacers. Fins or spacers are sandwiched between parallel plates or formed tubes. Examples of compact plate-heat exchangers are shown in Figure 9.2.[4] Fins are attached to the plates by brazing, soldering, adhensive bonding, welding, mechanical fit, or extrusion. Alternate fluid passages are connected in parallel by end heads to form two sides of a heat exchanger. Fins are employed on both sides in gas-to-gas heat exchangers.

In gas-to-liquid heat exchanger applications, fins are usually employed only on the gas side where the heat transfer coefficient is lower; if fins are employed on the liquid side, the fins provide a structural strength. The fins used in a plate heat exchanger may be plain and straight fins; plain but wavy fins; or interrupted fins such as strip, louver, and perforated.

9.1.2 Tube–Fin Heat Exchangers

In a tube–fin exchanger, round, rectangular, and elliptical tubes are used and fins are employed either on the outside or on the inside, or on both outside and inside of the tubes, depending on the application. In a gas-to-liquid heat exchanger, the gas-side heat transfer coefficient is very low compared with the liquid-side heat transfer coefficient; therefore no fins are needed on the liquid side. In some applications, fins are also used inside the tubes.

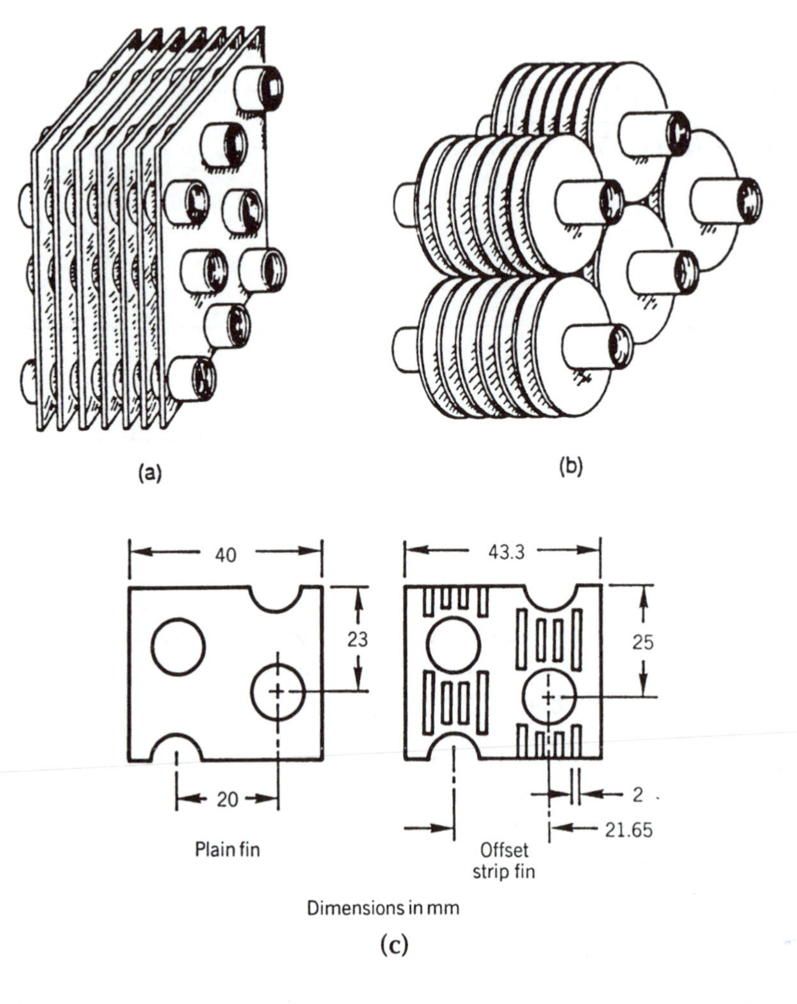

FIGURE 9.1
Finned-tube geometries used with circular tubes. (a) Plate fin-and-tube used for gases; (b) individually finned tubes; (c) plain fin and offset fin. (From Webb, R. L. [1994] *Principle of Enhanced Heat Transfer*, Wiley, New York.)

The liquids flow inside the tube, which can accommodate high pressures. Tube–fin heat exchangers are less compact than plate–fin heat exchangers. Examples of tube–fin heat exchangers are shown in Figure 9.3.[4]

Fins on the outside tubes may be categorized as (1) flat or continuous (plain, wavy, or interrupted) external fins on an array of tubes, (2) normal fins on individual tubes, and (3) longitudinal fins on individual tubes.

9.2 Heat Transfer and Pressure Drop

9.2.1 Heat Transfer

As outlined in the preceding section, compact heat exchangers are available in a wide variety of configurations of the heat transfer matrix. Their heat transfer and pressure drop characteristics have been studied by Kays and London.[3]

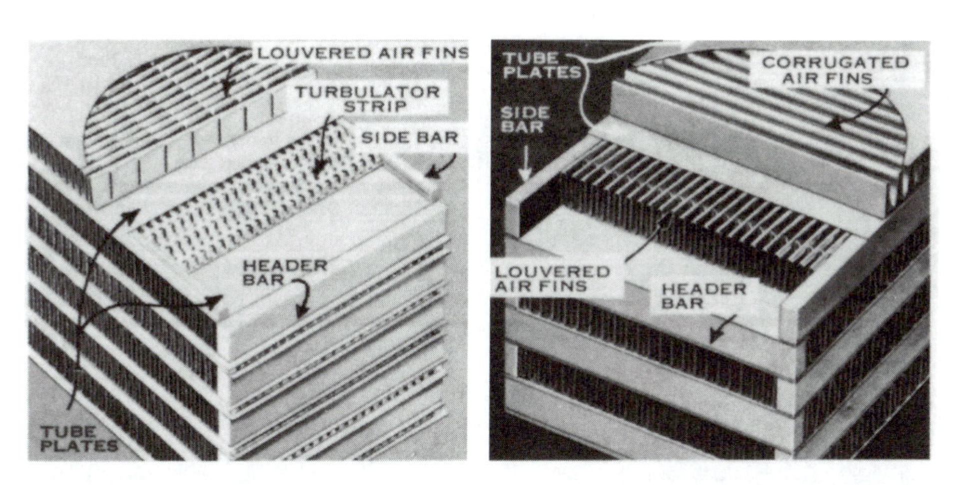

Bar and plate

Bar and plate

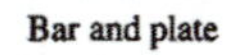

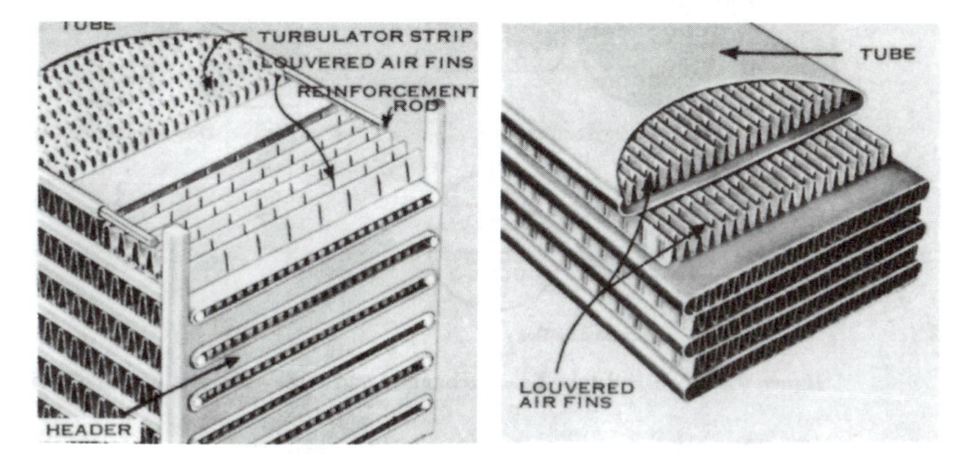

Formed plate fin

Formed plate fin

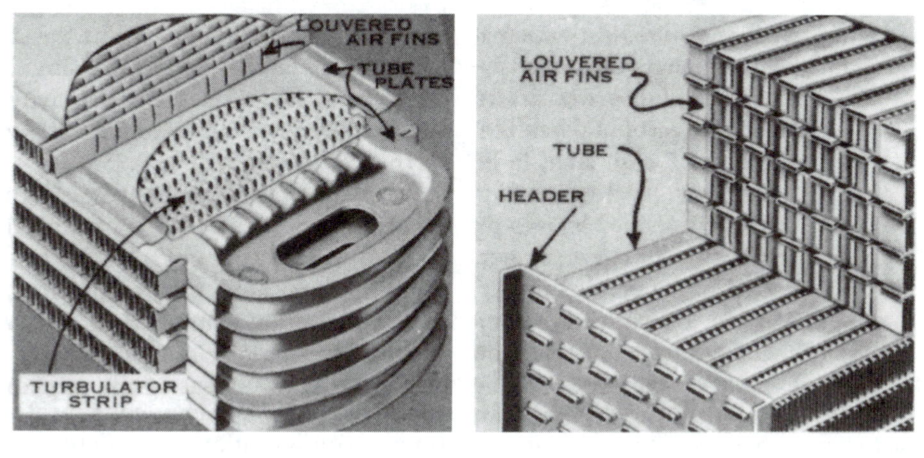

Drawn cup

Tube and center

FIGURE 9.2
Plate–fin heat exchangers. (Courtesy of Harrison Division, General Motors Corp., Lockport, NY.)

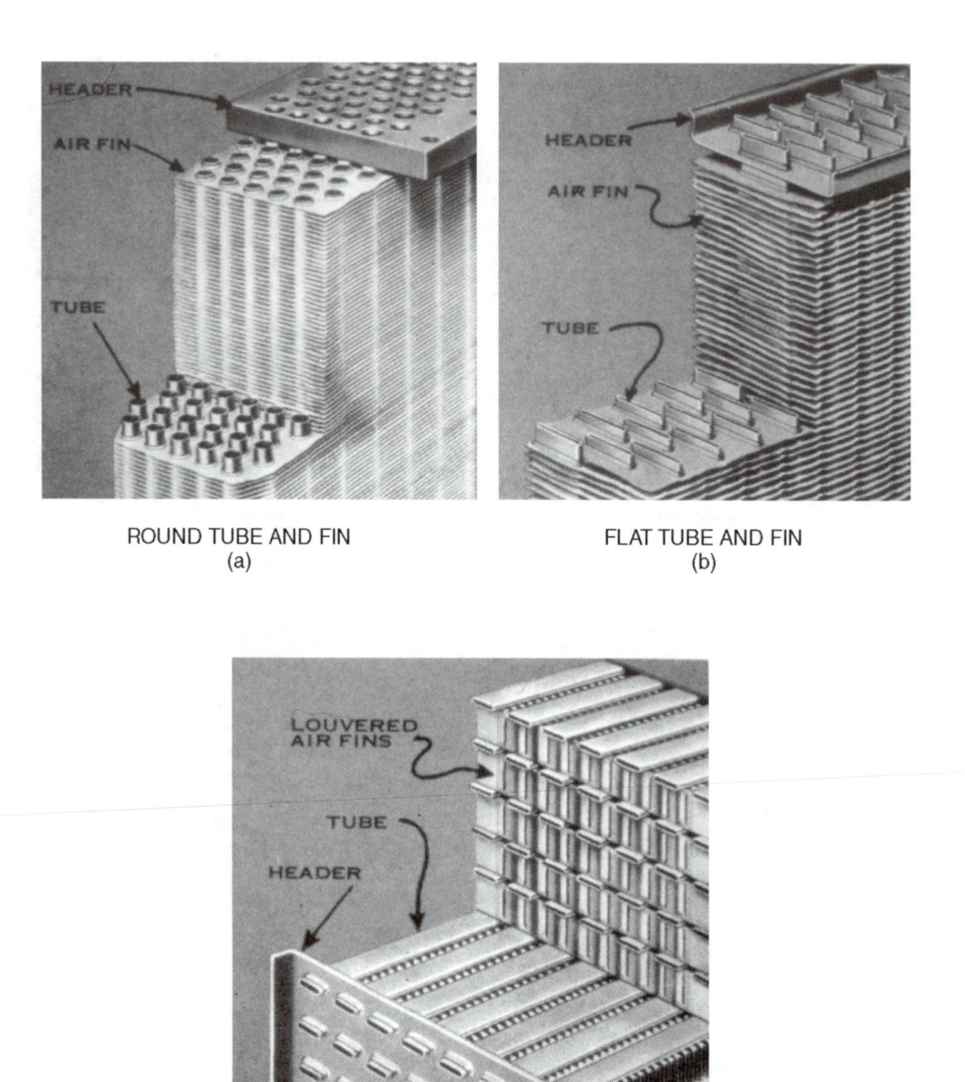

FIGURE 9.3
Tube fin heat exchangers. (Courtesy of Harrison Radiator Division, General Motor Corp., Lockport, NY.)

The heat transfer and pressure drop characteristics of various configurations for use as compact heat exchangers are determined experimentally. Figures 9.4 through 9.7 and Figure 9.9 show typical heat transfer and friction factor data for different configurations of compact heat exchangers.[3,5] For the fin–tube type of configuration, six different surface geometries are given in Table 9.1 and shown in Figure 9.8. Also data for the heat transfer and friction coefficient for five different plain plate–fins types are shown in Figure 9.9 and geometric data are given in Table 9.1.

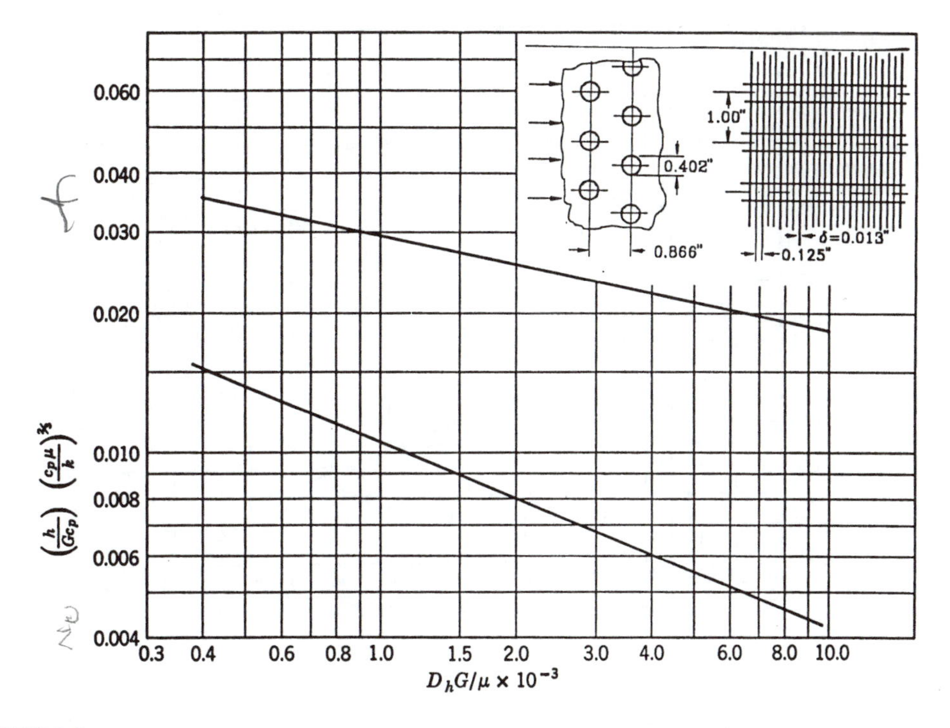

FIGURE 9.4
Heat transfer and friction factor for a circular tube-continuous fin heat exchanger, surface 8.0-3/8T. Surface 8.0-3/8T: tube O.D. = 1.02 cm; fin pitch = 3.15/cm; fin thickness = 0.033 cm; fin area/total area = 0.839; air-passage hydraulic diameter = 0.3633 cm; free-flow area/frontal area, σ = 0.534; heat transfer area/total volume = 587 m^2/m^3. (From Kays, W. M. and London, A. L. [1984] *Compact Heat Exchangers*, 3rd ed., McGraw-Hill, New York. With permission.)

Note that there are three dimensionless groups governing these correlations that are the Stanton, Prandtl, and Reynolds numbers:

$$St = \frac{h}{Gc_p} \quad Pr = \frac{c_p\mu}{k} \quad Re = \frac{GD_h}{\mu} \tag{9.1}$$

where, G is the mass velocity or mass flux, defined as:

$$G = \rho U_{max} = \frac{\dot{m}}{A_{min}} \tag{9.2}$$

$\dot{m}$ is the total mass flow rate of fluid, and A_{min} is the minimum free-flow cross-sectional area regardless of where this minimum occurs.

The value of the hydraulic diameter for each configuration is specified on Figures 9.4 to 9.6 and in Table 9.1.

The hydraulic diameter is defined as four times the flow passage volume divided by the total heat transfer area:

$$D_h = 4\frac{LA_{min}}{A} \tag{9.3}$$

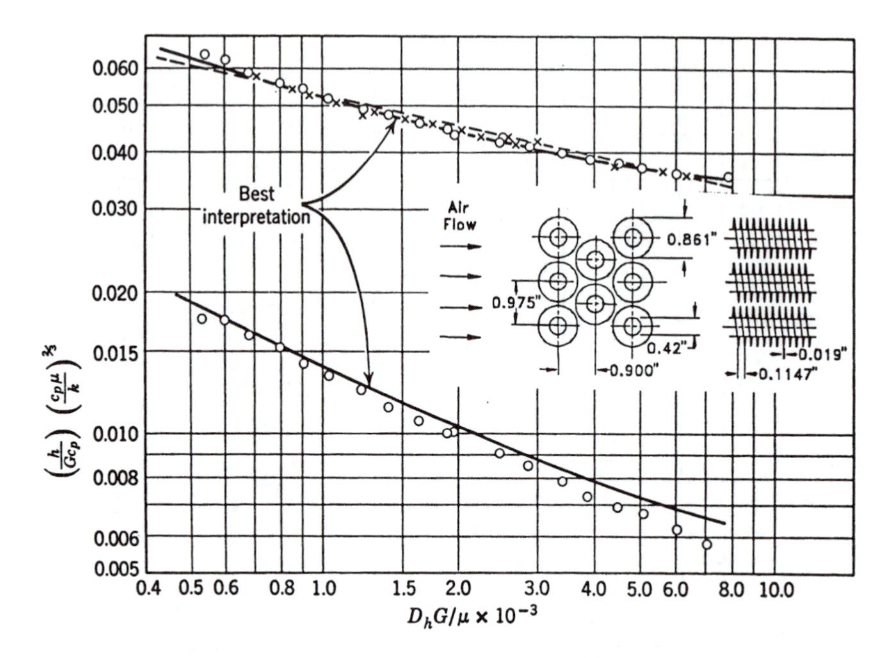

FIGURE 9.5
Heat transfer and friction factor for flow across circular finned-tube matrix. Surface CF-872(c): tube O.D. = 1.07 cm; fin pitch = 3.43/cm; fin thickness = 0.048 cm; fin area/total area = 0.876; air-passage hydraulic diameter, D_h = 0.443 cm; free-flow area/frontal area, σ = 0.494; heat transfer area/total volume = 446 m²/m³. (From Kays, W. M. and London, A. L. [1984] *Compact Heat Exchangers*, 3rd ed., McGraw-Hill, New York. With permission.)

where L is the flow length of the exchanger matrix, LA_{min} is the minimum free-flow passage volume, and A is the total heat transfer area.

Once the Reynolds number for flow is known, the heat transfer coefficient (Colburn modules, $J_H = St\, Pr^{2/3}$) and the friction factor, f, for flow across the matrix can be evaluated.

The overall heat transfer coefficient based on the gas–side surface area in a gas–liquid heat exchanger neglecting fouling effects can be written as:

$$\frac{1}{U_o} = \frac{A_t}{A_i}\frac{1}{h_i} + A_t R_w + \frac{1}{\eta_o h_o} \tag{9.4}$$

where η_o is the outside overall surface efficiency defined as:

$$\eta_o = 1 = \frac{A_f}{A_t}(1 - \eta_f)$$

and A_t (= $A_u + A_f$) is the total external air-side surface area.

The overall heat transfer coefficient based on the internal surface area becomes:

$$\frac{1}{U_i} = \frac{1}{h_i} + A_i R_w + \frac{A_i}{A_t}\frac{1}{\eta_o h_o} \tag{9.5}$$

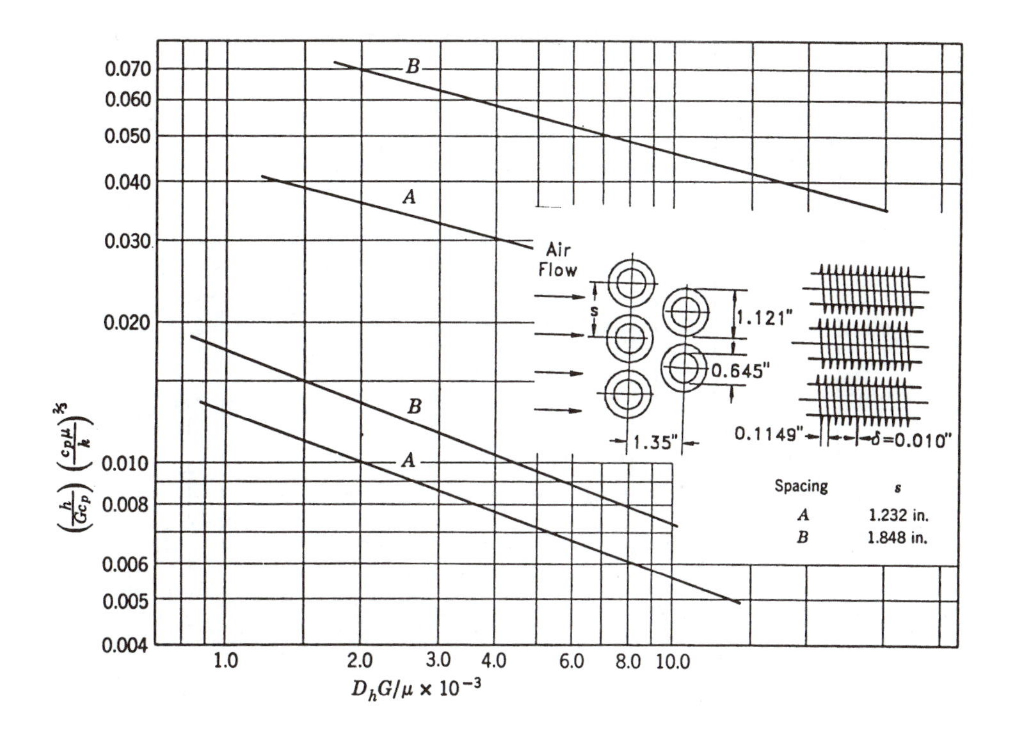

FIGURE 9.6
Heat transfer and friction factor for flow across finned-tube matrix. Surface CF-8.7-5/8 J: tube O.D. = 1.638 cm; fin pitch = 3.43/cm; fin thickness = 0.0254 cm; fin area/total area = 0.862; air-passage hydraulic diameter, D_h = 0.5477 cm (A), 1.1673 (B); free-flow area/frontal area, σ = 0.443 (A), 0.628 (B); heat transfer area/total volume = 323.8 m^2/m^3 (A), 215.6 m^2/m^3 (B). (From Kays, W. M. and London, A. L. [1984] *Compact Heat Exchangers*, 3rd ed., McGraw-Hill, New York. With permission.)

If both sides are finned, then Equation (9.5) becomes:

$$\frac{1}{U_i} = \frac{1}{\eta_i h_i} + A_i R_w + \frac{A_i}{A_t}\frac{1}{\eta_o h_o} \tag{9.6}$$

The air-side conductance is

$$K = \frac{\eta_o h_o A_t}{A_i} \tag{9.7}$$

An increase in the number of fins per centimeter will increase the conductance by increasing the ratio A_t/A_i. Also, the use of more closely spaced fins will increase the heat transfer coefficient, h_o, because of a smaller hydraulic diameter; or the use of a special fin configuration, such as wavy fin, will produce a higher heat transfer coefficient. The outside surface efficiency, η_o, is influenced by the fin thickness, thermal conductivity, and fin length. The fin efficiency, η_f, may be calculated from appropriate graphs or equations given in most heat transfer textbooks.

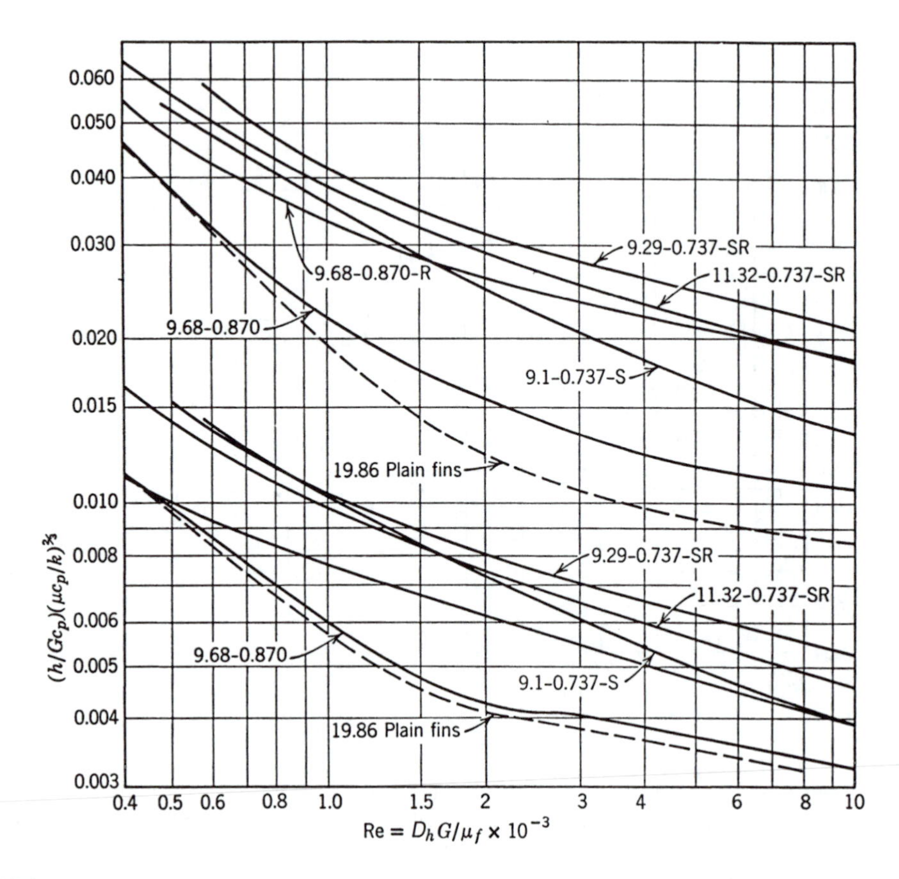

FIGURE 9.7
Heat transfer and friction factor for flow across finned-flat-tube matrix for the surfaces shown in Figure 9.8 and Table 9.1. (From Kays, W. M. and London, A. L. [1984] *Compact Heat Exchangers*, 3rd ed., McGraw-Hill, New York. With permission.)

9.2.2 Pressure Drop for Finned-Tube Exchangers

For flow normal to finned-tube banks such as illustrated in Figures 9.4 to 9.7, the total pressure drop, namely, the difference between the pressures at the inlet and outlet is given by:[3]

$$\Delta p = \frac{G^2}{2\rho_i}\left[f\frac{A_t}{A_{min}}\frac{\rho_i}{\rho} + (1+\sigma^2)\left(\frac{\rho_i}{\rho_o} - 1\right)\right] \tag{9.8}$$

where

$$\sigma = \frac{A_{min}}{A_{fr}} = \frac{\text{minimum free - flow area}}{\text{frontal area}} \tag{9.9}$$

$$\frac{A_t}{A_{min}} = \frac{4L}{D_h} = \frac{\text{total heat transfer area}}{\text{minimum flow area}} \tag{9.10}$$

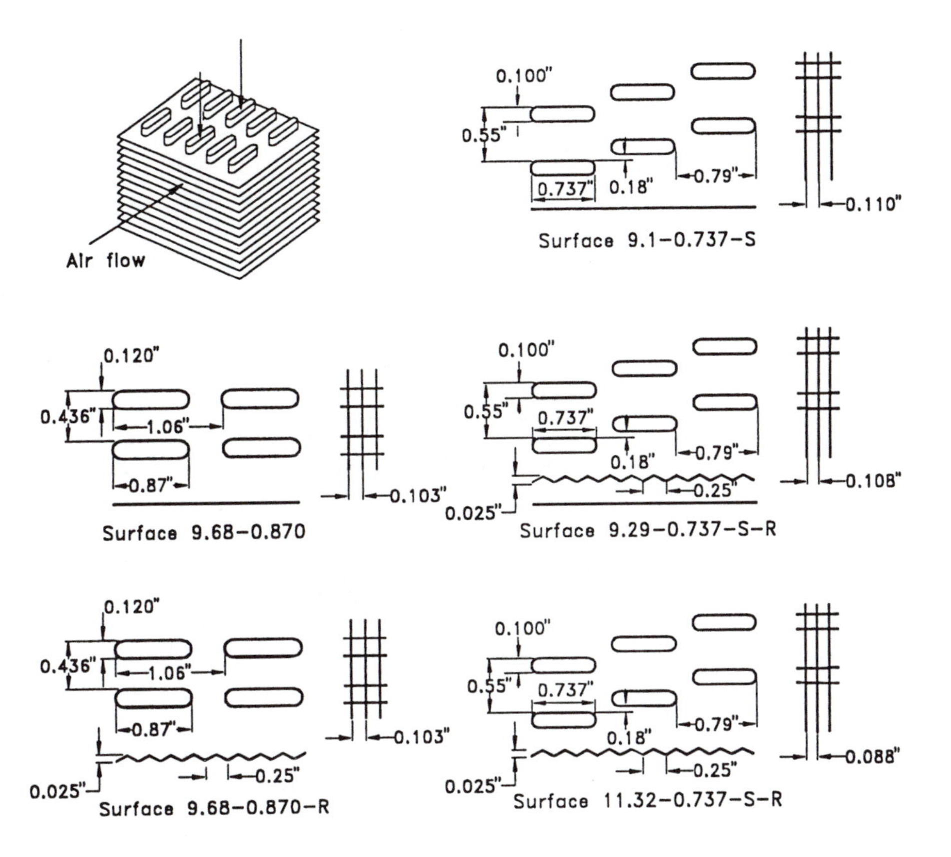

FIGURE 9.8
Various flattened tube plate–fin compact surfaces for which test data are presented in Figure 9.7. (From Kays, W. M. and London, A. L. [1984] *Compact Heat Exchangers*, 3rd ed., McGraw-Hill, New York. With permission.)

$$G = \frac{\rho U_\infty A_{fr}}{A_{min}} = \frac{\rho U_\infty A_{fr}}{A_{min}} = \frac{\rho U_\infty}{\sigma} \tag{9.11}$$

In this equation, ρ is the average density evaluated at the average temperature between the inlet and outlet, or it can also be estimated by averaging the fluid specific volume between the inlet and outlet as:

$$\frac{1}{\rho} = \frac{1}{2}\left(\frac{1}{\rho_i} + \frac{1}{\rho_o}\right) \tag{9.12}$$

The friction factor, *f*, for some tube–plate–fin heat exchangers has been found experimentally and plotted in Figures 9.4 to 9.7[3,5] and accounts for fluid friction against the solid wall and for entrance and exit losses. The second term on the right-hand side of Equation (9.8) accounts for the acceleration or deceleration of flow. This term is negligible for liquids for which the density is essentially constant.

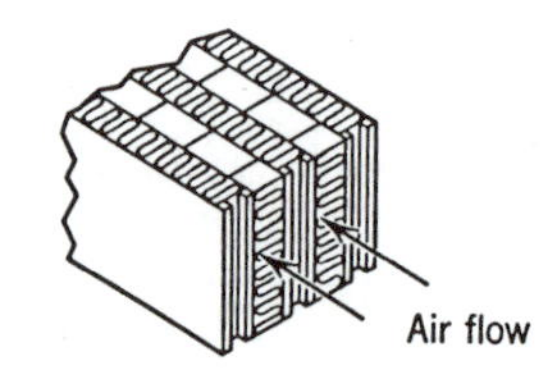

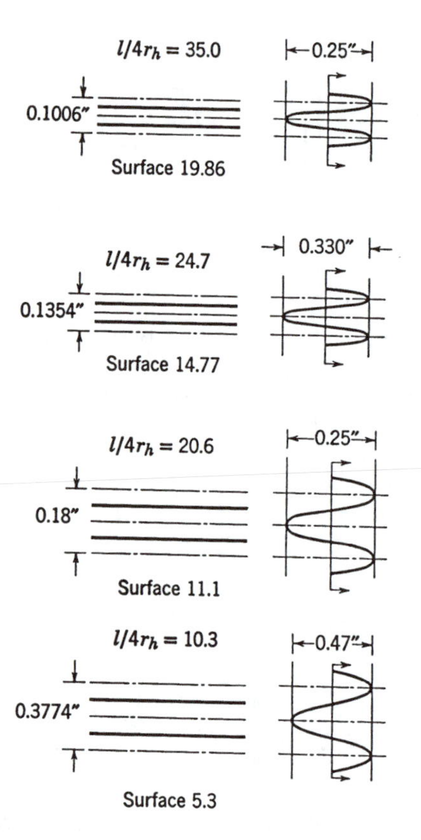

FIGURE 9.9
Heat transfer and friction factor for four plain plate–fin heat transfer matrices of Table 9.1. (From Kays, W. M. and London, A. L. [1984] *Compact Heat Exchangers*, 3rd ed., McGraw-Hill, New York. With permission.)

9.2.3 Pressure Drop for Plate–Fin Exchangers

Consider plate-fin exchanger surfaces as shown in Figure 9.8. The total pressure drop for flow across the heat exchanger matrix is given by:[3,6]

$$\Delta p = \frac{G^2}{2\rho_i}\left[(k_c + 1 - \sigma^2) + 2\left(\frac{\rho_i}{\rho_o} - 1\right) + f\frac{A}{A_{\min}}\frac{\rho_i}{\rho} - (1 - k_e - \sigma^2)\frac{\rho_i}{\rho_o}\right] \tag{9.13}$$

The first term inside the bracket shows the entrance effect; the second term, the flow acceleration effect; the third term, the core friction; and the last term, the exit effect.

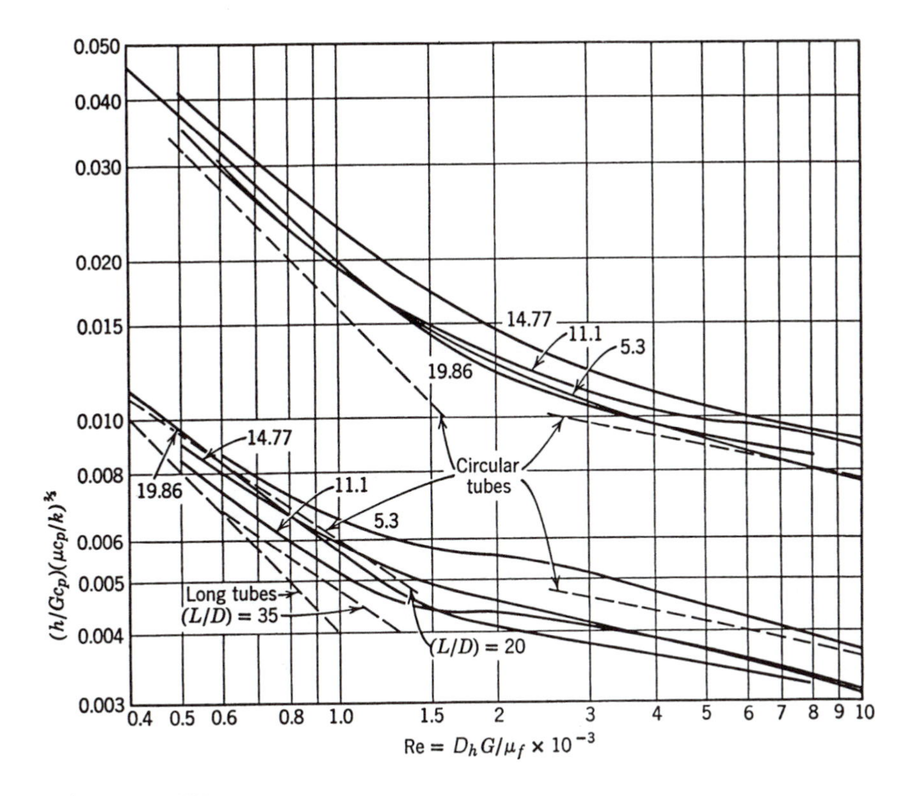

FIGURE 9.9 (CONTINUED)

In Equation (9.13), the frictional pressure drop generally dominates and accounts for about 90% or more of the total pressure drop across the core. The friction factor, f, has been found experimentally and plotted in Figure 9.9 for the surfaces shown in Figure 9.8.

The entrance and exit losses become important for short cores, small values of σ, high values of the Reynolds number, and gases. For liquids they are negligible. Typical values of the contraction loss coefficient, K_c, and the enlargement loss coefficient, K_e, are given in References 3 and 7.

Example 9.1

Air at 1 atm and 400 K and with a velocity of $U_\infty = 10$ m/s flows across a compact heat exchanger matrix having the configuration shown in Figure 9.4. Calculate the heat transfer coefficient, h, and frictional pressure drop for air side. The length of the matrix is 0.6 m.

SOLUTION

At 400 K and 1 atm, properties of air from Appendix B (Table B.1) are

$$\rho = 0.8825 \text{ kg/m}^3$$

$$\mu = 2.29 \times 10^{-5} \text{ kg/m·s}$$

$$c_p = 1013 \text{ J/kg·K}$$

$$Pr = 0.719$$

TABLE 9.1

Heat Transfer Matrix Geometries for Plate–Plain-Fin and Fin–Flat-Tube Types for Which Test Data are Presented in Figure 9.7 and 9.9

Surface Designation	Fins per cm	Hydraulic Diameter (D_h, cm)	Plate Spacing (b, cm)	Tube or Fin Thickness (cm)	Extended Total Area	Area of Volume Between Plates β (m^2/m^3)	Area of Core Volume β (m^2/m^3)	Free-Flow Frontal Area σ
Plate–Plain-Fin Type								
5.3	13.46	0.051	1.194	0.0152	0.719	511.8		
11.1	28.19	0.257	0.635	0.0152	0.730	1095.8		
14.77	37.52	0.215	0.838	0.0152	0.831	1210.6		
19.86	50.44	0.152	0.635	0.0152	0.833	1493.0		
Fin-Flat-Tube Type								
9.68-0.870	24.587	0.2997		0.0102	0.795		751.3	0.697
9.68-0.870-R	24.587	0.2997		0.0102	0.795		751.3	0.697
9.1-0.737-S	23.114	0.3565		0.0102	0.813		734.9	0.788
9.29-0.737-S-R	28.753	0.351		0.0102	0.845		885.8	0.788
11.32–0.737-S-R	23.596	0.3434		0.0102	0.814		748.0	0.780

From Kays, W. M. and London, A. L. (1984) *Compact Heat Exchangers*, 3rd ed., McGraw-Hill, New York. With permission.

From Figure 9.4, we have:

$$\frac{A_{min}}{A_{fr}} = \sigma = 0.534 \text{ and } D_h = 0.3633 \text{ cm}$$

Then

$$G = \frac{\rho U_\infty A_{fr}}{A_{min}} = \frac{\rho U_\infty}{\sigma} = \frac{0.8825 \times 10}{0.534} = 16.53 \text{ kg/(m}^2 \cdot \text{s)}$$

$$Re = \frac{GD_h}{\mu} = \frac{16.53 \times 0.3633 \times 10^{-2}}{2.29 \times 10^{-5}} = 2622$$

From Figure 9.4, for $Re = 2622$, we can obtain:

$$\frac{h}{Gc_p} Pr^{2/3} = 0.0071$$

$$h = 0.0071 \times \frac{Gc_p}{Pr^{2/3}} = 0.0071 \times \frac{16.53 \times 1013}{(0.719)^{2/3}}$$

$$= 148.1 \text{ W/m}^2 \cdot \text{K}$$

For Re = 2622, from Figure 9.4, $f = 0.025$:

$$\Delta p_f = f \frac{G^2}{2\rho_a} \frac{A_t}{A_{min}}$$

$$\frac{A_t}{A_{min}} = \frac{4 \times L}{D_h} = \frac{4 \times 0.6}{0.3633 \times 10^{-2}} = 660.6$$

Then

$$\Delta p_f = 0.025 \times \frac{16.53^2}{2 \times 0.8825} \times 660.6 = 2556 \ N$$

Example 9.2

Air at 2 atm and 500 K with a velocity of $U_\infty = 20$ m/s flows across a compact heat exchanger matrix having the configuration shown in Figure 9.8 (surface 11.36-0737-S-R). Calculate the heat transfer coefficient and the frictional pressure drop. The length of the matrix is 0.8 m.

SOLUTION

Air at 500 K and 2 atm, properties from Appendix B (Table B.1) are

$$\rho = 1.41 \text{ kg/m}^3$$

$$c_p = 1030 \text{ J/kg·K}$$

$$\mu = 2.69 \times 10^{-5} \text{ kg/m·s}$$

$$Pr = 0.718$$

The mass flux G is

$$G = \frac{\dot{m}}{A_{min}} = \frac{\rho U_\infty A_{fr}}{A_{min}} = \frac{\rho U_\infty}{\sigma} = \frac{1.41 \times 20}{0.78} = 36.15 \text{ kg/m}^2 \cdot \text{s}$$

The Reynolds number becomes:

$$Re = \frac{GD_h}{\mu} = \frac{36.15 \times 0.351 \times 10^{-2}}{2.69 \times 10^{-5}} = 4717$$

From Figure 9.6, for $Re = 4750$, we get:

$$\frac{h}{Gc_p} \cdot Pr^{2/3} = 0.0056$$

$$h = 0.0056 \frac{Gc_p}{Pr^{2/3}}$$

$$= 0.0056 \times \frac{36.15 \times 1030}{(0.718)^{2/3}} = 260.1 \text{ W/m}^2 \cdot \text{K}$$

$$\Delta p_f = f \frac{A_t}{A_{min}} \frac{\rho_i}{\rho} \frac{G^2}{2\rho_i}$$

$$\frac{A_t}{A_{min}} = \frac{4 \times L}{D_h} = \frac{4 \times 0.8}{0.351 \times 10^{-2}} = 911.68$$

For $Re = 4717$, from Figure 9.7, $f = 0.023$ and $\rho_i/\rho_o \approx 1$

$$\Delta p_f = 0.023 \times 911.68 \times \frac{(36.15)^2}{2 \times 1.41} = 9717.12 \ N/\text{m}^2$$

Example 9.3

Air enters at 1 atm and 30°C the core of a finned-tube heat exchanger of the type shown in Figure 9.4. The air flows at the rate of 1500 kg/h perpendicular to the tubes and exits with a mean temperature of 100°C. The core is 0.5 m long with a 0.25-m² frontal area. Calculate the total pressure drop between the air inlet and outlet and the average heat transfer coefficient on the air side.

SOLUTION

The air densities at inlet and outlet from Appendix B (Table B.1) are

$$\rho_i = 1.177 \text{ kg/m}^3$$

$$\rho_o = 0.954 \text{ kg/m}^3$$

The properties at the bulk temperature of (100 + 30)/2 = 65°C, from Appendix B (Table B.1) are

$$\rho = 1.038 \text{ kg/m}^3$$

$$Pr = 0.719$$

$$\mu = 2.04 \times 10^{-5} \text{ kg/m·s}$$

$$c_p = 1.007 \text{ kJ /kg·K}$$

The pressure drop is calculated from Equation (9.8) by:

$$\frac{A_t}{A_{min}} = \frac{\beta A_{fr} L}{A_{fr}\sigma} = \frac{\beta L}{\sigma}$$

where

$$\beta = \frac{A_t}{V} = 446 \text{ m}^2/\text{m}^3$$

$$\frac{A_t}{A_{min}} = \frac{\beta}{\sigma} L = \frac{446 \times 0.5}{0.494} = 451.4 \text{ kg/m} \cdot \text{s}$$

$$A_{min} = \sigma A_h = 0.494 \times 0.25 = 0.124 \text{ m}^2$$

$$G = \frac{\dot{m}}{A_{min}} = \frac{1500}{3600} \times \frac{1}{0.124} = 3.36 \text{ kg/m}^2 \cdot \text{s}$$

$$D_h = 4 \times \frac{L}{A_t / A_{min}} = 4 \times \frac{0.5}{451.4} = 4.43 \text{ mm}$$

$$Re = \frac{GD_h}{\mu} = \frac{3.36 \times 0.00443}{2.04 \times 10^{-5}} = 729$$

From Figure 9.4, for $Re = 729, f = 0.061$:

$$\Delta p = (3.36)^2 \frac{1}{2\times 1.177}\left[0.06\times 451.4\times\frac{1.177}{1.038}+(1+0.494^2)\left(\frac{1.177}{1.038}-1\right)\right]$$

$$= 150.5\ N/\text{m}^2$$

For the heat transfer coefficient, Colburn modulus $(h/Gc_p)\ Pr^{2/3}$ can be read from Figure 9.4 for $Re = 729$ as 0.018.

$$\frac{h}{Gc_p} Pr^{2/3} = 0.018$$

$$h = 0.018\times 3.36\times 1.007\times 10^3\times (0.719)^{-2/3}$$

$$\approx 75\ \text{W/m}^2\cdot\text{K}$$

Nomenclature

- A area, m^2
- A_f finned area, m^2
- A_{fr} frontal area, m^2
- A_{min} minimum free-flow area, m^2
- A_t total heat transfer area on the outside, m^2
- A_u unfinned area, m^2
- d diameter of the tube, m
- D_h hydraulic diameter, m
- f friction factor
- G mass velocity, kg/m^2·s
- k thermal conductivity, W/m·K
- h heat transfer coefficient, W/m^2·K
- U_∞ free stream velocity, m/s
- V volume of the core, m^3
- β heat transfer area density, m^2/m^3
- Δp pressure drop, N/m^2
- ΔT temperature difference, K, °C
- η_f fin efficiency
- η_o overall surface efficiency
- μ viscosity, kg/m·s
- ρ density, kg/m^3

References

1. Webb, R. L. (1994) *Principle of Enhanced Heat Transfer*, Wiley, New York.
2. Kakaç, S., Shah, R. K., and Mayinger, F. (Eds.), (1982) *Low Reynolds Number Flow Heat Exchangers*, Hemisphere, Washington, D.C.
3. Kays, W. M. and London, A. L. (1984) *Compact Heat Exchangers*, 3rd ed., McGraw-Hill, New York.
4. Kakaç, S., Bergles, A. E., and Mayinger, F. (Eds.), (1981) *Heat Exchangers: Thermal Hydraulic Fundamentals and Design*, Hemisphere, Washington, D.C.
5. Fraas, A. P. (1989) *Heat Exchanger Design*, 2nd ed., Wiley, New York.
6. Özisik, M. N. (1985) *Heat Transfer-A Basic Approach*, McGraw-Hill, New York.
7. Bejan, A. (1993) *Heat Transfer*, Wiley, New York.

Problems

9.1. Air at 1 atm and 400 K and with a velocity of 10 m/s flows across the compact heat exchanger shown in Figure 9.6, and exits with a mean temperature of 300 K. The core is 0.6 m long. Calculate the total frictional pressure drop between the air inlet and outlet, and the average heat transfer coefficient on the air side.

9.2. Air enters at 2 atm and 150°C the core of a finned-tube (aluminum) heat exchanger of the type shown in Figure 9.4. The air mass flow rate is 10 kg/s and flows perpendicular to the tubes. The core is 0.5 m long with a 0.30-m^2 frontal area. The height of the core is 0.5 m. Water at an inlet temperature of 15°C and a flow rate of 50 kg/s flows inside the tubes. Air-side data are given on Figure 9.4. For water-side data, assume that $\sigma_w = 0.129$, $D_h = 0.373$ m, and water-side heat transfer area/total volume = 138 m^2/m^3. This is a rating problem. Calculate:

a. Air-side and water-side heat transfer coefficients

b. Total heat transfer

c. Outlet temperatures of air and water

9.3. Hot air at 2 atm and 500 K at a rate of 8 kg/s flows across a circular finned-tube matrix configuration shown in Figure 9.6. The frontal area of the heat exchanger is 0.8 × 0.5 m and the core is 0.5 m long. Geometric configurations are shown in Figure 9.6. Calculate:

a. Heat transfer coefficient

b. Total frictional pressure drop between the air inlet and outlet

9.4. Repeat Problem 9.3 for a finned-tube matrix shown in Figure 9.5 and discuss the results.

9.5. Repeat Problem 9.3 for heat exchanger matrix configuration in Figure 9.8 for the matrix 9.1-0.737-S as given in Table 9.1.

9.6. Repeat Problem 9.3 for the heat exchanger matrix configuration shown in Figure 9.8 for the surface 11.32-0.737-S-R (see Table 9.1).

9.7. Air at 2 atm and 500 K and at a rate of 12 kg/s flows across a plain plate–fin matrix of configuration shown in Figure 9.8 and in Table 9.1 for the surface 9.68-0.870. The frontal area is 0.8 × 0.6 m and the length of the matrix is 0.6 m. Calculate:

a. Heat transfer coefficient

b. Friction coefficient

9.8. Repeat Problem 9.7 for a finned-tube matrix configuration shown in Figure 9.8 for the surface 9.68-0.870-R (see Figure 9.7) geometric configuration that is given in Table 9.1.

9.9. An air-to-water compact heat exchanger is to be designed to serve as an intercooler in a gas turbine plant. Geometric details of the proposed surface (surface 9.29-0.737-S-R) for the air side is given in Figure 9.8 and in Table 9.1. Hot air at 2 atm and 400 K with a flow rate of 20 kg/s enters the matrix. The outlet temperature of air is 300 K and the allowable pressure drop on the air side is 0.3 bar. Water at an inlet temperature of 17°C and a flow rate of 50 kg/s flows inside the flat tubes. Water velocity is 1.5 m/s. Water-side geometric details are

$$D_h = 0.373 \text{ cm}$$

$$\sigma_w = 0.129$$

$$\frac{\text{water - side heat transfer area}}{\text{total volume}} = 138 \text{ m}^2/\text{m}^3$$

Calculate:

a. Overall heat transfer coefficient

b. Air flow frontal area

c. Flow length and core size

9.10. Repeat Problem 9.7 for air flow rate of 15 kg/s while everything else remains the same.

9.11. An air-cooled refrigerant condenser is to be designed. A flattened tube with corrugated fins will be used. The surface selected for the matrix is similar in Figure 9.4. Cooling load (heat duty) is 125 kW. The refrigerant 134A condenses inside the tubes at 310 K. Air enters the condenser at 18°C and leaves at 26°C. The mean pressure is 2 atm. Calculate:

a. Air-side heat transfer coefficient

b. Tube-side condensation heat transfer coefficient

c. Overall heat transfer coefficient

d. Core dimensions based on the following parameters:

Air-side geometric configurations for the surface 7.75-5/8 T^3

Tube O.D. = 1.717 m

Tube arrangement = staggered

Fins/in. = 7.75

Fin type = plain

Fin thickness = 0.41×10^{-3} m

Minimum free-flow area/frontal area , $\sigma = 0.481$

Hydraulic diameter, $D_h = 3.48 \times 10^{-3}$ m

Heat transfer area/total volume, $\alpha = 554$ m^2/m^3

9.12. Design an air-to-water compact heat exchanger to serve as an intercooler for a gas turbine plant. The heat exchanger is to meet the following heat transfer and pressure drop performance specifications for air-side operating conditions:

Flow rate = 25 kg/s

Inlet temperature = 500°C

Outlet temperature = 350°C

Inlet pressure = 2×10^5 N/m^2

Pressure drop rate = 8%

and for water-side operating conditions:

Flow rate = 50 kg/s

Inlet temperature = 290 K

The heat exchanger surface for this heat exchanger is given in Figure 9.8 (surface 9.1-0.737-S) with flattened tube–fin compact surface. Fins are continuous aluminum fins. The geometric data for air side are given in Table 9.1. For the water side the flattened tube is 0.2 × 1.6 cm. The I.D. of the tubes before it was flattened was 1.23 cm, with a wall thickness of 0.025 cm. Water velocity inside is 1.5 m/s. The design should specify the core size and the core pressure drop.

Design Project 9.1

Cooling System and Radiator of a Truck

Some of the design specifications as applicable to a typical truck are given as:

Heat load	100 kW
Water inlet temperature	80°C
Water outlet temperature	70°C
Air inlet temperature	35°C
Air outlet temperature	46°C
Air pressure	100 kPa
Air pressure drop	0.3 kPa
Core matrix of the radiator	To be selected

Different compact surfaces must be studied for thermal and hydraulic analysis of the radiator and compared. A parametric study is expected to come up with an acceptable final design that includes the selection of pump, materials selection, mechanical design, drawings, and cost estimation.

Design Project 9.2

Air-Cooled Refrigeration Condenser

The process specification and the geometric parameters of the condenser are given in Problem 9.1. Two different compact surfaces given in this chapter including 7.75-5/8^3 will be analyzed and compared. The final design will include materials selection, mechanical design, technical drawings, and cost estimation.

10

Gasketed-Plate Heat Exchangers

10.1 Introduction

The gasketed-plate heat exchangers (plate and frame) were introduced in the 1930s, mainly for the food industries because of the ease of cleaning; and their design reached maturity in 1960s with the development of more effective plate geometries, assemblies, and improved gasket materials. The range of possible applications has widened considerably and at present, under specific and appropriate conditions, overlaps and successfully competes in areas historically considered to be the domain of tubular heat exchangers. They are capable of meeting an extremely wide range of duties in many industries. Therefore, they can be used as an alternative to tube-and-shell type of heat exchangers for low- and medium-pressure liquid-to-liquid heat transfer applications.

The design of plate heat exchangers is highly specialized in nature considering the variety of design available for the plates and arrangements that are possible to suit varied duties. Unlike tubular heat exchangers for which design data and methods are easily available, plate heat exchanger design continues to be proprietary in nature. Manufacturers have developed their own computerized design procedures applicable to the exchangers marketed by them.

10.2 Mechanical Features

A typical plate and frame heat exchanger is shown in an exploded view in Figures 10.1 and 1.16.[1] The elements of the frame are fixed plate, compression plate, pressing equipment, and connecting ports. The heat transfer surface is composed of a series of plates with parts for fluid entry and exit in the four corners (Figure 10.2). The flow pattern through a gasketed-plate heat exchanger is illustrated in Figure 10.3.[1]

10.2.1 Plate Pack and Frame

When a package of plates is pressed together, the holes at the corners of the plates form continuous tunnels or manifolds, leading the media from the inlets into the plate package, where they are distributed into the narrow channels between the plates. The plate pack is tightened by means of either a mechanical or a hydraulic tightening device that permits control of tightening pressure to the desired level. These passages formed between the plates and corner ports are arranged so that the two heat transfer media can flow through

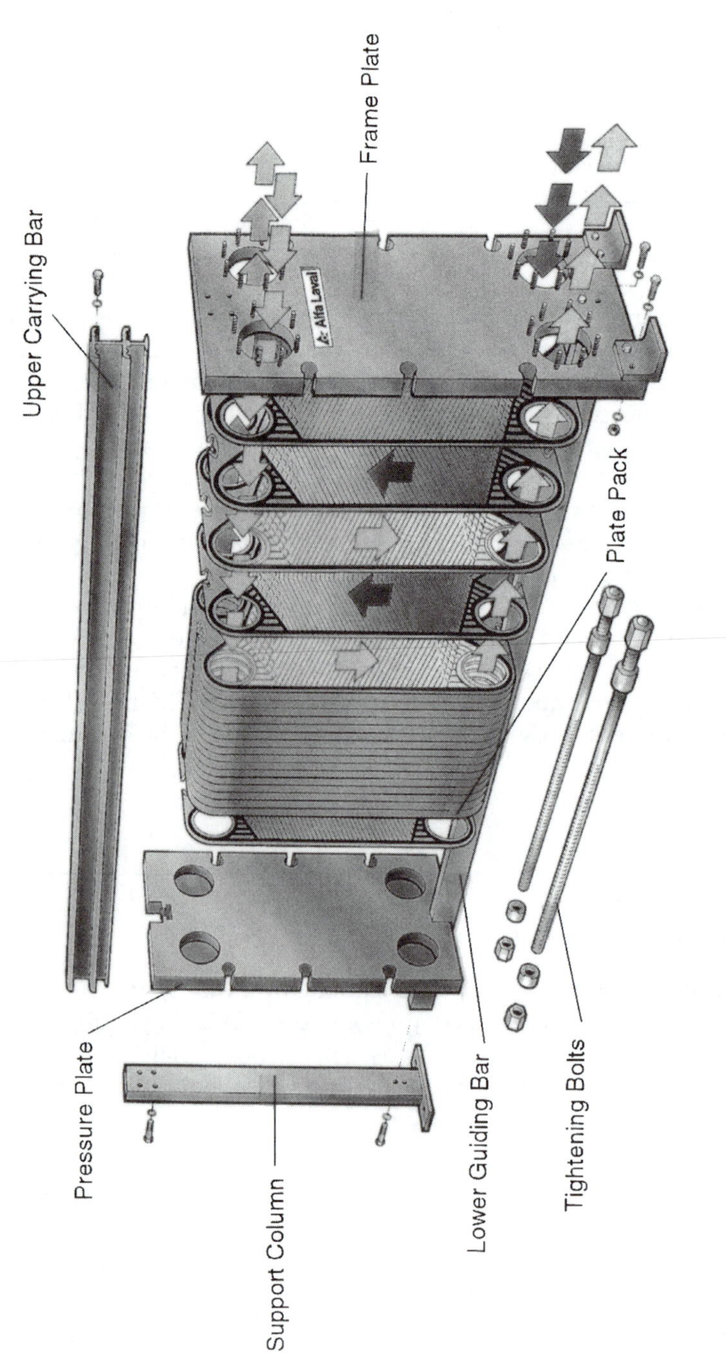

FIGURE 10.1
Gasketed-plate heat exchangers. (Courtesy of Alfa-Laval Thermal AB.)

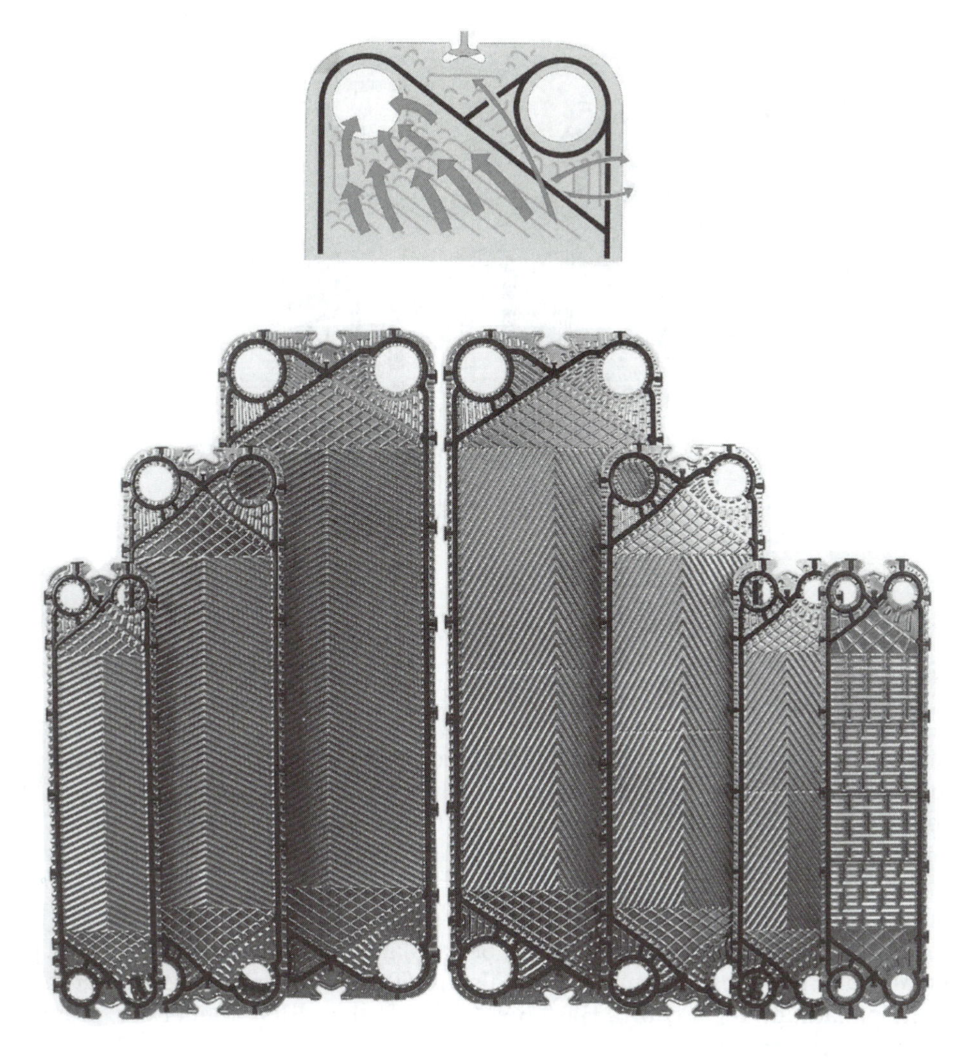

FIGURE 10.2
A Chevron-type heat exchanger plate. (Courtesy of Alfa-Laval Thermal AB.)

alternate channels, always in countercurrent flow. During the passage through the apparatus, the warmer medium will give some of its heat energy through the thin plate wall to the colder medium on the other side. Finally, the media are led into similar hole–tunnels as in the inlets at the other end of the plate package and discharged from the heat exchanger. The plates can be stacked up to several hundred in a frame and held together by the bolts that hold the stack in compression (Figure 10.4).[1]

The plate pack is compressed between the head plates using bolts. Plates and removable parts of the frame are suspended from the upper carrier bar and supported at their lower ends by a guide bar. The carrier and guide bars are bolted to the fixed part of the frame and, on all but smaller types, are attached to the end support.

The plate pack corresponds to the tube bundle in shell-and-tube exchangers with the important difference that the two sides of a plate heat exchanger are normally of identical hydrodynamic characteristics. The basic elements of the plate pack is the plate, a sheet of metal, precision pressed into a corrugated pattern as shown in Figure 10.2.[1] The largest single plate is of the order of 4.3 m high × 1.1 m wide. The heat transfer area for a single plate

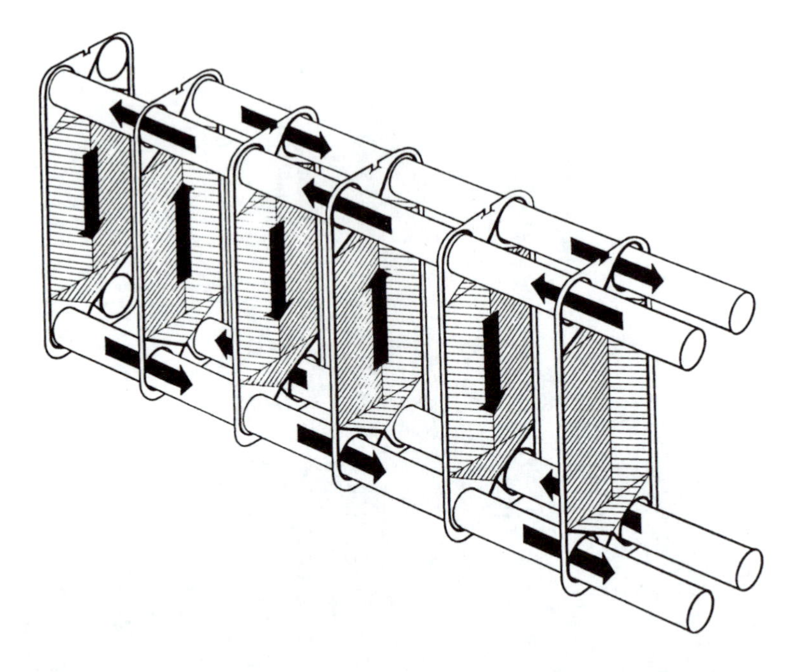

FIGURE 10.3
Flow diagram in a single-pass counterflow arrangement. (Courtesy of Alfa-Laval Thermal AB.)

lies in the range 0.01 to 3.6 m^2. It should be assured that the fluid is equally distributed over the full width of the plate. To avoid poor distribution of the fluid across the plate width, the minimum (length/width) ratio is of the order of 1.8. Plate thickness range between 0.5 and 1.2 mm and are spaced with nominal gaps of 2.5 to 5 mm, yielding hydraulic diameter for the flow channels of 4 to 10 mm.

Leakage from the channels between the plates to the surrounding atmosphere is prevented by the gasketing around the exterior of the plate. Plates can be made from all pressable materials. The most common materials are (Table 10.1): stainless steel, titanium, titanium–palladium, Incoloy 825, Hastelloy C-276, Diabon F100, Monel 400, aluminum, and aluminum brass.[2,3]

The number and size of the plates are determined by the flow rate, physical properties of the fluids, pressure drop, and temperature requirements.

10.2.2 Plate Types

A wide range of corrugation types are available in practical applications (Figure 10.2). Although most modern plate heat exchangers are the Chevron type, a number of commercial plates have a surface corrugated pattern called washboard.[4] Because of major variations in individual corrugation patterns of this plate type, the prediction methods rely on experimental data, particular to a specific pattern. In the washboard type, turbulence is promoted by a continuously changing flow direction and velocity of the fluids. In the Chevron type, adjacent plates are assembled such that the flow channel provides swirling motion to the fluids; the corrugated pattern has an angle β that is referred to as Chevron angle (Figure 10.6). The Chevron angle is reversed on adjacent plates so that when plates are clamped together, the corrugations provide numerous contact points (Figure 10.7). Because of the many support points of contact, the plates can be made from very thin

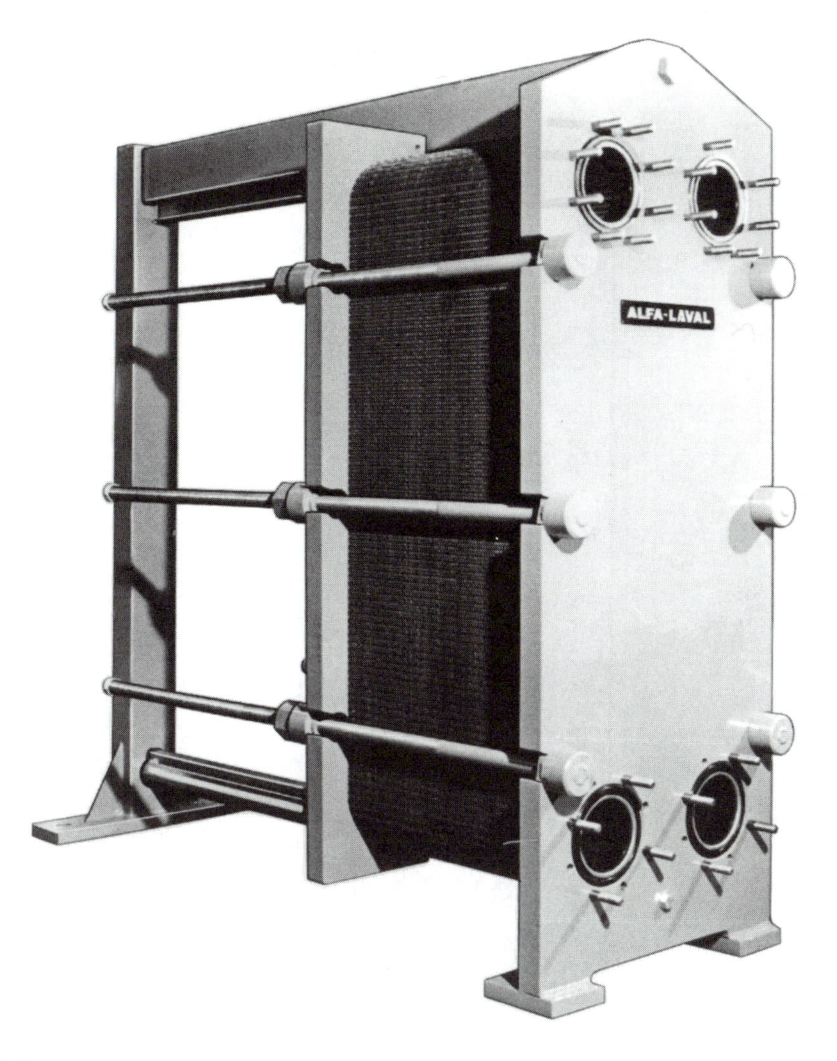

FIGURE 10.4
Gasketed-plate heat exchanger assembly. (Courtesy of Alfa-Laval Thermal AB.)

TABLE 10.1
Plate Materials

Material	Thermal Conductivity (W/m²·K)
Stainless steel (316)	16.5
Titanium	20
Inconel 600	16
Incolay 825	12
Hastelloy C-276	10.6
Monel 400	66
Nickel 200	66
9/10 Cupronickel	52
70/30 Cupronickel	35

From Raju, K. S. N. and Jagdish, C. B. (1983) In *Low Reynolds Number Flow Heat Exchangers*, Hemisphere, Washington, D.C. With permission.

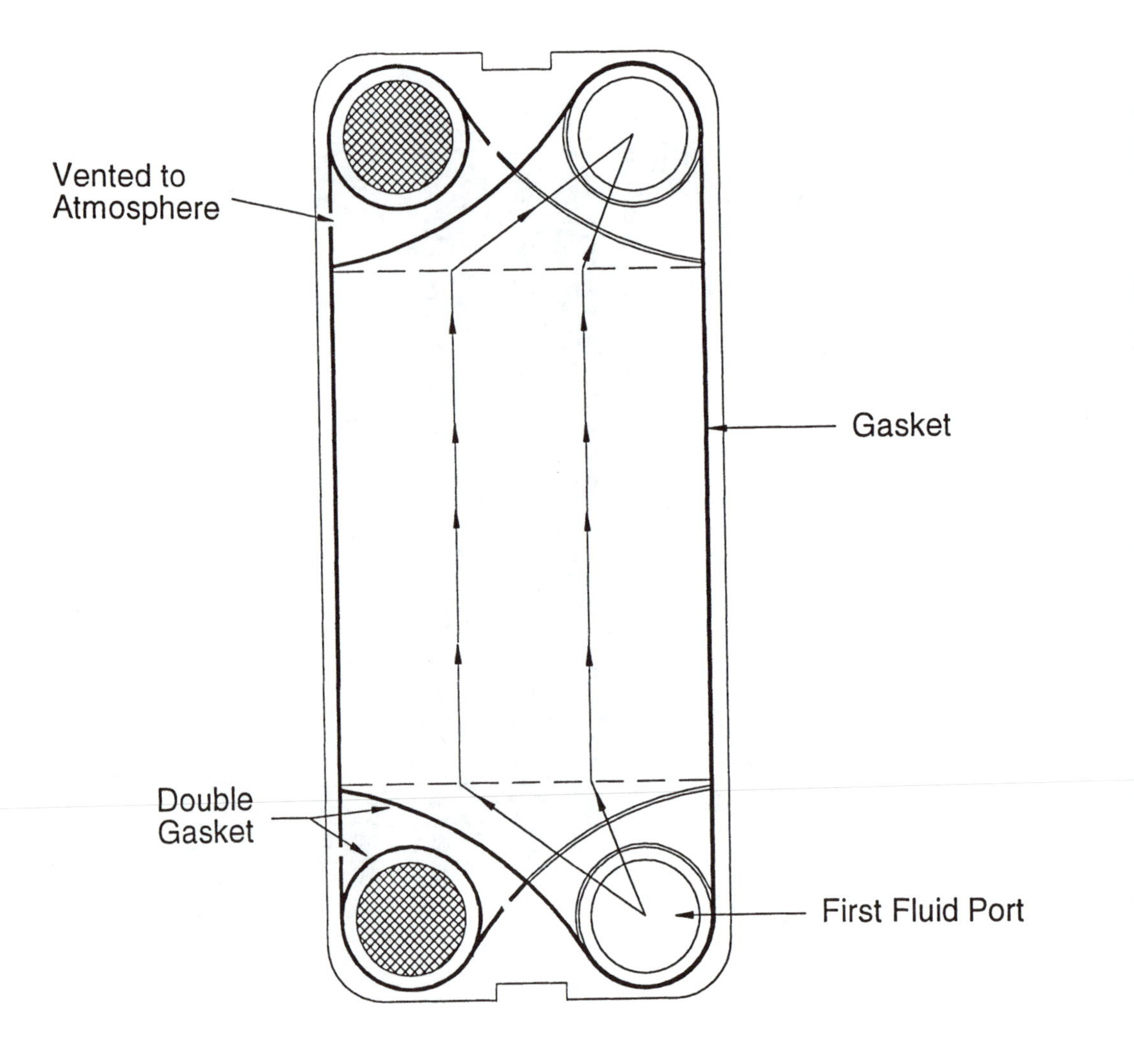

FIGURE 10.5
Gasket arrangements.

material, usually 0.6 mm. The Chevron angle varies between the extremes of about 65 and 25°C and determines the pressure drop and heat transfer characteristics of the plate.[4]

10.3 Operational Characteristics

The gasketed-plate heat exchanger can be opened for inspection, cleaning, maintenance, or rebuilding within the length of the frame (Figure 10.8).

10.3.1 Main Advantages

The gasket design minimizes the risk of internal leakage. Any failure in the gasket results in leakage to the atmosphere that is easily detectable on the exterior of the unit. The additional main advantages and benefits offered by the gasketed–plate heat exchangers are[1,3]

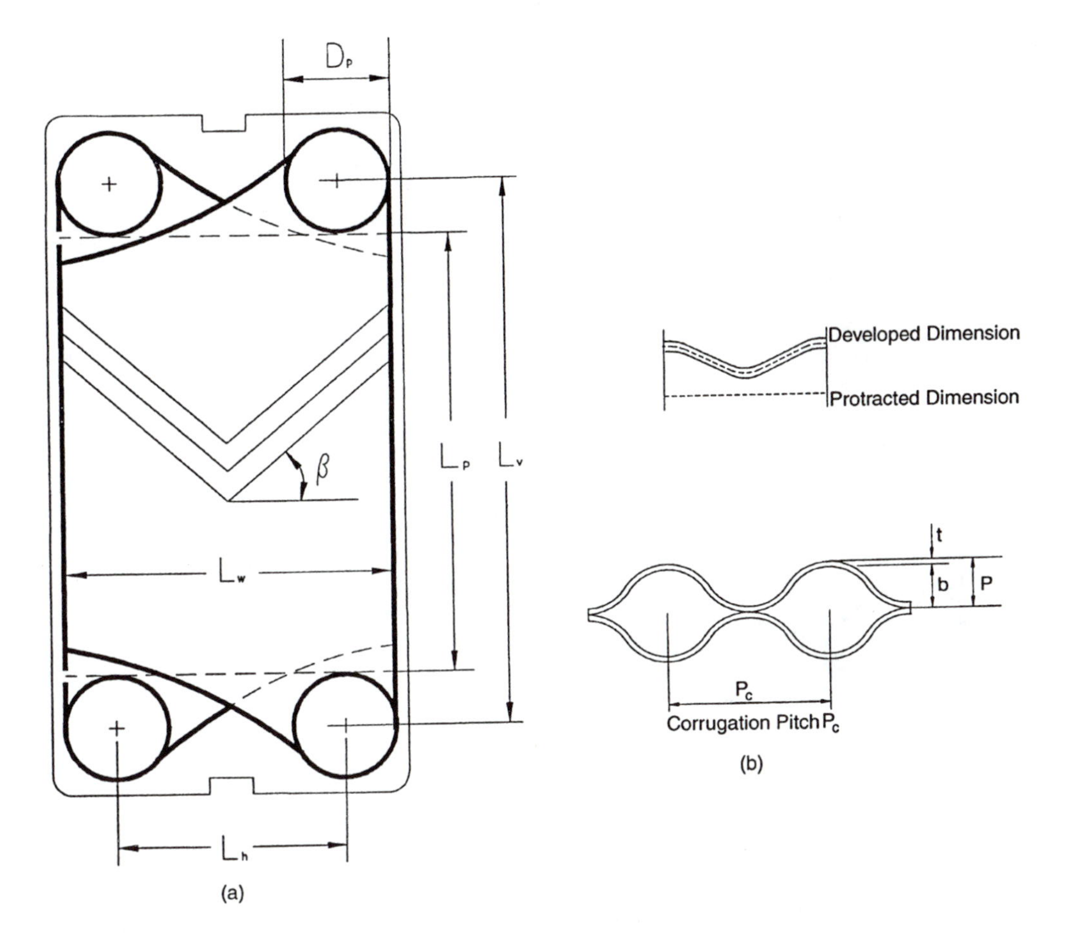

FIGURE 10.6
(a) Main dimensions of a Chevron plate; (b) developed and projected dimensions of a Chevron plate and cross section normal to the direction of troughs.

- Flexibility of design is through variety of plate size and pass arrangements.
- The heat transfer area is easily accessible, which permits changes in configuration to suit changes in process requirements through changes in the number of plates.
- Heat transfer is efficient; heat transfer coefficients are high for both fluids because of turbulence and small hydraulic diameter.
- They are very compact (large heat transfer area:volume ratio) and have low weight. In spite of their compactness, 2500 m^2 of surface is available in a single unit.
- Only the plate edges are exposed to the atmosphere. The heat losses are negligible and no insulation is required.
- Intermixing of the two fluids cannot occur under gasket failure.
- Plate units exhibit low-fouling characteristics due to high turbulence and low residence time.
- More than two fluids may be processed in a single unit.

The transition to turbulence flow occurs at low Reynolds number of 10 to 400. The high turbulence in a gasketed-plate heat exchanger leads to very high transfer coefficients,

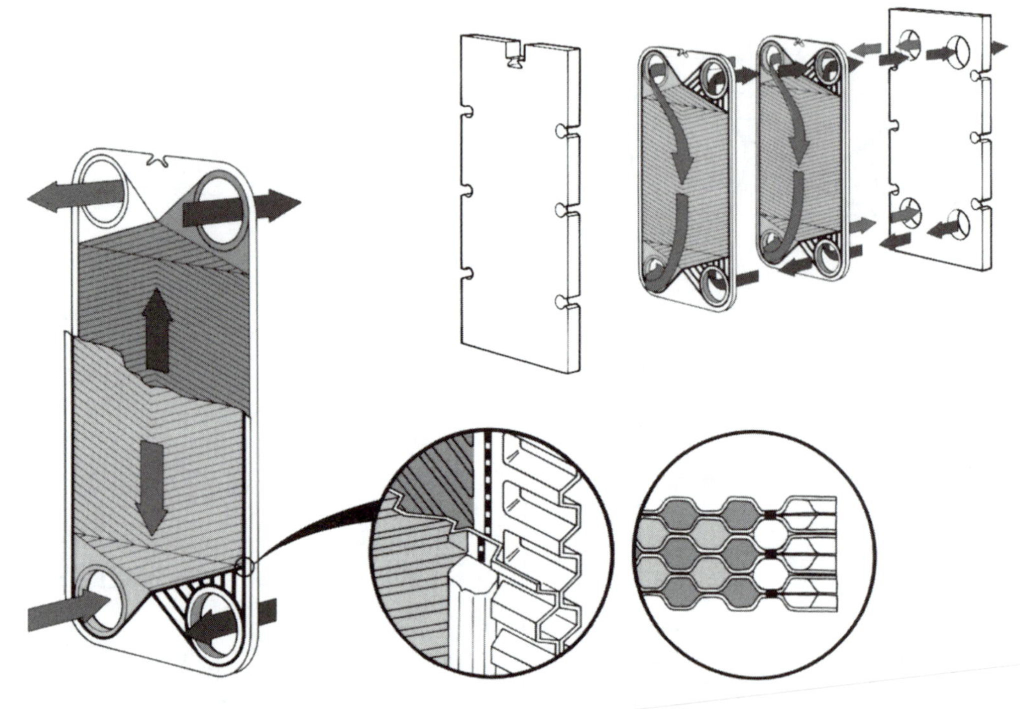

FIGURE 10.7
The Chevron angle is reversed on adjacent plates. (Courtesy of Alfa-Laval Thermal AB.)

low-fouling rates, and reduced size. The thin plates cut down the metal wall resistance to the minimum. These units can utilize up to about 82% of the theoretical log mean temperature difference (LMTD), while shell-and-tube units are capable of utilizing only 50% of it, mainly because of absence of crossflow in a plate unit. Over 90% of heat recovery is possible with a plate unit with very close temperature approaches.

The same frame can accommodate heat transfer duties for more than two liquids with the use of connecting plates. Costwise a plate unit competes favorably with a tubular one, if the tubes are to be of a costly material like stainless steel.

The plate unit is exceptionally compact and requires considerably less floor space than a tubular unit for the same duty. Low holdup volume; less weight; and lower costs for handling, transportation, and foundations are the other advantages involved in a plate unit.

10.3.2 Performance Limits

Capabilities of the gasketed-plate heat exchangers are limited due to the plates and the gasket. The main items can be summarized as follows:[5,6]

Maximum operating pressure	25 bar, with special construction 30 bar
Maximum temperature	160°C, with special gasket 250°C
Maximum flow rate	3600 m^3/h
Heat transfer area	0.1–2200 m^2
Heat transfer coefficient	3500–7500 $W/m^2 \cdot K$

FIGURE 10.8
Cleaning the plates. (Courtesy of Alfa-Laval Thermal AB.)

The gaskets impose restrictions on operating temperatures, on pressures, and on the nature of fluids that can be handled. Complex channel geometries result in the plate heat exchanges having high-shear characteristics. Friction factors are rather high for fully developed turbulent flow, but channel lengths are rather short and velocities are low so that pressure drops can be maintained within allowable limits for single-phase flow operations.

The main hindrance to the application of gasketed-plate heat exchangers in process industries is the upper limit on their size, which has been limited by the presses available to stamp out the plates from the sheet metal. Exchangers with sizes larger than 1500 m^2 are not normally available. Because the flow passages between the plates are thin, high liquid rates will involve excessive pressure drops, thus limiting the capacity. For specific cases, it is possible to have a maximum design pressure of up to 2.5 MPa; normally it is around 1 MPa. Operating temperatures are limited by the availability of its suitable gasket materials.

Gasketed-plate heat exchangers are not suitable as air coolers. They are also not quite suitable for air-to-air or gas-to-gas applications. Fluids with very high viscosity present some problems due to flow distribution effects, particularly when cooling is taking place. Flow velocities less than 0.1 m/s give low heat transfer coefficients and low heat exchanger efficiency. As such, velocities lower than 0.1 m/s are not used in plate heat exchangers.

Gasketed-plate heat exchangers are less suitable for condensing duties. This applies particularly to vapors under vacuum because the narrow plate gaps and induced turbulence result in appreciable pressure drops on the vapor side. Specially designed gasketed-plate heat exchangers are now available for duties involving evaporation and condensation systems.

10.4 Passes and Flow Arrangements

The term gasketed-plate heat exchanger "pass" refers to a group of channels in which the flow is in the same direction. Figure 10.9 shows a single-pass arrangement of the so-called "U" and "Z" arrangements. All four ports in U arrangement will be on the fixed-head plate, thus permitting disassembly of the exchanger for cleaning or repair without disturbing any external piping. In this arrangement, the flow distribution is less uniform than the Z arrangement.[7]

A multipass arrangement consists of passes connected in series, and Figure 10.10 shows an arrangement of a two-pass configuration (2/2 configuration) with three channels and four channels that are often abbreviated as $2 \times 3/2 \times 3$ and $2 \times 4/2 \times 4$. The system is in counterflow, except in the central plate where parallel flow prevails.

Figure 10.11 shows a two-pass/one-pass flow system (2/1 configuration) where one fluid flows in a single pass, while the other fluid flows in two passes in series. In this configuration one half of exchanger is in counterflow and the other half is in parallel flow (asymmetrical systems) if one of the fluids has a much greater mass flow rate or small allowable pressure drop.

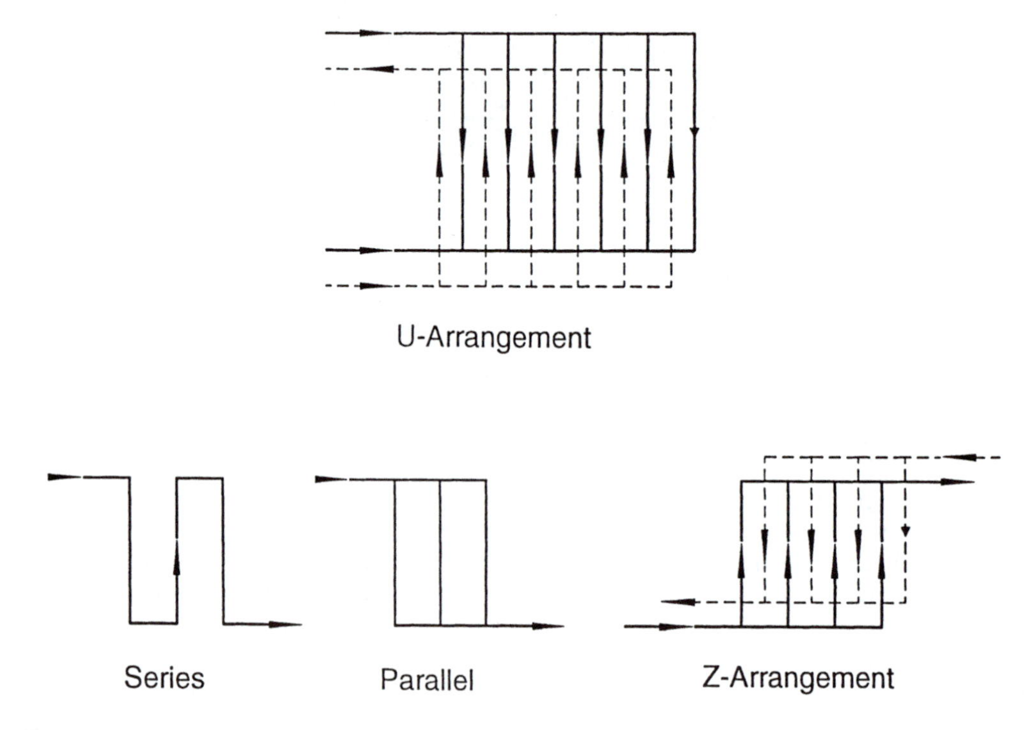

FIGURE 10.9
Flow pattern. (a) Schematic of a U-type arrangement — counterflow, single-pass flow ($1 \times 6/1 \times 6$); (b) Z-arrangement ($1 \times 4/1 \times 4$ configuration).

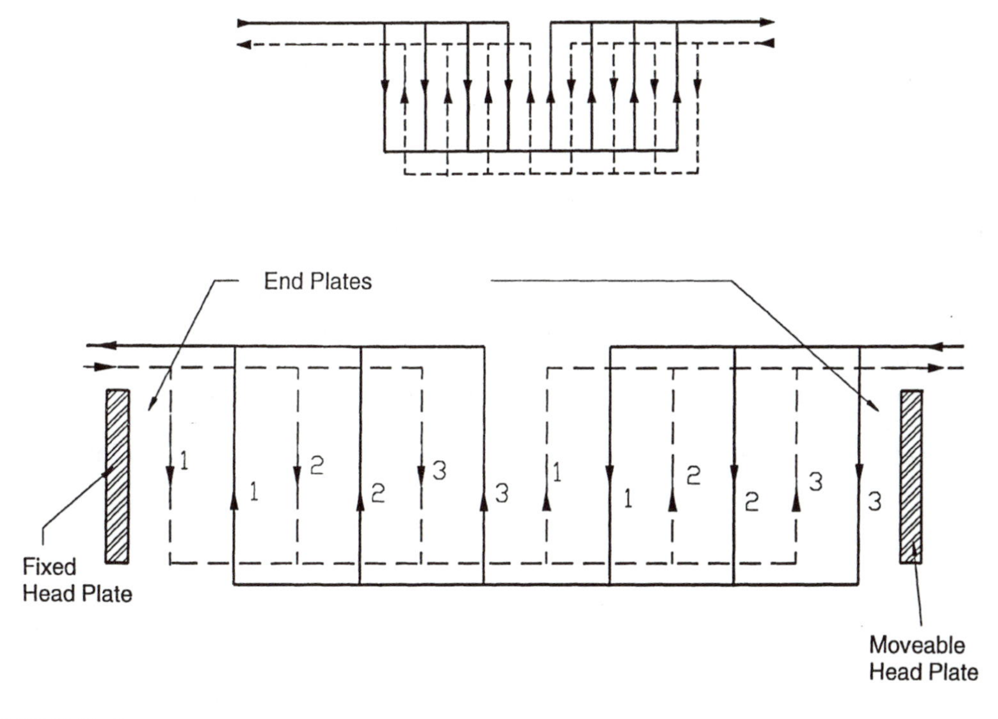

FIGURE 10.10
Schematic arrangement of a two-pass/two-pass flow system (2 × 3/2 × 3 configuration).

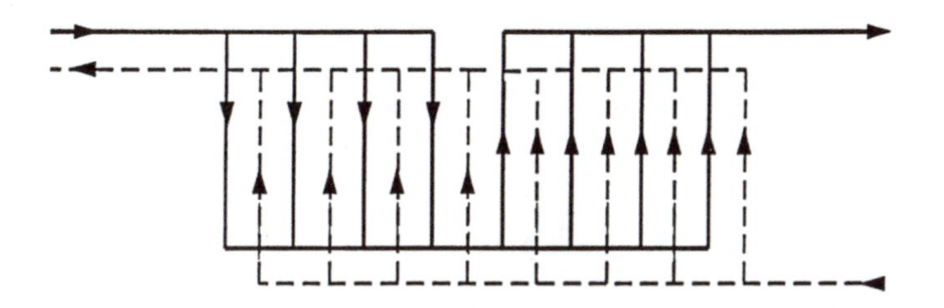

FIGURE 10.11
Schematic arrangement of a two-pass/one-pass flow system (2 × 4/1 × 8 configuration).

Multipass arrangements always require that the ports are located on both the fixed and movable head plate. Usually the number of passes and number of channels per pass are identical for the two fluids (symmetrically).

Multidistribution is a potential problem in any system of interconnected channels including gasketed-plate heat exchangers and must be considered carefully in their design.

10.5 Applications

Gasketed-plate heat exchangers are extensively used in food and dairy industries because of the ease for disassembling the heat exchanger for cleaning and sterilization to meet health and sanitation requirements; they are used as conventional process heaters and coolers, as well as condensers.[3,8]

Typical applications are mainly liquid-to-liquid turbulent flow situations. Of very great importance, especially at sea or in coastal locations, are central cooling systems using seawater, river water, or well water as the heat sink (e.g., in compressor intercooler systems). In a central cooling unit, a closed circuit of high-quality water that is used in the intercooler then is passed through gasketed-plate heat exchanger where heat is transferred from closed circuit water to water of lower quality, such as seawater or river water. Two closed circuit systems using a gasketed-plate heat exchanger as an intermediate exchanger are shown in Figure 10.12. Corrosion and fouling problems are transferred from the intercooler to the plate heat exchanger system. Equipment in the closed system (process plant equipment) can be fabricated out of low carbon steel, while the plate exchangers can be made out of titanium to withstand seawater corrosion. Fouling problems can be ultimately solved by the ability to disassemble the plate heat exchanger system for cleaning whenever it is necessary. Therefore, a gasketed-plate heat exchanger is ideal for offshore, and on-ship applications.

Some of the useful data on gasketed-plate heat exchangers are presented in Table 10.2.

10.5.1 Corrosion

When corrosion-resistant materials are essential, gasketed-plate heat exchangers become more attractive than conventional exchangers even though the plates are made out of expensive materials of construction. The range of materials of construction used has already been mentioned in Table 10.1. Because the plates are very thin compared with the tube thickness in shell-and-tube units, corrosion allowance normally recommended for tubular units becomes meaningless for gasketed-plate units. The corrosion allowance for gasketed-plate heat exchangers is much smaller than tubular units.

As a consequence of high turbulence over the entire plate surface, corrosion–erosion problems can be more significant in gasketed-plate heat exchangers. This requires the use of superior plate materials such as Monel or titanium (Table 10.3).

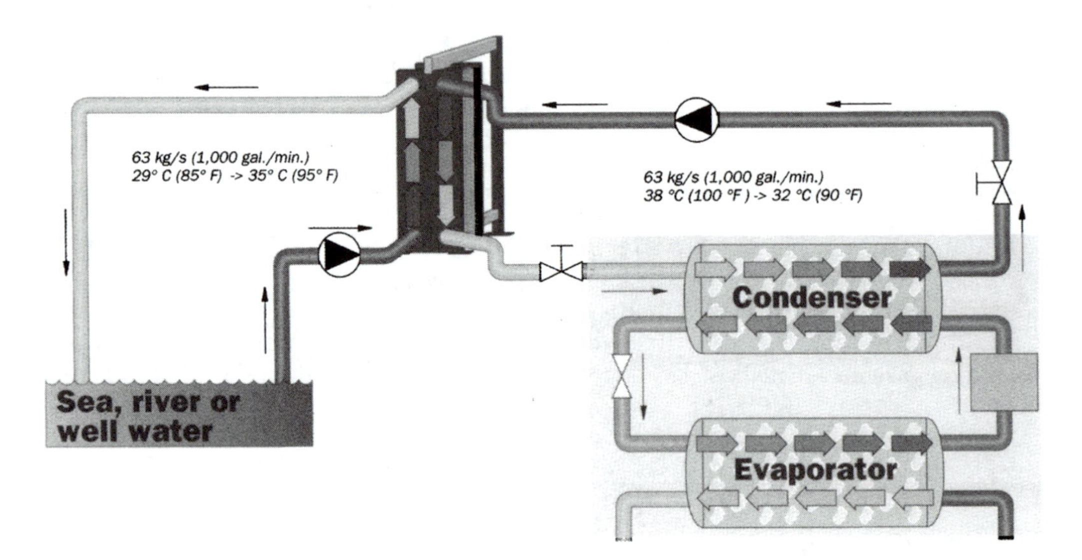

FIGURE 10.12
Closed-circuit cooling system. (Courtesy of Alfa-Laval Thermal AB.)

TABLE 10.2

Some Useful Data on Plate Heat Exchangers

Unit		
Largest size	1540 m^2	
Number of plates	Up to 700	
Port size	Up to 39 cm	
Plates		
Thickness	0.5–1.2 mm	
Size	0.03–2.2 m^2	
Spacing	1.5–5.0 mm	
Contact points	For every 1.5–20 cm^2	Depends on plate size and type of corrugations
Operation		
Pressure	0.1–1.5 MPa	Up to 2.5 MPa in special cases
Temperature	–25–150°C	With rubber gaskets
	–40–260°C	
Port velocities	5 m/s	With compressed asbestos fiber gaskets
Channel flow rates	0.05–12.5 m^3/h	
Maximum flow rates	2500 m3/h	
Performance		
Temperature approach	As low as 1°C	
Heat recovery	As high as 90%	
Heat transfer coefficients	3000–7000 W/m^2·°C	Water–water duties with normal fouling resistance
Number of transfer units	0.4–4.0	
Optimum pressure drops	30 kPa per NTU	

From Raju, K. S. N. and Jagdish, C. B. (1983) In *Low Reynolds Number Flow Heat Exchangers*, Hemisphere, Washington, D.C. With permission.

TABLE 10.3

Materials Selection

Application	Material
Natural cooling water, cooling tower water, or demineralized water	Stainless steel 316
Seawater or brackish water	Titanium
Dilute sulfuric and nitric acids up to 10% concentration, and for temperatures up to 70°C	Titanium, titanium–palladium alloy, Incoloy 825, Hastelloy
Chloride solution content	
< 200 ppm	Stainless steel
> 200 ppm	Titanium
Caustic solutions (50 to 70%)	Nickel
Wet chloride, chlorinated brines, hypochlorite solutions	Titanium
Copper sulfate solution in electrolyte refining	Stainless steel
Cooling hydrogen gas saturated with water vapor and mercury carryover in electrolysis plants	Incoloy

From Raju, K. S. N. and Jagdish, C. B. (1983) In *Low Reynolds Number Flow Heat Exchangers*, Hemisphere, Washington, D.C. With permission.

The increased concentration of a corrosive component in a scale deposit combined with the effect of high metal wall temperature under the deposit will enhance localized corrosion of heat transfer surface leading to eventual failure. In plate heat exchangers, this type of concentration corrosion is least because of low-fouling tendencies. The following example illustrates this point.

In a particular case, a gasketed-plate heat exchanger with stainless steel 316 plates operated for 5 years whereas a heating coil with a costlier material like Incolay 825 failed after 6 weeks of operation due to concentration corrosion under the heavy scale formed.[3]

As introduced in Chapter 5, the phenomenon of fouling is complex and very little information is available on the subject. Often it is necessary to conduct prolonged experimental studies under actual operating conditions to incorporate influence of fouling on the heat transfer rates. Careful choice of fouling factors while designing exchangers is necessary to minimize the chances of over- or underdesign. Besides adding to the costs, overdesign of a plate heat exchanger often reduces the flow velocities resulting in increased fouling rates.

Fouling is much less in a gasketed-plate heat exchanger than in a tubular unit because of the following reasons:

- High turbulence maintains solids in suspension.
- Velocity profiles across a plate are uniform with zones of low velocities absent.
- The plate surfaces are generally smooth and can be further electropolished.
- Deposits of corrosion products, to which fouling can adhere, are absent because of low-corrosion rates.
- In cooling duties, the high film coefficients maintain a moderately low metal wall temperature. This helps prevent crystallization growth of the inverse solubility compounds.
- Because of the ease with which plate units can be cleaned in place, deposits that grow with time can be kept to the minimum by frequent cleaning.

Table 10.4 gives recommended fouling factors for gasketed-plate heat exchangers assuming operation at the economic pressure drop of about 30 kPa per number of transfer units (NTU).[3,9] Manufacturer's recommendations based on test data should be a better guide while choosing fouling resistances. The general recommendations in this respect are 5% excess NTU for low fouling duties, 10% for moderate fouling, and 15 to 20% for high fouling.

TABLE 10.4

Recommended Fouling Factors for Plate Heat Exchangers

Service	Fouling factor ($m^2 \cdot K/W$)
Water	
Demineralized or distilled	0.0000017
Soft	0.0000034
Hard	0.0000086
Cooling tower (treated)	0.0000069
Sea (coastal) or estuary	0.000086
Sea (ocean)	0.000052
River, canal, tube well, etc.	0.000086
Engine jacket	0.0000103
Steam	0.0000017
Lubricating oils	0.0000034–0.0000086
Vegetable oils	0.0000017–0.0000052
Organic solvents	0.0000017–0.0000103
General process fluids	0.0000017–0.00000103

From Raju, K. S. N. and Jagdish, C. B. (1983) In *Low Reynolds Number Flow Heat Exchangers*, Hemisphere, Washington, D.C.; Cooper, A. et al. (1980) *Heat Transfer Eng.*, Vol. 1, no. 3, 50–55. With permission.

10.5.2 Maintenance

A gasketed-plate heat exchanger can be easily opened for inspection, mechanical cleaning, gasket replacement, extension or reduction to the number of plates, or other modifications of the duties (Figure 10.8). In-place cleaning by circulating suitable cleaning chemicals is relatively easy. Because the plate units are built from standard components that are interchangeable and replaceable, stocking of spare is minimum. Replacing the gasket does not pose much of a problem. In a single-pass unit, it is possible to locate all pipe connections in the fixed part of the frame. This is of great advantage because the unit can be opened for access to the heat transfer surface without opening any connections.

10.6 Heat Transfer and Pressure Drop Calculations

The design of gasketed-plate heat exchangers is highly specialized in nature considering the variety of designs available for the plates and arrangements that are possible to suit varied duties. Unlike tubular heat exchangers for which design data and methods are easily available, a gasketed-plate heat exchanger design continues to be proprietary in nature. Manufacturers have developed their own computerized design procedures applicable to the exchangers marketed by them.

Attempts have been made to develop heat transfer and pressure drop correlations for use with plate heat exchangers, but most of the correlations cannot be generalized to give a high degree of prediction ability.[6,10,11] In these exchangers, the fluids are much closer to countercurrent flow than in shell-and-tube heat exchangers. In recent years, some design methods have been reported. These methods are mostly approximate in nature to suit preliminary sizing of the plate units for a given duty. No published information is available on the rigorous design of plate heat exchangers.

10.6.1 Heat Transfer Area

The corrugations increase the surface area of the plate as compared with the original flat area. To express the increase of the developed length, in relation to projected length, a surface enlargement factor, ϕ, is then defined as the ratio of the developed length to the flat or projected length (Figure 10.6):

$$\phi = \frac{\text{developed length}}{\text{projected length}} \tag{10.1}$$

The value of ϕ is a function of the corrugation pitch and the corrugation depth or plate pitch (Figure 10.6) The enlargement factor varies between 1.15 and 1.25. The value of 1.17 can be assumed as a typical average.[2,11] The value of ϕ as given by Equation (10.1) is the ratio of the actual effective area as specified by the manufacturer, A_1, to the projected plate area A_{1p}:

$$\phi = \frac{A_1}{A_{1p}} \tag{10.2}$$

where A_{1p} can be approximated from Figure 10.6 as:

$$A_{1p} = L_p \cdot L_w \tag{10.3}$$

and L_p and L_w can be estimated from the port distance L_v and L_h and port diameter D_p as:

$$L_p \approx L_v - D_p \tag{10.4}$$

$$L_w \approx L_h + D_p \tag{10.5}$$

The value of ϕ is used to calculate the effective flow path.

10.6.2 Mean Flow Channel Gap

Flow channel is the conduit formed by two adjacent plates between the gaskets. The cross section of a corrugated surface being very complex, the mean channel spacing, b, is defined as shown in Figure 10.6b:

$$b = p - t \tag{10.6}$$

where p is the plate pitch or the outside depth of the corrugated plate and t is the plate thickness, b is also the thickness of a fully compressed gasket, as the plate corrugations are in metallic contact. Plate pitch is not to be confused by the corrugation pitch. Channel spacing b is required for calculation of the mass velocity and Reynolds number and is therefore a very important value that is usually not specified by the manufacturer. If not known or for existing units, the plate pitch p can be determined from the compressed plate pact (between the head plates) L_c, which is usually specified on drawings. Then p is determined as:

$$p = \frac{L_c}{N_t} \tag{10.7}$$

where N_t is the total number of plates.

10.6.3 Channel Equivalent Diameter

The equivalent diameter of the channel, D_e, is defined as:

$$D_h = \frac{4 \times \text{channel flow area}}{\text{wetted surface}} = \frac{4A_c}{P_w} \tag{10.8}$$

$$D_e = \frac{4(b)(L_w)}{2(b + L_w\phi)} \approx \frac{2b}{\phi} \tag{10.9}$$

with the approximation that $b \ll L_w$.

10.6.4 Heat Transfer Coefficient

Any attempt for the estimation of film coefficient of heat transfer in gasketed-plate heat exchangers involves extension of correlations that are available for heat transfer between flat flow passages. The conventional approach for such passages employs correlations applicable for tubes by defining an equivalent diameter for the noncircular passage, which is substituted for diameter, d, in the following correlation:[5,6,10]

$$\frac{hD_e}{k} = C_h \left(\frac{D_e G}{\mu}\right)^n \left(\frac{c_p \mu}{k}\right)^{1/3} \left(\frac{\mu_b}{\mu_w}\right)^{0.17} \tag{10.10}$$

where D_e is the equivalent diameter defined by Equation (10.8). Values of C_h and n depend on flow characteristics and Chevron angles. The transition to turbulance occurs at low Reynolds numbers and, as a result, the gasketed-plate heat exchangers give high heat transfer coefficients (Figure 10.13).

A fully developed documented method of estimating the heat transfer and pressure drop of gasketed-plate heat exchangers based on a wide range of commercially available Chevron-pattern plates are possible, which will be outlined here (Figure 10.2).

The Reynolds number, Re, based on channel mass velocity and the equivalent diameter, D_e, of the channel is defined as:

$$Re = \frac{G_c D_e}{\mu} \tag{10.11}$$

The channel mass velocity is given by:

$$G_c = \frac{\dot{m}}{N_{cp} b L_w} \tag{10.12}$$

where N_{cp} is the number of channel per pass and obtained from:

$$N_{cp} = \frac{N_t - 1}{2N_p} \tag{10.13}$$

where N_t is the total number of plates and N_p is the number of passes.

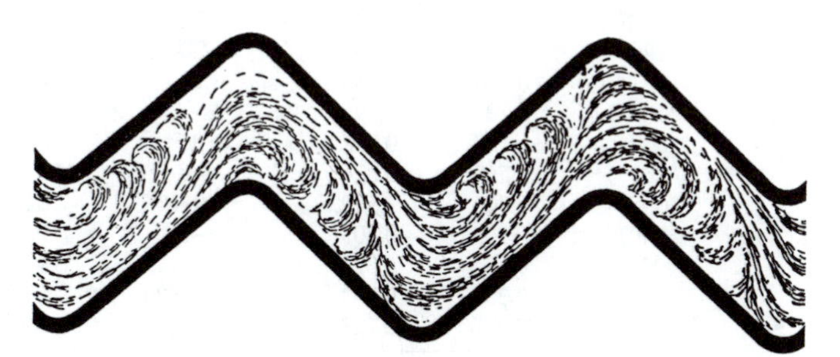

FIGURE 10.13
Flow regime between plates. (Courtesy of Alfa-Laval Thermal AB.)

A correlation in the form of Equation (10.9) has been proposed by Kumar,[10] and the values of constants C_h and n are given in Table 10.5.[7,8] In the literature, various correlations are available for plate heat exchangers for Newtonian and non-Newtonian fluids depending on flow characteristics and the geometry of plates.[11–13]

10.6.5 Channel Pressure Drop

The total pressure drop is composed of the frictional channel pressure drop, Δp_c and the port pressure drop Δp_p. The friction factor, f, is defined by the following equation for the frictional pressure drop Δp_c:[2,10,11]

$$\Delta p_c = 4f \frac{L_{eff} N_p}{D_e} \frac{G_c^2}{2\rho} \left(\frac{\mu_b}{\mu_w} \right)^{-0.17} \tag{10.14}$$

where L_{eff} is the effective length of the fluid flow path between inlet and outlet ports and it must take into account the corrugation enlargement factor ϕ; this effect is included in the definition of friction factor. Therefore $L_{eff} = L_v$, which is the vertical part distance as indicated on Figure 10.6.

The friction factor in Equation (10.14) is given by:

$$f = \frac{K_p}{Re^m} \tag{10.15}$$

Values of K_p and m are given in Table 10.5[2,10] as functions of Reynolds number for various values of Chevron angles. For various plate surface configurations, friction coefficient vs. Reynolds number must be provided by the manufacturer.

TABLE 10.5

Constants for Single-Phase Heat Transfer and Pressure Loss Calculation in Gasketed-plate Heat Exchangers

	Heat Transfer			Pressure Loss		
Chevron angle (degree)	Reynolds Number	C_h	n	Reynolds Number	K_p	m
≤30	≤10	0.718	0.349	<10	50	1
	>10	0.348	0.663	10–100	19.40	0.589
				>100	2.990	0.183
45	<10	0.718	0.349	<15	47	1
	10–100	0.400	0.598	15–300	18.29	0.652
	>100	0.300	0.663	>300	1.441	0.206
50	<20	0.630	0.333	<20	34	1
	20–300	0.291	0.591	20–300	11.25	0.631
	>300	0.130	0.732	>300	0.772	0.161
60	<20	0.562	0.326	<40	24	1
	20–400	0.306	0.529	40–400	3.24	0.457
	>400	0.108	0.703	>40	0.760	0.215
≥65	<20	0.562	0.326	50	24	1
	20–500	0.331	0.503	50–500	2.80	0.451
	>500	0.087	0.718	>500	0.639	0.213

From Saunders, E. A. D. (1988) *Heat Exchangers — Selection, Design, and Construction*, Wiley, New York; Kumar, H. (1984) 1st U.K. Natl. Conf. Heat Transfer, University of Leeds, July 3–5, *Inst. Chem. Symp.*, Ser. no. 86, p. 1275.

10.6.6 Port Pressure Drop

The pressure drop in the port ducts, Δp_p, can be roughly estimated as 1.4 velocity head:

$$\Delta p_p = 1.4 N_p \frac{G_p^2}{2\rho} \tag{10.16}$$

where

$$G_p = \frac{\dot{m}}{\dfrac{\pi D_p^2}{4}} \tag{10.17}$$

where $\dot{m}$ is the total flow rate in the port opening and D_p is the port diameter.

The total pressure drop is then:

$$\Delta p_t = \Delta p_c + \Delta p_p \tag{10.18}$$

10.6.7 Overall Heat Transfer Coefficient

The overall heat transfer coefficient for a clean surface:

$$\frac{1}{U_c} = \frac{1}{h_h} + \frac{1}{h_c} + \frac{t}{k_w} \tag{10.19}$$

and under fouling conditions (fouled or service overall heat transfer coefficient):

$$\frac{1}{U_f} = \frac{1}{h_h} + \frac{1}{h_c} + \frac{t}{k_w} + R_{fh} + R_{fc} \tag{10.20}$$

where h and c stand for hot and cold streams, respectively.

The relationship between U_c, fouled U_f, and the cleanliness factor, CF, can be written as:

$$U_f = U_c(CF) = \frac{1}{\dfrac{1}{U_c} + R_{fh} + R_{fc}} \tag{10.21}$$

10.6.8 Heat Transfer Surface Area

The heat balance relations in gasketed-plate heat exchangers are the same as for tubular heat exchangers. The required heat duty, Q_r, for cold and hot streams is

$$Q_r = (\dot{m}c_p)_c(T_{c_2} - T_{c_1}) = (\dot{m}c_p)_h(T_{h_1} - T_{h_2}) \tag{10.22}$$

On the other hand, the actually obtained heat duty, Q_f, for fouled conditions is defined as:

$$Q_f = U_f A_e F \Delta T_m \tag{10.23}$$

where A_e is the total developed area of all thermally effective plates, that is, $N_t - 2$ (see Figure 10.10) that accounts for the two plates adjoining the head plates. In multipass arrangements, the separating plate is in parallel flow that will have negligible effect on the mean temperature difference. If an entire pass is in parallel flow, correction factor, F, must be applied to obtain the true mean temperature difference, ΔT_m .[8,11,14–16] Otherwise, the true mean temperature difference, ΔT_m, is for the counterflow arrangement:

$$\Delta T_{lm,cf} = \frac{\Delta T_1 - \Delta T_2}{\ln \dfrac{\Delta T_1}{\Delta T_2}} \tag{10.24}$$

where ΔT_1 and ΔT_2 are the terminal temperature differences at the inlet and outlet.

A comparison between Q_r and Q_f defines the safety factor, C_s, of the design:

$$C_s = \frac{Q_f}{Q_r} \tag{10.25}$$

10.6.9 Performance Analysis

Assume that for a specific process, a manufacturer proposes a gasketed-plate heat exchanger. A stepwise calculation as documented in the following section will be used for rating (performance analysis of the proposed gasketed-plate heat exchanger as an existing heat exchanger).

Example 10.1

Cold water will be heated by a wastewater stream. The cold water stream with a 140 kg/s enters the gasketed-plate heat exchanger at 22°C and it will be heated to 42°C. The wastewater has the same flow rate entering at 65°C and leaving at 45°C. The maximum permissible pressure drop for each stream is 27 $lb_f/in.^2$.

Process Specifications:

Items	Hot Fluid	Cold Fluid
Fluids	Wastewater	Cooling water
Flow rates (kg/s)	140	140
Temperature in (°C)	65	22
Temperature out (°C)	45	42
Maximum permissible pressure drop ($lb_f/in.^2$)	27	27
Total fouling resistance ($m^2 \cdot K/W$)	0.00005	0
Specific heat (J/kg·K)	4183	4178
Viscosity ($N \cdot s/m^2$)	5.11×10^{-4}	7.68×10^{-4}
Thermal conductivity (W/m·K)	0.635	0.616
Density (kg/m^3)	986	995
Prandtl number	3.32	5.21

Constructional Data of the Proposed Plate Heat Exchanger

Plate material	SS304
Plate thickness (mm)	0.6
Chevron angle (degrees)	45
Total number of plates	105
Enlargement factor, ϕ	1.25
Number of passes	One pass
Overall heat transfer coefficient (clean/fouled) ($W/m^2 \cdot K$)	8,000/4,500
Total effective area (m^2)	110
All port diameters (mm)	200
Plate thickness (m)	0.0006
Compressed plate pack length, L_c, (m)	0.38
Vertical port distance, L_v, (m)	1.55
Horizontal port distance, L_h, (m)	0.43
Effective channel width, L_w, (m)	0.63
Thermal conductivity of the plate material (SS304), W/m·K	17.5

SOLUTION

Stepwise performance analysis — The required heat load can be calculated from heat balance as:

$$Q_{rh} = 140 \times 4183 \times (65 - 45) = 11712.4 \text{ kW}$$

$$Q_{rc} = 140 \times 4178 \times (42 - 22) = 11698.4 \text{ kW}$$

Compute the mean temperature difference by:

$$\Delta T_1 = \Delta T_2 = \Delta T_m = 23°\text{C}$$

therefore

$$\Delta T_{lm,cf} = \Delta T_{lm} = 23°\text{C}$$

The effective number of plates is

$$N_e = N_t - 2 = 103$$

Effective flow length between the vertical ports is

$$L_{eff} \approx L_v = 1.55 \text{ m}$$

Plate pitch can be determined from Equation (10.7):

$$p = \frac{L_c}{N_t} = \frac{0.38}{105} = 0.0036 \text{ m}$$

Determine the mean channel flow gap from Equation (10.6) by:

$$b = 0.0036 - 0.0006 = 0.0030 \text{ m}$$

One-channel flow area is calculated by:

$$A_{ch} = b \times L_w = 0.0030 \times 0.63 = 0.00189 \text{ m}^2$$

The single-plate heat transfer area is

$$A_1 = \frac{A_e}{N_e} = \frac{110}{103} = 1.067 \text{ m}^2$$

The projected plate area A_{1p} from Equation (10.3) is

$$A_{1p} = L_p \cdot L_w = (1.55 - 0.2) \times 0.63 = 0.85 \text{ m}^2$$

Enlargement factor has been specified by the manufacturer; but it can be verified from Equation (10.2):

$$\phi = \frac{1.067}{0.85} = 1.255$$

Calculate channel equivalent diameter from Equation (10.9):

$$D_e = \frac{2b}{\phi} = \frac{2 \times 0.0030}{1.255} = 0.00478 \text{ m}$$

Obtain the number of channels per pass, N_{cp} from Equation (10.13):

$$N_{cp} = \frac{N_t - 1}{2N_p} = \frac{105 - 1}{2} = 52$$

Heat transfer analysis — Calculate mass flow rate per channel by:

$$\dot{m}_{ch} = \frac{140}{52} = 2.69 \text{ kg/s}$$

Mass velocity, G_{ch} is

$$G_{ch} = \frac{1.97}{0.00189} = 1423.3 \text{ kg/m}^2 \cdot \text{s}$$

Hot and cold fluid Reynolds numbers are

$$Re_h = \frac{G_h D_e}{\mu_h} = \frac{1423.3 \times 0.00478}{5.11 \times 10^{-4}} = 13314$$

$$Re_c = \frac{G_c D_e}{\mu_c} = \frac{1423.3 \times 0.00478}{7.68 \times 10^{-4}} = 8859$$

Therefore, both fluids are in turbulent flow.

Hot fluid heat transfer coefficient, h_h — This can be obtained by referring to Table 10.5 C_h = 0.3, and n = 0.633, from the heat transfer coefficient from Equation (10.10):

$$Nu_h = \frac{h_h D_e}{k} = 0.3(Re)^{0.663}(Pr)^{1/3}\left(\frac{\mu_b}{\mu_w}\right)^{0.17}$$

By assuming $\mu_b \approx \mu_w$, then:

$$Nu_h = 0.3(13314)^{0.663}(3.32)^{1/3} = 242.8$$

$$h_h = \frac{Nu_h k}{D_e} = \frac{0.635 \times 242.8}{0.00478} = 32255 \, W/m^2 \cdot K$$

Cold fluid heat transfer coefficient, h_c — This can be calculated from Table 10.5 and Equation (10.10):

$$Nu_c = \frac{h_c D_e}{k} = 0.3(8859)^{0.663}(5.21)^{1/3} = 215.4$$

$$h_c = \frac{Nu_c k}{D_e} = \frac{0.616 \times 215.4}{0.00478} = 27759 \text{ W/m}^2 \cdot \text{K}$$

Overall heat transfer coefficient — Determine the clean overall heat transfer coefficient from Equation (10.19):

$$\frac{1}{U_c} = \frac{1}{h_c} + \frac{1}{h_h} + \frac{0.0006}{17.5}$$

$$U_c = 9870 \text{ W/m}^2 \cdot \text{K}$$

Fouled (or service) overall heat transfer coefficient is calculated from Equation (10.20):

$$\frac{1}{U_f} = \frac{1}{U_c} + 0.00005$$

$$U_f = 6609 \text{ W/m}^2 \cdot \text{K}$$

The corresponding cleanliness factor is

$$CF = \frac{U_f}{U_c} = \frac{6609}{9870} = 0.67$$

which is rather low because of the high fouling factor.

Actual heat duties for clean surface are

$$Q_c = U_c A_e \Delta T_m = 9870 \times 110 \times 23 = 24971 \text{ kW}$$

$$Q_f = U_f A_e \Delta T_m = 6609 \times 110 \times 23 = 16721 \text{ kW}$$

The safety factor is

$$C_s = \frac{Q_f}{Q_r} = \frac{16721}{11698} \approx 1.4$$

The percentage over surface design from Equation (5.20) is

$$OS = 100 U_c R_{ft} = 49.35\%$$

which is a rather large heat exchanger that could be smaller to satisfy the process specifications. A 30% oversurface design may be preferable to decrease the investment cost, and then the cleaning scheduling could be arranged accordingly. Therefore, the proposed design may be modified and re-rated.

Pressure drop analysis — For the hot and cold fluid friction coefficients from Equation (10.15) and Table 10.5:

$$f_h = \frac{1.441}{(Re_h)^{1.441}} = \frac{1.441}{(13314)^{0.206}} = 0.204$$

$$f_c = \frac{1.441}{(Re_c)^{1.441}} = \frac{1.441}{(8859)^{0.206}} = 0.222$$

The frictional pressure drop from hot and cold streams are calculated from Equation (10.14).

For hot fluid:

$$(\Delta p_c)_h = 4 \times 0.204 \times \frac{1.55 \times 1}{0.00478} \times \frac{(1423.3)^2}{2 \times 986} = 271819 \text{ Pa}$$

$$(\Delta p_c)_h = 39.4 \text{ lb}_f/\text{in.}^2$$

and for cold fluid:

$$(\Delta p_c)_c = 4 \times 0.222 \times \frac{1.55 \times 1}{0.00478} \times \frac{(1423.3)^2}{2 \times 995} = 293128 \text{ Pa}$$

$$(\Delta p_c)_c = 42.5 \text{ lb}_f/\text{in.}^2$$

The pressure drop in the port ducts is calculated from Equation (10.16).

By determining the port mass velocity as:

$$(G_p)_h = (G_p)_c = \frac{\dot{m}}{\pi\left(\frac{D_p^2}{4}\right)} = \frac{140}{\pi\left(\frac{0.2^2}{4}\right)} = 4458.60 \text{ kg/m}^2 \cdot \text{s}$$

for hot fluid:

$$(\Delta p_p)_h = 1.4 \times 1 \times \frac{G_p^2}{2 \times 986} = 14112.96 \text{ Pa}$$

$$(\Delta p_p)_h = 2.05 \text{ lb}_f/\text{in.}^2$$

and for cold fluid

$$(\Delta p_p)_c = 1.4 \times 1 \times \frac{G_p^2}{2 \times 995} = 13985.31 \text{ Pa}$$

$$(\Delta p_p)_c = 2.03 \text{ lb}_f/\text{in.}^2$$

Total pressure drop is calculated from Equation (10.18) for hot fluid:

$$(\Delta p_t)_h = (\Delta p_c)_h + (\Delta p_p)_h = 39.4 + 2.05 = 41.45 \text{ lb}_f/\text{in.}^2$$

and for cold fluid:

$$(\Delta p_t)_c = (\Delta p_c)_c + (\Delta p_p)_c = 42.5 + 2.03 = 44.53 \text{ lb}_f/\text{in.}^2$$

The calculation shows that the proposed unit satisfies the process required and pressure drop constraint, but it could be smaller unless it is used under heavy fouling conditions. Interesting design examples and performance analysis are also given in Reference 17.

10.7 Thermal Performance

To quantitatively assess the thermal performance of plates, it would be necessary to define the thermal duty of the plates. The plates are of two types, short duty or soft plates, and long duty or hard plates. The soft plates involve low-pressure drop and low heat transfer coefficients. Plates having a high Chevron angle provide this feature. A short and wide plate is of this type. The hard plates involve high-pressure drop and give high heat transfer coefficients. A plate having a low Chevron angle provides high heat transfer combined with high-pressure drop. Long and narrow plates belong to this category. The type of embossing, apart from the overall dimensions, has a significant effect on the thermal performance.

Manufacturers produce both hard and soft plates, so that a wide range of duties can be handled in the most efficient manner. In deciding the type of plate that will be most suitable for a given duty, the most convenient term to use is the NTU, defined as:

$$\mathrm{NTU}_c = \frac{UA}{(\dot{m}c_p)_c} = \frac{T_{c_2} - T_{c_1}}{\Delta T_m} \tag{10.26}$$

$$\mathrm{NTU}_h = \frac{UA}{(\dot{m}c_p)_h} = \frac{T_{h_1} - T_{h_2}}{\Delta T_m} \tag{10.27}$$

Manufacturers specify the plates having low values of Chevron angle as high-θ plates and plates having high values of Chevron angle as low-θ plates. A low Chevron angle is around 25° to 30°, while a high Chevron angle is around 60° to 65°.

The ε-NTU method is described in Chapter 2; the total heat transfer rate from Equation (2.43) is

$$Q = \varepsilon(\dot{m}c_p)_{min}(T_{h_1} - T_{h_2}) \tag{10.28}$$

Heat capacity rate ratio is given by Equation (2.35) as:

$$R = \frac{T_{h_1} - T_{h_2}}{T_{c_2} - T_{c_1}} \tag{10.29}$$

When $R < 1$:

$$(\dot{m}c_p)_c = (\dot{m}c_p)_{min} = C_{min} \tag{10.30}$$

$$\mathrm{NTU} = \frac{UA}{C_{min}} = \frac{UA}{(\dot{m}c_p)_c} \tag{10.31}$$

and when R > 1:

$$(\dot{m}c_p)_h = (\dot{m}c_p)_{min} = C_{min} \tag{10.32}$$

$$\text{NTU} = \frac{UA}{C_{min}} = \frac{UA}{(\dot{m}c_p)_h} \tag{10.33}$$

In calculating the value of NTU for each stream, the total mass flow rates of each stream must be used.

Graphs and tables of heat exchanger effectiveness, ε, and of correction factor, F, to ΔT_m as functions of R and NTU for various plate heat exchangers are available in the literature.[5,17,18]

The heat exchanger effectiveness for pure counterflow and for parallel flow is given by Equations (2.46) and (2.47), respectively. Heat exchanger effectiveness, ε, and $(\text{NTU})_{min}$ for counterflow can be expressed as:

$$\varepsilon = \frac{\exp[(1 - C_{min}/C_{max})\text{NTU}_{min}] - 1}{\exp[(1 - C_{min}/C_{max})\text{NTU}_{min}] - C_{min}/C_{max}} \tag{10.34}$$

and

$$\text{NTU}_{min} = \frac{\ln[(1 - C_{min}/C_{max})/(1 - \varepsilon)]}{(1 - C_{min}/C_{max})} \tag{10.35}$$

which are useful in rating analysis when outlet temperatures of both streams are not known.

One of the problems associated with plate units is with respect to the exact matching of the thermal duties. It is very difficult to achieve the required thermal duty and at the same time utilize the available pressure drop fully. A solution to this problem is to install more than one kind of channel in the same plate pack with differing NTU values.[13] The types could be mixed to produce any desired value of NTU between the highest and the lowest. The high NTU plate has Chevron corrugations with a large included angle giving a comparatively larger pressure drop, while the low NTU plate has Chevron corrugations with a much smaller included angle leading to a relatively low-pressure drop.

Nomenclature

A_t total effective plate heat transfer area, m^2

A_1 single-plate effective area, m^2

A_{1p} single-plate projected area, m^2

b mean mass channel gap

c_p specific heat of fluid, J/kg·K
CF cleanliness factor
C_h constant in Equation (10.10)
C_s safety factor defined by Equation (10.25)
D_e channel equipment diameter defined by Equation (10.9), m
D_p port diameter, m
f friction factor
K_p constant in Equation (10.15)
k thermal conductivity, W/m·k
G_c mass velocity through a channel, kg/m²·s
G_p heat transfer coefficient, W/m·K
L_e length of compressed plate pack, m
L_h horizontal port distance, m
L_p projected plate length, m
L_v vertical port distance (flow length in one pass), m
L_w plate width inside gasket, m
L_{eff} effective flow length between inlet and outlet port ($\approx L_v$), m
$\dot{m}$ mass flow rate, kg/s
NTU number of transfer units defined by Equation (10.26)
P plate pitch, m
Pr Prandtl number, $c_p\mu/k$
P_w wetted perimeter, m
Q duty, W
Q_c heat load under clean conditions, W
Q_f heat load under fouled conditions, W
Q_r heat transfer required, W
R_{fc} fouling factor for cold fluid, m²·K/W
R_{fh} fouling factor for hot fluid, m²·K/W
Re Reynolds number, G_cD_e/μ, or G_pD_p/μ
t plate thickness, m
U_c clean overall heat transfer coefficient, W/m²·K
U_f fouled (service) overall heat transfer coefficient, W/m²·K

Greek Symbols

β Chevron angle, degree
Δp_c channel flow pressure drop, kPa
Δp_p port pressure drop, kPa
ΔT_m mean temperature difference,°C, K
ρ fluid density, kg/m³
μ dynamic viscosity at average inlet temperature, N·s/m² or Pa·s

References

1. Alfa-Laval Thermal AB, *Thermal Handbook*, Lund, Sweden.
2. Saunders, E. A. D. (1988) *Heat Exchangers — Selection, Design, and Construction*, Wiley, New York.
3. Raju, K. S. N. and Jagdish C. B. (1983) Plate heat exchangers and their performance. In *Low Reynolds Number Flow Heat Exchangers*, S. Kakaç, R. K. Shah, and A. E. Bergles (Eds.), Hemisphere, Washington, D.C.
4. Cooper, A. and Usher, J. D. (1988) Plate heat exchangers. In *Heat Exchanger Design Handbook*, Chapter 3.7. Hemisphere, Washington, D.C.
5. Raju, K. S. N. and Jagdish C. B. (1983) Design of plate heat exchangers. In *Low Reynolds Number Flow Heat Exchangers*, S. Kakaç, R. K. Shah, and A. E. Bergles (Eds.), Hemisphere, Washington, D.C.
6. Edwards, M. F. (1983) Heat transfer in plate heat exchangers at low Reynolds numbers. In *Low Reynolds Number Flow Heat Exchangers*, S. Kakaç, R. K. Shah, and A. E. Bergles (Eds.), Hemisphere, Washington, D.C.
7. Wilkinson, W. L. (1974) Flow distribution in plate exchangers. *Chem. Engineer*, no. 285, 289–293.
8. Marriott, J. (1971) Where and how to use plate heat exchangers. *Chem. Eng.*, Vol. 78, no. 8, 127–134.
9. Cooper, A., Suitor, J .W., and Usher, J. D. (1980) Cooling water fouling in plate heat exchangers. *Heat Transfer Eng.*, Vol. 1, no. 3, 50–55.
10. Kumar, H. (1984) The plate heat exchanger: construction and design, 1st U.K. Nat. Conf. Heat Transfer, University of Leeds, July 3–5, *Inst. Chem. Symp.*, Ser. no. 86, 1275.
11. Edwards, M. F., Changal, A. A., and Parrott, D. L. (1974) Heat transfer and pressure drop characteristics of a plate heat exchanger using Newtonian and non-Newtonian liquids. *Chem. Engineer*, 286–293, May 1974.
12. McKillop, A. A. and Dunkley, W. L. (1960) Heat transfer. *Ind. Eng. Chem.*, Vol. 52, no. 9, 740–744.
13. Clark, D. F. (1974) Plate heat exchanger design and recent development. *Chem. Engineer*, no. 285, 275–279.
14. Buonopane, R. A., Troupe, R. A., and Morgan, J. C. (1963) Heat transfer design method for plate heat exchangers. *Chem. Eng. Prog.*, Vol. 59, 57.
15. Foote, M. R. (1967) Effective mean temperature difference in multi-pass plate heat exchangers, National Engineering Laboratory, U.K., Report no. 303.
16. Bell, J. K. (1981) Plate heat exchanger. In *Heat Exchangers — Thermal–Hydraulic Fundamentals and Design*, S. Kakaç, A. E. Bergles, and F. Mayinger (Eds.), Hemisphere, Washington, D.C.
17. Hewitt, G. F., Shirs, G. L., and Bott, T. R. (1994) *Process Heat Transfer*, CRC, Boca Raton, FL.
18. Shah, R. K. and Focke, W. W. (1988) Plate heat exchangers and their design theory. In *Heat Transfer Equipment Design*, Shah, R. K., Subbarao, E. C., and Mashelkar, R. A. (Eds.), Hemisphere, Washington, D.C.

Problems

10.1. The following constructional information is available for a gasketed-plate heat exchanger:

Chevron angle	50°
Enlargement factor	1.17
All port diameters	15 cm
Plate thickness	0.0006 m
Vertical port distance	1.50 m

Horizontal port distance	0.50 m
Plate pitch	0.0035 m

Calculate:

a. Mean channel flow gap
b. One-channel flow area
c. Channel equivalent diameter
d. Projected plate area
e. Effective surface area per plate

10.2. A gasketed-plate heat exchanger will be used for heating city water (R_{fc} = 0.00006 m²·K/W) using the wastewater available at 90°C. The vertical distance between the ports of the plate is 1.60 m and the width of the plate is 0.50 m with a gap of 6 mm between the plates. The enhancement factor is provided by the manufacturer as 1.17 and the Chevron angle is 50°. The plates are made of titanium (k = 20 W/m·K) with a thickness of 0.0006 m. The port diameter is 0.15 m. The cold water enters to the plate heat exchanger at 15°C and leaves at 45°C at a rate of 6 kg/s; and it will be heated by the hot water available at 90°C, flowing at a rate of 12 kg/s. By considering single-pass arrangements for both streams, calculate:

a. The effective surface area and the number of plates of this heat exchangers
b. The pressure drop for both streams

10.3. The performance of the single-pass heat exchanger given in Problem 10.1 could be improved by using plates with increased Chevron angle. Repeat Example 10.1 for Chevron angles β = 60° and 65°.

10.4. Repeat Example 10.1 for 30% or less oversurface design.

10.5. Solve Problem 10.2 for two-pass/two-pass arrangement.

10.6. A heat exchanger is required to heat treated cooling water with a flow rate of 60 kg/s from 10 to 50°C using the waste heat from water, cooling from 60 to 20°C with the same flow rate as the cold water. The maximum allowable pressure drop for both streams is 120 kPa. A gasketed-plate heat exchanger with 301 plates having a channel width of 50 cm, a vertical distance of 1.5 m between ports is proposed, and the plate pitch is 0.0035 m with an enlargement factor of 1.25. The spacing between them is 6 mm. Plates are made of stainless steel (316) (k = 16.5 W/m·K). For a two-pass/two-pass arrangement, analyze the problem to see if the proposed design is feasible. Could this heat exchanger be smaller or larger?

10.7. Repeat Example 10.1 for different types of water conditions:

a. Coastal ocean (R_{fc} = 0.0005 K·m²/W), demineralized closed loop (0.000001 K·m²/W)
b. Wastewater (0.00001 K·m²/W)/cooling tower (R_{fc} = 0.000069 m²·K/W)

Write your conclusions.

10.8. A one-pass countercurrent flow heat exchanger has 201 plates. The exchanger has a vertical port distance of 2 m and is 0.6 m wide, with a gap between the plates of 6 mm. This heat exchanger will be used for the following process: cold water from the city supply with an inlet temperature of 10°C is fed to the heat exchanger at a rate of 15 kg/s that will be heated to 75°C with a wastewater entering at an inlet temperature of 90°C. The flow rate of hot water is 30 kg/s, which is a distilled water. The other construction parameters are given as in

Problem 10.2. There is no limitation on the pressure drop. Is this heat exchanger suitable for this purpose (larger or smaller)?

10.9. In Example 10.1 a new mass flow rate is provided as 90 kg/s. Assume that the inlet temperatures of hot and cold fluid are changed to 120 and 20°C, respectively. If the flow arrangement is single pass for both streams, use the ε-NTU method to calculate the outlet temperatures for the specific heat exchanger. There may be a constraint on the outlet temperature of the cold stream; then this type of analysis is important to see if the required outlet temperature of the cold stream is satisfied.

Design Project 10.1

Comparative Study of Gasketed-Plate Heat Exchangers vs. Double-Pipe (Hairpin) Heat Exchangers

The main considerations involved in the choice of gasketed-plate heat exchangers for a specific process in comparison with double-pipe heat exchanger and shell-and-tube heat exchanger units may be made on the basis of weight and space limitations, temperature approach, operating temperatures, pressure drop limitations, maintenance requirements and capital, and operating costs that must be considered in a comparative study. For given specifications make a comparative study of a gasketed-plate heat exchanger with a double-pipe heat exchanger. Specifications given in design Project 6.1 or in Problem 10.5 can be used, and the constructional parameters given in Example 10.1 or an alternative can be selected.

Design Project 10.2

Comparative Study of Gasketed-Plate Heat Exchangers vs. Shell-and-Tube Heat Exchangers

Specifications given in Example 10.1 or in Problem 10.5 can be used for such a comparative study.

Design Project 10.3

Blood Heat Exchanger for an Open Heart Operation

The objective of a blood heat exchanger is to shorten the time normally required to cool a patient's blood prior to open heart surgery. Because blood behaves in a non-Newtonian manner, necessary assumptions were made to complete the calculations. The material in contact needs to have the smoothest possible surface; therefore a stainless steel with a specific nonwetting silicone resin shell-and-tube heat exchanger was chosen for design. For a sufficient design, the pressure drop must be minimal due to the fragile nature of blood, and the heat transfer rate must be high due to the required rapid temperature change. This design project will take into account, thermal, dynamic heat transfer, and parametric analysis for a blood heat exchanger.

The purpose of a blood heat exchanger is to shorten the time normally required to cool a patient's blood prior to open heart surgery. Assume that the patient's blood will be cooled from 37 to 27°C by cooling water available at 15°C. The mass flow rate of blood is 0.03 kg/s and the properties at 32°C are ρ = 1055 kg/m^3, μ = 0.00045 kg/m·s, k = 0.426 W/m·K, Pr = 3.52 and c_p = 3330 J/kg·K. Cooling water may be available at 0.20 kg/s. One can directly select a gasketed-plate heat exchanger, or a comparative study — as in Design Projects 10.1 and 10.2 — can be conducted.

11

Condensers and Evaporators

11.1 Introduction

A condenser is a two-phase flow heat exchanger in which heat is generated from the conversion of vapor into liquid (condensation) and the heat generated is removed from the system by a coolant.

Condensers may be classified into two main types: those in which the coolant and condensate stream are separated by a solid surface, usually a tube wall; and those in which the coolant and condensing vapor are brought into direct contact.

The direct contact type may consist of a vapor that is bubbled into a pool of liquid, a liquid that is sprayed into a vapor, or a packed-column in which the liquid flows downward as a film over a packing material against the upward flow of vapor.

Those in which the streams are separated may be subdivided into three main types: air-cooled, shell-and-tube, and plate condensers. In the air-cooled type, condensation occurs inside tubes with cooling being provided by air blown or sucked across the tubes. Fins, having a large surface area, are usually provided on the air side to compensate for the low air-side heat transfer coefficients.

In shell-and-tube condensers, the condensation may occur inside or outside the tubes. The orientation of the unit may be vertical or horizontal.

In the refrigeration and air-conditioning industry, various types of two-phase flow heat exchangers are used. They are classified according to whether they are coils or shell-and-tube heat exchangers. Evaporator and condenser coils are used when the second fluid is air because of the low heat transfer coefficient on the air side.

In the following sections, the basic type of condensers, air-conditioning evaporators, and condenser coils, and their rating and sizing problems are briefly discussed. Further information can be obtained from References 1 and 2.

11.2 Shell-and-Tube Condensers

11.2.1 Horizontal Shell-Side Condensers

The shell-and-tube condensers for process plants are covered by the Tubular Exchanger Manufacturers Association (TEMA); Figure 8.2 shows various TEMA shell types as described in Chapter 8. The most common condenser types are shown in Figure 8.3; these condensers may be used as shell-side condensers with the exception of the F-shell, which is unusual. The E-type is the simplest form of these types of condensers. In the shell-and-tube condensers, condensation may occur inside or outside the tubes, depending on the design requirements (Figures 11.1 and 11.2).[3] A very important feature of a condenser as compared with any other type of heat exchanger is that it must have a vent for removal of noncondensable gas. Thus, an E-type condenser will have two outlet nozzles, one for the vent and the other for the condensate outlet. Noncondensables have the effect of depressing the condensation temperature, and thus reducing the temperature difference between the streams. Therefore, it is clear that there is no chance of noncondensable accumulation during condenser operation. Hence the vent is provided.

There are some advantages and disadvantages of the different shell-type condensers, and they are briefly discussed as follows:

The J-shell has a great advantage over the E-shell in that it can be arranged with two nozzles, one at either end for the vapor inlet, and one small nozzle in the middle for the condensate outlet. One would normally have a small nozzle in the middle at the top to vent noncondensable gases. By having these two inlet nozzles, a large vapor volume coming into the condenser can be accommodated more easily. Also, by splitting the vapor flow into

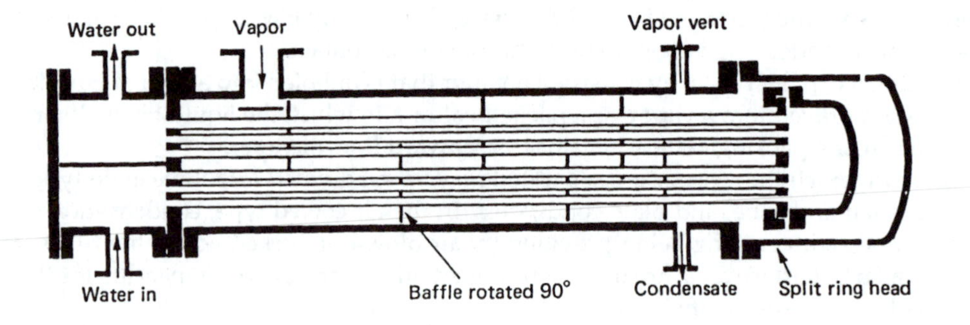

FIGURE 11.1
Horizontal shell-side condenser. (From Mueller, A. C. [1983] In *Heat Exchanger Design Handbook*, pp. 3.4.2–1. Hemisphere, Washington, D.C. With permission from Begell House, NY.)

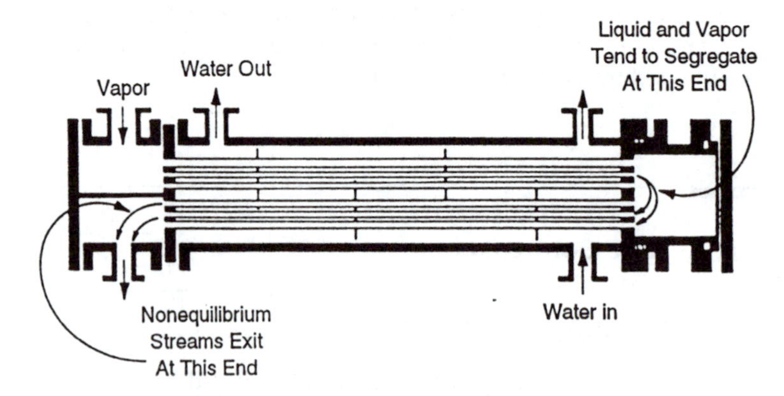

FIGURE 11.2
Horizontal in-tube condenser. (From Mueller, A. C. [1983] In *Heat Exchanger Design Handbook*, pp. 3.4.2–1. Hemisphere, Washington, D.C.; Breber, G. [1988] In *Heat Transfer Equipment Design*, pp. 477–496. Hemisphere, Washington, D.C. With permission from Begell House, NY.)

two and by halving the path length for vapor flow, the pressure drop may be reduced substantially over that for a similar size E-shell. It is a good practice with the J-shell to make sure that the heat loads in both halves of the exchanger are the same to prevent the possibility that noncondensed vapor coming from one end of the exchanger meets subcooled liquid from the other. This could give rise to periodic violent vapor collapse and possible exchanger damage. This problem usually means that the J-shell should not be designed with a single tube-side pass if there is a large temperature variation in the tube-side fluid as it flows from one end of the exchanger to the other. J-shells would normally have baffles similar to those found in E-shells, except that a full-circle tube support plate may be placed in the center of the exchanger (Figure 11.3).

The F-, G-, and H-shells can also have transverse baffles in addition to the longitudinal baffle (Figures 8.2 and 11.4). Full-circle tube support plates may be placed in line with the inlet nozzles, and for H-shells additional full-circle tube support plates can be placed halfway along the shell. An H-shell would therefore have three tube support plates along the lengths of the tubes, and it might be possible to avoid having any further segmented baffles supporting the tubes. In such circumstances, an H-shell gives a fairly low-pressure drop. The vent nozzles in G- and H-shells have to be placed in the side of the shell above the condensate outlet nozzles but, of course, below the longitudinal baffles. If there are multiple tube-side passes in G- and H-shells, these should be arranged so that the coldest pass is at the bottom and the warmest, at the top so that there is some degree of countercurrent flow.[1]

The crossflow, or X-type exchanger, is a very useful unit for vacuum operation. In such operations, large volumes of vapor must be handled, and it is possible to avoid the chance of tube vibration. The large flow area combined with the short flow path also means that pressure drops can be kept low. It is important, particularly, to keep pressure drops low in vacuum operation so as to avoid reducing the saturation temperature, and therefore losing temperature difference. Figure 11.5 shows a typical crossflow unit. This particular unit has

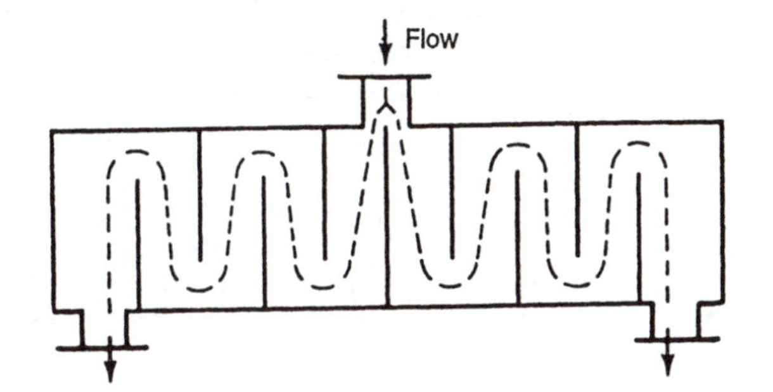

FIGURE 11.3
Schematic sketch of TEMA J-shell.

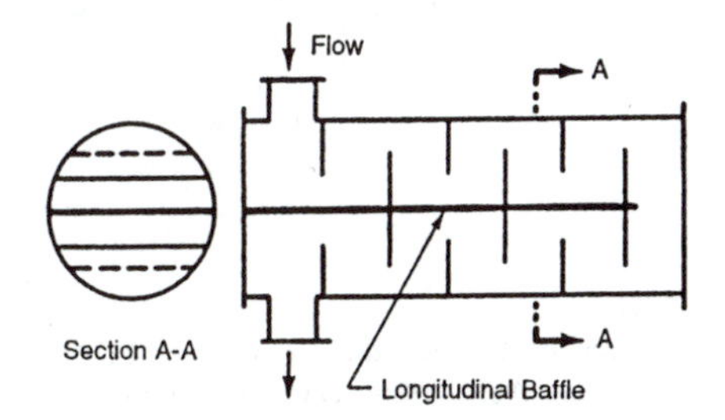

FIGURE 11.4
Schematic sketch of TEMA F-shell.

three inlet nozzles to avoid having a very large single inlet nozzle that may lead to difficulties in mechanical construction. A large space above the top of the bundle is necessary to give good vapor distribution along the exchanger length, and this may be assisted by the introduction of a perforated distributor inserted as is necessary to give sufficient tube support to prevent vibration. Noncondensable gases must be vented from as low as possible in the exchanger, as shown in Figure 11.5. Variations on the tube bundle layout are possible in crossflow condensers.[1]

In shell-and-tube condensers, a variety of alternative baffle arrangements like single segmental, double segmental, triple segmental, and no tube in the window are used. Baffles in condensers are usually arranged with the cut vertical so that the vapor flow is from side to side. The baffles are notched on the bottom to allow drainage of the condensate from one compartment to the next and finally to the condensate exit.

With most of the baffle types, having the baffles closer together increases the shell-side fluid velocities. This would be undesirable if vibration or high-pressure drop is a problem. The design with no tubes in the window, however, allows one to have additional intermediate baffles that support the tubes without having any significant effect on the flow. However, such a design is expensive because there is a large volume of empty shell.

It is almost a universal practice to use an impingement plate under the vapor inlet nozzle on a shell-side condenser to prevent tube erosion from the high velocity of the incoming vapor (Figure 11.6). Otherwise the outer tubes of the first row of the bundle could be subject to serious damage.[1]

When flow-induced vibration is a problem, an extra tube support plate may be inserted near the inlet nozzle, as shown in Figure 11.7.[1]

11.2.2 Vertical Shell-Side Condensers

A vertical E-shell as condenser is shown in Figure 11.8. The vent is shown near the condensate exit at the cold end of the exchanger. Vertical shell-side condensation has definite thermodynamic advantages when a wide-condensing-range mixture is to be condensed. There

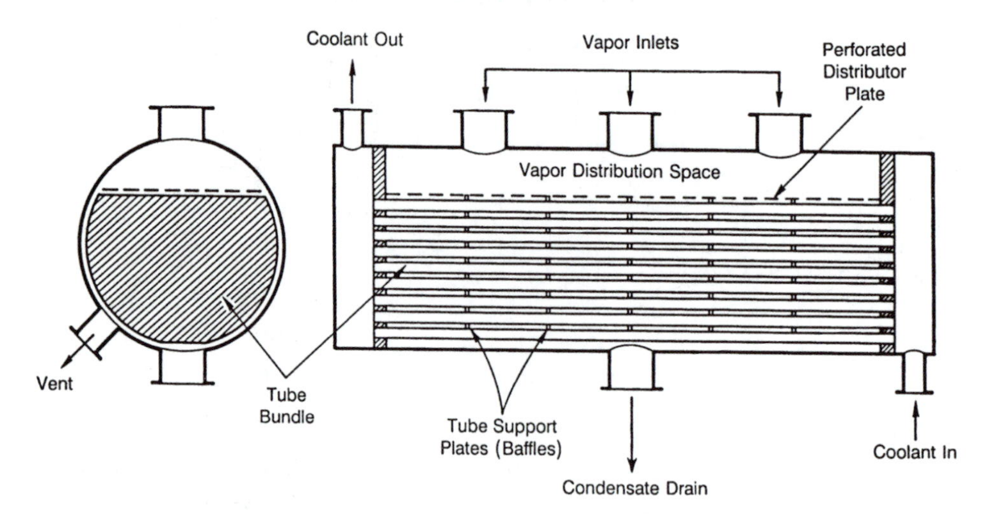

FIGURE 11.5
Main features of a crossflow condenser (TEMA X-type). (From Butterworth, D. [1991] In *Boilers, Evaporators and Condensers*, pp. 571–633. Wiley, New York; Butterworth, D. In *Two-Phase Flow Heat Exchangers*, Kluwer, Dordrecht, The Netherlands. With permission.)

will be very good mixing between condensate and vapor. In this arrangement, it may not be easy to clean the inside of the tubes. Thus this arrangement should not be used with a fouling coolant.

11.2.3 Vertical Tube-Side Condensers

These condensers are often designed with downflows for both the vapor and the condensate. Vapor enters at the top and flows down through the tubes with the condensate

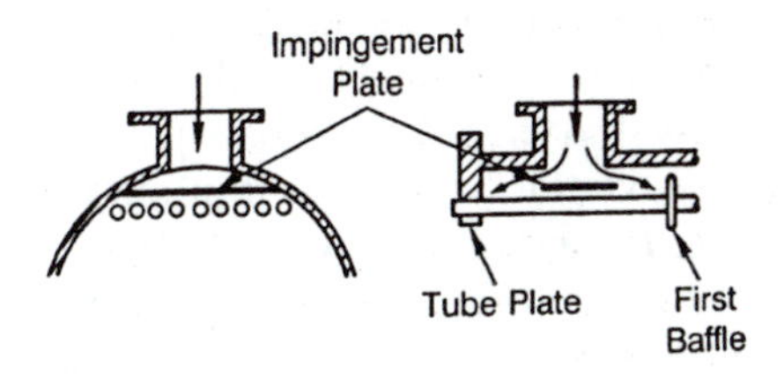

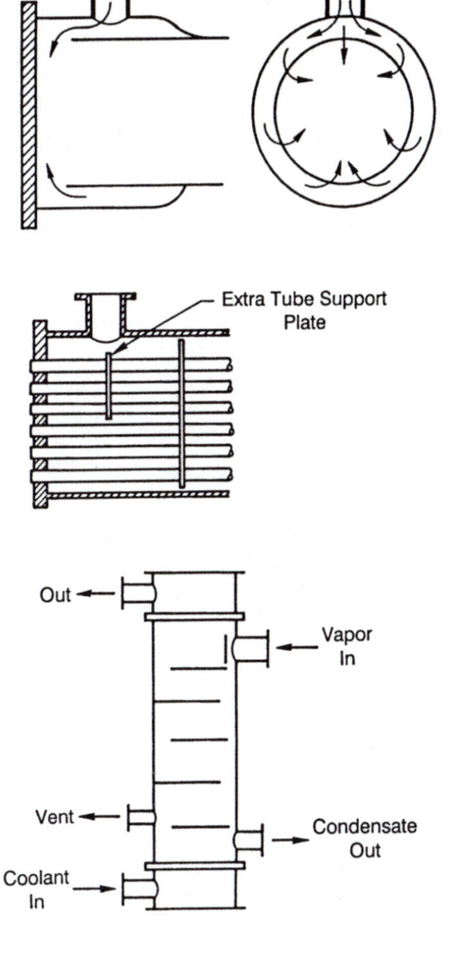

FIGURE 11.6
Impingement plate and vapor belt. (From Butterworth, D. [1991] In *Boilers, Evaporators and Condensers*, pp. 571–633. Wiley, New York; Butterworth, D. [1988] In *Two-Phase Flow Heat Exchangers*, Kluwer, Dordrecht, The Netherlands. With permission.)

FIGURE 11.7
Extra tube support plate to help prevent vibration of tubes near the inlet nozzle.

FIGURE 11.8
Vertical E-shell as condenser. (From Butterworth, D. [1991] In *Boilers, Evaporators and Condensers*, pp. 571–633. Wiley, New York; Butterworth, D. [1988] In *Two-Phase Flow Heat Exchangers*, Kluwer, Dordrecht, The Netherlands. With permission.)

draining from the tubes by gravity and vapor shear (Figure 11.9). If the shell-side cleaning can be done chemically, a fixed tube sheet construction can be designed. Two tube-side passes, using U-tubes, would be possible where there is upward flow in the first pass and downward flow in the second. Tube-side condensers must also have adequate venting. The vent line in vertical units with downflow should be placed in the oversized lower header above any possible pool of condensate formed here.

Vertical tube-side condensers may also be designed to operate in the reflux mode, with an upward flow of vapor but a downward counterflow of any condensate formed on the tube walls. Clearly, such units can only operate provided the flooding phenomena is avoided (Figure 11.10). Capacity is limited due to flooding of the tubes.

The vapor condenses on the cold tube walls, and the condensate film drains downward by gravity. Any noncondensable gases pass up the tube and can be evacuated through a vent.

The tube extends beyond the bottom tube sheet and is cut at a certain angle to provide drip points for the condensate. The shell should be vented through the upper tube sheet. Reflux condenser tubes are usually short, only 2 to 5 m long, with large diameters. In this type of condenser, it must be ensured that the vapor velocity entering the bottom of the tubes is low so that the condensate can drain freely from the tubes.[5]

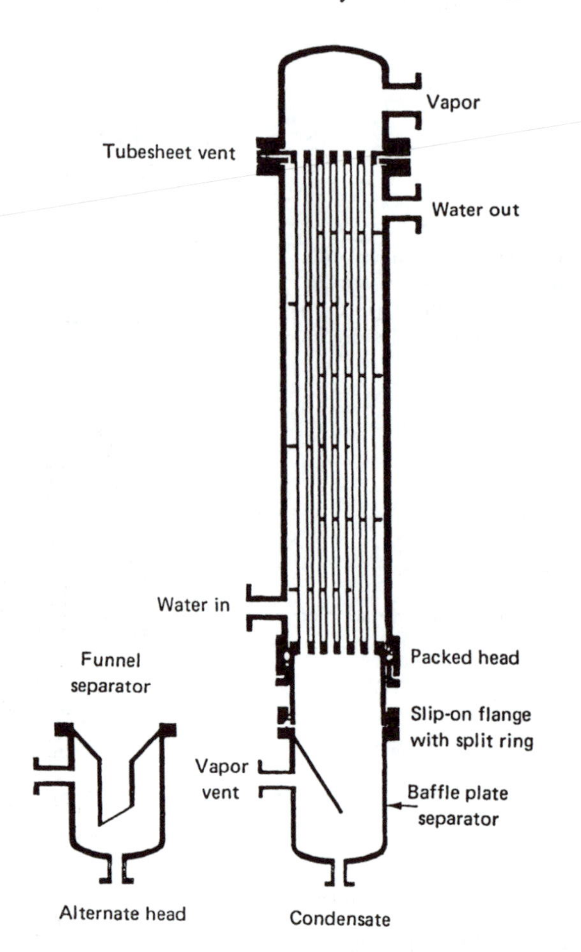

FIGURE 11.9
Vertical in-tube downflow condenser. (From Mueller, A. C. [1983] In *Heat Exchanger Design Handbook*, pp. 3.4.2–1. Hemisphere, Washington, D.C. With permission from Begell House, NY.)

11.2.4 Horizontal In-Tube Condensers

Horizontal tube-side condensation is most frequently used in air-cooled condensers and in kettle or horizontal thermosiphon reboilers.

Horizontal in-tube condensers can be in single-pass, multipass, or U-tube arrangements. More than two passes is unusual. In U-tube bundles, the different lengths between the inner and outer U-tubes will result in different condensing than the same area in a straight-tube, single-pass bundle.[3,4]

Figure 11.2 shows a typical two-pass arrangement for horizontal tube-side condensation. The vapor coming in is being partially condensed in the first pass in the top of the bundle. When the vapor–condensate mixture exits from the first pass into the turnaround header, it is difficult — by existing methods — to estimate the distribution of the liquid and the vapor among the tubes of the second pass. The condensate dropout between passes will cause maldistribution, and its effect on the condensation processes must be considered. The maldistribution of the condensate and vapor into certain tubes alters the heat transfer and fluid flow characteristics of these tubes, resulting in a net reduction of the average effective heat transfer coefficient for the entire pass. If a multicomponent mixture is being condensed, the separation of the liquid and vapor will result in a partial fractionation of the

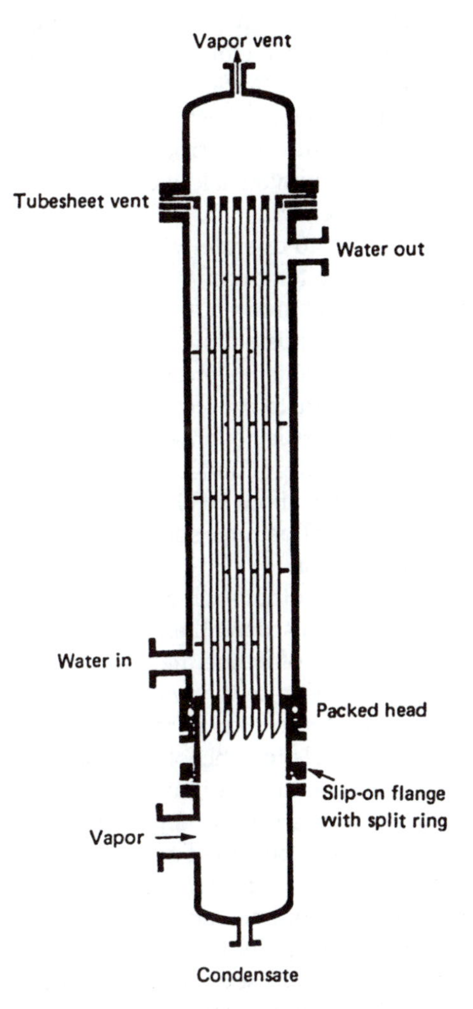

FIGURE 11.10
Reflux condenser. (From Mueller, A. C. [1983] In *Heat Exchanger Design Handbook*, pp. 3.4.2–1. Hemisphere, Washington, D.C.; Breber, G. [1988] In *Heat Transfer Equipment Design*, pp. 477–496. Hemisphere, Washington, D.C. With permission from Begell House, NY.)

mixture being condensed. It is possible in that case to have substantial quantities of non-condensed vapor exiting the condenser.

A vent must be located where the noncondensables are finally concentrated. Tube layouts depend on the coolant requirements. Because of the blanketing effect of condensate accumulation in the bottom portion of the tubes, this configuration is not very effective in gravity-controlled flow.

It is desirable to subcool any condensate leaving a condenser to prevent flashing in the piping circuit and equipment downstream of the condenser.

For condensation inside vertical tubes, subcooling can be achieved by having a liquid level control and running the tubes full of liquid up to a certain height. However, this is clearly only possible if there is a noncondensable present; otherwise there is no convenient way of venting such noncondensables.

11.3 Steam Turbine Exhaust Condensers

For historical reasons, steam turbine exhaust condensers are often referred to as surface condensers. In principle, they are no different from the shell-side condensers described in Section 11.2.1, in particular, the X-type (Figure 11.5). In practice, there are certain severe demands placed on these units that have been overcome by special design features. These special demands arise from the large heat duties that they must perform, and from the necessity to maintain a low-condensing temperature to achieve the highest possible power station efficiency.[1]

The aim is to operate with the condensing temperature only a few degrees above the cooling water temperature. Typically, the cooling water is about 20°C, with condensation taking place at around 30°C. Saturation pressure of water at this temperature is 0.042 bar absolute, which is a typical operating pressure for these condensers. Clearly, there is little pressure available for pressure drop through the unit. There is also small temperature difference to spare to overcome the effect of noncondensable gases. Hence, the design of surface condensers is governed by the need for good venting and low-pressure drop.

These very large condensers often have box-shaped shells, the smaller ones, with surface areas less than about 5000 m^2, may have cylindrical shells.[1]

Figure 11.5 illustrates some of the main features of a surface condenser that are common to most designs.

As with any other shell-and-tube unit, tubes in surface condensers must be supported at regular intervals along their length with tube support plates. Such support plates also have the advantage of deliberately preventing any axial flow of vapor, thus making it easier for designers to ensure that vapor flow paths through the bundles are relatively straightforward, giving rise to no recirculation pockets where noncondensables can accumulate.

It is desirable to subcool any condensate leaving a condenser to prevent flashing in the piping circuit and equipment downstream of the condenser. One solution is to design a separate unit as a subcooling exchanger. One can also overdesign the condenser; a subcooling section can be added as a second pass. This second section often has closer baffling to increase the heat transfer coefficient in this region. For example, the unit may be designed with all the required condensation achieved in the first pass. The vent line would then be placed in the header at the end of this first pass. The liquid level would be maintained in this header and the second pass would be running full of liquid. With this particular method of achieving subcooling, however, a much smaller number of tubes would normally be required in the second pass to give high velocities to ensure good heat transfer.

There is such a variety of different surface condenser designs that it is impossible to illustrate them all here, but many examples of modern condensers are described by Sebald.[6]

11.4 Plate Condensers

There are three main types of plate exchangers: plate-and-frame, spiral plate, and plate–fin exchangers, which are introduced in Chapters 1 and 10 (see Figures 1.13, 1.15, and 10.1).

Plate-and-frame (gasketed) heat exchangers are usually limited to fluid streams with pressures below 25 bar and temperatures below 250°C. Detailed information is given by Alfa Laval.[7]

Gasketed heat exchangers are mainly developed for single-phase flows. They are not well suited for condensers, because of the size of the parts in the plates that are small for handling large-volume flows of vapor. However, they are frequently used with service steam on one side to heat some process stream.

A *spiral* heat exchanger as a condenser is shown in Figure 11.11. This type can operate at pressures up to 20 bar and temperatures up to 400°C.

Plate–fin heat exchangers are discussed in Chapter 1. These heat exchangers are very compact units having heat transfer surface density around 2000 m^2/m^3.[1] They are often used in low-temperature (cryogenic) plants where the temperature difference between the streams is small (1 to 5°C).

The flow channels in plate–fin exchangers are small and often contain many irregularities. This means that the pressure drop is high with high flow velocities. They are not mechanically cleanable; therefore, their use is restricted to clean fluids.

Plate–fin exchangers can be arranged as parallel and counterflow or the streams may be arranged in crossflow (Figures 1.17 and 1.25c).

The corrugated sheets that are sandwiched between the plates serve both to give extra heat transfer and to provide structural support to the flat plate. The most common corrugated sheets are shown on Figure 1.28.

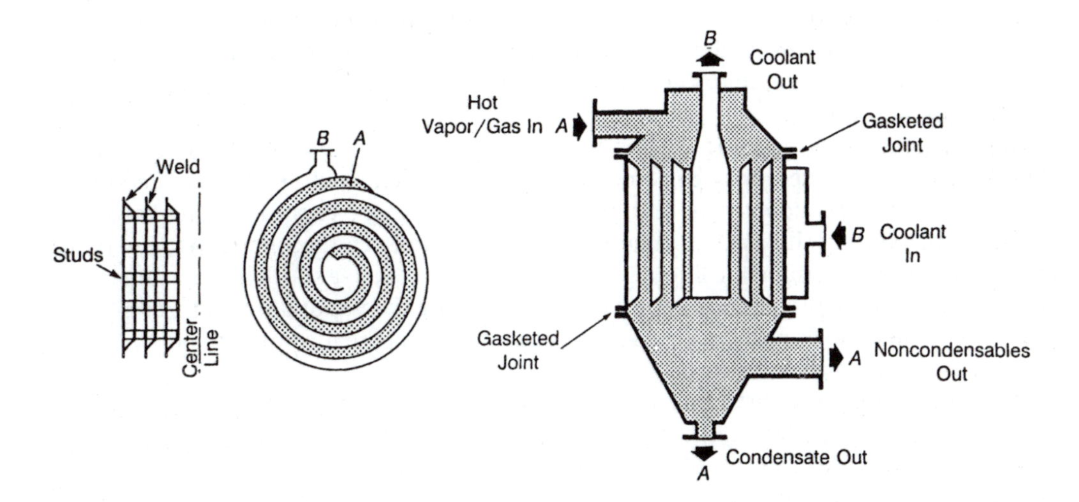

FIGURE 11.11
Spiral heat exchanger for condenser applications. (Courtesy of APV International.)

Uniform flow distribution is very essential for good performance of plate–fin exchangers as condensers. When the nominal flow rate of coolant in the adjacent channels is small, partial condensation can occur that decreases the unit efficiency and can lead to several operating problems.

11.5 Air-Cooled Condensers

Air-cooled condensers may be favored where water is in short supply. Condensation occurs inside tubes that are transversally finned on the air side for a low air-side heat transfer coefficient. Figure 1.23 illustrates a typical air-cooled heat exchanger. The unit shown is a forced draft unit because the air is blown across. The alternative design includes a fan at the top that is called induced draft. Air-cooled condensers can also become economical if condensation takes place at temperatures that are about 20°C higher than ambient temperature.[4]

Disadvantages of the air-cooled condensers are their requirements of a relatively large ground area and space around the unit, noise from the fans, and problems with freezing of condensate in climates with low temperatures.

Air-cooled heat exchangers as condensers normally have only a few tube rows, and the process stream may take one or more passes through the unit. With multipass condensers, a problem arises with redistributing the two-phase mixture on entry to the next pass. This can be overcome in some cases by using U-tubes or by having separate tube rows only for the condensate (Figure 1.23). There is only one arrangement practically possible: each successive pass must be located below the previous one to enable the condensate to continue downward.

11.6 Direct Contact Condensers

Direct contact condensers are inexpensive and simple to design, but have limited application because the condensate and coolant are mixed. The main advantages of these condensers besides their low cost is that they cannot be fouled and that they have very high heat transfer rates per unit volume. There are generally three types of direct contact condensers: pool, spray, and tray.

In pool condensers, vapor is injected into a pool of liquid. The coolant may be a process fluid to be heated. There may be some operational problems with this type of condenser. The first is that the condensation front may move back into the vapor inlet line, causing the liquid to be periodically ejected, often with some violence. The second is that a very large vapor bubble may form in the liquid pool, and this may collapse suddenly causing damage to the vessel. These problems may be avoided by having the vapor injected through a large number of small holes or by using special ejectors that mix the incoming vapor with liquid in a special mixing tube.[1,4]

The most common type of direct contact condenser is one in which subcooled liquid is sprayed into the vapor in a large vessel. This arrangement is illustrated in Figure 1.7a. Very often these units are used for condensing steam using water as a coolant. In these cases, the mixing of water with condensate presents no major problem. When condensing a vapor whose condensate is immiscible with the spray liquid, however, a separator is usually required after the condenser to recover the product. Alternatively, the condensate product may be cooled in a single-phase exchanger and some recycled as coolant spray. At first observation, there then seems little benefit in using a combination of a direct contact condenser and a conventional single-phase exchanger instead of using only one shell-and-tube condenser. The advantage appears, however, when the condenser is operating under vacuum. As has been seen already, tubular condensers for vacuum operation are large and complex. It can therefore sometimes be economical to replace such a condenser by a simple spray condenser and a compact single-phase cooler.[1]

Spray condensers cannot be used with dirty coolants because the spray nozzles may be blocked. In those circumstances, a tray condenser may be used, as illustrated in Figure 1.7b. The trays may be sloped slightly to prevent dirt accumulation on them. The tray arrangement can have a slight thermodynamic advantage over a spray unit because some degree of countercurrent flow may be achieved between the falling liquid and upward flowing vapor–gas mixture.

11.7 Thermal Design of Shell-and-Tube Condensers

Design and rating methods of a heat exchanger have already been introduced in Chapter 8. The rating problem is a performance analysis for a known exchanger geometry with given specifications. The condenser dimensions of the known condenser or tentative estimates of condenser dimensions are used as input data to the rating program. The thermal duty and outlet conditions of the condensate are calculated if the length is fixed. If the heat duty is fixed, the length of the condenser can be calculated (Figure 8.10).

Besides these parameters, the rating program will also calculate pressure drops for the condensing vapor side and the coolant side, and the outlet temperature of the coolant; and an advanced program will calculate many other useful parameters.

Some advanced programs provide design modifications. They take the output from the rating subroutines and modify the condenser parameters in such a way that the new condenser configuration will better satisfy the required process conditions.

Two approaches are used in designing heat exchangers, especially two-phase flow heat exchangers such as condensers. In the first approach, the heat exchanger is taken as a single control volume with average overall heat transfer coefficient and two inlets and two outlets (lumped analysis). In the second analysis, the heat exchanger is divided into segments or multiple control volumes, with the outlet of one control volume being the inlet to an adjacent control volume. This is a local analysis. The heat transfer rate (heat duty) for the heat exchanger is obtained by integrating the local values. The first approach is more common and simpler, and it generally provides acceptable results for design.

Let us treat the first approach (lumped analysis). Heat transfer rate can be expressed for two control volumes defined on the condensing side (hot fluid) and coolant side, respectively, as follows:

$$Q_h = \dot{m}_h (i_i - i_o) \tag{11.1}$$

$$Q_c = (\dot{m}c_p)_c (T_{c_2} - T_{c_1}) \tag{11.2}$$

The basic design equation for a heat exchanger in terms of mean quantities that are related by Equation (2.7) is

$$Q = U_m A_o \Delta T_{lm} \tag{11.3}$$

where U_m is the mean overall heat transfer coefficient and ΔT_{lm} is the mean temperature difference that is the well-known log mean temperature difference (LMTD) given by Equation (2.28):

$$\Delta T_{lm} = \frac{\Delta T_1 - \Delta T_2}{\ln(\Delta T_1 / \Delta T_2)} \tag{11.4}$$

The overall heat transfer coefficient, U_m, is built up of a number of components, including the individual heat transfer coefficients, and wall and fouling resistances; and is given by Equation (2.17) for a bare tube:

$$\frac{1}{U_m} = \frac{1}{h_o} + R_{fo} + A_o R_w + \left(R_{fi} + \frac{1}{h_i} \right) \frac{A_o}{A_i} \tag{11.5}$$

where U_m is based on the outside heat transfer surface area.

A simplified form of Equation (11.5) can also be used:

$$\frac{1}{U_m} = \frac{1}{h_o} + R_t + \frac{1}{h_i} \tag{11.6}$$

where R_t is the combined thermal resistance of the tube wall and fouling and U_m is defined as:

$$U_m = \frac{1}{A_o} \int_{A_o} U dA \tag{11.7}$$

where U is the local overall heat transfer coefficient and may vary considerable along the heat exchanger.

For variable U, local heat transfer from a differential surface area, dA, is given by:

$$\delta Q = U dA dT \tag{11.8}$$

The total heat transfer can be calculated by the procedure outlined in Chapter 2, Section 2.8.

If U varies linearly with A, Equation (11.7) can be integrated between the overall heat transfer coefficients at the inlet and outlet of the condenser.

$$U_m = \frac{1}{2}(U_1 + U_2) \tag{11.9}$$

For the overall heat transfer coefficient U and ΔT varying linearly with Q, Equation (2.61) can be used.

If both $1/U$ and ΔT vary linearly with Q, Equation (11.5) can be integrated with the aid of Equation (11.8) to give:

$$\frac{1}{U_m} = \frac{1}{U_1}\frac{\Delta T_{l,m} - \Delta T_2}{\Delta T_1 - \Delta T_2} + \frac{1}{U_2}\frac{\Delta T_1 - \Delta T_{l,m}}{\Delta T_1 - \Delta T_2} \tag{11.10}$$

Equation (11.9) is simple to use, but it is not valid if there are large differences between U_1 and U_2. It is difficult to decide which of the preceding equations for U_m is valid for a given circumstance.[1]

Example 11.1

A shell-and-tube type of condenser is to be designed for a coal-fired power station of 200 MW. Steam enters the turbine at 5 MPa and 400°C (i_i = 3195.7 kJ/kg). The condenser pressure is 10 kPa (0.1 bar). The thermodynamic efficiency of the turbine is $\eta_t = 0.85$. The actual enthalpy of steam entering the condenser at 0.1 bar is calculated to be i_e= 2268.4 kJ/kg with 0.80% quality. The condenser is to be designed without subcooling. A single-tube pass is used and the cooling water velocity is assumed as 2 m/s. Cooling water is available at 20°C and can exit the condenser at 30°C. Allowable total pressure drop on the tube side is 35 kPa.

SOLUTION

Mass flow rate of wet steam entering the condenser can be found from the first law of thermodynamics:

$$\dot{m}_s = \frac{W_T}{\Delta i} = \frac{200 \times 10^3}{3195.7 - 2268.4} = 215.68 \text{ kg/s}$$

Cooling water properties at the mean temperature of 25°C from Appendix B:

c_{pc} = 4180 J/kg·K
μ_c = 0.00098 N·s/m^2
k_c = 0.602 W/m·K
ρ_c = 997 kg/m^3
Pr_c = 6.96

Saturated liquid properties of the condensed water at 10 kPa, T_s = 45.8°C, from Appendix B:

i_{lg} = 2392 kJ/kg
μ_l = 0.000588 N·s/m^2
k_l = 0.635 W/m·K
ρ_l = 990 kg/m^3
i_l = 191.8 kJ/kg
O.D. of the tubes, d_o= 0.0254 m
I.D. of the tubes, d_i = 0.02291 m
tube wall thermal conductivity, k = 111 W/m·K

The cooling water velocity $U_c = 2\text{m/s}$, which is chosen as a compromise between fouling that increases at low velocities, and erosion that occurs at high velocities. Fouling resistances chosen from TEMA tables in Chapter 5 are

$$R_{fi} = 0.00018 \text{ m}^2\cdot\text{K/W}$$

$$R_{fo} = 0.00009 \text{ m}^2\cdot\text{K/W}$$

Condenser heat load, Q is

$$\begin{aligned} Q &= \dot{m}(i_{in} - i_l) \\ &= 215.68(2268.4 - 191.8) \text{ kW} \\ &= 447.9 \text{ MW} \end{aligned}$$

Cooling water mass flow rate, $\dot{m}_c$ is calculated by:

$$\begin{aligned} \dot{m}_c &= \frac{Q}{(T_{c_2} - T_{c_1})c_{pc}} \\ &= \frac{(4.479 \times 10^5) \times 10^3}{(30 - 20) \times 4180} \\ &= 10717.4 \text{ kg/s} \end{aligned}$$

The number of tubes, N_T, is determined from the fixed cooling water velocity U_c, as follows (one-tube pass):

$$\dot{m}_c = U_c \rho_c \left(\frac{\pi d_i^2}{4} \right) N_T$$

or

$$\begin{aligned} N_T &= \frac{4\dot{m}_c}{U_c \rho_c \pi d_i^2} \\ &= \frac{4 \times 1.07174 \times 10^4}{997 \times 2 \times \pi \times (0.02291)^2} \\ &= 13038 \text{ tubes} \end{aligned}$$

To calculate the condensing-side heat transfer coefficient, an estimate of the average number of tube rows in a vertical column, N_T, is needed. For typical condenser tube layouts, this was estimated as 70.

The coolant heat transfer coefficient, h_i, can be estimated from single-phase, in-tube heat transfer correlations given in Chapter 2.

Tube-side Reynolds number is calculated first:

$$Re = \frac{U_c \rho_c d_i}{\mu_c}$$

$$= \frac{0.02291 \times 2 \times 997}{0.00098}$$

$$= 46614.8$$

Therefore, the flow inside the tubes is turbulent. The Petukhov–Kirillov correlations, for example, can be used to calculate the heat transfer coefficient:

$$Nu = \frac{(f/2) Re\, Pr}{1.07 + 12.7(f/2)^{1/2}(Pr^{2/3} - 1)}$$

where

$$f = (1.58 \ln Re - 3.28)^{-2}$$

$$= [11.58 \ln(46614.8) - 3.28]^{-2}$$

$$= 0.00532$$

Thus

$$f/2 = 0.00266$$

$$Nu = \frac{0.00266 \times 46614.8 \times 6.96}{11.07 + 12.7(0.00266)^{1/2}[(6.96)^{2/3} - 1]} = 304.4$$

Hence

$$h_i = \frac{Nu \cdot k_c}{d_i}$$

$$= \frac{304.4 \times 0.602}{0.02291}$$

$$= 8027 \text{ W/m}^2 \cdot \text{K}$$

The condenser is to be designed without subcooling, so that:

$$\Delta T_{in} = (45.8 - 20) = 25.8°\text{C}$$

and

$$\Delta T_{out} = (45.8 - 30) = 15.8°\text{C}$$

Hence the LMTD is

$$\Delta T_{lm} = \frac{\Delta T_{in} - \Delta T_{out}}{\ln(\Delta T_{in} / \Delta T_{out})} = \frac{25.8 - 15.8}{\ln(25.8/15.8)} = 20.4°\text{C}$$

Next, the shell-side heat transfer coefficient is determined to obtain the overall heat transfer coefficient. This coefficient depends on the local heat flux and hence an iteration are necessary. The equation required in this iteration is developed first. The overall heat transfer coefficient, U, based on the tube O.D., is given by:

$$\frac{1}{U} = R_t + \frac{1}{h_o}$$

where h_o is the coefficient outside the tubes (shell side) and R_t is the sum of all other thermal resistances given by Equation (11.3):

$$R_t = R_{fo} + \left[\frac{1}{h_i} + R_{fi}\right]\frac{d_o}{d_i} + \frac{t_w}{k_w}\frac{d_o}{D_m}$$

where D_m is approximated as:

$$D_m = \frac{d_o - d_i}{\ln(d_o / d_i)} \approx \frac{1}{2}(d_o + d_i)$$

and t_w is the wall thickness.

Hence

$$R_t = 0.00009 + \left[\frac{1}{8027} + 0.00018\right]\frac{0.0254}{0.0229} + \frac{0.0013}{111}\cdot\frac{0.0254}{0.0242}$$

$$= 4.39 \times 10^{-4}$$

and

$$\frac{1}{U} = 4.39 \times 10^{-4} + \frac{1}{h_o} \quad (1)$$

The condensing-side heat transfer coefficient, h_o, may be calculated by the Nusselt method with the Kern correction for condensate inundation (Equations 7.2 and 7.10). Hence

$$h_o = 0.728\left[\frac{\rho_l^2 g i_{lg} k_l^3}{\mu_l \Delta T_w d_o}\right]^{1/4}\frac{1}{N^{1/6}}$$

where ΔT_w is the difference between the saturation temperature and the temperature at the surface of the fouling. Because $\rho_l \gg \rho_g$, Equation (7.2) has been simplified. By inserting the values, we get:

$$h_o = 0.728\left[\frac{(990)^2(9.81)(2392\times 10^3)(0.635)^3}{(5.66\times 10^{14})\Delta T_w(0.0254)}\right]^{1/4}\frac{1}{70^{1/6}} \tag{2}$$
$$= 8990/\Delta T_w^{1/4}$$

The temperature difference, ΔT_w, is given by:

$$\Delta T_w = \Delta T - R_t q''$$

where ΔT is the local temperature difference between the streams; R_t is the sum of all other resistances; and q'' is the local heat flux, which is given by:

$$q'' = U\Delta T$$

Hence

$$\Delta T_w = \Delta T(1 - R_t U)$$
$$= \Delta T(1 - 4.39\times 10^{-4} U) \tag{3}$$

A suggested iteration at the inlet and outlet is therefore:

1. Guess ΔT_w
2. Calculate h_o from Equation (2)
3. Calculate U from Equation (1)
4. Recalculate ΔT_w from Equation (3)
5. Repeat the calculations from step 2 and continue the iteration until U converges.

Tables 11.1 and 11.2 summarize the results of this iteration for the inlet and outlet of the condenser when ΔT_{in} = 25.8°C and ΔT_{out} = 15.8°C, respectively. The initial guess of ΔT_w is 10°C.

TABLE 11.1

Iteration for Overall Coefficient at the Inlet of the Power Condenser

ΔT_w, °C (Eq. [3])	h_o, W/m²·K (Eq. [2])	U, W/m²·K (Eq. [1])
10	5049	1569
9.47	5118	1575
9.45	5121	1575
9.44	5122	1575

TABLE 11.2

Iteration for Overall Coefficient at the Outlet of the Power Condenser

ΔT_w, °C (Eq. [3])	h_o, W/m²·K (Eq. [2])	U, W/m²·K (Eq. [1])
6.0	5737	1629
5.89	5764	1631
5.86	5771	1632
5.84	5776	1632

The mean overall heat transfer coefficient can then be determined by taking the average of the inlet and outlet overall heat transfer coefficients:

$$U_m = \frac{1575 + 1632}{2} = 1603 \text{ W/m}^2 \cdot \text{K}$$

The required surface area is therefore calculated from Equation (11.3):

$$Q = U_m A_o \Delta T_{lm}$$

$$A_o = \frac{Q}{U_m \Delta T_{lm}}$$

$$= \frac{447.9 \times 10^6}{1603 \times 20.4} = 1.370 \times 10^5 \text{ m}^2$$

The required length is found by the following equation:

$$A_o = N_T \pi d_o L$$

or

$$L = \frac{A_o}{N_T \pi d_o} = \frac{1.370 \times 10^5}{15340 \times 0.0254} = 13.2 \text{ m}$$

Now, the thermal resistances of the heat exchanger will be analyzed. This is useful when cleaning practices must be considered for the exchanger. Table 11.3 shows this for the inlet of the condenser.

As can be seen from Table 11.3, a considerable amount of fouling is due to the inside tubing. Therefore, it is necessary to concentrate on the tube side when cleaning the exchanger. Several possible methods of cleaning can be used. One method of cleaning will include the cooling water chemistry. The second method involves passing balls down the tubes.

Now that the length of the power condenser has been determined, the shell size of the condenser, the tube-side pressure drop, and the tube-side pumping power can all be calculated.

TABLE 11.3

Comparison of Thermal Resistances at the Inlet of the Condenser

Item	Given by	Value (m²·K/kW)	%
Tube-side fluid	$\frac{d_o}{h_i d_i}$	0.138	25
Tube-side fouling	$\frac{R_{fi} d_o}{d_i}$	0.200	36
Tube wall	$\frac{t_w d_o}{k_w D_m}$	0.012	2
Shell-side fouling	R_{fo}	0.09	16
Shell-side fluid	$\frac{1}{h_o}$	0.195	35

The expression of the shell diameter as a function of heat transfer area, A_o; tube length, L; and tube layout dimensions, P_T, PR, and d_o as parameters can be estimated from Equation (8.10):

$$D_s = 0.637\sqrt{\frac{CL}{CPT}}\left[\frac{A(PR)^2 d_o}{L}\right]^{1/2} \tag{4}$$

where CL is the tube layout constant, $CL = 1.00$ for 90° and 45°; and $CL = 0.87$ for 30° and 60°.

CPT accounts for the incomplete coverage of the shell diameter by the tubes:

$CPT = 0.93$ for 1-tube pass
$CPT = 0.9$ for 2-tube pass
$CPT = 0.85$ for 3-tube pass
P_T (the tube pitch) is 0.0381 m.
PR (the tube pitch ratio) is $P_T/d_o = 0.0381/0.0254 = 1.501$.

Then using the values given for the preceding parameters in Equation (4), the shell diameter is determined as:

$$D_s = 0.637\sqrt{\frac{1}{0.93}}\left[\frac{13696.6 \times 1.501^2 \times 0.0254}{13.2}\right]^{1/2}$$

$$= 5.1 \text{ m}$$

The pressure drop on the tube side is calculated as follows:

$$\Delta p_{tot} = \Delta p_t + \Delta p_r$$

The pressure drop through the tubes can be calculated from Equation (8.19):

$$\Delta p_t = 4f \frac{LN_p}{D_e} \frac{G^2}{2\rho}$$

where

$$f = 0.046\, Re^{-0.2}$$
$$= 0.046 \times (46.6 \times 10^3)^{-0.2} = 0.00535$$

also

$$N_p = \text{number of passes}$$
$$D_e = d_i$$
$$G = U_c \rho$$

Therefore

$$\Delta p_t = 4 \times 0.00535 \times \frac{13.2 \times 1}{0.02291} \times \frac{(2 \times 997)^2}{2 \times 997}$$
$$= 24559.3 \text{ Pa}$$

The pressure drop due to the return is given by Equation (8.21):

$$\Delta p_r = 4N_p \frac{\rho U_c^2}{2}$$
$$= 4 \times 1 \times \frac{997 \times 2^2}{2}$$
$$= 7976 \text{ Pa}$$

Therefore, the total pressure drop on the tube side is determined by:

$$\Delta p_{tot} = \Delta p_t + \Delta p_r$$
$$= 24559.3 + 7976$$
$$= 32535.3 \text{ Pa}$$

The pumping power is proportional to the pressure drop across the condenser:

$$P = \frac{\dot{m} \Delta p_{tot}}{\rho \eta_p}$$

where η_p is the efficiency of the pump, assumed to be 85%.

$$P = \frac{10717.4 \times 56602.1}{997 \times .85} = 829597 \text{ W}$$

This procedure was done for cooling water velocities of 1.5 and 2.5 m/s, and the results are given in Table 11.4.

During the rating procedures, the thermal performance and the pressure drop for both streams are calculated for this design. The better the starting point estimate is, the sooner the final design will be done. On a computer, however, it is usually faster to let the computer select a starting point — usually a very conservative case — and use its enormous computational speed to move toward the suitable design. If the calculation shows that the required amount of heat cannot be transferred or if one or both allowable pressure drops are exceeded, it is necessary to select new geometric parameters and to recalculate the condenser. On the other hand, if pressure drops are much smaller than allowable, a better selection of parameters may result in a smaller and more economical condenser while utilizing more of the available pressure drop.

The best economic choice then is a result of the additional step that is an economic evaluation of several designs. The final design must meet several conditions:

1. The condenser must be operable in a range of conditions that include the design conditions.
2. The design must meet all specific physical limitations that may be imposed, such as weight, size, coolant process constraints, and metallurgical constraints.
3. The design must allow standard or economical fabrication.
4. Operation and maintenance expenses must not exceed economical limits.

11.8 Design and Operational Considerations

Mueller[3] and Taborek[8] described a number of requirements that should be considered during selection and design practices of condensers.

Condensation modes — Condensers are designed under filmwise condensation conditions. Although dropwise condensation gives higher heat transfer coefficients, it is not possible to sustain this condensation mode for a long period of time on individual condensers.

TABLE 11.4
Effect of Coolant Velocity on Design

	Cooling Water Velocity		
Parameters	U_c = 1.5 m/s	U_c = 2.0 m/s	U_c = 2.5 m/s
Number of tubes, N_t	17385	13038	10430.7
Reynolds no., *Re* (coolant)	34961.2	46614.8	58268.5
Heat transfer coefficient (coolant), h_c (W/m²·K)	6307.6	8026.7	9685.3
Heat transfer coefficient (shell), h_o (W/m²·K)	5441	5441	5443
Overall heat transfer coefficient, U_m (W/m²·K)	1476	1603.5	11666.5
Heat transfer area, A_o (m²)	14525.5	13696.6	13178.8
Length, L (m)	10.5	13.2	15.8
Shell diameter, D_s (m)	5.9	5.1	4.6
Pressure drop (tubes), ΔP_t (kPa)	16.1	32.5	56.6
Pumping power (tubes, η_p = 85%), P_t (kW)	378.5	829.6	1540.4

Condensation Regimes — Depending on the flow characteristics of the vapor and condensate, the designer must determine the flow regime applicable along the vapor flow path. At low vapor velocities the so-called "gravity controlled" or Nusselt flow regime exists. At high vapor velocities, the "vapor shear controlled" regime will predominate (see Section 7.4.2).

De-super-heating — Some vapor stream enters the condenser superheated. If the wall temperature is below the dew point, condensation will take place and the de-super-heating duty must be calculated as a single-phase process.

Subcooling — It is sometimes desired to subcool the condensate slightly before further processing. This can be accomplished by raising the level of the condensate so that it will be in contact with the cool tubes. For larger subcooling heat applications it is more efficient to allocate a separate unit.

Construction considerations — Some guidelines and practices must be observed for design of well-functioning condensers:

1. Vertical in-tube condensation is very effective, but the tube length is limited because it may fill up with condensate. Thus the size of such condensers is restricted, because otherwise large shell diameters would be required.
2. Horizontal tube-side condensation is less effective and much more difficult to calculate because of the stratification of the condensate. Positive tube inclination must be used.
3. Horizontal shell-side condensation is very popular because it is well predictable and permits use of large surfaces; and the extremely low-pressure drop required for vacuum operations can be obtained by proper unit selection (TEMA X- shell).

If noncondensable gases are present in vapors, the condensing coefficient will vary an order of magnitude between vapor inlet and outlet. Stepwise calculation is unconditionally required. A detailed stepwise analysis is given in Reference 1. Furthermore, the unit construction must be suitable for effective removal (venting) of the gases that would otherwise accumulate and render the condenser inoperative.

Some of the main reasons of failure of condenser operation are given in Reference 9 and briefly summarized here:

1. The tubes may be fouled more than expected — a problem not unique to condensers.
2. The condensate may not be drained properly, causing tubes to be flooded. This could mean that the condensate outlet is too small or too high.
3. Venting of noncondensables may be inadequate.
4. The condenser was designed on the basis of end temperatures without noticing that the design duty would involve a temperature cross in the middle of the range.
5. Flooding limits have been exceeded for condensers with a backflow of liquid against upward vapor flow.
6. Excessive fogging may occur. This can be a problem when condensing high-molecular-weight vapors in the presence of noncondensable gas.
7. There is a possibility of severe maldistribution in parallel condensing paths, particularly with vacuum operation. This occurs because there can be two flow rates that satisfy the imposed pressure drops. An example might be that in which the pressure drop is zero. One channel may have a very high vapor inlet flow

but achieve a zero pressure drop because the momentum pressure recovery cancels out the frictional pressure loss. The next channel may achieve the zero pressure drop by having no vapor inlet flow. To be stable in this case the channel would be full of noncondensables. This problem may occur with parallel tubes in a tube-side condenser or with whole condensers when they are arranged in parallel.[1]

11.9 Condensers for Refrigeration and Air-Conditioning

Several types of heat exchangers are used in refrigeration and air-conditioning applications. They are two-phase flow heat exchangers — one side is refrigerant and the other side is air or liquid. In general, condensers in refrigeration and air-conditioning industry are of two types: (1) condenser coil where the refrigerant flows through the tubes and air flows over the finned tubes; and (2) shell-and-tube condenser where refrigerant flows on the shell side and the liquid flows through the tubes. The common heat exchanger types used in the refrigeration and air-conditioning industry are shown in Figure 11.12. An illustration of a commercial two-stage vapor compression refrigeration cycle is given in Figure 11.13.

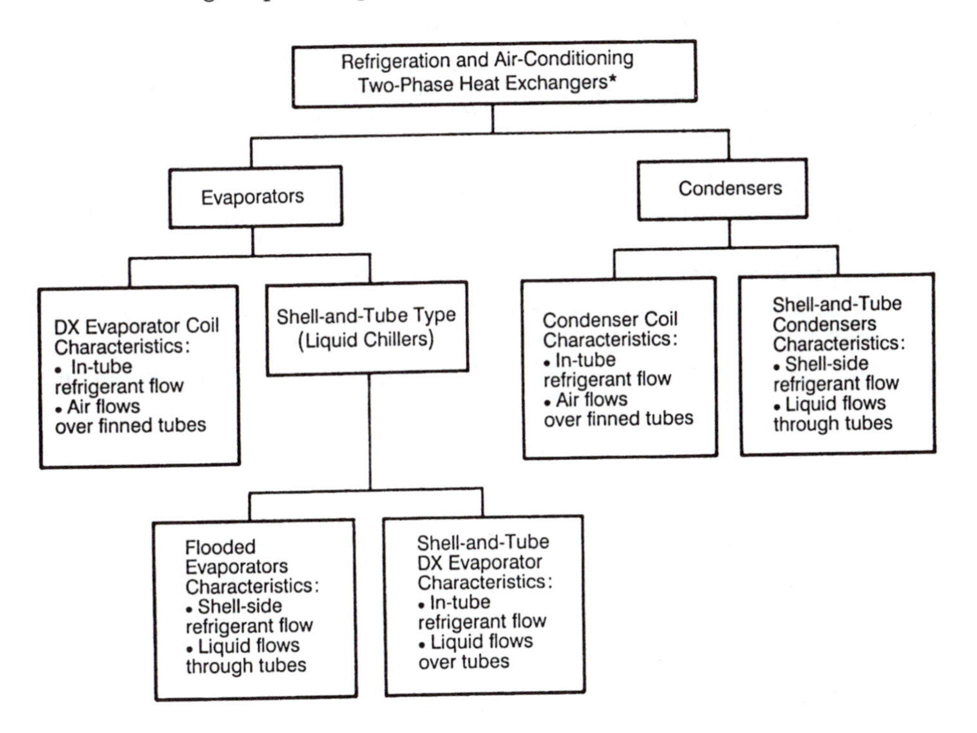

FIGURE 11.12
Common heat exchanger types used in refrigeration and air-conditioning industry. (From Pate, M. B. [1991] In *Boilers, Evaporators and Condensers*, Wiley, New York. With permission.)

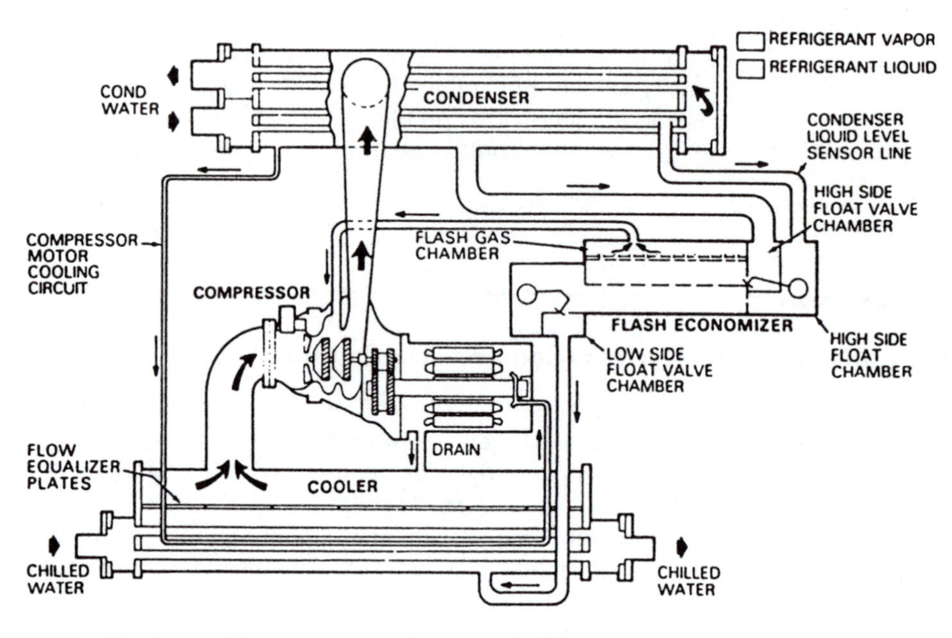

FIGURE 11.13
Illustration of a two-stage vapor compression refrigeration cycle. (Courtesy of Carrier Corp.)

The main heat transfer components of the refrigeration system are the evaporator and condenser that absorb and reject heat, respectively. The evaporator vaporizes the refrigerant and also heat, while the condenser rejects heat and discharges high-pressure and -temperature liquid refrigerant.

Figure 11.13 shows a schematic of a commercial air-conditioning system cycle with all its components.[13]

The maximum efficiency in a refrigeration cycle is ultimately obtained by the combined performance of all the major components; evaporator and condenser (along with the compressor) make up the bulk of the system in terms of size and cost.

The common forms of condensers may be classified on the basis of the cooling medium as:

1. Water-cooled condensers
   - Horizontal shell and tube
   - Vertical shell and tube
   - Shell and coil
   - Double pipe
2. Air-cooled condensers
3. Evaporative condensers (air cooled and water cooled)

11.9.1 Water-Cooled Condensers

Water-cooled condensers are shell-and-tube type (horizontal and vertical), shell-and-U type, shell-and-coil type, and double-pipe type. The examples of shell-and-tube horizontal and vertical condensers that can be used with refrigerant condensing on the shell side and with cooling water circulating inside the tubes are given in Chapter 8 and in Section 11.2.

The shell-and-tube condenser with superheated refrigerant vapor entering and subcooled liquid–refrigerant exiting is the most common type of water-cooled condenser. This type of unit is usually mounted in the horizontal position to facilitate condensate drainage from the tube surface. Shell-and-tube heat exchangers are introduced in Chapter 8, and the geometric parameters are already discussed. In air-conditioning and refrigeration, copper tubes are usually used in shell-and-tube condensers with sizes ranging from 5/8- to 1-in. O.D., with 3/4-in. O.D. being the most popular. Ammonia requires carbon steel tubes. Equivalent triangular pitches are normally used for tube arrangements in the shell. Water is circulated through the tubes in single to multipass circuits.

When fixed-tube sheet, straight-tube construction is used, provisions and space for tube cleaning and replacement must be provided at one end of the condenser. Location of the gas and liquid outlet nozzles should be carefully considered; they should be located far enough apart to allow the entering superheated vapor to be exposed to the maximum tube surface area. As the refrigerant vapor is condensed on the outside of the tubes, it drips down onto the lower tubes and collects at the bottom of the condenser. In some cases, the condenser bottom is designed to act as a reservoir (receiver) for storing refrigerant. It must be ensured that excessive amounts of liquid refrigerant are maintained to seal the outlet nozzle from inlet gas flow.

A shell-and-tube type of vertical condenser with an open water circuit is widely used in ammonia refrigeration systems. Shell-and-tube condensers with U-tubes with a single tube sheet are also manufactured. A shell-and-coil condenser has cooling water circulated through one or more continuous or assembled coils contained within the shell.

A double-pipe or hairpin condenser consists of one or more assemblies of two tubes, one within the other (see Chapter 6). The refrigerant vapor is condensed in either the annular space or the inner tube.

11.9.2 Air-Cooled Condensers

Air-cooled condensers are made of finned tubes. The air side is finned and the liquids flow through the tubes. The heat transfer processes in an air-cooled condenser has three main phases: (1) de-super-heating, (2) condensing, and (3) subcooling.

Condensing takes place in approximately 85% of the condensing area at a relatively constant temperature. As a result of the frictional pressure drop, there will be a small condensing temperature drop along the surface. The geometric parameters of air-cooled condensers are the same as compact heat exchangers, some examples of which are shown in Chapter 9.

Coils are commonly constructed of copper, aluminum, or steel tubes, ranging from 0.25 to 0.75 in. (6.35 to 19.05 mm) in diameter. Copper, the most expansive material, is easy to use in manufacturing and requires no protection against corrosion. Aluminum requires exact manufacturing methods, and special protection must be provided if aluminum-to-copper joints are made. Steel tubing requires weather protection.

Fins are used to improve the air-side heat transfer. Most fins are made of aluminum, but copper and steel are also used.

Air-cooled condensers may be classified as remote from the compressor or as part of the compressor (condensing unit). A more specific subdivision may be forced flow or free convection air flow. In this type, air is passed over the coils where in-tube condensation occurs.

11.9.3 Evaporative Condensers

An evaporative condenser is a type of water-cooled condenser that offers means of conserving water by combining the condenser and the cooling tower into one piece of equipment. The coil containing the refrigeration gas is wet outside; and then the cooling air is forced over the coils, or may flow freely over the coils (free convection), at which time some of the water evaporates, in turn, cooling and condensing the tube-side refrigerant (Figures 11.14 and 11.15). Principal components of an evaporative condenser include the condensing coil, fan, spray water pump, water distribution system, cold water pump, drift eliminators, and water makeup assembly. The heat from a condensing vapor is ejected into the environment. Evaporative condensers eliminate the need for the pumping and chemical treatment of large quantities of water associated with cooling tower and refrigerant condenser systems. They require substantially less fan power than air-cooled condensers of comparable capacity.[14] As can be seen from Figures 11.14 and 11.15, vapor to be condensed is circulated through a condensing coil that has water continually on the outside by recirculating water systems. Air is simultaneously directed over the coil, causing a small portion of the recalculated water to evaporate. This evaporation results in the removal of heat from the coil, thus cooling and condensing the vapor in the coil. The high rate of energy transfer from the wetted external surface to the air eliminates the need for extended surface design. Therefore, evaporative condensers are more easily cleaned. Coil materials are steel, copper, iron, or stainless steel.

Coil water is provided by circulating water by the pump to the spray nozzles located above the coil. The water descends through the air circulated by the fan, over the coil surface, and eventually returns to the pan sump. The water distribution system is designed for complete and continuous wetting of full coil surface. This ensures the high rate of heat transfer achieved with fully wetted tubes and prevents excessive scaling, which is more likely to occur on intermittently or partially wetted surfaces. Water lost through evaporation and

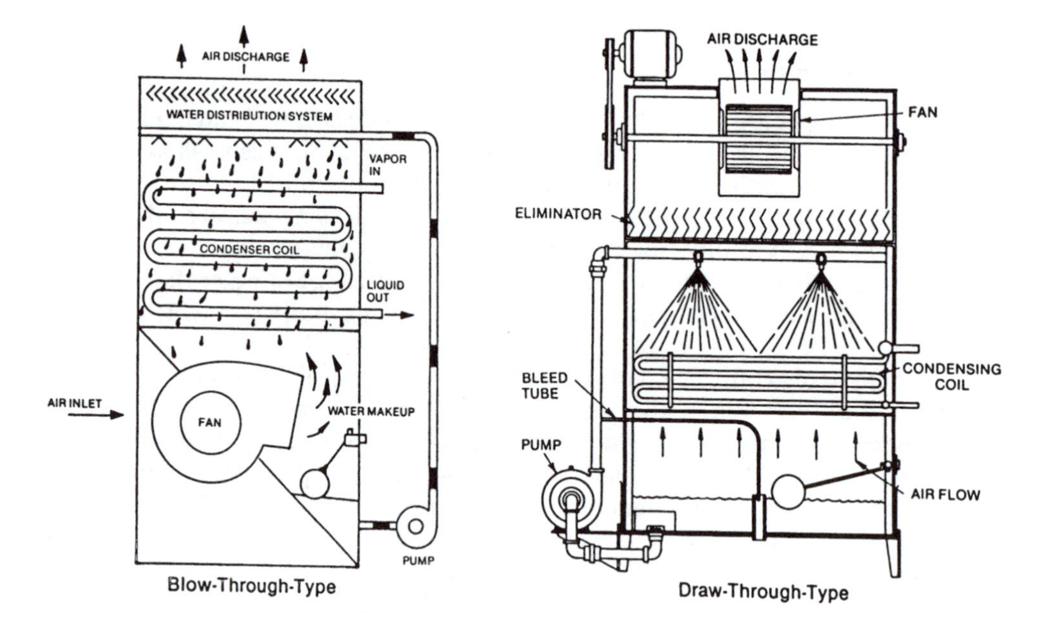

FIGURE 11.14
Operation of evaporative condenser. (From the 1983 ASHRAE Handbook — Equipment [1983] Chapter 16, Condensers, ASHRAE, Atlanta, GA. With permission.)

blowdown from the cold-water pump is replaced through an assembly that typically consists of a mechanical float valve or solenoid valve and float switch combination.

Most evaporative condensers employ fan(s) to either blow or draw air through the unit. Typically, the fans are either the centrifugal or propeller type, depending on the external pressure needs, permissible sound level, and energy usage requirements.

Drift eliminators are used to recover entrained moisture from the air stream. These eliminators strip most of the water from the discharge air system; however, a certain amount is discharged as drift.

Evaporative condensers require substantially less fan horsepower than air-cooled condensers of comparable capacity. Perhaps most importantly, however, system utilizing evaporative condensers can be designed for lower condensing temperatures, and subsequently for lower compressor energy input than system utilizing conventional or water-cooled condensers.

The evaporative condenser allows for a lower condensing temperature than does an air-cooled condenser because the heat rejection is limited by the ambient wet-bulb temperature, which is normally 14 to 25°F (8 to 14°C) lower than the ambient dry-bulb temperature. Though less obvious, the evaporative condenser also provides lower condensing temperature than does the cooling tower or water-cooled condenser because of the reduction of heat-transfer and mass-transfer steps from two steps (the first step is between the refrigerant and cooling water and the second step is between the water and ambient air) to one step (refrigerant directly to ambient air).[13]

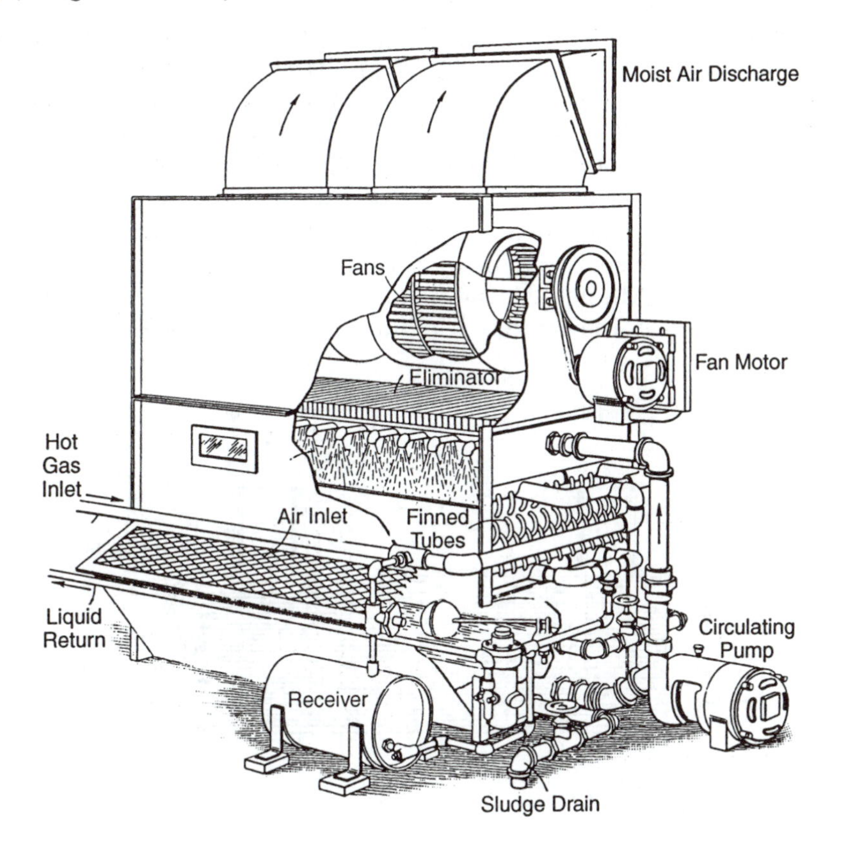

FIGURE 11.15
Evaporative condenser. (From Trane Air-Conditioning Manual [1979] The Trane Co., LaCrosse, WI.)

11.10 Evaporators for Refrigeration and Air-Conditioning

As can be seen from Figure 11.12, two-phase flow heat exchangers in the refrigeration and air-conditioning industry can be classified as coils when the refrigerant boils inside the tubes and air follows over finned tubes. Shell-and-tube evaporators are used when the second fluid is a liquid that flows either through the tubes (shell-and-tube flooded evaporators) or over the tubes (shell-and-tube direct expansion evaporators). Flooded evaporators and direct expansion (DX) evaporators perform similar functions. The five types of heat exchangers shown in Figure 11.12 represent the majority of heat exchanger types used in the refrigeration and air-conditioning industry. However, other types such as plate–fin and double-pipe heat exchangers are also used for automotive air-conditioning systems and low-tonage liquid cooling systems, respectively.

11.10.1 Water-Cooling Evaporators (Chillers)

As indicated in the preceding section, evaporators, while the refrigerant boils inside the tubes, are the most common commercial evaporators; boiling in the shell, outside the tubes is an important class of liquid-chilling evaporators standard in centrifugal compressor applications. Examples of liquid-cooling-flooded evaporators are shown in Figures 11.16 and 11.17.

Direct expansion evaporators are usually of the shell-and-tube type with the refrigerant inside the tubes, cooling the shell-side fluid (Figures 11.16 and 11.17). Two crucial items in the direct expansion shell and tube evaporators are the number of tubes (refrigerant passes) and the type of tubes.

Another type is the flooded shell-and-tube evaporator that cools liquids flowing through tubes by transferring heat to the evaporating refrigerant on the shell side; the tubes are covered (flooded) with a saturated mixture of liquid and vapor (Figure 11.18).

Baudelot evaporators, also used in industrial applications, are used for cooling a liquid to near its freezing point. The liquid is circulated over the outside of a number of horizontal tubes one above the other. The liquid is distributed uniformly along the length of the tube

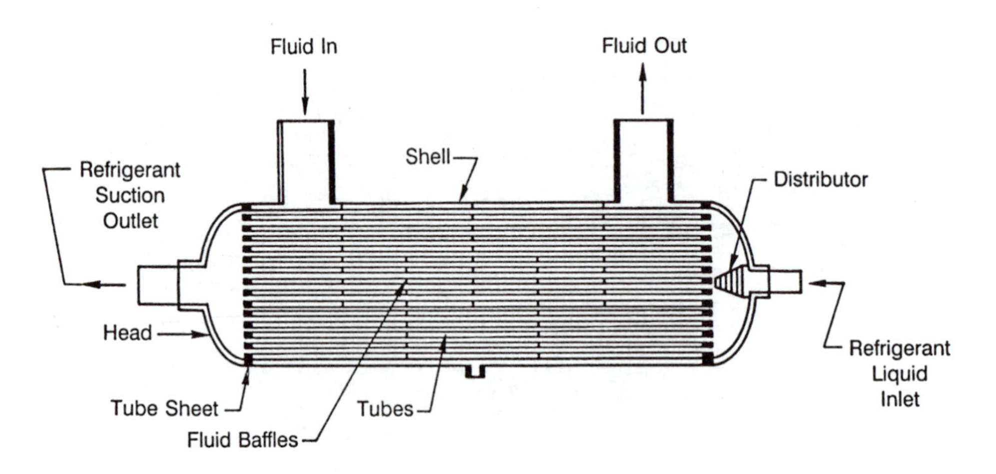

FIGURE 11.16
Direct (straight-tube type) expansion evaporator. (From the 1983 ASHRAE Handbook — Equipment [1983] Chapter 16, Condensers, ASHRAE, Atlanta, GA. With permission.)

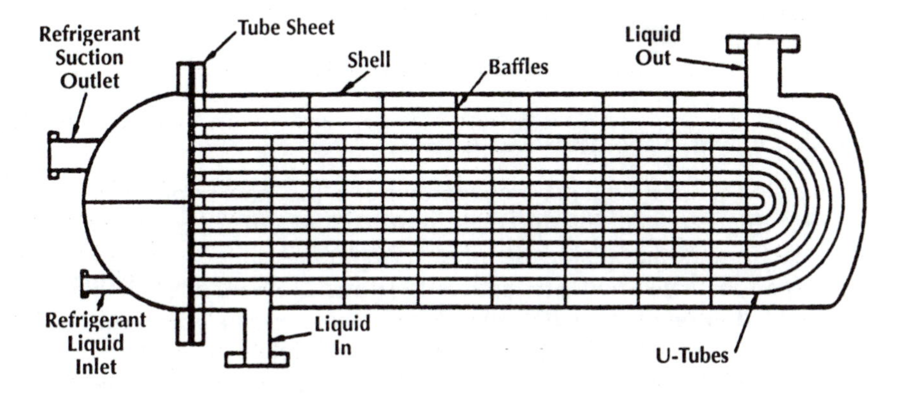

FIGURE 11.17
Direct (U-type) expansion evaporator. (From the 1983 ASHRAE Handbook — Equipment [1983] Chapter 16, Condensers, ASHRAE, Atlanta, GA. With permission.)

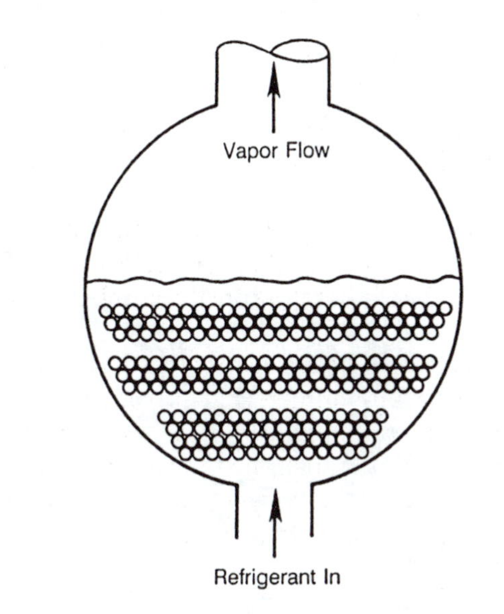

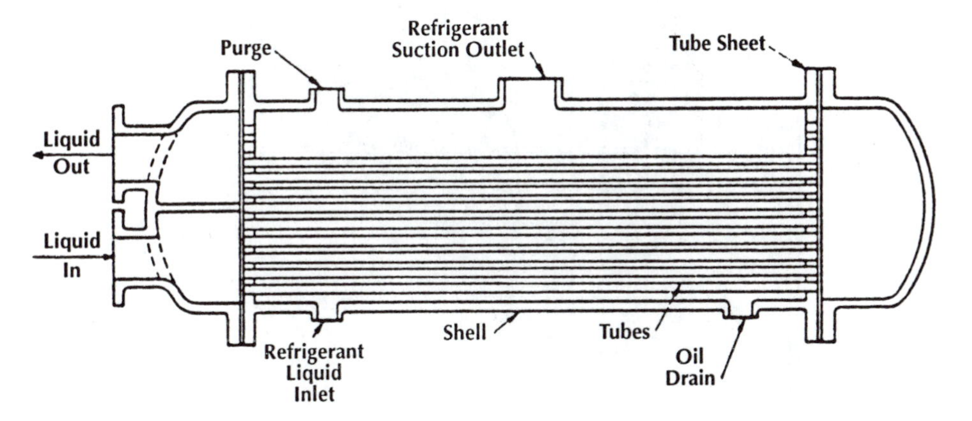

FIGURE 11.18
Flooded shell and tube evaporator (liquid cooler). (From the 1983 ASHRAE Handbook — Equipment [1983] Chapter 16, Condensers, ASHRAE, Atlanta, GA. With permission.)

and flows by gravity to the tubes below. The cooled liquid is collected by a pan from which it is pumped back up to the source of heat and then to the cooler again.[13]

11.10.2 Air-Cooling Evaporators (Air Coolers)

The dry expansion coil is a type of air evaporator. It is the most popular and widely used air cooler. The refrigerant entering the coil is mostly liquid and as it passes through the coil picking up heat from the air outside the tubes, it evaporates and exits as mostly vapor. The typical dry expansion coils shown in Figures 11.19 and 11.20 are multicircuited (from 2 to 22 circuits) and fed by one expansion valve. The air cooler can be forced convection implementing the use of a fan, or can rely on natural convection.

The most common air-conditioning evaporator and condenser is the type that air flows over a circular tube bank that has been finned with continuous plates as shown in Figure 11.20; hence it is a plate finned-tube heat exchanger. The evaporating or condensing refrigerant flows through tubes that are mounted perpendicular to the air flow and arranged in staggered rows. The end views in Figure 11.20 show that the tubes can be connected and coiled to form any number of passes, rows, and parallel paths, hence the name evaporator and condenser coil. Evaporator coils use round tubes for the most part. Typical tube sizes representing a wide range of applications are outside diameters of 5/16, 3/8, 1/2, 5/8, 3/4,

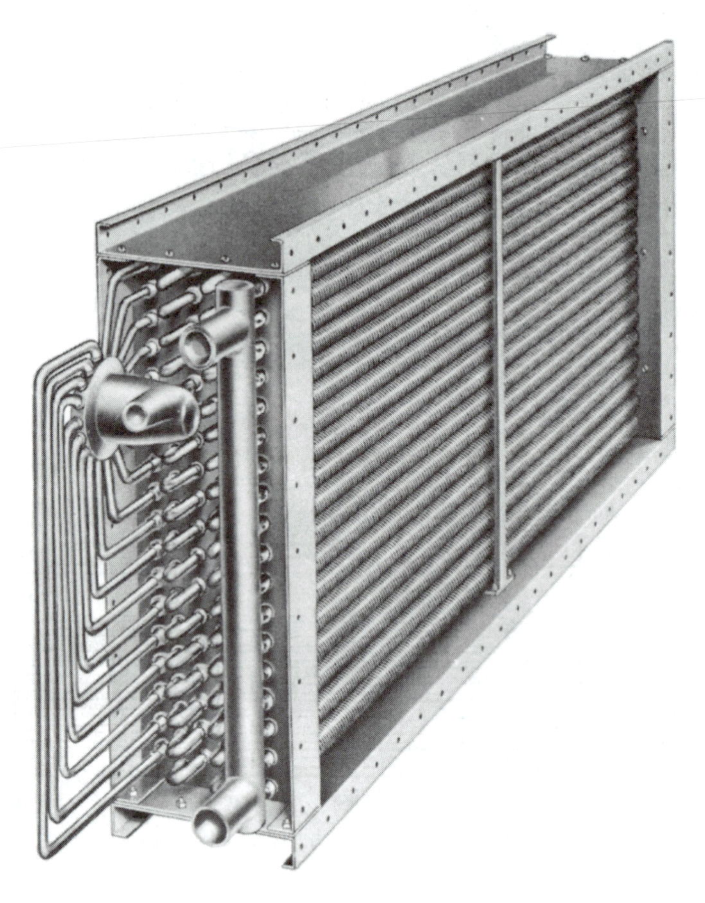

FIGURE 11.19
Dry expansion coil-finned Tube Bank for an Air-Conditioning System. (Courtesy of Aerofin Corp.)

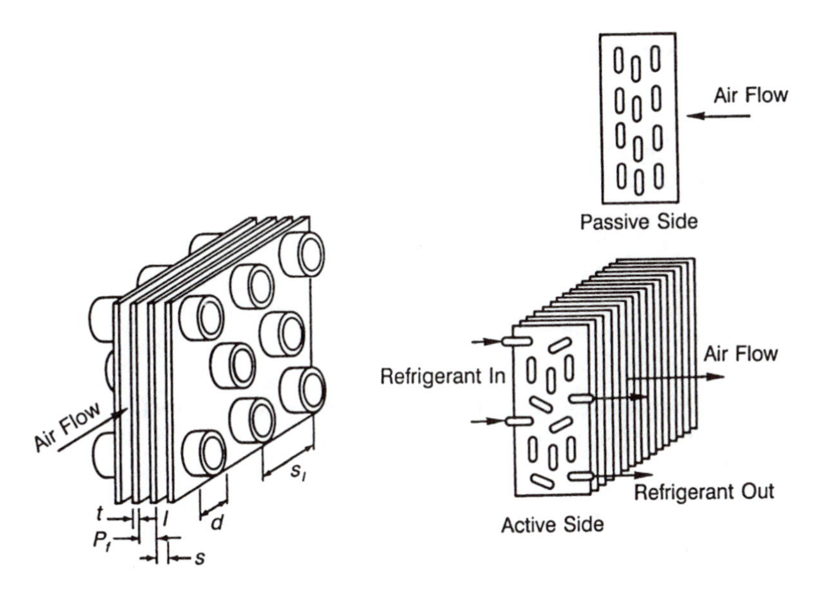

FIGURE 11.20
Dry expansion coil (air evaporator). (From Pate, M. B. [1991] In *Boilers, Evaporators and Condensers*, Wiley, New York. With permission.)

and 1 in. Oval tubes are also used for special applications. The fins and tubes are generally made of aluminum and copper, respectively; however, sometimes both components are manufactured from the same material. The fins are connected to the tubes by inserting the tubes into holes stamped in the fins and then expanding the tubes by either mechanical or hydraulic means.

11.11 Thermal Analysis

The thermal design methods of condensers and evaporators for air-conditioning and refrigeration systems are the same as single-phase applications or as power and process industries that are discussed in previous chapters and in Section 11.7. The only difference will be in the selection of proper correlations to calculate heat transfer coefficients. In refrigeration and air-conditioning, refrigerants like R-12, R-22, R-502, and R-134A are most commonly used. The correlations given to calculate heat transfer coefficient for refrigerant are often different from other fluids.

Three recent correlations for in-tube flow boiling refrigerants are given by Shah,[15] Kandlikar,[16] and Güngör and Winterton;[17] these correlations are introduced and discussed in Chapter 7. These correlations have been verified by experiment. For the completeness of this chapter, and to make the chapter comprehensible by itself, these three correlations are briefly repeated here.

11.11.1 Shah Correlation

The Shah correlation is applicable to nucleate, convection, and stratified boiling regions; and it uses four dimensionless parameters given by the following equations:

$$\Psi = \frac{h_{TP}}{h_L} \tag{11.11}$$

where h_{TP} is the two-phase heat transfer coefficient, and h_L is the all-liquid superficial convection heat transfer coefficient for liquid phase and is calculated by the Dittus–Boelter equation as:

$$h_L = 0.023\left[\frac{G(1-x)d_i}{\mu_l}\right]^{0.8} Pr^{0.4}\frac{k_l}{d_i} \tag{11.12}$$

The other three dimensionless numbers are the convection number (Co), boiling number (Bo), and Froude number (Fr_L) which are

$$Co = \left(\frac{1}{x}-1\right)^{0.8}\left(\frac{\rho_g}{\rho_l}\right)^{0.5} \tag{11.13}$$

$$Bo = \frac{q''}{Gi_{fg}} \tag{11.14}$$

$$Fr_L = \frac{G^2}{\rho_l^2 g d_i} \tag{11.15}$$

When $Co > 1$, it is in the nucleate boiling regime where ψ is independent of Co and depends on Bo number. For horizontal tubes, the surface is fully wet only if $Fr_L \geq 0.04$; for $Fr_L < 0.04$, part of the tube surface is dry and the heat transfer coefficient is lower than in the vertical tubes. We define a dimensionless parameter N_s as follows:

For horizontal tubes with $Fr_L \leq 0.04$:

$$N_s = 0.38\, Fr_L^{-0.3}\, Co \tag{11.16}$$

For vertical tubes at all values of Fr_L and for horizontal tubes with $Fr_L \geq 0.04$

$$N_s = Co \tag{11.17}$$

For $N_s > 1$

$$\Psi_{cb} = \frac{1.8}{N_s^{0.8}} \tag{11.18}$$

$$\Psi_{nb} = 230\, Bo^{0.5} \qquad Bo > 0.3\times10^{-4} \tag{11.19}$$

$$\Psi_{nb} = 1 + 46\,Bo^{0.5} \quad Bo < 0.3 \times 10^{-4} \tag{11.20}$$

ψ is the larger of ψ_{nb} and ψ_{cb}. Thus if $\psi_{nb} > \psi_{cb}$, $\psi = \psi_{nb}$. If $\psi_{cb} > \psi_{nb}$, $\psi = \psi_{cb}$.

For $0.1 < N_s \leq 1.0$:

$$\Psi_{bs} = F\,Bo^{0.5} \exp\,(2.74 N_s^{-0.1}) \tag{11.21}$$

ψ_{cb} is calculated with Equation (11.18), ψ equals the larger of ψ_{bs} and ψ_{cb}.

For $N_s \leq 0.1$:

$$\Psi_{bs} = F\,Bo^{0.5} \exp\,(2.47 N_s^{-0.15}) \tag{11.22}$$

ψ_{cb} is calculated with Equation (11.18) and ψ equals the large of ψ_{cb} and ψ_{bs}. The constant F in Equations (11.21) and (11.22) is as follows:

$$Bo \geq 11 \times 10^{-4}, \qquad F = 14.7 \tag{11.23}$$

$$Bo < 11 \times 10^{-4}, \qquad F = 15.43 \tag{11.24}$$

For horizontal tubes with $Fr_L < 0.04$, these equations are recommended only for $Bo \geq 11 \times 10^{-4}$. For horizontal tubes with $Fr_L \geq 0.04$, and vertical tubes with any Fr_L values, these equations are recommended for all values of Bo without restriction.[15]

As expected, at low qualities, nucleate boiling dominates, while at high qualities, convective boiling dominates.

11.11.2 Kandlikar Correlation

The two-phase boiling heat transfer coefficient, h_{TP}, was expressed as the sum of the convective and nucleate boiling terms:

$$h_{TP} = h_{cov} + h_{nucl} \tag{11.25}$$

Kandlikar represented the convective boiling region by the convection number, Co, and the nucleate boiling region by the boiling number, Bo, as defined by Equations (11.13) and (11.14), respectively.

The final form of the proposed correlation is

$$\frac{h_{TP}}{h_L} = C_1 (Co)^{c_2} (25 Fr_L)^{c_5} + C_3 Bo^{C_4} F_{fl} \tag{11.26}$$

where h_L is the single-phase liquid-only heat transfer coefficient and is calculated from Equation (11.12). F_{fl} is the fluid-dependent parameter and is given in Table 11.5. The values of constants C_1 to C_5 are given next.

TABLE 11.5

Fluid-Dependent Parameter F_{fl} in the Proposed Correlation

Fluid	F_{fl}
Water	1.00
R-11	1.30
R-12	1.50
R-13B1	1.31
R-22	2.20
R-113	1.30
R-114	1.24
R-152a	1.10
Nitrogen	4.70
Neon	3.50

$Co < 0.65$ — convective boiling region, then:

$$C_1 = 1.1360, \quad C_4 = 0.7$$
$$C_2 = -0.90, \quad C_5 = 0.3$$
$$C_3 = 667.\ 2$$

$Co > 0.65$ — nucleate boiling region, then:

$$C_1 = 0.6683, \quad C_4 = 0.7$$
$$C_2 = -0.2, \quad C_5 = 0.3$$
$$C_3 = 1058.0$$

for vertical tubes, and for horizontal tubes with $Fr_L > 0.04$.

The correlation, Equation (11.26), is valid for horizontal and vertical smooth tubes.

11.11.3 Güngör and Winterton Correlation

A general correlation for forced convection boiling has been developed by Güngör and Winterton, and the basic form of the correlation is

$$h_{TP} = Eh_L + Sh_{pool} \tag{11.27}$$

where the all-liquid convection heat transfer coefficient, h_L, is calculated from Equation (11.12); E and S are the enhancement and suppression factors, respectively, and are given by the following expectations:

$$E = 1 + 2.4 \times 10^4\, Bo^{1.16} + 1.37\left(\frac{1}{X_{tt}}\right)^{0.86} \tag{11.28}$$

and

$$S = \frac{1}{1 + 1.15 \times 10^{-6} E^2 Re_L^{1.17}} \tag{11.29}$$

where Bo is given by Equation (11.14) and the X_{tt} is the well-known Martinelli parameter:

$$X_{tt} = \left(\frac{1-x}{x}\right)^{0.9} \left(\frac{\rho_g}{\rho_l}\right)^{0.5} \left(\frac{\mu_L}{\mu_g}\right)^{0.1} \tag{11.30}$$

X_{tt} is similar to the convective number, Co, used in Shah's correlation.

For the pool boiling term, h_{pool}, the correlation given by Cooper[18] is used:

$$h_{pool} = 55\, Pr^{0.12} (-\log_{10} Pr)^{-0.55} M^{-0.5} (q'')^{0.67} \tag{11.31}$$

If the tube is horizontal and the Froude number is less than 0.05, E should be multiplied by:

$$E_2 = Fr_L^{(0.1 - 2Fr_L)} \tag{11.32}$$

and S should be multiplied by:

$$S_2 = \sqrt{Fr_L} \tag{11.33}$$

As the equations stand it is assumed that heat flux, q'', is known, in which case it is straight forward to calculate T_w. If T_w is known, then — as with many other correlations — an iteration is required.

Boiling in annuli is treated by means of an equivalent diameter that depends on the annular gap:

$$\begin{aligned} D_e &= \frac{4 \times \text{ flow area}}{\text{wetted perimeter}} \qquad \text{for gap } > 4 \text{ mm} \\ D_e &= \frac{4 \times \text{ flow area}}{\text{heated perimeter}} \qquad \text{for gap } < 4 \text{ mm} \end{aligned} \tag{11.34}$$

Note that in the data only one of the annulus walls was heated.

In subcooled boiling, the driving temperature differences for nucleate boiling and for forced convection are different, so Equation (11.27) is replaced by:

$$q'' = h_L (T_w - T_b) + S h_{pool} (T_w - T_s) \tag{11.35}$$

There is no enhancement factor because there is no net vapor generation, but the suppression factor remains effective (calculated according to Equations [11.28] and [11.29]). It could be argued that there should still be an enhancement factor because there is still local vapor generation, but this approach gives a worse fit to the data.

11.12 Standards for Evaporators and Condensers

The designer of air-conditioning evaporators and condensers should be aware of standards that are used to rate heat exchangers once they are built according to the recommend design. In the United States, standards are published by both the Air-Conditioning and Refrigerating Institute (ARI) and the American Society of Heating, Refrigerating, and Air-Conditioning Engineers (ASHRAE). For example, the ASHRAE standard for testing air-conditioning evaporator coils is standard 33-78, *Methods of Testing Forced Circulation Air Cooling and Air Heating Coils,* while the standard for condenser coils is standard 20-70, *Methods of Testing for Rating Remote Mechanical-Draft Air-Cooled Refrigerant Condensers.* Both of these standards are presently being reviewed and updated so that revised standards may be forthcoming. The ARI standards are very similar to the ASHRAE standards.

The standards noted previously prescribed laboratory methods including procedures, apparatus, and instrumentation that are to be followed when determining the capacity of the heat exchanger coil. Only then can be the ratings that are published by manufacturers be compared on a common basis.

Example 11.2

Consider the following data to design a water-cooled, shell-and-tube freon condenser for the given heat duty:

Cooling load of the condenser	125 kW
Refrigerant	R-22
	Condensing temperature, 37°C
	In-tube condensation
Coolant	City water
	Inlet temperature, 18°C
	Outlet temperature, 26°C
	Mean pressure, 0.4 MPa
Heat transfer matrix:	3/4-in. O.D., 20 Birmingham wire gauge (BWG)
	Brass tubes

SOLUTION

For preliminary design — Assume values for the shell-and-tube condenser are

$$N_p = 1 \text{ tube pass}$$

$$D_s = 15.25'' = 0.387 \text{ m}$$

$$N_T = 137 \text{ tubes}$$

$$P_T = 1'' \text{ square pitch}$$
$$B = 0.35 \text{ m baffle spacing}$$

Properties of the shell-side coolant fluid, city water, are

$$T_{c1} = 18°C \text{ (inlet temperature)}$$
$$T_{c2} = 26°C \text{ (outlet temperature)}$$
$$d_o = 0.75 \text{ in.} = 0.01905 \text{ m}$$

Bulk temperature is

$$T_b = (T_{c_1} + T_{c_2})/2$$
$$= (18 + 26)/2 = 22°C$$

Properties at bulk temperature from Appendix B are

$$v_l = 0.001002 \text{ m}^3/\text{kg}, \quad c_{pl} = 4.181 \text{ kJ/kg}\cdot\text{K}$$
$$k_l = 0.606 \text{ W/m}\cdot\text{K}, \quad \mu_l = 959 \times 10^{-6} \text{Pa}\cdot\text{s}$$
$$Pr = 6.61$$

The cooling water flow rate (shell side) is

$$Q = (\dot{m}c_p)_c \Delta T_c = (\dot{m}c_p)_h \Delta T_h$$
$$\dot{m}_h = \frac{(\dot{m}c_p)_c \Delta T_c}{c_p \Delta T_h} = \frac{1.384 \times 4.1794 \times 10^3 \times 15}{4.268 \times 10^3 \times 15} = 1.36 \text{ kg/s}$$

The properties of the tube-side-condensing fluid, R-22, are

$$T_h = T_{sat} = 37°C \text{ (saturation temperature)}$$
$$d_i = 0.68 \text{ in.} = 0.01727 \text{ m}$$
$$A_c = 0.362 \text{ in.}^2 = 2.34 \times 10^{-4} \text{ m}^2$$

Properties at saturation temperature from Appendix B are

$$p_{sat} = 14.17 \text{ bar} \qquad c_{pl} = 1.305 \text{ kJ/kg}\cdot\text{K}$$
$$v_l = 8.734 \times 10^{-4} \text{ m}^3/\text{kg} \qquad v_g = 0.01643 \text{ m}^3/\text{kg}$$
$$\mu_l = 0.000186 \text{ Pa}\cdot\text{s} \qquad \mu_g = 0.0000139 \text{ Pa}\cdot\text{s}$$
$$k_l = 0.082 \text{ W/m}\cdot\text{K} \qquad Pr = 2.96$$
$$i_{fg} = 169 \text{ kJ/kg}$$

Refrigerant flow rate (in-tube) is

$$\dot{m}_R = \frac{Q_c}{i_{fg}} = \frac{125}{169} = 0.737 \text{ kg/s}$$

Shell-side calculations — Determine A_s = shell-side area by obtaining a fictitious cross-sectional area (see Section 8.4.1):

$$A_S = \frac{D_S C B}{P_T}$$

$$A_S = \frac{(0.387)(0.0254 - 0.01905)(0.35)}{(0.0254)} = 0.0339 \text{ m}^2$$

Mass flux is

$$G_s = \frac{\dot{m}_c}{A_s} = \frac{3.73}{0.0339} = 110.03 \text{ kg/m}^2 \cdot \text{s}$$

Equivalent diameter, D_e, by assuming square pitch, is then:

$$D_e = \frac{4(P_T^2 - \pi d_o^2/4)}{\pi d_o}$$

$$D_e = \frac{4[(0.0254)^2 - \pi(0.01905)^2/4]}{\pi(0.01905)} = 0.0241 \text{ m}$$

Calculate Reynolds number, shell side by

$$Re_S = \frac{G_S D_e}{\mu}$$

$$Re_S = \frac{(110.03)(0.0241)}{(959 \times 10^{-6})} = 2772$$

so flow is turbulent on the shell side.

Shell-side heat transfer coefficient for $2000 < Re < 10^6$ is obtained by:

$$\frac{h_0 D_e}{k} = 0.35 \, Re_S^{0.55} \, Pr_S^{1/3} \left(\frac{\mu_b}{\mu_w} \right)^{0.14}$$

where μ_b/μ_w is assumed to be 1 because of the small temperature difference:

$$h_0 = 0.35(2772)^{0.55}(6.61)^{1/3}(0.606/0.0241) = 1329 \text{ W/m}^2 \cdot \text{K}$$

Tube-side calculations — These can be made by determining mass flux (refrigerant):

$$G_R = \frac{\dot{m}_R}{A_R}$$

$$A_R = \frac{A_C N_T}{N_P} = \frac{(0.00234)(137)}{1} = 0.0321\ \text{m}^2$$

where A_c is the cross-sectional tube area.

$$G_R = \frac{(0.737)}{(0.0321)} = 22.97\ \text{kg/m}^2 \cdot \text{s}$$

and the Reynolds number:

$$Re_L = \frac{G_R(1-x)d_i}{\mu_l}$$

For a quality of $x = 0.5$:

$$Re_L = \frac{(22.97)(1-0.5)(0.0173)}{186 \times 10^{-6}} = 1068$$

$$Re_v = \frac{G_R x d_i}{\mu_l}$$

$$Re_v = \frac{(22.97)(0.5)(0.0173)}{14 \times 10^{-6}} = 14192$$

Re_L is the liquid-only Reynolds number, and Re_v is the vapor-only Reynolds number.

The average heat transfer coefficient for condensation in tube-side, h_{TP}, is given by the Cavallini and Zecchin correlation,[26] Equation (7.31):

$$h_{TP} = 0.05\, Re_{eq}^{0.8}\, Pr^{1/3}\, k_L / d_i$$

where the equivalent Reynolds number, Re_{eq}, is defined as:

$$Re_{eq} = Re_v \left(\frac{\mu_v}{\mu_L}\right)\left(\frac{\rho_L}{\rho_v}\right)^{0.5} + Re_L$$

$$Re_{eq} = (14192)\left(\frac{14}{186}\right)\left(\frac{0.01648}{8.73 \times 10^{-4}}\right)^{0.5} + 1068 = 5709$$

so,

$$h_{TP} = (0.05)(5709)^{0.8}(2.96)^{1/3}(0.082/0.0173) = 345\ \text{W/m}^2 \cdot \text{K}$$

Overall heat transfer coefficient is calculated by:

$$U_0 = \left[\frac{d_0}{h_i d_i} + \frac{d_0 R_{fi}}{d_i} + \frac{d_0 \ln(d_0/d_i)}{2k} + R_{fo} + \frac{1}{h_0} \right]^{-1}$$

where the fouling resistances are:

$R_{fi} = 0.000176$ m²·K/W for refrigerant liquids

$R_{fo} = 0.000176$ m²·K/W for city water

k (brass) = 111 W/m·K

$h_i = h_{TP}$, the inside heat transfer coefficient

$$U_0 = \left[\frac{(0.191)}{(345)(.0173)} + \frac{(0.0191)(.000176)}{(.0173)} + \frac{(0.0191)\ln(0.0191/0.0173)}{2(111)} + 0.000176 + \frac{1}{1283} \right]^{-1}$$

$$U_0 = 230 \text{ W/m}^2 \cdot \text{K}$$

Log mean temperature difference (LMTD) is

$$\Delta T_{lm} = \frac{\Delta T_1 - \Delta T_2}{\ln(\Delta T_1 / \Delta T_2)}$$

where $\Delta T_1 = T_h - T_{c1} = 19°\text{C}$, $\Delta T_2 = T_h - T_{c2} = 11°\text{C}$:

$$\Delta T_{lm} = \frac{19 - 11}{\ln(19/11)} - 14.64\,°\text{C}$$

Determine mean temperature difference by:

$$\Delta T_m = \Delta T_{LMTD} \cdot F = 14.64°\text{C}$$

where $F = 1$ for a condenser.

Calculate the heat transfer surface area by

$$Q = U_o A_o \Delta T_m$$

$$A_o = N_t \pi d_o L$$

$$= \frac{Q}{U_0 \Delta T_m}$$

$$= \frac{125000}{(230)(14.64)} = 37.12 \text{ m}^2$$

and the length of tube by:

$$L = \frac{A_0}{N_T \pi d_0}$$

$$L = \frac{(37.12)}{(137)\pi(0.01905)} = 4.53 \text{ m}$$

Nomenclature

A	heat transfer area, m^2
Bo	boiling number, q''/Gi_{fg}
Co	convection number, $[(1-x)/x]^{0.8}(\rho_g/\rho_l)^{0.5}$
C_1 *to* C_5	constants in Equation (11.26)
c_p	specific heat at constant pressure, J/kg·K
d	tube diameter, m
D_e	equivalent diameter, m
E	enhancement factor
F	constant given in Equation (11.22)
F_{fl}	fluid-dependent parameter in Equation (11.26)
Fr_L	Froude number, $G^2/(\rho_l^2 gd)$
g	acceleration of gravity, m/s^2
G	mass velocity (flux), $kg/m^2 \cdot s$
h	heat transfer coefficient, $W/m^2 \cdot K$
h_L	liquid heat transfer coefficient, $W/m^2 \cdot K$
h_{TP}	local two-phase heat transfer coefficient, $W/m^2 \cdot K$
i	specific enthalpy, J/kg
i_{lg}	latent heat of vaporization, J/kg
k	thermal conductivity, $W/m^2 \cdot K$
L	tube length, m
M	molecular weight
$\dot{m}$	mass flow rate, kg/s
N	average number of tubes in a vertical column
N_s	parameter defined by Equations (11.16) and (11.17)
N_T	number of tubes in a bundle
Nu	Nusselt number, hD_e/k, hd_o/k, hd_i/k
p	pressure, Pa
Pr	Prandtl number, $c_p\mu/k$

Q	cumulative heat release rate, W
q''	heat flux, W/m^2
Re	Reynolds number, GD_e/μ, Gd/μ
Re_L	liquid Reynolds number, $Gd_i\,(1-x)/\mu$
S	suppression factor
t	thickness of the wall, m
T	temperature, °C, K
ΔT	temperature difference, $(T_s - T_w)$,°C, K
U	overall heat transfer coefficient, W/m^2·K
x	vapor quality
X_{tt}	Lockhart–Martinelli parameter, $(1-x/x)^{0.9}\,(\rho_g/\rho_l)^{0.5}$
μ_g	dynamic viscosity of vapor, N_s/m^2
μ_l	dynamic viscosity of liquid, N_s/m^2
ρ_g	density of vapor, kg/m^3
ρ_l	density of liquid, kg/m^3
Ψ	h_{TP}/h_L
Ψ_{bs}	value of Ψ in the bubble suppression regime
Ψ_{cb}	value of Ψ in the pure convective boiling regime
Ψ_{nb}	value of Ψ in the pure nucleate boiling regime

References

1. Butterworth, D. (1991) Steam power plant and processes condensers. In *Boilers, Evaporators and Condensers*, S. Kakaç (Ed.), pp. 571–633. Wiley, New York.
2. Butterworth, D. (1988) Condensers and their design. In *Two-Phase Flow Heat Exchangers*, S. Kakaç, A. E. Bergles, and E. O. Fernandes (Eds.), Kluwer, Dordrecht, The Netherlands.
3. Mueller, A. C. (1983) Selection of condenser types. In *Heat Exchanger Design Handbook*, pp. 3.4.2–1. Hemisphere, Washington, D.C.
4. Breber, G. (1988) Condenser design with pure vapor and mixture of vapors. In *Heat Transfer Equipment Design*, R. K. Shah, E. C. Subbarao, and R. A. Mashelkar (Eds.), pp. 477–496. Hemisphere, Washington, D.C.
5. Bell, K. J. and Mueller, A. C. (1971) *Condensation Heat Transfer Condenser Design*, AIChE Today Series, American Institute of Chemical Engineers, New York.
6. Sebald, J. F. (1980) A development history of steam surface condensers in the electrical utility industry. ASME/AIChE National Heat Transfer Conference, San Diego, *Heat Transfer Eng.*, Vol. 1, no. 3, 80–87; Vol. 1, no. 4, 76–81.
7. Alfa Laval, *Thermal Handbook*, Lund, Sweden.
8. Taborek, J. (1991) Industrial heat transfer exchanger design practices. In *Boilers, Evaporators and Condensers*, S. Kakaç (Ed.), pp. 143–177. Wiley, New York.
9. Steinmeyer, D. E. and Muller, A. C. (1974) Why condensers don't operate as they are supposed to. *Chem. Eng. Prog.*, Vol. 70, 78–82.
10. Pate, M. B. (1991) Evaporators and condensers for refrigeration and air- conditioning systems. In *Boilers, Evaporators and Condensers*, S. Kakaç (Ed.), Wiley, New York.
11. Althouse, A. D., Turnquist, C. H., and Bracciano, A. F. (1968) Modern R Refrigeration and Air-Conditioning, Goodheat-Willcox Co.

12. Trane Air Conditioning Manual (1979) Trane Company, Lacrosse, WI.
13. Equipment Handbook (1988) ASHRAE, Atlanta, GA.
14. Equipment Handbook (1983) Chapter 16, Condensers, ASHRAE, Atlanta, GA.
15. Shah, M. M. (1982) Chart correlation for saturated boiling heat transfer: equation and further study. *ASHRAE Trans.*, Vol. 88, 185–196.
16. Kandlikar, S. G. (1990) A general correlation for saturated two-phase flow boiling heat transfer inside horizontal and vertical tubes. *J. Heat Transfer*, Vol. 11, 219–228.
17. Güngör, K. E. and Winterton, R. H. S. (1986) A general correlation for flow boiling in tubes and annuli. *Int. J. Heat Mass Transfer*, Vol. 29, 351–358.
18. Cooper, M. G. (1949) Saturated nucleate pool boiling, a simple correlation of data for isothermal two-phase two-component flow in pipes. *Chem. Eng. Prog.*, Vol. 45, no. 39, 39–48.

Problems

11.1. (See also Example 11.1) Design a shell-and-tube type power condenser for a 250 MWe coal-fired power station. Steam enters the turbine at 500°C and 4 MPa. The thermodynamic efficiency of the turbine is 0.85. Assume that the condenser pressure is 10 kPa. Cooling water for the operation is available at 15°C. Assume a laminar film condensation on the shell side.

11.2. In a power plant, a shell-and-tube type heat exchanger is used as a condenser. This heat exchanger consists of 20,000 tubes and the fluid velocity through the tubes is 2 m/s. The tubes are made of Admiralty metal and have 18 BWG, 7/8-in. O.D. The cooling water enters at 20°C and exits at 30°C. The average temperature of the tube walls is 55°C and the shell-side heat transfer coefficient is 4000 W/m^2·K. Fouling on both sides is neglected. By making acceptable engineering assumptions, perform the thermal design of this condenser.

11.3. A surface condenser is designed as a two-tube-pass, shell-and-tube type of heat exchanger at 10 kPa (h_{fg} = 2007.5 kJ/kg, T_s = 45°C). Coolant water enters at 15°C and leaves at 25°C inside the tubes. Coefficient is 300 W/m^2·K. The tubing is thin walled, has a 5-cm I.D. and is made of carbon steel; and the length of the heat exchanger is 2 m.

Calculate:

a. LMTD
b. Steam mass flow rate
c. Surface area of the condenser
d. Number of tubes
e. Effectiveness, ε

11.4. A shell-and-tube steam condenser is to be constructed of 18 BWG Admiralty tubes with 7/8-in. O.D. Steam will condense outside these single-pass horizontal tubes at T_s = 60°C. Cooling water enters each tube at T_i =17°C, with a flow rate of $\dot{m}$ = 0.7 kg/s per tube and leaves at T_o = 29°C. The heat transfer coefficient for the condensation of steam is h_s = 4500 W/m^2·K. Steam-side fouling is neglected. The fouling factor for the cooling water is 1.76×10^{-4} m^2·K/W. Assume that the tube wall temperature is 50°C. The tube material is carbon steel.

11.5. A shell-and-tube type of condenser will be designed for refrigerant-22 in a refrigeration system that provides a capacity of 100 kW for air-conditioning. The condensing temperature is 47°C at design conditions. The tubes are copper (386 W/m·K) and are 14-mm I.D. and 16-mm O.D. Cooling water enters the condenser tubes at 30°C with velocity of 1.5 m/s and leaves at 35°C. The condenser will be two tube passes. In-line and triangular pitch arrangements will be considered. Make acceptable engineering assumption and size this condenser.

11.6. Repeat Problem 11.5 for the refrigerant-134A.

11.7. A shell-and-tube type of condenser will be designed for a refrigerant-134A in a refrigeration system that provides a capacity of 100 kW for air-conditioning. The condensing temperature is 47°C at design conditions. The tubes are copper and are 14-mm I.D. and 16-mm O.D. Cooling water enters the condenser tubes at 20°C with a velocity of 2 m/s and leaves at 26°C. The condenser will be in two tube passes with a 30° triangular pitch arrangement. Assuming acceptable engineering design parameters, size the condenser.

Design Project 11.1

Compact Air-Cooled Refrigerant Condenser

Specifications are

Cooling load (heat duty)	Q = 125 kW
Refrigerant	R-134A Condensing inside tubes at T_s = 37°C (310 K)
Coolant	Air
	Inlet temperature, T_{c1} = 18°C
	Inlet temperature, T_{c1} = 26°C
	Mean pressure p = 2 atm (0.2027 MPa)
Heat transfer matrix	To be selected from Chapter 9

Different (at least two) surfaces must be studied for thermal and hydraulic analysis and compared; the primary attention must be given to obtaining the smallest possible heat exchanger. A parametric study is expected to come up with a suitable final design. Final design will include material selection, mechanical design, technical drawings, and cost estimation.

Design Project 11.2

Air-Conditioning Evaporator

Specifications are

Cooling load	4-ton air-conditioning system
Evaporator temperature	10°C

Refrigerant	Freon-134A
Operates at	10°C
Air inlet temperature	30°C
Outlet temperature	18°C

Finned brass tubes will be used for the heat transfer matrix, and the core matrix will be selected from one of the compact surfaces given in Chapter 9.

The final report will include thermal, hydraulic analysis, material selection, and mechanical design together with cost estimation and other considerations as studied in Design Project 11.1.

Design Project 11.3

Water-Cooled, Shell-and-Tube Type of Freon Condenser

Specifications are

Cooling load	200 kW
Refrigerant	Freon-134A
Temperature	27°C
Mean pressure	0.702 MPa
Coolant	Water
Inlet temperature	18°C
Outlet temperature	26°C
Mean pressure	0.4 MPa
Heat transfer matrix	To be selected from Chapter 9

Condensation will be on the shell-side of the heat exchanger. A shell type must be selected. The size will be estimated first and then it will be rated as described in Chapter 8. Final report will include thermal–hydraulic analysis, optimization, material selection, mechanical design, drawings, sizing, and cost estimation.

Design Project 11.4

Steam Generator for a Pressurized Water Reactor (PWR) - 600 MWe

Specifications are

Power to be generated from a single generator	150 MWt
Reactor coolant inlet temperature	300°C

Reactor coolant outlet temperature	337°C
Reactor coolant pressure	15 MPa
Secondary circuit water inlet temperature	200°C
Secondary circuit steam outlet temperature	285°C
Secondary fluid pressure	6.9 MPa

One may find the number of steam generators needed (see Figure 1.27). Thermal–hydraulic analysis must be conducted on the selection of the type of steam generator. The final report will include a parametric study, material selection, mechanical design, technical drawings, and cost estimation.

Appendix A

Physical Properties of Metals and Nonmetals

Nomenclature

b	normal boiling point
c	critical point
c_p	specific heat at constant pressure, kJ/(kg·K)
c_v	specific heat at constant volume, kJ/(kg·K)
c_{pf}	specific heat at constant pressure of the saturated liquid, kJ/(kg·K)
c_{pg}	specific heat at constant pressure of the saturated vapor, kJ/(kg·K)
h	specific enthalpy, kJ/(kg·K)
h_f	specific enthalpy of the saturated liquid, kJ/(kg·K)
h_g	specific enthalpy of the saturated vapor, kJ/(kg·K)
k	thermal conductivity, W/m·K)
k_f	thermal conductivity of the saturated liquid, W/(m·K)
k_g	thermal conductivity of the saturated vapor, W/(m·K)
M	molecular weight, kg/k mol
P	pressure, bar
P_f	saturation vapor pressure, liquid, bar
P_g	saturation vapor pressure, gas, bar
Pr	Prandtl number, $c_p\mu/k$
Pr_f	Prandtl number of the saturated liquid, $(c_p\mu)_f/k_f$
Pr_g	Prandtl number of the saturated vapor, $(c_p\mu)_g/k_g$
s	specific entropy, kJ/(kg·K)
s_f	specific entropy of the saturated liquid, kJ/(kg·K)
s_g	specific entropy of the saturated vapor, kJ/(kg·K)
T	temperature, °C, K
T_b	normal boiling point temperature, °C, K
T_m	normal melting point temperature, °C, K
v	specific volume, m^3/kg
v_f	specific volume of the saturated liquid, m^3/kg
v_g	specific volume of the saturated vapor, m^3/kg
$\overline{v}_s$	velocity of sound, m/s
Z	compressibility factor

Greek Symbols

α	thermal expansion coefficient
β	isothermal compressibility coefficient
γ	ratio of principal specific heat, c_p/c_v
μ	viscosity, Pa·s
μ_f	viscosity of the saturated liquid, Pa·s
μ_g	viscosity of the saturated vapor, Pa·S
σ	surface tension, N/m
ρ	density, kg/m^3

TABLE A.1

Thermophysical Properties of Metals

Metal	Temperature Range T (°C)	Density ρ (g/cm³)	Specific Heat c (kJ/kg·K)	Thermal Conductivity k (W/m·K)	Emissivity ε
Aluminum	0–400	2.72	0.895	202–250	0.04–0.06 (Polished) 0.07–0.09 (Commercial) 0.2–0.3 (Oxidized)
Brass (70% Cu, 30% Zn)	100–300	8.52	0.38	104–147	0.03–0.07 (Polished) 0.2–0.25 (Commercial) 0.45–0.55 (Oxidized)
Bronze (75% Cu, 25% Sn)	0–100	8.67	0.34	26	0.03–0.07 (Polished) 0.4–0.5 (Oxidized)
Constantan (60% Cu, 40% Ni)	0–100	8.92	0.42	22–26	0.03–0.06 (Polished) 0.2–0.4 (Oxidized)
Copper	0–600	8.95	0.38	385–350	0.02–0.04 (Polished) 0.1–0.2 (Commercial)
Iron (C = 4%, cast)	0–1000	7.26	0.42	52–35	0.2–0.25 (Polished) 0.55–0.65 (Oxidized) 0.6–0.8 (Rusted)
Iron (C ≃ 0.5%, wrought)	0–1000	7.85	0.46	59–35	0.3–0.35 (Polished) 0.9–0.95 (Oxidized)
Lead	0–300	11.37	0.13	35–30	0.05–0.08 (Polished) 0.3–0.6 (Oxidized)
Magnesium	0–300	1.75	1.01	171–157	0.07–0.13 (Polished)
Mercury	0–300	13.4	0.125	8–10	0.1–0.12
Molybdenum	0–1000	10.22	0.251	125–99	0.06–0.10 (Polished)
Nickel	0–400	8.9	0.45	93–59	0.05–0.07 (Polished) 0.35–0.49 (Oxidized)
Platinum	0–1000	21.4	0.24	70–75	0.05–0.03 (Polished) 0.07–0.11 (Oxidized)
Silver	0–400	10.52	0.23	410–360	0.01–0.03 (Polished) 0.02–0.04 (Oxidized)
Steel (C ≃ 1%)	0–1000	7.80	0.47	43–28	0.07–0.17 (Polished)

TABLE A.1 (CONTINUED)

Thermophysical Properties of Metals

Metal	Temperature Range T (°C)	Density ρ (g/cm³)	Specific Heat c (kJ/kg·K)	Thermal Conductivity k (W/m·K)	Emissivity ε
Steel (Cr ≃ 1%)	0–1000	7.86	0.46	62–33	0.07–0.17 (Polished)
Steel (Cr 18%, Ni 8%)	0–1000	7.81	0.46	16–26	0.07–0.17 (Polished)
Tin	0–200	7.3	0.23	65–57	0.04–0.06 (Polished)
Tungsten	0–1000	19.35	0.13	166–76	0.04–0.08 (Polished) 0.1–0.2 (Filament)
Zinc	0–400	7.14	0.38	112–93	0.02–0.03 (Polished) 0.10–0.11 (Oxidized) 0.2–0.3 (Galvanized)

Kakaç, S. and Yener, Y. (1995) *Convective Heat Transfer*, 2nd ed., CRC Press, Boca Raton, FL.

TABLE A.2

Thermophysical Properties of Nonmetals

Metal	Temperature Range T (°C)	Density ρ (g/cm³)	Specific Heat c (kJ/kg·K)	Thermal Conductivity k (W/m·K)	Emissivity ε
Asbestos	100–1000	0.47–0.57	0.816	0.15–0.22	0.93–0.97
Brick, rough red	100–1000	1.76	0.84	0.38–0.43	0.90–0.95
Clay	0–200	1.46	0.88	1.3	0.91
Concrete	0–200	2.1	0.88	0.81–1.4	0.94
Glass, window	0–600	2.2	0.84	0.78	0.94–0.66
Glass wool	23	0.024	0.7	0.038	
Ice	0	0.91	1.9	2.2	0.97–0.99
Limestone	100–400	2.5	0.92	1.3	0.95–0.80
Marble	0–100	2.60	0.79	2.07–2.94	0.93–0.95
Plasterboard	0–100	1.25	0.84	0.43	0.92
Rubber (hard)	0–100	1.2	1.42	0.15	0.94
Sandstone	0–300	2.24	0.71	1.83	0.83–0.9
Wood (oak)	0–100	0.6–0.8	2.4	0.17–0.21	0.90

Kakaç, S. and Yenner, Y. (1995), *Convective Heat Transfer*, 2nd ed., CRC Press, Boca Raton, FL.

Appendix B

Physical Properties of Air, Water, Liquid Metals, and Refrigerants

TABLE B.1

Properties of Dry Air at Atmospheric Pressure

Temperature (°C)	ρ (kg/m^3)	c_p (kJ/kg·K)	k (W/m·K)	$\beta \times 10^3$ (1/K)	$\mu \times 10^5$ (kg/m·s)	$\nu \times 10^6$ (m^2/s)	$\alpha \times 10^6$ (m^2/s)	Pr
–150	2.793	1.026	0.0120	8.21	0.870	3.11	4.19	0.74
–100	1.980	1.009	0.0165	5.82	1.18	5.96	8.28	0.72
–50	1.534	1.005	0.0206	4.51	1.47	9.55	013.4	0.715
0	1.2930	1.005	0.0242	3.67	1.72	013.30	18.7	0.711
20	1.2045	1.005	0.0257	3.43	1.82	15.11	21.4	0.713
40	1.1267	1.009	0.0271	3.20	1.91	16.97	23.9	0.711
60	1.0595	1.009	0.0285	3.00	2.00	18.90	26.7	0.709
80	0.9908	1.009	0.0299	2.83	2.10	20.94	29.6	0.708
100	0.9458	1.013	0.0314	2.68	2.18	23.06	32.8	0.704
120	0.8980	1.013	0.0328	2.55	2.27	25.23	36.1	0.70
140	0.8535	1.013	0.0343	2.43	2.35	27.55	39.7	0.694
160	0.8150	1.017	0.0358	2.32	2.43	29.85	43.0	0.693
180	0.7785	1.022	0.0372	2.21	2.51	32.29	46.7	0.69
200	0.7475	1.026	0.0386	2.11	2.58	34.63	50.5	0.685
250	0.6745	1.034	0.0421	1.91	2.78	41.17	60.3	0.68
300	0.6157	1.047	0.0390	1.75	2.95	47.85	70.3	0.68
350	0.5662	1.055	0.0485	1.61	3.12	55.05	81.1	0.68
400	0.5242	1.068	0.0516	1.49	3.28	62.53	91.9	0.68
450	0.4875	1.080	0.0543	—	3.44	70.54	103.1	0.685
500	0.4564	1.092	0.0570	—	3.86	70.48	114.2	0.69
600	0.4041	1.114	0.0621	—	3.58	95.57	138.2	0.69
700	0.3625	1.135	0.0667	—	4.12	113.7	162.2	0.70
800	0.3287	0.156	0.0706	—	4.37	132.8	185.8	0.715
900	0.321	1.172	0.0741	—	4.59	152.5	210	0.725
1000	0.277	1.185	0.0770	—	4.80	175	235	0.735

From Kakaç, S. and Yener, Y. (1995) *Convective Heat Transfer*, 2nd ed., CRC Press, Boca Raton, FL.

TABLE B.2

Thermophysical Properties of Saturated Ice-Water-Steam

P (bar)	T (K)	v_f* (10^{-3} m³/kg)	v_g (m³/kg)	h* (kJ/kg)	h_g (kJ/kg)	μ_g (10^{-4} Pa·s)	k_f* (W/(m·K)	k_g (W/(m·K)	Pr_f*	Pr_g
0.001	252.84	1.0010	1167	–374.9	2464.1	0.0723	2.40	0.0169		
0.002	260.21	1.0010	600	–360.1	2477.4	0.0751	2.35	0.0174		
0.003	265.11	1.0010	408.5	–350.9	2486.0	0.0771	2.31	0.0177		
0.004	267.95	1.0010	309.1	–344.4	2491.9	0.0780	2.29	0.0179		
0.005	270.74	1.0010	249.6	–337.9	2497.3	0.0789	2.27	0.0180		
0.006	273.06	1.0010	209.7	–333.6	2502	0.0798	2.26	0.0182		
0.0061	273.15	1.0010	206.0	–333.5	2502	0.0802	2.26	0.0182		
0.0061	273.15	1.0002	206.0	0.0	2502	0.0802	0.566	0.0182	13.04	0.817
0.008	276.73	1.0001	159.4	21.9	2508	0.0816	0.568	0.0184	11.66	0.823
0.010	280.13	1.0001	129.2	29.4	2513.4	0.0829	0.578	0.0186	10.39	0.828
0.02	290.66	1.0013	67.00	73.5	2532.7	0.0872	0.595	0.0193	7.51	0.841
0.03	297.24	1.0028	45.66	101.1	2544.8	0.0898	0.605	0.0195	6.29	0.854
0.04	302.13	1.0041	34.80	121.4	2553.6	0.0918	0.612	0.0198	5.57	0.865
0.05	306.04	1.0053	28.19	137.8	2560.6	0.0933	0.618	0.0201	5.08	0.871
0.06	309.33	1.0065	23.74	151.5	2566.6	0.0946	0.622	0.0203	4.62	0.877
0.08	314.68	1.0085	18.10	173.9	2576.2	0.0968	0.629	0.0207	4.22	0.883
0.10	318.98	1.0103	14.67	191.9	2583.9	0.0985	0.635	0.0209	3.87	0.893
0.20	333.23	1.0172	7.65	251.5	2608.9	0.1042	0.651	0.0219	3.00	0.913
0.30	342.27	1.0222	5.23	289.3	2624.6	0.1078	0.660	0.0224	2.60	0.929
0.40	349.04	1.0264	3.99	317.7	2636.2	0.1105	0.666	0.0229	2.36	0.941
0.5	354.50	1.0299	3.24	340.6	2645.4	0.1127	0.669	0.0233	2.19	0.951
0.6	359.11	1.0331	2.73	359.9	2653.0	0.1147	0.673	0.0236	2.06	0.961
0.8	366.66	1.0385	2.09	391.7	2665.3	0.1176	0.677	0.0242	1.88	0.979
1.0	372.78	1.0434	1.6937	417.5	2675.4	0.1202	0.6805	0.0244	1.735	1.009
1.5	384.52	1.0530	1.1590	467.1	2693.4	0.1247	0.6847	0.0259	1.538	1.000
2.0	393.38	1.0608	0.8854	504.7	2706.3	0.1280	0.6866	0.0268	1.419	1.013
2.5	400.58	1.0676	0.7184	535.3	2716.4	0.1307	0.6876	0.0275	1.335	1.027
3.0	406.69	1.0735	0.6056	561.4	2724.7	0.1329	0.6879	0.0281	1.273	1.040
3.5	412.02	1.0789	0.5240	584.3	2731.6	0.1349	0.6878	0.0287	1.224	1.050
4.0	416.77	1.0839	0.4622	604.7	2737.6	0.1367	0.6875	0.0293	1.185	1.057
4.5	421.07	1.0885	0.4138	623.2	2742.9	0.1382	0.6869	0.0298	1.152	1.066
5	424.99	1.0928	0.3747	640.1	2747.5	0.1396	0.6863	0.0303	1.124	1.073
6	432.00	1.1009	0.3155	670.4	2755.5	0.1421	0.6847	0.0311	1.079	1.091

7	438.11	1.1082	0.2727	697.1	2762.0	0.1443	0.6828	0.0319	1.044	1.105
8	445.57	1.1150	0.2403	720.9	2767.5	0.1462	0.6809	0.0327	1.016	1.115
9	448.51	1.1214	0.2148	742.6	2772.1	0.1479	0.6788	0.0334	0.992	1.127
10	453.03	1.1274	0.1943	762.6	2776.1	0.1495	0.6767	0.0341	0.973	1.137
12	461.11	1.1386	0.1632	798.4	2782.7	0.1523	0.6723	0.0354	0.943	1.156
14	468.19	1.1489	0.1407	830.1	2787.8	0.1548	0.6680	0.0366	0.920	1.175
16	474.52	1.1586	0.1237	858.6	2791.8	0.1569	0.6636	0.0377	0.902	1.191
18	480.26	1.1678	0.1103	884.6	2794.8	0.1589	0.6593	0.0388	0.889	1.206
20	485.53	1.1766	0.0995	908.6	2797.2	0.1608	0.6550	0.0399	0.877	1.229
25	497.09	1.1972	0.0799	962.0	2800.9	0.1648	0.6447	0.0424	0.859	1.251
30	506.99	1.2163	0.0666	1008.4	2802.3	0.1684	0.6347	0.0449	0.849	1.278
35	515.69	1.2345	0.0570	1049.8	2802.0	0.1716	0.6250	0.0472	0.845	1.306
40	523.48	1.2521	0.0497	1087.4	2800.3	0.1746	0.6158	0.0496	0.845	1.331
45	530.56	1.2691	0.0440	1122.1	2797.7	0.1775	0.6068	0.0519	0.849	1.358
50	537.06	1.2858	0.0394	1154.5	2794.2	0.1802	0.5981	0.0542	0.855	1.386
60	548.70	1.3187	0.0324	1213.7	2785.0	0.1854	0.5813	0.0589	0.874	1.442
70	558.94	1.3515	0.0274	1267.4	2773.5	0.1904	0.5653	0.0638	0.901	1.503
80	568.12	1.3843	0.0235	1317.1	2759.9	0.1954	0.5499	0.0688	0.936	1.573
90	576.46	1.4179	0.0205	1363.7	2744.6	0.2005	0.5352	0.0741	0.978	1.651
100	584.11	1.4526	0.0180	1408.0	2727.7	0.2057	0.5209	0.0798	1.029	1.737
110	591.20	1.4887	0.0160	1450.6	2709.3	0.2110	0.5071	0.0159	1.090	1.837
120	597.80	1.5268	0.0143	1491.8	2689.2	0.2166	0.4936	0.0925	1.163	1.963
130	603.98	1.5672	0.0128	1532.0	2667.0	0.2224	0.4806	0.0998	1.252	2.126
140	609.79	1.6106	0.0115	1571.6	2642.4	0.2286	0.4678	0.1080	1.362	2.343
150	615.28	1.6579	0.0103	1611.0	2615.0	0.2373	0.4554	0.1307	1.502	2.571
160	620.48	1.7103	0.0093	1650.5	2584.9	0.2497	0.4433	0.1280	1.688	3.041
170	625.41	1.7696	0.0084	1691.7	2551.6	0.2627	0.4315	0.1404	2.098	3.344
180	630.11	1.8399	0.0075	1734.8	2513.9	0.2766	0.4200	0.1557	2.360	3.807
190	634.58	1.9260	0.0067	1778.7	2470.6	0.2920	0.4087	0.1749	2.951	8.021
200	638.85	2.0370	0.0059	1826.5	2410.4	0.3094	0.3976	0.2007	4.202	12.16

TABLE B.2 (CONTINUED)

Thermophysical Properties of Saturated Ice-Water-Steam

P (bar)	s_f^* (kJ/kg·K)	s_g (kJ/kg·K)	c_{pf}^* (kJ/kg·K)	c_{pg} (kJ/kg·K)	μ_l (10^{-4} Pa·s)	γ_f	γ_g	$\bar{v}_{sf}$ (m/s)	$\bar{v}_{sg}$ (m/s)	σ^* (N/m)
0.001	−1.378	9.848	1.957							
0.002	−1.321	9.585	2.015							
0.003	−1.280	9.456	2.053							
0.004	−1.260	9.339	2.075							
0.005	−1.240	9.250	2.097	1.851						
0.006	−1.222	9.160	2.106	1.854						
0.0061	−1.221	9.159	2.116	1.854						
0.0061	0.0000	9.159	4.217	1.854	17.50					0.0756
0.008	0.0543	9.0379	4.206	1.856	15.75					0.0751
0.010	0.1059	8.9732	4.198	1.858	14.30					0.0747
0.02	0.2605	8.7212	4.183	1.865	10.67					0.0731
0.03	0.3543	8.5756	4.180	1.870	9.09					0.0721
0.04	0.4222	8.4724	4.179	1.874	8.15					0.0714
0.05	0.4761	8.3928	4.178	1.878	7.51					0.0707
0.06	0.5208	8.3283	4.178	1.881	7.03					0.0702
0.08	0.5925	8.2266	4.179	1.887	6.35					0.0693
0.10	0.6493	8.1482	4.180	1.894	5.88					0.0686
0.20	0.8321	7.9065	4.184	1.917	4.66					0.0661
0.30	0.9441	7.7670	4.189	1.935	4.09					0.0646
0.40	1.0261	7.6686	4.194	1.953	3.74					0.0634
0.5	1.0912	7.5928	4.198	1.967	3.49					0.0624
0.6	1.1454	7.5309	4.201	1.978	3.30					0.0616
0.8	1.2330	7.4338	4.209	2.015	3.03					0.0605
1.0	1.3027	7.3598	4.222	2.048	2.801	1.136	1.321	438.74	472.98	0.0589
1.5	1.4336	7.2234	4.231	2.077	2.490	1.139	1.318	445.05	478.73	0.0566
2.0	1.5301	7.1268	4.245	2.121	2.295	1.141	1.316	449.51	482.78	0.0548
2.5	1.6071	7.0520	4.258	2.161	2.156	1.142	1.314	452.92	485.88	0.0534
3.0	1.6716	6.9909	4.271	2.198	2.051	1.143	1.313	455.65	488.36	0.0521
3.5	1.7273	6.9392	4.282	2.233	1.966	1.143	1.311	457.91	490.43	0.0510
4.0	1.7764	6.8943	4.294	2.266	1.897	1.144	1.310	459.82	492.18	0.0500
4.5	1.8204	6.8547	4.305	2.298	1.838	1.144	1.309	461.46	493.69	0.0491
5	1.8604	6.8192	4.315	2.329	1.787	1.144	1.308	462.88	495.01	0.0483
6	1.9308	6.7575	4.335	2.387	1.704	1.144	1.306	465.23	497.22	0.0468

7	1.9918	6.7052	4.354	2.442	1.637	1.143	1.304	467.08	498.99	0.0455
8	2.0457	6.6596	4.372	2.495	1.581	1.142	1.303	468.57	500.55	0.0444
9	2.0941	6.6192	4.390	2.546	1.534	1.142	1.302	469.78	501.64	0.0433
10	2.1382	6.5821	4.407	2.594	1.494	1.141	1.300	470.76	502.64	0.0423
12	2.2161	6.5194	4.440	2.688	1.427	1.139	1.298	472.23	504.21	0.0405
14	2.2837	6.4651	4.472	2.777	1.373	1.137	1.296	473.18	505.33	0.0389
16	2.3436	6.4175	4.504	2.862	1.329	1.134	1.294	473.78	506.12	0.0375
18	2.3976	6.3751	4.534	2.944	1.291	1.132	1.293	474.09	506.65	0.0362
20	2.4469	6.3367	4.564	3.025	1.259	1.129	1.291	474.18	506.98	0.0350
25	2.5543	6.2536	4.640	3.219	1.193	1.123	1.288	473.71	507.16	0.0323
30	2.6455	6.1837	4.716	3.407	1.143	1.117	1.284	472.51	506.65	0.0300
35	2.7253	6.1229	4.792	3.593	1.102	1.111	1.281	470.80	505.66	0.0200
40	2.7965	6.0685	4.870	3.781	1.069	1.104	1.278	468.72	504.29	0.0261
45	2.8612	6.0191	4.951	3.972	1.040	1.097	1.275	466.31	502.68	0.0244
50	2.9206	5.9735	5.034	4.168	1.016	1.091	1.272	463.67	500.73	0.0229
60	3.0273	5.8908	5.211	4.582	0.975	1.077	1.266	457.77	496.33	0.0201
70	3.1219	5.8162	5.405	5.035	0.942	1.063	1.260	451.21	491.31	0.0177
80	3.2076	5.7471	5.621	5.588	0.915	1.048	1.254	444.12	485.80	0.0156
90	3.2867	5.6820	5.865	6.100	0.892	1.033	1.249	436.50	479.90	0.0136
100	3.3606	5.6198	6.142	6.738	0.872	1.016	1.244	428.24	473.67	0.0119
110	3.4304	5.5595	6.463	7.480	0.855	0.998	1.239	419.20	467.13	0.0103
120	3.4972	5.5002	6.838	8.384	0.840	0.978	1.236	409.38	460.25	0.0089
130	3.5616	5.4408	7.286	9.539	0.826	0.956	1.234	398.90	453.00	0.0076
140	3.6243	5.3803	7.834	11.07	0.813	0.935	1.232	388.00	445.34	0.0064
150	3.6859	5.3178	8.529	13.06	0.802	0.916	1.233	377.00	437.29	0.0053
160	3.7471	5.2531	9.456	15.59	0.792	0.901	1.235	366.24	428.89	0.0043
170	3.8197	5.1855	11.30	17.87	0.782	0.867	1.240	351.19	420.07	0.0034
180	3.8765	5.1128	12.82	21.43	0.773	0.838	1.248	336.35	410.39	0.0026
190	3.9429	5.0332	15.76	27.47	0.765	0.808	1.260	320.20	399.87	0.0018
200	4.0149	4.9412	22.05	39.31	0.758	0.756	1.280	298.10	387.81	0.0011

* Above the solid line, solid phase; below the line, liquid.

From Liley, P. E. (1987) Thermophysical properties, in Boilers, Evaporators, and Condensers, S. Kakaç (Ed.), Wiley, New York.

TABLE B.3

Thermophysical Properties of Steam at 1-bar Pressure

T (K)	ν (m³/kg)	h (kJ/kg)	s (kJ/kg·K)	c_p (kJ/kg·K)	c_v (kJ/kg·K)	γ	Z	$\bar{V}_s$ (m/s)	μ (10^{-5} Pa·s)	k (W/m·K)	Pr
373.15	1.679	2676.2	7.356	2.029	1.510	1.344	0.9750	472.8	1.20	0.0248	0.982
400	1.827	2730.2	7.502	1.996	1.496	1.334	0.9897	490.4	1.32	0.0268	0.980
450	2.063	2829.7	7.741	1.981	1.498	1.322	0.9934	520.6	1.52	0.0311	0.968
500	2.298	2928.7	7.944	1.983	1.510	1.313	0.9959	540.3	1.73	0.0358	0.958
550	2.531	3028	8.134	2.000	1.531	1.306	0.9971	574.2	1.94	0.0410	0.946
600	2.763	3129	8.309	2.024	1.557	1.300	0.9978	598.6	2.15	0.0464	0.938
650	2.995	3231	8.472	2.054	1.589	1.293	0.9988	621.8	2.36	0.0521	0.930
700	3.227	3334	8.625	2.085	1.620	1.287	0.9989	643.9	2.57	0.0581	0.922
750	3.459	3439	8.770	2.118	1.653	1.281	9.9992	665.1	2.77	0.0646	0.913
800	3.690	3546	8.908	2.151	1.687	1.275	0.9995	685.4	2.98	0.0710	0.903
850	3.921	3654	9.039	2.185	1.722	1.269	0.9996	705.1	3.18	0.0776	0.897
900	4.152	3764	9.165	2.219	1.756	1.264	0.9996	723.9	3.39	0.0843	0.892
950	4.383	3876	9.286	2.253	1.791	1.258	0.9997	742.2	3.59	0.0912	0.886
1000	4.614	3990	9.402	2.286	1.823	1.254	0.9998	760.1	3.78	0.0981	0.881
1100	5.076	4223	9.625	2.36			0.9999	794.3	4.13	0.113	0.858
1200	5.538	4463	9.384	2.43			1.0000	826.8	4.48	0.130	0.837
1300	5.999	4711	10.032	2.51			1.0000	857.9	4.77	0.144	0.826
1400	6.461	4965	10.221	2.58			1.0000	887.9	5.06	0.160	0.816
1500	6.924	5227	10.402	2.65			1.0002	916.9	5.35	0.18	0.788
1600	7.386	5497	10.576	2.73			1.0004	945.0	5.65	0.21	0.735
1800	8.316	6068	10.912	3.02			1.0011	999.4	6.19	0.33	0.567
2000	9.263	6706	11.248	3.79			1.0036	1051.0	6.70	0.57	0.445

From Kakaç, S. and Yener, Y. (1995) *Convective Heat Transfer*, 2nd ed., CRC Press, Boca Raton, FL.

TABLE B.4

Thermophysical Properties of Water-Steam at High Pressures

T (K)	ν (m^3/kg)	h (kJ/kg)	s (kJ/kg·K)	c_p (kJ/kg·K)	c_v (kJ/kg·K)	γ	Z	$\bar{V}_s$ (m/s)	μ (10^{-5} Pa·s)	k (W/m·K)	Pr
P = 10 bar											
300	1.003.–3[a]	113.4	0.392	4.18	4.13	1.01	0.0072	1500	8.57.–4[b]	0.615	5.82
350	1.027.–3	322.5	1.037	4.19	3.89	1.08	0.0064	1552	3.70.–4	0.668	2.32
400	1.067.–3	533.4	1.600	4.25	3.65	1.17	0.0058	1509	2.17.–4	0.689	1.34
450	1.123.–3	749.0	2.109	4.39	3.44	1.28	0.0054	1399	1.51.–4	0.677	0.981
500	0.221	2891	6.823	2.29	1.68	1.36	0.957	535.7	1.71.–5[c]	0.038	1.028
600	0.271	3109	7.223	2.13	1.61	1.32	0.987	592.5	2.15.–5	0.047	0.963
800	0.367	3537	7.837	2.18	1.70	1.28	0.994	686.2	2.99.–5	0.072	0.908
1000	0.460	3984	8.336	2.30	1.83	1.26	0.997	759.4	3.78.–5	0.099	0.881
1500	0.692	5224	9.337	2.66			1.000	917.2	5.35.–5	0.18	0.80
2000	0.925	6649	10.154	3.29			1.002	1050	6.70.–5	0.39	0.57
P = 50 bar											
300	1.001.–3	117.1	0.391	4.16	4.11	1.01	0.0362	1508	8.55.–4	0.618	5.76
350	1.025.–3	325.6	1.034	4.18	3.88	1.08	0.0317	1561	3.71.–4	0.671	2.31
400	1.064.–3	536.0	1.596	4.24	3.64	1.16	0.0288	1519	2.18.–4	0.691	1.34
450	1.120.–3	751.4	2.103	4.37	3.43	1.27	0.0270	1437	1.52.–4	0.681	0.975
500	1.200.–3	976.1	2.575	4.64	3.25	1.43	0.0260	1246	1.19.–4	0.645	0.856
600	0.0490	3013	6.350	2.85	1.94	1.47	0.885	560.5	2.14.–5	0.054	1.129
800	0.0713	3496	7.049	2.31	1.74	1.32	0.966	674.5	3.03.–5	0.075	0.929
1000	0.0911	3961	7.575	2.35	1.85	1.27	0.987	756.5	3.81.–5	0.102	0.880
1500	0.1384	5214	8.589	2.66			1.000	918.8	5.37.–5	0.18	0.81
2000	0.1850	6626	9.398	3.12			1.002	1053	6.70.–5	0.33	0.64
P = 100 bar											
300	9.99.–4	121.8	0.390	4.15	4.09	1.01	0.0722	1516	8.52.–4	0.622	5.69
350	1.022.–3	329.6	1.031	4.17	3.87	1.08	0.0633	1571	3.73.–4	0.675	2.31
400	1.061.–3	539.6	1.590	4.23	3.64	1.16	0.0575	1532	2.20.–4	0.694	1.34
450	1.116.–3	754.1	2.097	4.35	3.43	1.27	0.0537	1452	1.53.–4	0.685	0.975
500	1.193.–3	977.3	2.567	4.60	3.24	1.42	0.0517	1269	1.21.–4	0.651	0.853

TABLE B.4 (CONTINUED)

Thermophysical Properties of Water-Steam at High Pressures

T (K)	v (m³/kg)	h (kJ/kg)	s (kJ/kg·K)	c_p (kJ/kg·K)	c_v (kJ/kg·K)	γ	Z	$\bar{v}_s$ (m/s)	μ (10^{-5} Pa·s)	k (W/m·K)	Pr
600	0.0201	2820	5.775	5.22	2.64	1.97	0.726	502.3	2.14.–5	0.073	1.74
800	0.0343	3442	6.685	2.52	1.82	1.38	0.929	662.4	3.08.–5	0.081	0.960
1000	0.0449	3935	7.233	2.44	1.88	1.30	0.973	753.3	3.85.–5	0.107	0.876
1500	0.0692	5203	8.262	2.68			1.000	921.1	5.37.–5	0.18	0.82
2000	0.0926	6616	9.073	3.08			1.003	1057	6.70.–5	0.31	0.67
P = 250 bar											
300	9.93.–3	135.3	0.385	4.12	4.06	1.02	0.1792	1542	8.48.–4	0.634	5.50
350	1.016.–3	341.7	1.022	4.14	3.84	1.08	0.1572	1599	3.78.–4	0.686	2.28
400	1.053.–3	550.1	1.578	4.20	3.62	1.16	0.1426	1568	2.24.–4	0.704	1.33
450	1.105.–3	762.4	2.078	4.30	3.41	1.26	0.1330	1496	1.57.–4	0.696	0.969
500	1.175.–3	981.9	2.541	4.50			0.1273	1331	1.24.–4	0.666	0.838
600	1.454.–3	1479	3.443	5.88	4.22	1.40	0.1313	896.9	8.63.–5	0.532	0.952
800	0.0120	3261	6.086	3.41			0.813	627.3	3.29.–5	0.109	1.03
1000	0.0173	3845	6.741	2.69	1.97	1.36	0.935	745.9	3.98.–5	0.125	0.856
1500	0.0277	5186	7.827	2.73			1.000	929.1	5.40.–5	0.18	0.819
2000	0.0372	6608	8.642	3.04			1.008	1068			
P = 500 bar											
300	9.83.–4	157.7	0.378	4.06	3.98	1.02	0.3549	1583	8.45.–4	0.650	5.28
350	1.005.–3	361.8	1.007	4.10	3.81	1.08	0.3112	1644	3.87.–4	0.700	2.27
400	1.041.–3	567.8	1.557	4.14	3.59	1.15	0.2820	1623	2.31.–4	0.719	1.33
450	1.088.–3	776.9	2.050	4.23	3.39	1.25	0.2618	1561	1.62.–4	0.714	0.960
500	1.151.–3	991.5	2.502	4.37			0.2493	1418	1.29.–4	0.689	0.822
600	1.362.–3	1456	3.346	5.08	3.72	1.37	0.2459	1080	9.34.–5	0.588	0.808
800	4.576.–3	2895	5.937	5.84	2.79	2.10	0.620	597.8	4.04.–5	0.178	1.33
1000	8.102.–3	3697	6.302	3.17	1.81	1.76	0.878	742.1	4.28.–5	0.150	0.905
1500	0.0139	5157	7.484	2.82			1.004	943.6			
2000	0.0188	6595	8.310	3.04			1.018	1086			

[a] Notation – 3 signifies × 10^{-3}.
[b] Notation – 4 signifies × 10^{-4}.
[c] Notation – 5 signifies × 10^{-5}.

TABLE B.5

Properties of Liquid Metals

Liquid Metal and Melting Point	Temperature, °C	k W/(m·K)	ρ kg/m³	c kJ(kg·K)	$\mu \times 10^4$ kg/(m·s)
Bismuth (288°C)	315	16.40	10000	0.1444	16.22
	538	15.70	9730	0.1545	10.97
	760	15.70	9450	0.1645	7.89
Lead (327°C)	371	18.26	10500	0.159	24.0
	482	19.77	10400	0.155	19.25
	704	—	10130	—	13.69
Mercury (–39°C)	10	8.14	13550	0.138	15.92
	149	11.63	13200	0.138	10.97
	315	14.07	12800	0.134	8.64
Potassium (64°C)	315	45.0	804	0.80	3.72
	427	39.5	740	0.75	1.78
	704	33.1	674	0.75	1.28
Sodium (98°C)	93	86.0	930	1.38	7.0
	371	72.5	860	1.29	2.81
	704	59.8	776	1.26	1.78
56% Na, 44% K (19°C)	93	25.6	885	1.13	5.78
	371	27.6	820	1.06	2.36
	704	28.8	723	1.04	5.94
22% Na, 78% K (–11°C)	93	24.4	850	0.95	4.94
	400	26.7	775	0.88	2.07
	760	—	690	0.88	1.46
44.5% Pb, 55.5% Bi (125°C)	315	9.0	10500	0.147	—
	371	11.9	10200	0.47	15.36
	650	—	9820	—	11.47

From Kakaç, S. and Yener, Y. (1995) *Convective Heat Transfer*, 2nd ed., CRC Press, Boca Raton, FL.

TABLE B.6

Thermophysical Properlies of Saturated Refrigerant 12

p (bar)	T (K)	v_f (10^{-4} m³/kg)	v_g (m³/kg)	h_f (kJ/kg)	h_g (kJ/kg)	s_f (kJ/kg·K)	s_g (kJ/kg·K)
0.10	200.1	6.217	1.365	334.8	518.1	3.724	4.640
0.15	206.3	6.282	0.936	340.1	521.0	3.750	4.627
0.20	211.1	6.332	0.716	344.1	523.2	3.769	4.618
0.25	214.9	6.374	0.582	347.4	525.0	3.785	4.611
0.30	218.2	6.411	0.491	350.2	526.5	3.798	4.606
0.4	223.5	6.437	0.376	354.9	529.1	3.819	4.598
0.5	227.9	6.525	0.306	358.8	531.2	3.836	4.592
0.6	231.7	6.570	0.254	362.1	532.9	3.850	4.588
0.8	237.9	6.648	0.198	367.6	535.8	3.874	4.581
1.0	243.0	6.719	0.160	372.1	538.2	3.893	4.576
1.5	253.0	6.859	0.110	381.2	542.9	3.929	4.568
2.0	260.6	6.970	0.0840	388.2	546.4	3.956	4.563
2.5	266.9	7.067	0.0681	394.0	549.2	3.978	4.560
3.0	272.3	7.183	0.0573	399.1	551.6	3.997	4.557
4.0	281.3	7.307	0.0435	407.6	555.6	4.027	4.553
5.0	288.8	7.444	0.0351	414.8	558.8	4.052	4.551
6.0	295.2	7.571	0.0294	421.1	561.5	4.073	4.549
8.0	306.0	7.804	0.0221	431.8	565.7	4.108	4.546
10	314.9	8.022	0.0176	440.8	569.0	4.137	4.544
15	332.6	8.548	0.0114	459.3	574.5	4.193	4.539
20	346.3	9.096	0.0082	474.8	577.5	4.237	4.534
25	357.5	9.715	0.0062	488.7	578.5	4.275	4.527
30	367.2	10.47	0.0048	502.0	577.6	4.311	4.517
35	375.7	11.49	0.0036	515.9	574.1	4.347	4.502
40	383.3	13.45	0.0025	532.7	564.1	4.389	4.471
41.2[a]	385.0	17.92	0.0018	548.3	548.3	4.429	4.429

p (bar)	c_{pf} (kJ/kg·K)	c_{pg} (kJ/kg·K)	u_f (10^{-4} Pa·s)	μ_g (10^{-5} Pa·s)	k_f (W/m·K)	k_g (W/m·K)	Pr_f	Pr_g	σ (N/m)
0.10	0.855		6.16		0.105	0.0050	5.01		
0.15	0.861		5.61		0.103	0.0053	4.69		
0.20	0.865		5.28		0.101	0.0055	452		
0.25	0.868		4.99		0.099	0.0056	4.38		
0.30	0.872		4.79		0.098	0.0057	4.26		
0.4	0.876		4.48		0.097	0.0060	4.05		0.0189
0.5	0.880	0.545	4.25	1.00	0.095	0.0062	3.94	0.89	0.0182
0.6	0.884	0.552	4.08	1.02	0.094	0.0063	3.84	0.88	0.0176
0.8	0.889	0.564	3.81	1.04	0.091	0.0066	3.72	0.88	0.0167
1.0	0.894	0.574	3.59	1.06	0.089	0.0069	3.61	0.88	0.0159
1.5	0.905	0.600	3.23	1.10	0.086	0.0074	3.40	0.89	0.0145
2.0	0.914	0.613	2.95	1.13	0.083	0.0077	3.25	0.90	0.0134
2.5	0.922	0.626	2.78	1.15	0.081	0.0081	3.16	0.91	0.0125
3.0	0.930	0.640	2.62	1.18	0.079	0.0083	3.08	0.91	0.0118
4.0	0.944	0.663	2.40	1.22	0.075	0.0088	3.02	0.92	0.0106
5.0	0.957	0.683	2.24	1.25	0.073	0.0092	2.94	0.93	0.0096
6.0	0.969	0.702	2.13	1.28	0.070	0.0095	2.95	0.95	0.0087
8.0	0.995	0.737	1.96	1.33	0.066	0.0101	2.95	0.97	0.0074
10	1.023	0.769	1.88	1.38	0.063	0.0107	3.05	1.01	0.0063
15	1.102	0.865	1.67	1.50	0.057	0.0117	3.23	1.11	0.0042
20	1.234	0.969	1.49	1.69	0.053	0.0126	3.47	1.30	0.0029
25	1.36	1.19	1.33		0.047	0.0134	3.84		0.0019
30	1.52	1.60	1.16		0.042	0.014	4.2		0.0009
35	1.73	2.5			0.037	0.016			0.0005
40									0.0001
41.2									0.0000

[a] Critical point.
From Kakaç, S. and Yener, Y. (1995) *Convective Heat Transfer*, 2nd ed., CRC Press, Boca Raton, FL.

TABLE B.7

Thermophysical Properties of Refrigerant 12 at 1-bar Pressure

T (K)	ν (m^3/kg)	h (kJ/kg)	s (kJ/kg·K)	μ (10^{-5} Pa·s)	c_p (kJ/kg·K)	k (W/m·K)	Pr
300	0.2024	572.1	4.701	1.26	0.614	0.0097	0.798
320	0.2167	584.5	4.741	1.34	0.631	0.0107	0.788
340	0.2309	597.3	4.780	1.42	0.647	0.0118	0.775
360	0.2450	610.3	4.817	1.49	0.661	0.0129	0.760
380	0.2590	623.7	4.853	1.56	0.674	0.0140	0.745
400	0.2730	637.3	4.890	1.62	0.684	0.0151	0.730
420	0.2870	651.2	4.924	1.67	0.694	0.0162	0.715
440	0.3009	665.3	4.956	1.72	0.705	0.0173	0.703
460	0.3148	697.7	4.987	1.78	0.716	0.0184	0.693
480	0.3288	694.3	5.018	1.84	0.727	0.0196	0.683
500	0.3427	709.0	5.048	1.90	0.739	0.0208	0.674

From Kakaç, S. and Yener, Y. (1995) *Convective Heat Transfer*, 2nd ed., CRC Press, Boca Raton, FL.

TABLE B.8

Thermophysical Properties of Saturated Refrigerant 22

T (K)	p (bar)	v_f (m³/kg)	v_g (m³/kg)	h_f (kJ/kg)	h_g (kJ/kg)	s_f (kJ/kg·K)	s_g (kJ/kg·K)	c_{pf} (kJ/kg·K)	c_{pg} (kJ/kg·K)
150	0.0017	6.209.–4[a]	83.40	268.2	547.3	3.355	5.215	1.059	
160	0.0054	6.293.–4	28.20	278.2	552.1	3.430	5.141	1.058	
170	0.0150	6.381.–4	10.85	288.3	557.0	3.494	5.075	1.057	
180	0.0369	6.474.–4	4.673	298.7	561.9	3.551	5.013	1.058	
190	0.0821	6.573.–4	2.225	308.6	566.8	3.605	4.963	1.060	
200	0.1662	6.680.–4	1.145	318.8	571.6	3.675	4.921	1.065	0.502
210	0.3116	6.794.–4	0.6370	329.1	576.5	3.707	4.885	1.071	0.544
220	0.5470	6.917.–4	0.3772	339.7	581.2	3.756	4.854	1.080	0.577
230	0.9076	7.050.–4	0.2352	350.6	585.9	3.804	4.828	1.091	0.603
240	1.4346	7.195.–4	0.1532	361.7	590.5	3.852	4.805	1.105	0.626
250	2.174	7.351.–4	0.1037	373.0	594.9	3.898	4.785	1.122	0.648
260	3.177	7.523.–4	0.07237	384.5	599.0	3.942	4.768	1.143	0.673
270	4.497	7.733.–4	0.05187	396.3	603.0	3.986	4.752	1.169	0.703
280	6.192	7.923.–4	0.03803	408.2	606.6	4.029	4.738	1.193	0.741
290	8.324	8.158.–4	0.02838	420.4	610.0	4.071	4.725	1.220	0.791
300	10.956	8.426.–4	0.02148	432.7	612.8	4.113	4.713	1.257	0.854
310	14.17	8.734.–4	0.01643	445.5	615.1	4.153	4.701	1.305	0.935
320	18.02	9.096.–4	0.01265	458.6	616.7	4.194	4.688	1.372	1.036
330	22.61	9.535.–4	9.753.–3[b]	472.4	617.3	4.235	4.674	1.460	1.159
340	28.03	1.010.–3	7.479.–3	487.2	616.5	4.278	4.658	1.573	1.308
350	34.41	1.086.–3	5.613.–3	503.7	613.3	4.324	4.637	1.718	1.486
360	41.86	1.212.–3	4.036.–3	523.7	605.5	4.378	4.605	1.897	
369.3[c]	49.89	2.015.–3	2.015.–3	570.0	570.0	4.501	4.501	∞	∞

TABLE B.8 (CONTINUED)

Thermophysical Properties of Saturated Refrigerant 22

T (K)	μ_f (10^{-4} Pa·s)	μ_g (10^{-4} Pa·s)	k_f (W/m·K)	k_g (W/m·K)	$\bar{v}_{sf}$ (m/s)	$\bar{v}_{fg}$ (m/s)	Pr_f	Pr_g	σ (N/m)
150			0.161						
160			0.156						
170	7.70		0.151			142.6	5.39		
180	6.47		0.146			146.1	4.69		
190	5.54		0.141			149.4	4.16		
200	4.81		0.136		1007	152.6	3.77		0.024
210	4.24		0.131		957	155.2	3.47		0.022
220	3.78		0.126		909	157.6	3.24		0.021
230	3.40	0.100	0.121	0.0067	862	159.7	3.07	0.89	0.019
240	3.09	0.104	0.117	0.0073	814	161.3	2.92	0.89	0.017
250	2.82	0.109	0.112	0.0080	766	162.5	2.83	0.89	0.0155
260	2.60	0.114	0.107	0.0086	716	163.1	2.78	0.89	0.0138
270	2.41	0.118	0.102	0.0092	668	163.4	2.76	0.90	0.0121
280	2.25	0.123	0.097	0.0098	622	162.1	2.77	0.93	0.0104
290	2.11	0.129	0.092	0.0105	578	161.1	2.80	0.97	0.0087
300	1.98	0.135	0.087	0.0111	536	160.1	2.86	1.04	0.0071
310	1.86	0.141	0.082	0.0117	496	157.2	2.96	1.13	0.0055
320	1.76	0.148	0.077	0.0123	458	153.4	3.14	1.25	0.0040
330	1.67	0.157	0.072	0.0130	408	148.5	3.39	1.42	0.0026
340	1.51	0.171	0.067	0.0140	355	142.7	3.55	1.60	0.0014
350	1.30		0.060		290	135.9	3.72		0.0008
360	1.06								
369.3									

[a] Notation – 4 signifies × 10^{-4}.
[b] Notation – 3 signifies × 10^{-3}.
[c] Critical point.

From Kakaç, S. and Yener, Y. (1995) *Convective Heat Transfer*, 2nd ed., CRC Press, Boca Raton, FL.

TABLE B.9

Thermophysical Properties of Refrigerant R22 at Atmospheric Pressure

T (K)	v (m^3/kg)	h (kJ/kg)	s (kJ/kg·K)	c_p (kJ/kg·K)	Z	$\bar{V}_s$ (m/s)	μ (10^{-6} Pa·s)	k (W/m·K)	Pr
232.3	0.2126	586.9	4.8230	0.608	0.9644	160.1	10.1	0.0067	0.893
240	0.2205	591.5	4.8673	0.6117	0.9682	163.0	10.4	0.0074	0.860
260	0.2408	604.0	4.8919	0.6255	0.9760	169.9	11.2	0.0084	0.838
280	0.2608	616.8	4.9389	0.6431	0.9815	176.2	12.0	0.0094	0.820
300	0.2806	630.0	4.9840	0.6619	0.9857	182.3	12.8	0.0106	0.804
320	0.3001	643.4	5.0274	0.6816	0.9883	188.0	13.7	0.0118	0.790
340	0.3196	657.3	5.0699	0.7017	0.9906	193.5	14.4	0.0130	0.777
360	0.3390	671.7	5.1111	0.7213	0.9923	198.9	15.1	0.0142	0.767
380	0.3583	686.5	5.1506	0.7406	0.9936	204.1	15.8	0.0154	0.760
400	0.3775	701.5	5.1892	0.7598	0.9945	209.1	16.5	0.0166	0.755
420	0.3967	717.0	5.2267	0.7786	0.9953	214.0	17.2	0.0178	0.753
440	0.4159	732.8	5.2635	0.7971	0.9961	218.8	17.9	0.0190	0.752
460				0.8150		223.5	18.6	0.0202	0.751
480				0.8326		227.9	19.3	0.0214	0.751
500				0.8502			19.9	0.0225	0.750

From Kakaç, S. and Yener, Y. (1995) *Convective Heat Transfer*, 2nd ed., CRC Press, Boca Raton, FL.

TABLE B.10

Thermophysical Properties of Saturated Refrigerant R134a

T (K)	p (bar)	v_f (m^3/kg)	v_g (m^3/kg)	h_f (kJ/kg)	h_g (kJ/kg)	s_f (kJ/kg·K)	s_g (kJ/kg·K)	c_{pf} (kJ/kg·K)	c_{pg} (kJ/kg·K)	μ_f (10^{-4} Pa·s)	μ_g (10^{-4} Pa·s)	k_f (W/m·K)	kg (W/m·K)	Pr_f	Pr_g	σ (N/m)
200	0.070	0.000661	2.32	−36.0	201.0	−0.1691	1.0153									
210	0.187	0.000674	0.906	−26.5	208.1	−0.1175	0.9941									
220	0.252	0.000687	0.698	−15.3	214.5	−0.0664	0.9758									
230	0.438	0.000701	0.416	−3.7	220.8	−0.0158	0.9602	1.113	0.732							
240	0.728	0.000716	0.258	8.1	227.1	0.0343	0.9471	1.162	0.764	4.25	0.095	0.099	0.008	4.99	0.90	
250	1.159	0.000731	0.167	20.3	233.3	0.0840	0.9363	1.212	0.798	3.70	0.099	0.095	0.008	4.72	0.96	0.0145
260	1.765	0.000748	0.112	32.9	239.4	0.1331	0.9276	1.259	0.835	3.25	0.104	0.091	0.008	4.49	1.02	0.0131
270	2.607	0.000766	0.077	45.4	244.8	0.1817	0.9211	1.306	0.876	2.88	0.108	0.087	0.009	4.31	1.08	0.0117
280	3.721	0.000786	0.055	59.2	251.1	0.2299	0.9155	1.351	0.921	2.56	0.112	0.083	0.009	4.17	1.14	0.0103
290	5.175	0.000806	0.040	72.9	256.6	0.2775	0.9114	1.397	0.972	2.30	0.117	0.079	0.010	4.07	1.20	0.0090
300	7.02	0.000821	0.029	87.0	261.9	0.3248	0.9080	1.446	1.030	2.08	0.121	0.075	0.010	4.00	1.27	
310	9.33	0.000865	0.022	101.5	266.8	0.3718	0.9050	1.497	1.104	1.89	0.125	0.071	0.010	3.98	1.34	
320	12.16	0.000895	0.016	116.6	271.2	0.4189	0.9021	1.559	1.198	1.72	0.129	0.068	0.011	3.98	1.57	
330	15.59	0.000935	0.012	132.3	275.0	0.4663	0.8986	1.638	1.324	1.58	0.133	0.064	0.011	3.94	1.44	
340	19.71	0.000984	0.0094	148.9	277.8	0.5146	0.8937	1.750	1.520	1.45	0.137	0.060	0.012	4.23	1.74	
350	24.60	0.00105	0.0071	166.6	279.1	0.5649	0.8861	1.931	1.795	1.34	0.14	0.056	0.012	4.62	2.09	
360	30.40	0.00115	0.0051	186.5	277.7	0.6194	0.8721	2.304	2.610	1.20	0.16	0.054	0.013	5.16	3.21	
370	37.31	0.00139	0.0035	216.0	270.0	0.6910	0.8370			0.95	0.26					
374.3[a]	40.67	0.00195	0.0020	248.0	248.0	0.7714	0.7714									

[a] Critical point.

From Liley, P.E. (1987) Thermophysical properties. In *Boilers, Evaporators, and Condensers*, S. Kakac (Ed.), Wiley, New York.

TABLE B.11

Properties of Refrigerant 134a at Atmospheric Pressure

T (K)	v (m^3/kg)	h (kJ/kg)	s (kJ/kg·K)	c_p (kJ/kg·K)	Z	$\bar{V}_s$ (m/s)
247	0.1901	231.5	0.940	0.787	0.957	145.9
260	0.2107	241.8	0.980	0.801	0.965	150.0
280	0.2193	258.1	1.041	0.827	0.974	156.3
300	0.2365	274.9	1.099	0.856	0.980	162.1
320	0.2532	292.3	1.155	0.885	0.984	167.6
340	0.2699	310.3	1.209	0.915	0.987	172.8
360	0.2866	328.8	1.263	0.945	0.990	177.6
380	0.3032	347.8	1.313	0.976	0.992	182.0
400	0.3198	367.2	1.361	1.006	0.994	186.0

From Liley, P. E. (1987) Thermophysical properties. In *Boilers, Evaporators, and Condensers*, S. Kakaç (Ed.), Wiley, New York.

TABLE B.12

Thermophysical Properties of Unused Engine Oil

T (K)	ν_f (m^3/kg)	c_{pf} (kJ/kg·K)	μ_f (W/m·K)	k_f (W/m·K)	Pr_f	α_f (m^2/s)
250	1.093–3[a]	1.72	32.20	0.151	367,000	9.60–8[b]
260	1.101–3	1.76	12.23	0.149	144,500	9.32–8
270	1.109–3	1.79	4.99	0.148	60,400	9.17–8
280	1.116–3	1.83	2.17	0.146	27,200	8.90–8
290	1.124–3	1.87	1.00	0.145	12,900	8.72–8
300	1.131–3	1.91	0.486	0.144	6,450	8.53–8
310	1.139–3	1.95	0.253	0.143	3,450	8.35–8
320	1.147–3	1.99	0.141	0.141	1,990	8.13–8
330	1.155–3	2.04	0.084	0.140	1,225	7.93–8
340	1.163–3	2.08	0.053	0.139	795	7.77–8
350	1.171–3	2.12	0.036	0.138	550	7.62–8
360	1.179–3	2.16	0.025	0.137	395	7.48–8
370	1.188–3	2.20	0.019	0.136	305	7.34–8
380	1.196–3	2.25	0.014	0.136	230	7.23–8
390	1.205–3	2.29	0.011	0.135	185	7.10–8
400	1.214–3	2.34	0.009	0.134	155	6.95–8

[a] Notation – 3 signifies $\times 10^{-3}$.
[b] Notation – 8 signifies $\times 10^{-8}$.
From Kakaç, S. and Yener, Y. (1995) *Convective Heat Transfer*, 2nd ed., CRC Press, Boca Raton, FL.

TABLE B.13

Conversion Factors

Area: 1 m^2 = 1550.0 in.^{2} = 10.7639 ft^2 = 1.19599 yd^2 = 2.47104 × 10^{-4} acre = 1 × 10^{-4} ha = 10^{-6} km^2 = 3.8610 × 10^{-7} mi^2

Density: 1 kg/m^3 = 0.06243 lb$_m$/ft^3 = 0.01002 lb$_m$/U.K. gal = 8.3454 × 10^{-3} lb$_m$/U.S. gal = 1.9403 × 10^{-3} slug/ft^3 = 10^{-3}g/cm^3

Energy: 1 kJ = 737.56 ft·lb$_f$ = 238.85 cal = 0.94783 Btu = 3.7251 × 10^{-4} hp·h = 2.7778 × 10^{-4} kW·h

Heat transfer coefficient: 1 W/(m^2·K) = 0.8598 kcal/(m^2·h·°C) = 0.1761 Btu/(ft^2·h·°F) = 10^{-4} W/(cm^2·K) = 0.2388 × 10^{-4} cal/(cm^2·s·°C)

Inertia: 1 kg·m^2 = 3.41717 × 10^3 lb·in.^{2} = 0.73756 slug·ft^2

Length: 1 m = 10^{10} Å = 39.370 in. = 3.28084 ft = 4.971 links = 1.0936 yd = 0.54681 fathoms = 0.04971 chain = 4.97097 × 10^{-3} furlong = 10^{-3} km = 5.3961 × 10^{-4} U.K. nautical miles = 5.3996 × 10^{-4} U.S. nautical miles = 6.2137 × 10^{-4} mi

Mass: 1 kg = 2.20462 lb$_m$ = 0.06852 slug = 1.1023 × 10^{-3} U.S. ton = 10^{-3} t = 9.8421 × 10^{-4} U.K. ton

Mass flow rate: 1 kg/s = 2.20462 lb/s = 132.28 lb/min = 7936.64 lb/h = 3.54314 long ton/h = 3.96832 short ton/h

Power: 1 W = 44.2537 ft·lb$_f$/min = 3.41214 Btu/h = 1 J/s = 0.73756 ft·lb$_f$/s = 0.23885 cal/s = 0.8598 kcal/h

Pressure: 1 bar = 10^5 N/m^2 = 10^5 Pa = 750.06 mm Hg at 0°C = 401.47 in. H_2O at 32°F = 29.530 in. Hg at 0°C = 14.504 lb$_f$/in.^{2} = 14.504 psia = 1.01972 kg/cm^2 = 0.98692 atm = 0.1 MPa

Specific energy: 1 kJ/kg = 334.55 ft·lb$_f$/lb$_m$ = 0.4299 Btu/lb$_m$ = 0.2388 cal/g

Specific energy per degree: 1 kJ/(kg·K) = 0.23885 Btu/(lb$_m$·°F) = 0.23855 cal/(g·°C)

Surface tension: 1 N/m = 5.71015 × 10^{-3} lb$_f$/in.

Temperature: T (K) = T (°C) + 273.15 = [T (°F) + 459.67]/1.8 = T (°R)/1.8

Temperature difference: ΔT (K) = ΔT (°C) = ΔT (°F)/1.8 = ΔT (°R)/1.8

Thermal conductivity: 1 W/(m·K) = 0.8604 kcal/(m·h·°C) = 0.5782 Btu/(ft·h·°F) = 0.01 W/(cm·K) = 2.390 × 10^{-3} cal/(cm·s·°C)

Thermal diffusivity: 1 m^2/s = 38,750 ft^2/h = 3600 m^2/h = 10.764 ft^2/s

Torque: 1 N·m = 1.41.61 oz·in. = 8.85073 lb$_f$·in. = 0.73756 lb$_f$·ft = 0.10197 kg$_f$·m

Velocity: 1 m/s = 100 cm/s = 196.85 ft/min = 3.28084 ft/s = 2.23694 mi/h = 2.23694 mph = 3.6 km/h = 1.94260 U.K. knot = 1.94384 Int. knot = 1.94384 Int. knot

Viscosity, dynamic: 1 (N·s)/m^2 = 1 Pa·s = 10^7 μP = 2419.1 lb$_m$/(ft·h) = 10^3 cP = 75.188 slug/(ft·h) = 10 P = 0.6720 lb$_m$/(ft·s) = 0.02089 (lb$_f$·s)/ft^2

Viscosity, kinematic: (see thermal diffusivity)

Volume: 1 m^3 = 61,024 in.^{3} = 1000 l = 219.97 U.K. gal = 264.17 U.S. gal = 35.3147 ft^3 = 1.30795 yd^3 = 1 stere = 0.81071 × 10^{-3} acre-foot

Volume flow rate: 1 m^3/s = 35.3147 ft^3/s = 2118.9 ft^3/min = 13198 U.K. gal/min = 791,891 U.K. gal/h = 15,850 U.S. gal/min = 951,019 U.S. gal/h

From Liley, P. E. (1987) Thermophysical properties, in *Boilers, Evaporators, and Condensers*, Kakaç, S. (ed.), Wiley, New York.

References

1. Liley, P. E. (1987) Thermophysical properties. In *Handbook of Single-Phase Convective Heat Transfer*, Kakaç, S., Shah, R. K., and Aung, W. (Eds.), Chapter 22. Wiley, New York.
2. Liley, P. E. (1987) Thermophysical properties. In *Boilers, Evaporators, and Condensers*, Kakaç, S. (Ed.), Wiley, New York.
3. Kakaç, S. and Yener, Y. (1995) *Convective Heat Transfer*, 2nd Ed., CRC Press, Boca Raton, FL.
4. Kakaç, S. and Yener, Y. (1993) *Heat Conduction*, 3rd Ed., Taylor & Francis, London.

Index

D

E

H

I

J

K

L

M

N

O

P

R

S

T

U